Co-Engineering and Participatory Water Management

Organisational Challenges for Water Governance

Effective participatory water management requires effective co-engineering – the collective process whereby organisational decisions are made on how to bring stakeholders together.

This trans-disciplinary book highlights the challenges involved in the collective initiation, design, implementation and evaluation of participatory water planning and management processes. It also demonstrates how successful management typically requires the effective handling of two participatory processes: the stakeholder water management process and the co-engineering process required to organise this. The book provides practical methods for supporting improved participatory processes, including the application of theory and models to aid decision making. Case studies of these applications from Australia and Europe, with additional examples from all over the world, including Africa, are used to examine negotiations and leadership approaches, and their effects on the participatory stakeholder processes.

This international review of participatory water governance and its organisational challenges forms an important resource for academic researchers in hydrology, environmental management and water policy, and also practitioners and policy makers working in water management.

KATHERINE A. DANIELL is a Research Fellow in the Australian National University's Centre for Policy Innovation. Her work focuses on resolving the challenges associated with implementing multi-level participatory processes to bring about coordinated policy, adaptation strategies and local action for sustainable development. Her other research interests include developing decision-aiding theory for 'multi-accountable' groups and encouraging effective inter-organisational collaborations. She also teaches executive development courses for the Australian National Institute for Public Policy (ANIPP) on multi-level governance. Dr Daniell is a guest editor for the journal *Ecology and Society*, and she has received many awards and honours for her work, including a General Sir John Monash Award, a prize for best paper presentation at the 2011 IAHR World Congress and election as a Fellow of the Peter Cullen Water and Environment Trust.

INTERNATIONAL HYDROLOGY SERIES

The **International Hydrological Programme** (IHP) was established by the United Nations Educational, Scientific and Cultural Organization (UNESCO) in 1975 as the successor to the International Hydrological Decade. The long-term goal of the IHP is to advance our understanding of processes occurring in the water cycle and to integrate this knowledge into water resources management. The IHP is the only UN science and educational programme in the field of water resources, and one of its outputs has been a steady stream of technical and information documents aimed at water specialists and decision makers.

The **International Hydrology Series** has been developed by the IHP in collaboration with Cambridge University Press as a major collection of research monographs, synthesis volumes and graduate texts on the subject of water. Authoritative and international in scope, the various books within the series all contribute to the aims of the IHP in improving scientific and technical knowledge of fresh-water processes, in providing research know-how and in stimulating the responsible management of water resources.

TITLES IN PRINT IN THIS SERIES

M. Bonell, M. M. Hufschmidt and J. S. Gladwell *Hydrology and Water Management in the Humid Tropics: Hydrological Research Issues and Strategies for Water Management*

Z. W. Kundzewicz *New Uncertainty Concepts in Hydrology and Water Resources*

R. A. Feddes *Space and Time Scale Variability and Interdependencies in Hydrological Processes*

J. Gibert, J. Mathieu and F. Fournier *Groundwater/Surface Water Ecotones: Biological and Hydrological Interactions and Management Options*

G. Dagan and S. Neuman *Subsurface Flow and Transport: A Stochastic Approach*

J. C. van Dam *Impacts of Climate Change and Climate Variability on Hydrological Regimes*

D. P. Loucks and J. S. Gladwell *Sustainability Criteria for Water Resource Systems*

J. J. Bogardi and Z. W. Kundzewicz *Risk, Reliability, Uncertainty, and Robustness of Water Resource Systems*

G. Kaser and H. Osmaston *Tropical Glaciers*

I. A. Shiklomanov and J. C. Rodda *World Water Resources at the Beginning of the Twenty-First Century*

A. S. Issar *Climate Changes during the Holocene and their Impact on Hydrological Systems*

M. Bonell and L. A. Bruijnzeel *Forests, Water and People in the Humid Tropics: Past, Present and Future Hydrological Research for Integrated Land and Water Management*

F. Ghassemi and I. White *Inter-Basin Water Transfer: Case Studies from Australia, United States, Canada, China and India*

K. D. W. Nandalal and J. J. Bogardi *Dynamic Programming Based Operation of Reservoirs: Applicability and Limits*

H. S. Wheater, S. Sorooshian and K. D. Sharma *Hydrological Modelling in Arid and Semi-Arid Areas*

J. Delli Priscoli and A. T. Wolf *Managing and Transforming Water Conflicts*

H. S. Wheater, S. A. Mathias and X. Li *Groundwater Modelling in Arid and Semi-Arid Areas*

L. A. Bruijnzeel, F. N. Scatena and L. S. Hamilton *Tropical Montane Cloud Forests*

S. Mithen and E. Black *Water, Life and Civilisation: Climate, Environment and Society in the Jordan Valley*

K. A. Daniell *Co-Engineering and Participatory Water Management*

Co-Engineering and Participatory Water Management

Organisational Challenges for Water Governance

Katherine A. Daniell

CAMBRIDGE
UNIVERSITY PRESS

CAMBRIDGE
UNIVERSITY PRESS

University Printing House, Cambridge CB2 8BS, United Kingdom

One Liberty Plaza, 20th Floor, New York, NY 10006, USA

477 Williamstown Road, Port Melbourne, VIC 3207, Australia

4843/24, 2nd Floor, Ansari Road, Daryaganj, Delhi - 110002, India

79 Anson Road, #06-04/06, Singapore 079906

Cambridge University Press is part of the University of Cambridge.

It furthers the University's mission by disseminating knowledge in the pursuit of education, learning and research at the highest international levels of excellence.

www.cambridge.org
Information on this title: www.cambridge.org/9781108446495

First published 2012
First paperback edition 2017

A catalogue record for this publication is available from the British Library

Library of Congress Cataloging in Publication data
Daniell, Katherine A., 1981–
Co-engineering and participatory water management: organisational challenges for water governance / Katherine A. Daniell.
p. cm.
(International hydrology series)
Includes bibliographical references and index.
ISBN 978-1-107-01231-8
1. Watershed management–International cooperation. 2. Water resources development–International cooperation. I. Title.
TC409.D26 2012
628.1–dc23 2011046219

ISBN 978-1-107-01231-8 Hardback
ISBN 978-1-108-44649-5 Paperback

Contents

Acknowledgements

Co-engineering and Participatory Water Management is based partly on the author's PhD thesis submitted to the Australian National University (ANU) and AgroParisTech-ENGREF (Daniell, 2008), and on ongoing work published since then. Approval for the research was obtained from the CRES Local Ethics Sub-Committee of the Australian National University and was based on the *NHMRC/AVCC National Statement on Ethical Conduct in Research Involving Humans*: Protocol No. 2006/91.

In an attempt to reduce the environmental impact of this research, in particular due to the unavoidable plane flights, 50 tonnes of carbon offsets were purchased from CO_2 Australia. These will take the form of mallee plantings, which offer further benefits to the Australian environment because of their carbon sequestering capacity. This includes mitigating land salinisation and erosion, as well as aiding the improvement of catchment environmental quality.

This book is the result of a journey through life and work in Europe and Australia. The large amount of collaborative work that took place and the many close personal relationships that were formed made the research work a thoroughly enjoyable and fulfilling experience. It is through work with others that the contents of this investigation into – and accounts of – participatory water management theory and practice around the world have been enriched. For this experience, I have many people to thank.

First of all, thanks go to my panel of research directors. To Ian White, for his continuous support, time, discussions, friendship and meticulous and thought-provoking comments on my work, as well as for encouraging and supporting me in publishing this work in the International Hydrology Series of Cambridge University Press and UNESCO-IHP. To Alexis Tsoukiàs, for helping me to find an (inter)disciplinary home in the French research system, for his suggestions on research directions and literature that led to my interest in operationalising decision-aiding theory for water management practice, and for his ongoing enthusiasm for the development of new research together. To Pascal Perez, for initially spawning my interest in participatory processes and in undertaking a PhD, as well as for believing in my capacities and helping me to navigate the French–Australian administrative systems for the cotutelle programme. To Stewart Burn, for helping to keep my research practically grounded and useful to the Australian water management community, as well as for his administrative support, which has given me the opportunity to network and work with a much more diverse range of researchers than would otherwise have been possible. And last but not least to Nils Ferrand, for investing so much time and energy in setting up and contributing to the direction of this research project on a professional and personal level, for being such a wonderful source of intellectual stimulation and for inspiring me to work towards clarifying my thoughts and developing a critically reflective attitude to my research. I have learnt an enormous amount in such a short time and look forward to working on many more collaborative projects in the future.

To Peter Coad from the Hornsby Shire Council in NSW, Australia, I have many thanks for his willingness to develop a case study with me, for his vision to improve the estuary's management and environmental quality and for daring to think big and try many innovative things, and for his and his family's friendship. The work in the Lower Hawkesbury Estuary could not have taken place without much support from many other people: in particular, Philip Haines, Verity Rollason and Michelle Fletcher from BMT WBM, Michael Baker from SJB Planning, Kristy Guise and Ross McPherson from the HSC, Natalie Jones and Ian White from the ANU, as well as all of the participants of the process, who gave so much of their time, work efforts and enthusiasm.

I thank Irina Ribarova from UACEG in Sofia, Bulgaria, very much for warmly accepting me into the team partway through the project, for trusting us and trying to establish and manage such a broad-scale participatory water management process in her country, and for her support and friendship through and following the process. Moreover, the work in the Iskar Basin could not have taken place without the efforts of many other people; in particular, I would like to thank Albena Popova, Svetlana Vasileva, Petar Kalinkov, Galina Dimova and Anna Denkova from UACEG; Nils Ferrand, Jean-Emmanuel Rougier, Geraldine Abrami and Dominique Rollin from Cemagref; Matt Hare from

Seecon Deutschland GmBH; Plamen Ninov from the Bulgarian Bureau of Meteorology; and all of the participants of the process, for their time, work efforts and enthusiasm.

For the pilot trial participatory process that took place in Montpellier, France, in 2005, I must add my thanks to Nils Ferrand, Nicolas Jahier, Innocent Bakam, Pascal Finaud-Guyot, Yorck von Korff, Pieter Bots and Quentin Leseney, as well as the nine participants and staff at Le Jogging for taking part in our experiment.

For the Dhünn case, I thank Sabine Moellenkamp, for introducing me to this case and for our discussions on aspects of collaborative participatory process organisation, as well as all of her colleagues who worked on the case and did such a thorough job of documenting it.

For the Mitidja case, I am greatly indebted to Amar Imache, for sharing his work with me over a number of years and building my understanding of the challenges and opportunities present in Algeria for participatory water management practice. I equally thank Mathieu Dionnet, Sami Bouarfa and the Lisode team for their support of the development of this case, as well as all the members of the Communauté Montpelliéraine de la Participation and other participants, organisers and supporters of this work.

For the Tarawa case, thanks must go to all of the people who have shared their perspectives on island hydrology and participatory water management in the Republic of Kiribati over the years, including Ian White, Tony Falkland, Trevor Daniell, Pascal Perez, Anne Dray, Magnus Moglia, Patrick D'Aquino and Natalie Jones.

None of this work would have been possible without the financial support of a number of sources. Firstly, I have been greatly honoured to receive substantial support of both a financial and a personal nature from the General Sir John Monash Foundation and its sponsors. Their support has provided great freedom and endowed me with the responsibility to pursue the research paths I believed of most importance to Australians and other international communities, and it has given me the opportunity to meet many inspirational people.

I also thank my current employer, the Centre for Policy Innovation at the ANU, for support while the book project was being completed. Further financial support from the following sources is also greatly appreciated: a CSIRO Postgraduate Scholarship (Land and Water Division); operating costs and a research contract from Cemagref UMR G-EAU; a Publishing Fellowship, operating costs and an ANU Miscellaneous Scholarship from the Fenner School of Environment and Society; a Cotutelle Scholarship from the French Embassy in Australia; and the John Crampton Travelling Scholarship. The Bulgarian part of this work was supported financially by the European Commission, 6th Framework programme, AquaStress Project, Contract GOCE, Contract No. 511231–2.

The contents of this book are the sole responsibility of the author and can under no circumstances be regarded as reflecting the position of the European Union. Administrative support from many people at the ANU, Cemagref, AgroParisTech-ENGREF and CSIRO, including Thierry Rieu, Alain Delacourt, Jean-Philippe Tortorotot, Patrice Garin, Elodie Borg, Christine Moretti, Marie-Claude Lafforgue, Augustin Luxin, Nikki Hughes, Roz Smith, Noel Chan and Phil Greaves, is much appreciated.

This book could not have been developed without the support of UNESCO-IHP and Cambridge University Press, so I thank Shahbaz Khan, Chris Hudson, Susan Francis, Kirsten Bot, Abigail Jones, Alison Lees and Laura Clark, their publishing support teams and the two anonymous reviewers, whose comments have helped to shape its contents and quality. Anne Daniell's proofreading of the manuscript in its various versions has been greatly appreciated. Many thanks also to Geoff Syme, Valerie Belton, Eddy Moors and Flavie Cernesson for their first reviews of the manuscript as part of my thesis examination and support of its publication.

Research life in both Europe and Australia was greatly enriched by the vibrant research communities of the UMR G-EAU and the Fenner School, as well as the AquaStress project and other external associations, such as the General Sir John Monash Foundation, the Peter Cullen Environment and Water Trust, BEQUEST and the ANUVC. Many thanks to all my colleagues, fellow PhD students and friends for enriching my life and work, including Ariella Helfgott, Ashley Kingsborough, Steve Dovers, Pat Troy, Quentin Grafton, Karen Hussey, Jamie Pittock, Daniel Connell, Andrew Ross, Anthony Hogan, Adrian Kay, Tony Jakeman, Peter Coombes, Julien Lepetit, Ken Crompton, Peter Binks, Deane Terrell, Will Steffen, Helen Allison, Su Wild River, Mark Matthews, Adam Graycar, David Marsh, Suzy Marsh, Justin Iu, Hannah French, Matthew Rosenberg, Daniel Murfet, Catherine Gross, Rado Faletič, Jean-François Desvignes-Hicks, Merrilyn Fitzpatrick, Kerrie Glennie, Bev Biglia, Lain Dare, Trish Mercer, Wendy Jarvie, Sophie Thoyer, Pieter Valkering, Rianne Bijsma, Bernie Foley, Holger Maier, Heath Sommerville, Nick Fleming, Bridget Vincent, Matt Baker, Sarah Milne, David Matthews, Satis Arnold, Olivier Barreteau, Chabane Mazri, Philippe Ker Rault, Alexandre Gaudin, Marwan Ladki, Laure Isnard, Martine Antona, Cécile Barnaud, Jean-Yves Jamin, Rémi Barbier, Pierre Maurel, David Crevoisier, Caroline Sart, Myriam Taoussi, Pauline Bremond, Frédéric Grelot, Audrey Richard-Ferroudji, Clémence Bedu, Christelle Gramaglia, Gabrielle Bouleau, Katrin Erdlenbruch, Clément Geney, Nicolas Desquinabo, Christelle Pezon, Gilian Cadic, François Molle, Névine Kocher, Kaddour Raissi, Yann Chabin, Cécile Jeannin, as well as Mike, Kim, Nina, Hélène, Gwen, Kristen, Lucy, Margaret, Will, Chris, Katie, Sophie, Huy, Gus, Aziliz, Gus, Anne-Sophie, Jeff, Emmanuel, Sandrine,

Daniel, Brigitte, Mathieu, Patrick, Annick, Thomas, Marilyne, Claire, Nicolas, Aurélie, Jean, Maren and many others.

Most of all, thank you to my family – Anne, Trevor and James Daniell, Margaret Crisp and extended French, German, Canberra and Adelaide family – for so much love and support of many kinds through this phase of my life and the previous ones. I dedicate this book to you.

Enfin, à Quentin de m'avoir accompagnée à travers le monde et de m'avoir soutenue jusqu'au bout de ce travail. J'ai hâte de parcourir les prochaines étapes du voyage avec toi…

Glossary

Commons in the environmental management sense refers to natural assets that belong to or support a group of people; for example, common water, air or land resources.

Messes are dynamic situations that consist of complex systems of interacting and changing problems (Ackoff, 1979).

Non-government organisations (NGOs): this appellation includes citizen or local action groups, as well as not-for-profit local, national and international organisations and associations.

Organisation is considered in the broadest possible manner as a group, association, business, institution, government or any other appellation of at least two people who share something in common (i.e. have the same interest). This can include individual citizens, as they can be considered as representatives of their country or region.

Problem situation can be described as a context in which decisions need to be made.

Stakeholders are considered as people, institutions or organisations that have a stake in the outcome of decisions related to water management, as they are directly affected by the decisions made or have the power to block or influence the decision-making process (Nandalal and Simonovic, 2003).

Stakes refer to the stakeholders' interests or those issues or problems with which they are concerned.

Values are considered to take one of two of the following definitions: firstly, the type of values that are 'held', such as principles, morals, beliefs or other ideas that serve as guides to individual and collective action; and secondly, the type of values that are 'assigned' in reference to the qualities and characteristics seen in objects or people, especially positive characteristics (actual and potential) or those that are considered worthwhile or desirable (Mason, 2002).

Risk in water management can be considered as a function of: hazard; the probability of occurrence or likelihood of certain impacts resulting from a hazard event; and vulnerability defined as the magnitude of potential consequences or impacts resulting from an event's occurrence (Dwyer *et al.* 2004, Standards Australia, 2004; 2006).

Vulnerability (in this definition of risk) is often considered as both a function of susceptibility or exposure to hazards and of resilience, which is defined as the adaptive capacity of systems to respond and cope in the face of hazard events (DIFD, 2004; Dwyer *et al.* 2004; Kundzewicz and Schellnhuber, 2004).

Part I
Framing the context

1 Introduction

1.1 ORGANISING THE STRUGGLE TO GOVERN THE COMMONS

The recent worldwide push for broad-scale multi-level participatory processes to aid the adaptive management of socio-ecological systems has led to the emergence of important and rarely investigated actors – those who 'organise' the struggle to govern the commons.

In their 2003 *Science* article, entitled 'The struggle to govern the commons', Dietz, Ostrom and Stern (Dietz *et al.*, 2003) described the creation of effective governance systems for the world's critical environmental problems as 'akin to a co-evolutionary race'. They suggested that adaptive and robust governance mechanisms to deal with these problems are most likely to succeed if the following strategies are pursued:

- The development of 'analytic deliberation' or well-structured dialogues between 'interested parties, officials and scientists';
- The 'nesting' of institutional arrangements to maintain complexity and redundancy;
- Employing 'institutional variety' or a mix of institutional types and different types of rules; and
- The promotion of 'designs that facilitate experimentation, learning, and change' (Dietz *et al.*, 2003).

In recent years there has been a rapid increase in the number of attempts to promote governance mechanisms that fit some of these characteristics.

At a local level, hundreds of thousands of community stewardship groups have been established around the world since the beginning of the 1990s and these have supported marked improvements in environmental quality and maintenance of livelihoods (Pretty, 2003). Commonly, NGOs and governments provide funding support for facilitators and skills training for local people, to encourage ongoing self-governance (Pretty, 2003). This community group establishment has been coupled with an increasing realisation that such groups, researchers and other public and private actors should move away from 'treating the problems' to first asking what the problems or issues are and collectively structuring both them and visions for the future (Rosenhead and Mingers, 2001b; Adams *et al.*, 2003; Libicki and Pfleeger, 2004).

Attempts are also being made to understand and address the broader-scale challenges of today's rapidly changing and interconnected world, such as climate change, ocean management, biodiversity, resource budgets (food, energy, water, nutrients, etc.) and the need to structure problems and governance mechanisms for their treatment. These attempts include growing efforts to establish and maintain inter-organisational networks to govern socio-ecological systems adaptively and to improve their resilience (Jackson and Stainsby, 2000; Adger *et al.*, 2005; Fayesse, 2006). Multi-level participatory processes are also increasingly being organised to help overcome the scale mismatch issues, and to facilitate increased learning and changes in governance systems to meet the new challenges of today's increasingly interconnected world (Cumming *et al.*, 2006).

The emergence of these governance structures means that we are again entering a phase in the evolution of common pool resource management, where increasingly important research questions must be addressed, in particular:

- Who is responsible for designing and implementing the structures of these new participatory systems, processes and networks?
- How are they organised?
- Who chooses or designs the methods that are used to aid decision making through these processes?
- Who chooses the participants to be included or the scope of issues to be addressed?
- Are these organising processes monitored by anyone?
- Do the organisers have the knowledge and legitimacy required to organise these processes effectively?
- To what extent could organisation be improved to obtain better participatory process and socio-ecological system outcomes?
- Just how important are the roles of these organisers in helping to meet the challenges linked to the world's critical environmental problems?

1.2 WATER: A KEYSTONE OF COMMONS GOVERNANCE

Water and its management is an integral part of almost all of these interwoven challenges, as it is a fundamental need for life. Neither we, nor the entirety of the world's diverse ecosystems, can survive without an adequate quantity and quality of water for our basic needs. Many of the aforementioned international-level networks and community groups are rapidly growing around specific centres of interest, including a variety of different aspects of water management. However, their capacity to effect on-the-ground action and improve human living standards appears minimal (Gleick *et al.*, 2006), particularly for the almost 1 billion people who lack access to clean drinking water and the 2.6 billion who lack access to adequate forms of sanitation (UNDP, 2006).

General water 'scarcity' issues (Rijsberman, 2006) in many parts of the world and conflicts between competing water uses for potable water, sanitation, food production, industry, energy production and many other uses (social, recreational and spiritual), do not help the plight of these billions. Such drivers as population growth, climate change, technological innovations and past water management choices, including the construction of engineering structures and introduction of planning regulations, are to some extent all responsible for these issues that are linked to human behaviour. These drivers are also partially responsible for the increasing risk of damages and loss of life caused by 'natural' hazards, such as floods, droughts, storms, earthquakes and ecological shifts, such as algal blooms or fish kills (Kundzewicz and Takeuchi, 1999; Abramovitz, 2001).

1.3 PROBLEM STATEMENT

Current water management and planning, including their associated decision making processes, are commonly characterised by interconnecting and complex problems that exhibit high levels of conflict and uncertainty. This results from overlapping legislative requirements, multiple decision makers and managers, competing interests, unequally distributed water resources and social and environmental impacts of their development, as well as uncertainties about the future in a more interconnected and rapidly changing world. In such contexts, the decision making process for the selection and implementation of water management strategies becomes a major challenge. 'Traditional' methods of water management and planning are usually insufficient (Gleick, 2000a), as are 'traditional' or 'objective' forms of risk assessment (Klinke and Renn, 2002). The pertinence of expert-created integrated water models designed to inform policy decisions, or quantitative risk analyses to determine levels of 'acceptability', has been more broadly questioned due to the unrepresentative nature of these experts' values-based decisions (Fischer, 2000; Rayner, 2007). In such water management and planning contexts, it is unusual that one institution possesses all of the relevant knowledge and is in control of all the resources required to implement its own decisions successfully. This means that water engineers and managers are increasingly obliged to work with other institutions, stakeholders, experts and the general public to create more acceptable models and plans and to implement management solutions (Loucks, 1998). Therefore, there is a widely recognised and increasing need for the development of improved approaches to aid inter-organisational decision making in the water sector, in order to ensure the sustainable and equitable development of water resources and their dependent societies and environments.

Decision-aiding for water management and planning has long focused on the building of models by experts, which can be used to inform managers' decisions. However, it is considered that in many current inter-organisational water management and planning contexts, decision-aiding through the use of such expert-created models is problematic: in particular, model transparency, scope of the problem treated, uncertainty related to model inputs and outputs, and expert legitimacy are just some of the aspects that can come under attack when this type of decision-aiding practice is pursued, particularly when stakeholders who may be adversely impacted are not involved in the decision process (Fischer, 2000; Rayner, 2007). To address such issues, 'participatory modelling' has been mooted as a potential solution. Participatory modelling is a process that allows a number of different points of view to be explicitly represented and collectively reflected upon by a group of stakeholders through a series of semi-structured decision cycles (Ferrand, 1997). The potential for participatory modelling to be used as a process for inter-organisational decision-aiding in the water sector remains under-evaluated and in need of further investigation.

Such inter-organisational decision-aiding processes for water planning and management are typically organised or 'co-engineered' by several agencies or actors, owing to their size and complexity, meaning that participatory processes are co-initiated, co-designed and co-implemented by a number of people. Co-engineering has also received scant attention in studies of participatory decision making and remains a large gap in current knowledge.

1.4 UNDERLYING HYPOTHESES

The initial hypothesis that guided this research project is that:

> *Situations exist where it is useful to use a participatory modelling approach to aid inter-organisational decision making for water planning and management.*

Linked to this hypothesis, it is assumed that:

The increasing complexity of water-related problems has contributed to the need for improved inter-organisational decision-aiding for water planning and management.

It is then further assumed that:

Participatory modelling processes used for inter-organisational decision-aiding in complex water management contexts are co-engineered.

This then leads to the central hypothesis of this research, that:

Co-engineering can critically impact on the participatory modelling processes and their outcomes.

1.5 BOOK AIM AND OBJECTIVES

To examine these hypotheses, the aim of this study is:

To investigate the impact of co-engineering of participatory modelling processes for inter-organisational decision-aiding in water planning and management.

To fulfil this aim and investigate the listed hypotheses, the book has the following objectives:

1. To critically review past and current water governance systems, their management priorities and strategies to examine whether water management has become increasingly complex.
2. To critically review decision-aiding theory and methods, including participatory modelling, and the way in which they could be used to improve water planning and management.
3. To develop a definition of, and critically review, the concept of co-engineering as it relates to the organisation of participatory modelling processes for water management. This is to allow the identification of priority gaps in knowledge that require further research.
4. To formulate an intervention research programme and evaluation protocol for investigating co-engineering of participatory modelling processes, for inter-organisational decision-aiding in water planning and management.
5. To outline the lessons learnt through individual and comparative intervention case analysis, so as to determine to what extent co-engineering can critically impact on participatory modelling processes and their outcomes.
6. To propose suggestions for future best practice, new perspectives and priority areas in need of further research in co-engineering participatory modelling processes for inter-organisational decision-aiding in water planning and management.

1.6 SCOPE OF THE STUDY

The study will address the aim and objectives from the perspective of water governance at the international, Australian and European levels. An in-depth comparison of the co-engineering of two inter-organisational participatory modelling processes in the Australian and Bulgarian cultural and institutional contexts will be provided. Further examples of co-engineering practices being developed in other contexts, including Algeria, will also be outlined. The main focus of the book will therefore be based on the co-engineering of participatory processes for water management, which can be considered to be situated in the arenas of constitutional and collective choice (Ostrom, 1990), relative to the 'on-the-ground' socio-ecological process systems and day-to-day operational choices of water managers. The focus area is represented in Figure 1.1 as part of an idealised representation of the interconnected feedback systems.

Investigating the co-engineering of participatory processes for inter-organisational decision-aiding in water planning and management requires an understanding of previous theory and practice in a range of relevant academic disciplines. In this transdisciplinary book, a choice has been made to limit the range of literature and examples discussed to those that are relevant to on-the-ground practitioners and researchers, managers, consulting engineers and professional facilitators who are working towards the improvement of water planning and management. Disciplines with a focus on practice and action, such as water engineering, operational research or management science, regional planning and environmental policy, are therefore drawn upon to a greater extent than other academic disciplines with long theoretical and methodological histories, such as sociology, economics, anthropology or psychology.

Although elements of philosophical thought and theories of participation and democracy will be touched upon, this book is not directed towards advancing these bodies of knowledge. Other recent theses with bases in political, sociological and development theory develop these aspects of participation in water planning and management and are available to complement this enquiry (i.e. Barnaud, 2008; Ker Rault, 2008; Richard-Ferroudji, 2008). Similarly, this book is not an in-depth study into socio-ecological water processes, looking at water use behaviours, distribution, hydrological processes and so on; rather, it examines how coordinating decisions over the broader scale water governance aspects can be aided.

In other words, the research in this book stems from the observation that broad-scale participatory processes are becoming increasingly common in the water sector, and that practical, actionable knowledge of how better to organise them is needed. In particular, this work aims to highlight the practical need of water planners, policy makers, engineers, community workers and scientists for a greater understanding of how they could work

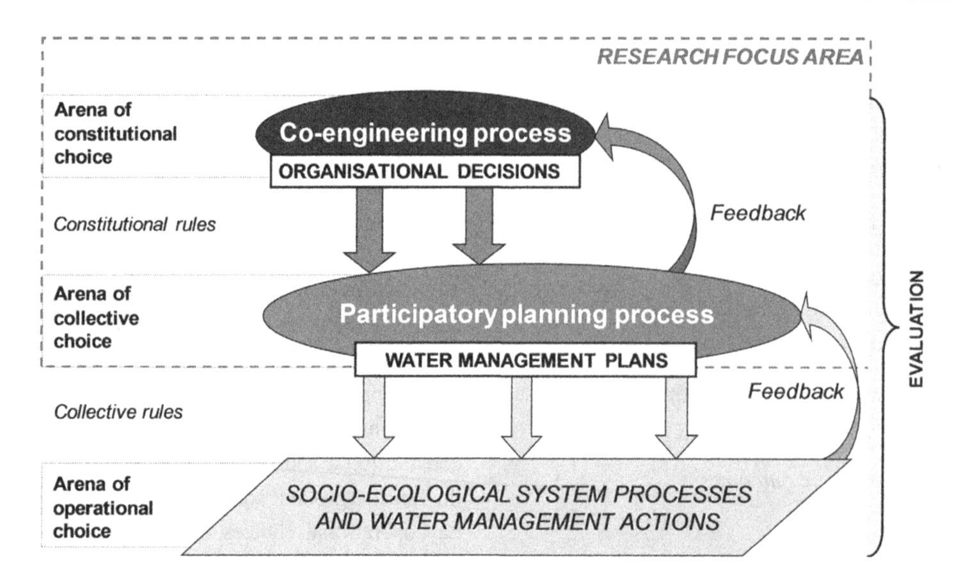

Figure 1.1 Linked systems definition of co-engineering processes for participatory water management relative to Ostrom's (1990) institutional analysis levels (left column).

with others using participatory modelling methodologies, so as to better manage the complex problems they face in today's world.

1.7 BOOK OVERVIEW

To achieve the aim and objectives, this book has been constructed with two principal parts. Part I, consisting of Chapters 1, 2, 3, 4 and 5, frames the context of the research work by presenting a critical review of literature to identify knowledge gaps and the development of research protocols which, when applied, could help to fill these gaps. Part II, consisting of Chapters 6, 7, 8, 9 and 10, then highlights the lessons learnt through research interventions and evaluation protocol application, as well as *ex-post* comparative analysis, extension, discussion, conclusions and areas of required future research. An outline of each of these parts is presented in more detail.

1.7.1 Part I: Framing the context

To introduce further the general context of this study, following Chapter 1's introduction of the water problem context and outline of the scope of the book, Chapter 2 reviews a range of current governance systems, issues and priorities for water planning and management internationally and in Europe and Australia, to provide sufficient background on the water governance contexts of the two intervention research case studies and other co-engineering examples in Part II. A brief review of the lessons learnt from failed management approaches is provided, to identify the need for alternative approaches. Reflecting on these reviews, a number of future needs and opportunities are highlighted, including the *need to develop and implement improved methods of aiding decision making in water planning and management*, in particular for inter-organisational decision-aiding.

Chapter 3 outlines the concept of decision-aiding and its use in relation to the water sector. A critical review of the literature discusses the origins and evolution of decision-aiding practices with a specific focus on theory and management practice from engineering, operational research and management science, and environmental and public policy literature, as well as decision-aiding and its relevance to the inter-organisational water management and planning. How decision-aiding models can be put into practice in this context is then examined. Participatory structure design is also examined, and a comparison of participatory modelling methods is made. An example of integrated participatory modelling for inter-organisational decision-aiding in water planning and management is proposed. The choice or design of different methods for use in such methodologies based on contextual needs and constraints is highlighted as a knowledge gap, as well as what happens when a number of analysts and decision makers are required to co-engineer the participatory modelling processes.

Chapter 4 fills these gaps by critically reviewing a number of current approaches designed to aid the choice, mixture or creation of methods in participatory interventions and determine the remaining gaps in knowledge. The large gap in the understanding of the co-engineering of participatory modelling processes for decision-aiding is then analysed. A definition of the co-engineering process is given, followed by a critical interdisciplinary review of literature to gain insights on the concept. A research agenda on the co-engineering of participatory processes is then outlined, including the need for an appropriate research approach and evaluation protocol to aid the comparative assessment and learning on inter-organisational decision-aiding in the water sector.

Based on these needs, Chapter 5 presents the research protocols to be used as part of a 'participatory intervention research process' for *investigating co-engineering of participatory modelling processes for inter-organisational decision-aiding in water planning and management.* The principal objects of interest within this process, the 'co-engineering process' and the internal 'participatory modelling process', are delimited and the choice of setting these research boundaries is discussed. An adaptation of the Tsoukiàs' (2007) decision-aiding process model to the inter-organisational context is proposed for use as the base for constructing participatory modelling methodologies. An outline of the kind of evaluation protocol and methods that could be used to monitor and develop further insights on the co-engineering of participatory modelling processes for inter-organisational decision-aiding for water planning and management then follows. Finally, the validation and legitimisation of research insights obtained through an intervention research approach are outlined.

1.7.2 Part II: Learning through intervention

Drawing on the research needs and theoretical framework identified in Part I, Part II presents a selection of intervention cases. Two in-depth case studies from the water sector, which focus on estuary management in Australia and flood and drought risk management in Bulgaria, are then described. The results of the evaluation procedures are reported and an outline is given of a range of lessons learnt. The discussion is further extended by examining co-engineering practices in other recent cases, including for agricultural water management in Algeria.

Commencing Part II, Chapter 6 outlines the practical intervention cases used to create actionable knowledge through interventions of co-engineering participatory modelling processes for inter-organisational decision-aiding in water planning and management. Information is provided on the case selection and a brief background to the cases, including data sources and interpretation schemes. An overview of the lessons learnt from the pilot intervention case carried out in Montpellier, France, that were used to inform the next two interventions in Australia and Bulgaria is also provided. Elements outlined include some adaptations to the evaluation protocol and learning about whether participatory modelling processes for decision-aiding require simulation models.

Chapter 7 presents the Australian intervention case based on the adaptation of a participatory modelling methodology to a 'participatory values-based risk management approach', which was used for collective decision-aiding in the creation of the Lower Hawkesbury Estuary Management Plan in New South Wales (NSW), Australia. This process, driven by local government, and using the Australian and New Zealand Risk Management Standard (AS/NZS 4630:2004), included three interactive stakeholder workshops with a range of stakeholders from state and local governments, the water and sanitation authority, local industries, community associations and residents. Evaluation results demonstrate that the process was efficient from a time and budgetary perspective and has a number of other potential benefits, including broad agency support, which are outlined, together with some lessons learnt and questions arising in need of future research.

Chapter 8 presents the intervention – 'Living with Floods and Droughts' – in the Upper Iskar Basin in Bulgaria to be used for building collective capacity in flood and drought risk management. This year-long process, driven by a number of researchers and regional stakeholders, included two phases of interviews and 15 workshops organised into series for six groups of paid stakeholders from national-level policy makers and experts to municipal-level Government representatives and citizens from around the region. The process was co-engineered to include qualitative participatory modelling activities on: stating expectations, modelling systems and actors, eliciting visions and values using cognitive mapping and causal modelling techniques; developing management options and strategies, framing and assessing strategies using option cards and multi-criteria analysis; and robustness testing of scenarios, risk response project planning and process evaluation. The co-engineering of this participatory modelling process is presented and discussed along with a range of participatory modelling process evaluation results, lessons learnt and areas of interest for further research.

Following the descriptions and results of the case studies presented in Chapters 7 and 8, Chapter 9 presents a comparative discussion of the two case studies, with a focus on: context effects; participatory modelling methodologies; the co-engineering team processes and effect of divergent objectives and leadership; participatory process ethics; and participant evaluation results. It then looks at the validation of the models and protocols used through the intervention research methodology, including an inter-organisational decision-aiding model, a participatory structure model, the evaluation protocol and the legitimisation of the intervention research findings. The discussion is also enriched with further examples of co-engineering practices from other participatory processes around the world, including cases from: the Dhünn Basin in Germany, on ecological river restoration; the Mitidja Plain in Algeria, on understanding and managing agricultural water use behaviours; and

Tarawa in the Republic of Kiribati, on water and sanitation development. Suggestions are then made for future best practice in the use of participatory modelling for inter-organisational decision-aiding in the water sector.

Chapter 10 gives final conclusions relating to the extent to which *co-engineering can critically impact on the participatory modelling processes and their outcomes* and the other research hypotheses. The key contributions of the research are summarised and related to the book aim and objectives. Finally, a range of priority areas and questions that require further research is outlined.

1.7.3 Book overview

The structure and flow diagram of the book is presented in Figure 1.2. The solid arrows represent direct linkages in the book structure, and the dashed arrows represent conceptual or indirect linkages.

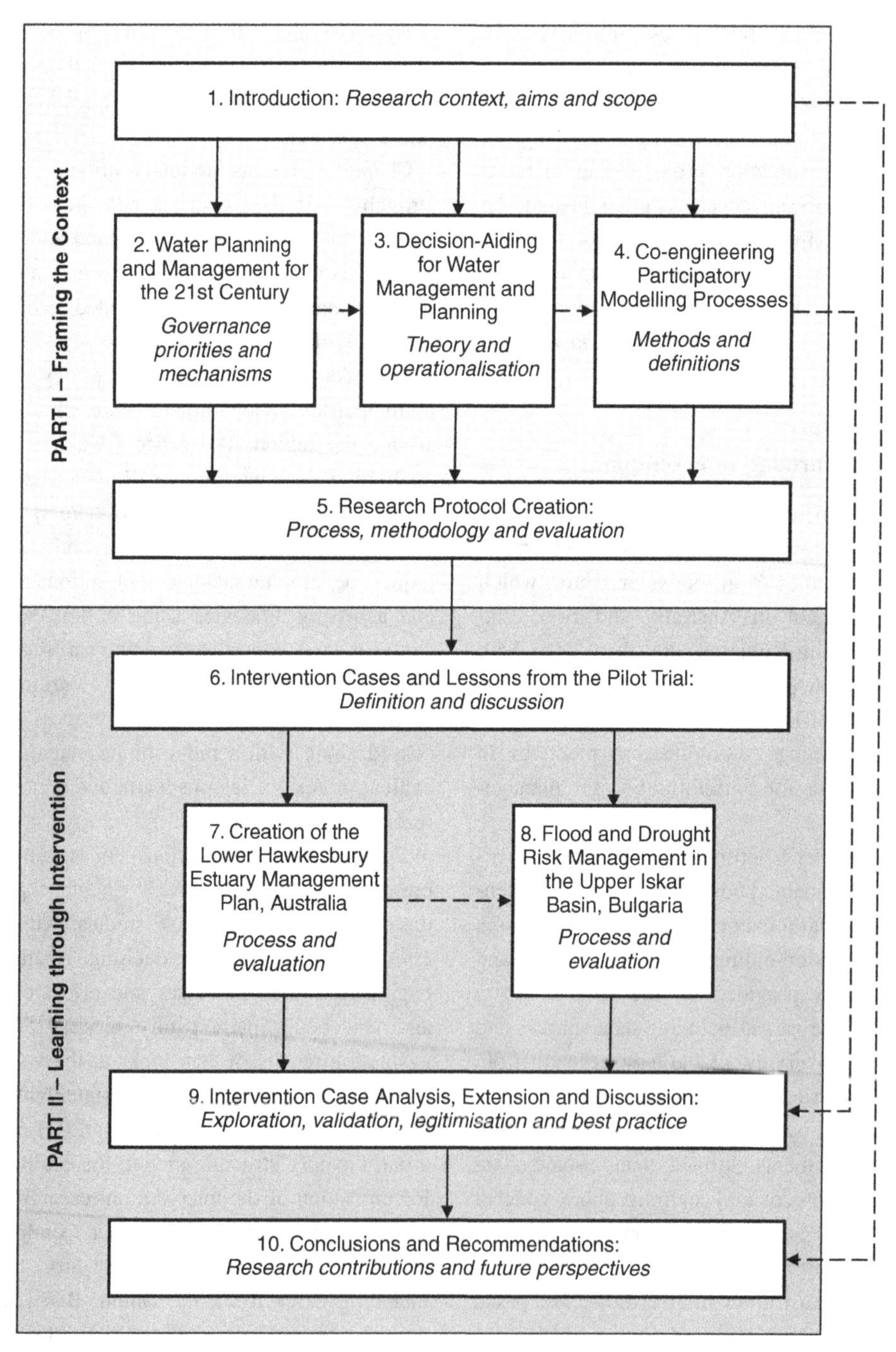

Figure 1.2 Book flow diagram summary.

2 Water planning and management for the twenty-first century

The world is heading for a water crisis that is unprecedented in human history; water development and management will change more in the next 20 years than in the past 2000 years.

ASIT BISWAS *Kyoto World Water Forum (2003)*

This chapter aims to outline the current context of water management, through the critical review and comparison of Australian, European and International water governance systems, and the challenges that remain for the future, as well as to provide a brief review of management approaches that could be considered as history's mistakes and from which we have much to learn. The purpose of this review is to propose responses to a number of larger questions, including:

- What are the principal challenges facing water planners and managers in the twenty-first century?
- What governance mechanisms are currently being put in place for managing these challenges and do they appear adequate?
- What can we learn from history that may help us to manage today's and tomorrow's water issues better?

2.1 CURRENT GOVERNANCE SYSTEMS FOR WATER PLANNING AND MANAGEMENT

Systems for water planning and governance vary widely around the world. The priorities of the water governance systems are closely linked to the main concerns and resources of their surrounding political systems and differ significantly, depending on the scale of management being addressed, its location and its hydrological and socio-economic context. This section will highlight just three examples and give a brief comparison of such systems, including their main challenges and management priorities: water governance at the Australian national level; the European Union's supra-national level; and an international level.

2.1.1 International frameworks and priorities

Access to water for life is a basic human need and a fundamental human right.

Human Development Report 2006 (UNDP, 2006)

As innocuous and reasonable as this statement in the 2006 Human Development report may seem, finding unanimous support for it from national ministers in international forums is just one of a significant number of challenges with which governance of water issues on an international scale must cope (Gleick *et al.*, 2004). Despite water and sanitation recently being acknowledged explicitly as a fundamental right and prerequisite for the realisation of all other human rights in the United Nations General Comment No. 15 (United Nations, 2002) and specifically in a legally enforceable resolution adopted by the General Assembly of the United Nations (United Nations, 2010), opponents of acknowledging water rights do so potentially to avoid it being considered by international lawyers as part of 'customary international law', in order to shirk legal, financial and moral responsibilities (Gleick *et al.*, 2004).

Legislating that water is a 'human right' and is not just considered as a 'need' is seen by many in the global community as a key step in highlighting how essential water is for the full enjoyment of life and all other human rights, and to incite increased action to uphold this right for the almost 1 billion people who currently lack access to clean and safe drinking water and the 2.6 billion who lack access to adequate forms of sanitation (UNDP, 2006). The direct results of the current lack of water and sanitation include the deaths of between 14 000 and 30 000 people per day, the majority of whom are children and the elderly (Gleick, 2000b), and daily disease-related problems for the equivalent of about half those living in the developing world (United Nations, 1997). Clearly, considering these statistics, efforts to improve water management on a worldwide scale have a long way to go, despite concerted efforts to address these kinds of issues for almost half a century.

The acknowledged need for a concerted international effort to address water issues effectively commenced in the years following the establishment of the United Nations after the

Second World War. It started with the creation of a number of water-related scientific and political international associations and the UNESCO-run 'International Hydrological Decade' from 1965–1974 (Varady, 2004). This initiative was then developed into the International Hydrological Programme (UNESCO-IHP), which has played an important role in international water initiatives ever since, including aiding the publication of the first *World Water Balance and Assessment of Water Resources of the Earth* in 1978 (Varady, 2004). During this time, the rise in worldwide social and environmental movements (for example, The 'Club of Rome') and calls for citizen participation in decision making corresponded with a range of United Nations 'mega-conferences' to address such issues (see Biswas (2004) for details). One of these conferences in 1977 was the United Nations Water Conference in Argentina, the first high-level political meeting of its type, where the 'Mar del Plata Action Plan' for water development and resources was drafted (Gleick *et al.*, 2006).

Even by today's standards, the plan is considered a remarkable and insightful political declaration that considered water in a holistic and comprehensive manner (Biswas, 2004). The opening statement of the conference gives an image of the objectives of the meeting:

> *It is hoped that the Water Conference would mark the beginning of a new era in the history of water development in the world and that it would engender a new spirit of dedication to the betterment of all peoples; a new sense of awareness of the urgency and importance of water problems; a new climate for better appreciation of these problems; higher levels of flow of funds through the channels of international assistance to the course of development; and, in general, a firmer commitment on the parts of all concerned to establish a real breakthrough so that our planet will be a better place to live in.*
>
> (Mageed (1978), in Biswas (2004)).

Although a small number of issues that were neglected, including international transboundary water management issues and the financial aspects of how the action plan could be successfully implemented (Gleick, 2000b), the plan outlined a number of important principles and recommendations ranging across the areas of: evaluation and assessment; water use and efficiency for development and sectorial needs; environment, health and fighting pollution; politics, planning, management and institutional aspects; teaching, education and research; natural disasters; and regional and international cooperation (United Nations, 1982). These included making the first explicit declaration of the human right to water:

> *All peoples, whatever their stage of development and their social and economic conditions, have the right to have access to drinking water in qualities and quantities and of a quality equal to their basic needs.*
>
> United Nations (1977)

The period after this conference saw water issues slipping largely off the world stage again until the 1992 International Conference on Water and the Environment in Dublin, Ireland (with the notable exception of on-ground work to implement actions from the Mar del Plata Action Plan through the 'International Drinking Water Supply & Sanitation Decade from 1981–1990' (WWAP, 2007)). The Dublin conference was prepared as a precursor to the Earth Summit in Rio de Janeiro, Brazil, which was to be held four months later, although the Dublin conference was predominantly an expert meeting with no inter-governmental committee, unlike the Mar del Plata conference. This omission meant that the recommendations, now known as the 'Dublin Principles', were not allowed to be officially considered at the UN Earth Summit where the Section 18 on water of Agenda 21 (United Nations, 1992) was drafted (Biswas, 2004). Despite this hurdle, as many of the experts were present at both the expert meeting and the Earth Summit drafting sessions, the information was still partially taken into account (Daniell, 2008; personal communication).

The recommendations of the Dublin conference have since been accepted into the common-knowledge sphere of many water professionals and policy makers. The Dublin Principles are (ICWE, 1992):

1. *Fresh water is a finite and vulnerable resource, essential to sustain life, development and the environment*;
2. *Water development and management should be based on a participatory approach, involving users, planners and policy makers at all levels*;
3. *Women play a central role in the provision, management and safeguarding of water; and*
4. *Water has an economic value in all its competing uses and should be recognised as a common good.*

Some water experts argue that these principles, and those of Section 18 in Agenda 21, are not a large improvement on the original Mar del Plata Action Plan, especially the economic recommendation, which moves away from the carefully defined recommendation of adopting *'appropriate pricing policies with a view to encourage efficient water use, and finance operation cost with due regard to social objectives'*, and omits the issues of equity and poverty (Biswas, 2004). However, a number of other authors have been more vocal in their support of the principles, in particular of Principle No. 2, on the need for a participatory approach to water development and management (FAO, 2000; Rahaman *et al.*, 2004), as it is a base element of the 'Integrated Water Resources Paradigm', which is heavily promoted on an international scale (Ker Rault, 2008).

Towards the end of the 1990s, a real movement started to occur, with the development of many new associations and initiatives, including the Global Water Partnership and the World Water Council in 1996, the World Commission on Dams in 1997, the World Commission on Water for the 21st Century in 1998, the Global International Waters Assessment (GIWA) and the Hydrology for Environment, Life and Policy (HELP) in 1999, the World Water Assessment Programme (WWAP) in 2000, the Dialogue on Water & Climate/Cooperative Program on Water & Climate (DWC/CPWC) in 2001 and the Global Water System Project (GWSP) in 2002 (Gleick, 1998; Varady, 2004). Although each of these groups has a slightly different range of objectives and constitution of participants, there are still a number of overlaps. For instance, the World Water Council and Global Water Partnership cover similar territory as they are open forums for a range of bodies, including UN bodies, national governments, other water associations, public and private companies, and NGOs, and they both spend a good deal of resources organising international meetings and forums (Gleick, 2000b). The World Water Council's series of 'World Water Forums' (Marrakech, Morocco, 1997; The Hague, Netherlands, 2000; Kyoto, Japan, 2003; and Mexico City, Mexico, 2006) have provided spaces for inter-governmental meetings and inter-ministerial declarations, which were traditionally organised by the UN, and are attracting increasing numbers of participants, as well as criticisms over cost effectiveness and bland and unoriginal statements (Biswas, 2004; Gleick *et al*., 2006).

In 2000, the United Nations also released their 'millennium development goals' (MDGs), including one on water (updated in 2002 at the World Summit on Sustainable Development in Johannesburg), which was to: 'Halve by 2015 the proportion of people without sustainable access to safe drinking water and basic sanitation (Goal 7, Target 10)' (UNDP, 2003). This was one of the first clear statements of intent with a measurable target for water management at an international scale. However, progress since the goals' definition has been unequal across regions, with a range of problems being identified, including a lack of political will, structural governance problems, inadequate financing and conflicting agendas. Progress towards the MDGs in sub-Saharan Africa has been particularly limited. Unless there is a change of direction very soon to overcome these blockages, it is now unlikely that the millennium goals will be met, especially for sanitation (Gleick *et al*., 2006).

Statements this century from the World Water Forums have reiterated a range of generally agreed priorities about how to 'provide water security in the twenty-first century', including (United Nations, 2000, 2003, 2006; WWC, 2009):

- Working to meet the Millennium Development Goals for basic water and sanitation needs;
- Empowering people, especially women, through participatory approaches to water management;
- Securing food supply, particularly for the poor and vulnerable to eradicate hunger, by supporting efficient and productive agricultural water use;
- Protecting ecosystems and preventing pollution: combating deforestation, desertification, biodiversity loss and land degradation;
- Implementing demand-management strategies and innovative development projects in both rural and urban areas, such as drought-resistant crops, hydropower production, desalination, water reuse and local technologies for water harvesting, water saving and water storage;
- Peacefully sharing water resources for all uses through effective transboundary cooperation, river basin organisation or other appropriate approaches;
- Managing risks and preventing disasters linked to water-related and climate change hazards (i.e. droughts, floods, pollution) through improved cooperation, preparedness risk assessment, infrastructure to increase storage and drainage capacity, community awareness, resilience and response;
- Valuing water: to reflect its economic, social, environmental and cultural values in all its uses, promoting full cost recovery and polluter-pays, equity and consideration of the poor and vulnerable;
- Coordinating monitoring systems, assessments and research production through all management scales (local, basin, national, international); and
- Governing water wisely in an integrated, accountable, transparent and participatory manner with input from all stakeholders, including public and private institutions, NGOs, research institutions and civil society: this should include focusing on local community-driven approaches, equitable benefit sharing, and due respect for the perspectives of the poor, women and the most vulnerable, as well as working to clarify roles, rights and responsibilities of all actors.

In retrospect, some authors question the extent to which inroads have been made in global water problems since the Mar del Plata Conference in 1977, and are becoming increasingly sceptical about the usefulness of the multiplicity of large water forums and conferences of recent years, apart from providing increased networking opportunities and visibility of issues (Biswas, 2004; Gleick *et al*., 2006). It is estimated that just the cost of the organisation and conference fees of the Third World Water Forum was equal to twice the United States Agency for International Development's contribution to water supply and sanitation for the continent of Africa (Gleick and Lane, 2005). Would some of this money have been better spent implementing projects to meet the millennium goals or will the 'talk-fests' encourage transformatory processes in the governments of the attendees? Despite this criticism, others see positive changes in the focus of international initiatives, e.g. Varady (2004):

- Moving away from thinking about water as an important resource, to recognising that potable water and sanitation access is a key to alleviating problems and aiding development;
- Moving to international scale multi-lateral initiatives to tackle water issues from sectorial and technical aspirations of improving scientific bases of knowledge and communication; and
- Moving beyond data acquisition and theory and broad-scale programmes with overarching goals, towards trying to improve on-site conditions through local targeted action programmes focused on the equitable and efficient use of water.

However, even if the international community has voiced its priorities for water management, and has made some attempts over recent decades to move from harder technological paths for treating water scarcity and development to softer paths of working with people, education and alleviating poverty, there are many obstacles and challenges that remain to be overcome, including:

- Fragmentation and multiplicity of international institutions: clear responsibilities still remain to be defined to avoid financial losses and time wasting;
- Diametrically opposed national positions on priorities for water management and use in development initiatives: for example, that a number of nations see water management as a low international priority, especially in terms of providing water as a right, and thus do not provide sufficient financial aid, and that there is a cleavage between general positions on the best paths forward (i.e. using markets, public–private partnerships, efficient centralised systems; the 'soft' path run predominantly by local communities with NGOs and governments; or mixes of these);
- Translation of high-level coordination and talk to on-ground implementation that makes a real difference to the receiving communities;
- Distributed information systems: much data required to aid the implementation of projects on-the-ground, especially at national levels, remains difficult to access due to regionally stored or privately owned data;
- A general lack of long-term visions for water futures at the international level; and
- Changing political priorities and human behaviours.

Considering the initial issue of water rights in this larger context, one of the key underlying challenges of the future in international water management will be to determine whether the two strong positions for the future of water management are reconcilable: the current conflict between treating water as a public good to be provided as a right for meeting basic needs and treating water as a resource and commodity to be regulated through markets. This will require in-depth analysis of the underlying value systems that have led to these positions and whether there might be water management options that could allow these value-systems to cohabit peacefully or be transformed to ensure a better future and governance of water for all members of current and future generations.

2.1.2 The Australian system and priorities

Patter, patter ... Boolcoomatta,
Adelaide and Oodnadatta,
Pepegonna, parched and dry
Laugh beneath a dripping sky.

C. J. DENNIS *A Song of Rain* (1918)

Australia is a land of extremes and paradoxes. It is the driest inhabited continent and has the highest stream-flow variability in the world (Finlayson and McMahon, 1988; McMahon, 1988). The country is physically isolated and per capita there are bountiful renewable national water resources with about 2500 m^3 to 3500 m^3 per capita per annum depending on the yearly climatic variations (NWC, 2007), which is well outside of the typical physical standards of water scarcity of 1000 m^3 per capita per annum, as described in Section A.1.3. The equivalent of approximately 1000 m^3 is consumed per capita per annum (NWC, 2007), with about 70% of the total being used for agricultural production (Chartres, 2006). However, Australia's small population of around 22 million is poorly distributed relative to the available water resources. Far north Australia receives 52% of the average annual runoff for 2% of the total population and Southern Australia receives 27% of the total average runoff for 82% of the total population (Nix (1988) cited in Hugo (2007)). Moreover, Australia is highly urbanised and, of the total population, 85% live within 50 km of the coastline (ABS, 2007), which presents Australia with a challenging range of water management considerations.

Australia is also a particularly young country in terms of written history – with settlement of Europeans from the First Fleet from Britain occurring in 1788 – but has an extraordinary and long oral history from the indigenous populations who have inhabited the territory for over 40 000 years (Diamond, 1999). Over this time the Australian indigenous populations have developed intimate knowledge of the Australian environment, their 'country', and hold strong spiritual and cultural bonds with it (Rose, 1996; Yu, 2000; Jackson *et al.*, 2005). Many indigenous cultural management systems and traditions were used to nurture and sustain the environmental and spiritual health of the water sources and lands (Lingiari Foundation, 2002). However, since the arrival of the European colonists, Australian landscapes have been quickly and dramatically altered. In a little over 200 years, much of the temperate and semi-arid land in the south was deforested and turned into farmland, cities were built and the Aboriginal populations of those regions were devastated by disease (Diamond, 1999). Land use changes and the introduction of exotic species also caused widespread havoc and environmental degradation in a country that has many unique ecosystems and some of the oldest and nutritionally poorest soils in the world (Flannery, 1994). Even before the Federation of the Australian States in 1900, numerous water-related and environmental problems were observed, including: water scarcity that restricted population growth, agricultural and industrial development, especially in the Southern Australian colonies; devastating droughts; conflicts over the use of shared water resources; and land degradation (Hutton and Connors, 1999; Connell, 2007).

Transboundary water use, conflict and governance agreements have long been a feature and are important issues in the Australian psyche, especially in the Murray-Darling Basin, Australia's 'Food Bowl' or 'Bread Basket', where approximately 50% of Australia's food is currently produced from only 6.1% of the country's runoff (Chartres, 2006). Water management and especially water allocations in Australia have always been governed by the States and Territories, and not at a national level. In the lead-up to Federation, intense negotiation surrounded the creation of Section 100 of the Australian Constitution on water use, in an attempt to leave water management decisions largely to policy makers and not lawyers (Connell, 2007). As can be seen, the power for water allocation was left with the States:

> *The Commonwealth shall not, by any law or regulation of trade or commerce, abridge the right of a State or of the residents therein to the reasonable use of waters of rivers for conservation or irrigation.*
>
> (Commonwealth of Australia Constitution Act, 1900)

Leaving water under State control has had a number of important impacts on water management in Australia, including: the development of different bodies of water-related legislation and management systems in each State, adapted to the goals and priorities of each State, which were sometimes in competition with each other; and the need to cooperate over a certain number of issues, in particular over the management of the cross-boundary Murray-Darling Basin, where a number of inter-jurisdictional initiatives were implemented which included the *River Murray Waters Agreement* and the development of the *River Murray Commission* in 1915 to oversee integrated planning works and water-use allocations between the States.

Coinciding with the development of worldwide hydrological programmes after the Second World War, Australia also embarked on an intensive campaign of 'Nation Building' to tame or divert many of its rivers, with the general aim of providing water security for agricultural production, industry and burgeoning urban populations. These projects included: the Snowy Mountains Hydro-Electric Scheme, constructed from 1949 to 1974 in the Murray-Darling Basin (ABS, 1986); the Ord River Irrigation Scheme, constructed from 1968–1973 in Northern Western Australia (Kittel, 2005); a series of major pipelines from the Murray to South Australian settlements, constructed from 1940 to 1973 (SA Water, 2008); and the construction of many major dams around the continent to supply urban centres and irrigation areas with adequate drinking water and irrigation supplies. The results of such developments have been that Australia has created the highest water storage capacity in the world compared with its rivers' mean annual flows (Finlayson and McMahon, 1988).

However, by the early 1970s a number of significant environmental issues had emerged, including widespread land salinisation (Sexton, 2003) and ecosystem damage (Hutton and Connors, 1999). Environmental and social movements gained increased strength; people in these groups carried out protests against engineering projects such as the Lake Pedder Dam in Tasmania (Smith and Handmer, 1991) and called for the protection of many sensitive sites from continued development, as well as more local participation in decision making (Handmer *et al.*, 1991; Hutton and Connors, 1999). Despite increased environmental social activism during this period, it was some time before this translated into Federal Government Policy.

The 1980s and early 1990s were a much more productive period for water reform in Australia with the:

- Development of the *Salinity and Drainage Strategy*, which was finally implemented in 1988 to combat the issues related to salinisation (Sexton, 2003; Connell, 2007);
- The Murray-Darling Basin Agreement of 1987 and installation of a new Murray-Darling Basin Commission, Community Advisory Committee and Inter-Ministerial Council, whose role was: 'to promote and coordinate effective planning and management for the equitable, efficient and sustainable use of the land, water and environmental resources of the Murray-Darling Basin' (Powell, 2002; Connell, 2007);
- A number of participatory water planning and management initiatives, including 'Total Catchment Management' or 'Integrated Catchment Management' schemes around the country, especially to treat rural management issues (SCEH, 2000) and involve urban residents in water conservation and long-term planning, such as in the country's capital city, Canberra (Penman, 1988; CSIRO ASSERT, 1992);
- Establishment of the *Coalition of Australian Governments* (CoAG) in 1992 and its adoption of the *National Strategy for Ecologically Sustainable Development* (ESD) (Australia's Equivalent of Agenda 21 (United Nations, 1992)), the *Natural Resources Management (Financial Assistance) Act 1992*, which included the installation of the *Australian Landcare Council* with its Water Reform Framework in

1994 to legally protect freshwater ecosystems, under which the states are required to allocate water to environmental needs (Gleick *et al.*, 2006; Allan, 2007);

- Development of the *National Water Quality Management Strategy* (NWQMS) by the *Australian and New Zealand Environment Conservation Council* (ANZECC) (DEWHA, 2007); and
- Introduction of the Murray-Darling Basin '*Cap*' in 1995 on water extractions to attempt to improve riverine health – as the Audit of water use in the Basin showed the riverine ecosystems and surrounding lands were under severe stress as a result of flow regime changes – and to meet the objectives of the CoAG environmental reforms (MDBC, 2004; Connell, 2007).

This period also saw a new wave of economic liberalism emerge to enhance national productivity, efficiency and economic growth, which included many micro-economic reforms and a review of Australia's competition policy for electricity, gas, water and road transport, which resulted in the development of the *National Competition Policy Reform Act* in 1995 and the *Australian Competition and Consumer Commission* (ACCC) and *National Competition Council* in the same year, to promote competition and regulate fair trade under the *Trade Practices Act* of 1974 (Kain *et al.*, 2003). These reforms had a major impact on water management in Australia, involving the following general trends:

- Separation of water supply management from regulation and planning functions;
- Corporatisation or privatisation of water suppliers; and
- Introduction and push for water markets and water trading to aid a move to 'high-value' uses of water.

By the late 1990s, it was increasingly noted that the severity of environmental degradation in Australia was having major economic impacts. The lack of coordination between numerous bodies at local, state, and national levels in part responsible for management issues, which now included private water companies and community associations, was seen to be one of the largest hurdles to improving environmental and water management in the country (SCEH, 2000). In an attempt to 'rescue' the country from further environmental and related economic decline, the Federal Government set about instituting a large range of reform and coordinating programmes, including the:

- National Heritage Trust (NHT) programme through the Natural Heritage Trust of Australia Act 1997, which included carrying out the National Land and Water Resources Audit in order:

  (a) *To estimate the direct and indirect causes and effects of land and water degradation on the quality of the Australian environment and to estimate the effects of land and water degradation on Australia's economy*;

  (b) *To provide a baseline for the purposes of carrying out assessments of the effectiveness of land and water degradation policies and programmes.*

  (COMLAW, 1997);

- National Action Plan for Salinity and Water Quality in 2000 (CoAG, 2000); and the
- National Water Initiative (NWI) of 2004, which included the creation of two new national water management bodies: the National Water Commission (NWC) under the National Water Commission Act 2004 and the Natural Resource Management Ministerial Council (NRMMC).

The NWI remains one of the overarching policies of water management in Australia and provides an indication of current water management priorities in the country (Hussey and Dovers, 2007). It is an ambitious policy, spanning eight major work areas of water management –

- *Water access entitlements and planning framework*;
- *Water markets and trading*;
- *Best practice water pricing*;
- *Integrated management of water for environmental and other public benefit outcomes*;
- *Water resource accounting*;
- *Urban water reform*;
- *Knowledge and capacity building; and*
- *Community partnerships and adjustment*;

with an implementation agenda until 2014 (CoAG, 2004). The initiative has the stated objectives of achieving (CoAG, 2004):

1. *Clear and nationally compatible characteristics for secure water access entitlements*;
2. *Transparent, statutory-based water planning*;
3. *Statutory provision for environmental and other public benefit outcomes, and improved environmental management practices*;
4. *Complete the return of all currently over-allocated or overused systems to environmentally sustainable levels of extraction*;
5. *Progressive removal of barriers to trade in water and meeting other requirements to facilitate the broadening and deepening of the water market, with an open trading market to be in place*;
6. *Clarity around the assignment of risk arising from future changes in the availability of water for the consumptive pool*;
7. *Water accounting which is able to meet the information needs of different water systems in respect to planning, monitoring, trading, environmental management and on-farm management*;

8. *Policy settings which facilitate water use efficiency and innovation in urban and rural areas*;
9. *Addressing future adjustment issues that may impact on water users and communities; and*
10. *Recognition of the connectivity between surface and groundwater resources and connected systems managed as a single resource.*

Despite the general appeal of the initiative, the mechanisms for achieving these objectives and the tasks outlined under each of the key work areas to be physically carried out were less well defined. Many of the outcomes and actions in the initiative require forms of collaboration and the development of relatively innovative projects, which have very few precedents in Australia or elsewhere in the world. This led to Land and Water Australia investigating the key areas that required research, to aid the implementation of the NWI. These areas were defined as follows (LWA, 2006):

1. ***Integrated assessment of impacts*** *of policy and water allocation changes across social, economic and environmental dimensions.*
2. ***Water plans and accreditation*** *in regard to content requirements and processes.*
3. ***Linkages between rural and urban water systems****, including in peri-urban areas.*
4. ***Indigenous perspectives*** *on water management, reforms and implementation.*
5. ***New frameworks for law and regulation****, and current settings as enablers or constraints on reform implementation.*
6. ***Values*** *attached to water and their shaping of understanding and communication of reform objectives and implementation.*
7. ***Auditing and review*** *of policy and water plans for effectiveness, and appropriate performance measures for impact detection and management.*
8. ***Water markets, pricing, trading*** *and transaction costs, and their establishment and functioning.*
9. ***Environmental water allocations*** *and their governance.*
10. ***Institutional roles, responsibilities and capacities*** *in reform implementation.*

The National Water Commission also carried out another such analysis, the results of which are outlined in Chartres (2006).

By 2007, the Federal Government was disappointed in the lack of progress on the initiative's implementation and increasingly worried by the impacts of one of Australia's longest droughts. The Government's reaction was to create a new *National Plan for Water Security*, which would include 10 billion Australian dollars of funding over 10 years, and the *Water Act 2007*. This Act also included the installation of the *Murray-Darling Basin Authority*, a new independent basin governance authority. The National Plan had a focus on improving the sustainability of rural water use, but was also supposed to support the implementation of the NWI and potentially lead to a Federal takeover of water management responsibility from the States if progress could not be made. The objectives of the Plan were as follows (DPMC, 2007):

1. *A nationwide investment in Australia's irrigation infrastructure to line and pipe major delivery channels*;
2. *A nationwide programme to improve on-farm irrigation technology and metering*;
3. *The sharing or water savings on a 50:50 basis between irrigators and the Australian Government leading to greater water security and increased environmental flows*;
4. *Addressing once and for all water over-allocation in the Murray-Darling Basin*;
5. *A new set of governance arrangements for the Murray-Darling Basin*;
6. *A sustainable cap on surface and groundwater use in the Murray-Darling Basin*;
7. *Major engineering works at key sites in the Murray-Darling Basin such as the Barmah Choke and Menindee Lakes*;
8. *Expanding the role of the Bureau of Meteorology to provide the water data necessary for good decision making by governments and industry*;
9. *A taskforce to explore future land and water development in northern Australia; and*
10. *Completion of the restoration of the Great Artesian Basin.*

Unlike the previous reforms, which had moved to treating environmental, economic and social issues in a more holistic manner, this plan appeared to put the emphasis back on centralised authority, ensuring water supply security and, in the process, Australian economic security. However, with the December 2007 change of Government, changes were again made to the direction of Australia's water policy with the adaptation of the previous Government's plan into the Government's own 12.9 billion dollar *Water for the Future Plan*.

Although maintaining many of the initiatives of the previous plan, the priorities were re-badged and the tone of the plan moved away from a technocratic and directive management style to a more integrated and cooperative management style. As stated by the then Minister for Climate Change and Water, the Hon. Penny Wong, at the presentation of the plan:

> *It is imperative for Commonwealth, state and local government to share a common understanding of the problems in water and respond in a comprehensive and coordinated way.*
>
> (Wong, 2008)

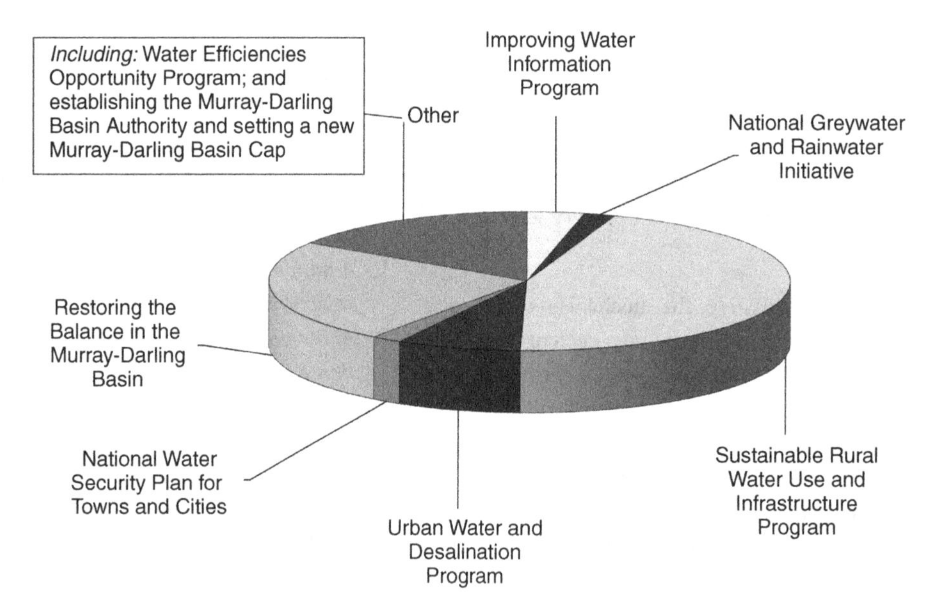

Figure 2.1 Summary of the programmes for the new 'Water for the Future Plan' and proportionate funding allocation of 12.9 billion dollars over the next ten years. Derived from data provided by the Australian Government (2008).

The four key priorities that these levels of government should then strive towards as part of the water plan were outlined as:

- *Taking action on climate change,*
- *Using water wisely,*
- *Securing water supplies; and*
- *Supporting healthy rivers.*

Each of these priorities then had a number of initiatives and a portion of the 12.9 billion dollars in funding attached, as outlined in Figure 2.1

The majority of the funding was therefore still allocated for investment in rural infrastructure and water use programmes to increase the efficiency and productivity of water use, and in large amounts in buying back water for the environment and health of the river systems in the Murray-Darling Basin. However, this plan also presents some innovative elements from an Australian Policy perspective, as over 1.5 billion dollars have also been allocated to urban water use programmes. Cities and urban infrastructure have typically been governed by the state and local governments with minimum policy direction from the Federal Government; so this plan represents an attempt at a national level for integrated water management, which may help to overcome the current Australian urban–rural water management divide. Funding has also been allocated to water information and management coordination efforts.

Only a few years into these reforms, progress has remained mixed. Many of the urban centres have invested in desalination plants, whether or not this made financial or hydrological sense compared with other decentralised options, and almost 1000 Gl of water entitlements have been bought by the Government for their Murray-Darling Basin buy-back programme to curb over-allocation. However, the process used to develop the plan for the Murray-Darling Basin by its new Federal Authority has got off to a very shaky start, with many stakeholders furious over the technocratic approach taken and lack of meaningful stakeholder engagement, and the head of the Authority resigning after citing irreconcilable differences with the Government over the understanding and implementation of the *Water Act 2007*. Recent serious floods across much of Australia, including in the Murray-Darling Basin, are also likely to add to the challenges of the reform agenda for the new head of the Murray-Darling Basin Authority and the current Federal Government.

Whether the 12.9 billion dollars of funding will be spent as planned and the planned objectives of the Water for the Future Plan are achieved remains to be seen; but, in any case, Australia has some formidable challenges ahead, aspects of which include managing:

- The impacts of extreme natural variability and climate change;
- Water quality and land degradation associated with land and water use behaviours;
- Salinisation, eutrophication and algal blooms, acid soils, sodic soils and erosion;
- Ecological impacts of river regulation and environmental flows;
- The implementation of sustainable technologies and required behavioural change to support more efficient and innovative technologies, including water reuse and recycling;

- The foreseeable negative impacts of some recent management choices, including desalination and water markets in terms of ecological, equity, energy, financial and other aspects;
- The lack of large investment in the sector since the 1970s and the large losses of corporate memory that have occurred after privatisation and institutional changes;
- Skill shortages, especially of engineers and facilitators with sufficient technical knowledge of water and its links to other systems to manage the complexity, uncertainties and conflicts present in the sector;
- Leadership required for the inter-organisational and inter-scale coordination between the multiplicity of committees and government departments with distributed resources, authority, knowledge and skills; and
- Adaptation to the new and uncertain world of the future.

2.1.3 The European system and priorities

Water management in Europe has a long written history, owing to the early societal developments, which are further outlined in Appendix A. The civilisations on the European continent had developed a number of water governance systems and bodies of law which evolved to regulate and allocate water for the most important uses of the times, such as navigation, fish passage, irrigation, and water mills, as well as to provide compensation to other users or property owners if their actions damaged others, such as flood damage due to a lack of infrastructure maintenance (Cech, 2005). With each change of empire or regime, the laws and management systems have been adapted, up until the current system of individual law and management systems developed by each national government on the European continent.

The history of the modern European Union follows on from the Second World War as a collaboration of western European countries who wanted to prevent the horrors of war happening again and instead promote peace, trade and human rights. The *Council of Europe* was created by ten signatory countries in 1949 (Council of Europe, 2008) and then after the creation of a number of trade and human rights agreements, six of these countries signed the *Treaty of Rome* in 1957 to create the *European Economic Community* (EEC). These six countries (France, Germany, Italy, Belgium, the Netherlands and Luxembourg) then created a 'common market' where people and goods could be transferred within the community, a customs union for regulation and common policies for economic development (EUROPA, 2008). Under this treaty, the European Commission, a parliament, a council of ministers and a court of justice were established as the principal decision making institutions to manage both national and community interests.

Through the 1960s, the EEC evolved its policies and cooperation and underwent successful economic growth. The 1970s saw an evolution in the sphere of policy issues treated in the community, which increased to include environmental and social concerns, with implementation of the 'polluter-pays' principle, the 'regional development fund' to improve solidarity between the richer and poorer regions of the community (EUROPA, 2008) and a number of water quality and pollution directives to protect human health (Ginestct, 2008). From 1972, a number of accession treaties were signed to allow other countries to join the EEC. Three countries were accorded the right in 1972 (Ireland, the United Kingdom and Denmark), one in 1979 (Greece), two in 1985 (Spain and Portugal), three in 1994 (Austria, Finland and Sweden), nine in 2003 (Cyprus, Estonia, Hungary, Latvia, Lithuania, Poland, the Czech Republic, Slovakia and Slovenia) and two more in 2005 (Bulgaria and Romania). From 2007, the European Union has been constituted of 27 Member States and a population of 490 million (EUROPA, 2008).

In 1992, the *Treaty on the European Union* (also known as the 'Maastricht Treaty') was signed to rename the 'European Economic Community' the 'European Community' or European Union (EU). This name change was associated with the broadened powers of the community beyond economic matters and the establishment of new decision making institutions. Issues on environmental policy, including powers of the European Union over water resources management, were present, with their objectives stated as ('Treaty on the European Union', 1992, Article 130r.1):

- *Preserving, protecting and improving the quality of the environment;*
- *Protecting human health;*
- *Prudent and rational utilisation of natural resources; and*
- *Promoting measures at international level to deal with regional or worldwide environmental problems.*

Throughout the 1990s, much information that was collected and processed related to Europe's environmental condition, including issues of water quality and quantity by the European Environment Agency, the EU's Member States and other international and research groups. As in many other areas of the world, there were many reasons for concern about the state of Europe's water resources and their ecosystems. Of particular concern was the pollution, which included pesticides, nitrates, organic matter, phosphates and acidification, and overuse of both surface waters and groundwaters (Stanners and Bourdeau, 1995). A lack of coordination for water management in and between many of the Member States, required to implement improvements and adhere to the 20 or so European Directives relating to water, was also of great concern.

To deal with some of these issues, the European Water Framework Directive (WFD) 2000/60/EC of the European Parliament and Council was passed in 2000 as a framework to aid community action in the field of water policy (EU, 2000). The self-stated purpose of the Directive is to (EU, 2000):

> *Establish a framework for the protection of inland surface waters, transitional waters, coastal waters and groundwater which:*
>
> (a) *Prevents further deterioration and protects and enhances the status of aquatic ecosystems and, with regard to their water needs, terrestrial ecosystems and wetlands directly depending on the aquatic ecosystems;*
> (b) *Promotes sustainable water use based on a long-term protection of available water resources;*
> (c) *Aims at enhanced protection and improvement of the aquatic environment, inter alia, through specific measures for the progressive reduction of discharges, emissions and losses of priority substances and the cessation or phasing-out of discharges, emissions and losses of the priority hazardous substances;*
> (d) *Ensures the progressive reduction of pollution of groundwater and prevents its further pollution; and*
> (e) *Contributes to mitigating the effects of floods and droughts.*

The commonly cited aim of the Water Framework Directive is for all water bodies (surface and groundwater; and natural, artificial or heavily modified) to achieve 'good' status or potential, both in terms of ecological and chemical parameters, by 2015 (Steyaert and Ollivier, 2007). The directive also dictates that management must take place at the water-basin scale and that these basins and their characteristics be defined and water basin management plans developed by 2009 with public information, consultation and involvement (EU, 2002):

> *Member States shall encourage the active involvement of all interested parties in the implementation of this Directive, in particular in the production, review and updating of the river basin management plans.*
>
> (EU, 2000, Article 14).

Exactly how this requirement should be interpreted and transferred into practice is far from evident (EU, 2002; Ker Rault, 2008; Ker Rault and Jeffrey, 2008). A number of specifications of economic management are also given in the WFD, including that: economic analyses of present and future water use in each basin should be carried out; full cost recovery of water services is pursued; and the 'polluter-pays', which could be through dissuasive penalties (van Ast and Boot, 2003). Analyses of the water bodies should also include analyses of pressures and impacts on the systems, as well as risk analyses (CEC, 2007).

The WFD currently concentrates much more on water quality than on water quantity issues, assuming that water bodies under water quantity stress are most likely to exhibit quality stress as well. However, there have been recent developments to add further specifications to the WFD with the addition of: the Groundwater Daughter Directive (2006/118/EC) in accordance with Article 17 of the WFD; a list of 33 priority substances (Decision 2455/2001/EC); the development of the Water Information System for Europe (WISE); Directive 2007/60/EC on the assessment and management of flood risks; the Marine Strategy Framework Directive (2008/56/EC); and proposed analyses of the potential impacts of water scarcities and droughts (CEC, 2007).

The European water governance system is therefore based on a largely economically rational, 'top-down' and incremental approach to improving the state of water and ecological resources, which is laid out in the water-related directives. Monitoring and reporting will be required to ensure compliance, with fines or other disincentives enforced for the case of non-compliance. It remains to be seen whether such governance will aid the European countries in more successfully meeting the challenges of water management for the future; these challenges are likely to be driven by population growth, climate change and competing economic and social priorities. Progress and review of these directives will form part of the *Blue Print for Safeguarding European Waters*, the next major European water policy document planned for 2012.

From the 2007 update assessment report of WFD implementation, it was shown that, although there is some progress, the capacity of Member States to implement the Directive varies greatly. In some cases, it was not the newest Member States that had the most difficulties, presumably as they are often willing to implement the required reform processes, but rather the older members (CEC, 2007). This may be due to the difficulties of administrative reform in countries with a long history of particular administrative systems that do not match the scales or domains of those dictated in the WFD. It is also estimated that at least 40% of water bodies are at risk of not meeting their 'good quality' status by 2015 for a number of reasons; clearly, the EU still has many challenges ahead (CEC, 2007).

2.1.4 Governance system comparison

The European approach to managing messy water situations varies significantly from other approaches currently being taken around the world, especially in its framing of the necessity to achieve certain types of ecological status as a means of obtaining sustainable water management. At the international level and in Australia there is a broadening to include more visions and interpretations of the problem situations and preferred paths of development for the future (through broad-scale participatory approaches), while the European Union seems to have decided on just one main path that could potentially be implemented,

Table 2.1 *Comparison of International, Australian and European water governance systems and priorities*

Comparison area	International	Australia	European Union
Decision-making loci	UN, inter-organisational networks, Member State influence, NGOs, private companies	State governments (& CoAG), National Water Commission, federal government influence, MDBA/CMAs, private companies, local governments and user groups	European Commission, Member States, Basin Institutions and associated groups
Focal challenges	Poverty, health, food production, water scarcities (all types), population growth, environmental degradation, pollution, transboundary management, coordination and financing for on-ground implementation	Extreme climatic variability and change, water allocation, ecological degradation, salinity, coordinating multiplicity of management institutions and initiatives, co-management of water and other resources (i.e. land, energy)	Pollution, health, ecological degradation, urban development, climatic variability (floods and droughts), healthy and productive marine environments
Main water user	Agriculture (70%), industry (including energy production and other uses) (22%), domestic (8%)	Agriculture (70%), industry (including energy production and other uses) (19%), domestic (11%)	Industry (including energy production and other uses) (56%), agriculture (30%), domestic (14%)
Governance initiatives	Millennium Development Goals, Agenda 21(Art. 18), UN and World Water Forum Statements	Water Act 2007, National Water Initiative, National Strategies for Salinity, Quality and Security, Murray-Darling Basin Agreement, State water legislation	EU Water Framework Directive, about 20 other water-related EU directives, Member State water legislation
Environmental priorities	Maintenance of ecosystem integrity, pollution prevention and control	Environmental water allocations, salinity management	'Good water status' by 2015 (ecological and chemical components)
Economic priorities	Valuation for all uses, cost recovery, polluter pays	Water markets and other pricing regimes, user pays, cost recovery	Full cost recovery, polluter pays
Social priorities	Poverty alleviation, health	Allocation between competing uses	Health and personal protection (e.g. from floods)
Technological priorities	Ensuring basic water needs (for drinking, sanitation and food production): hard and soft paths	Adaptation to climate change and other risks, hard and soft paths; water information systems (BoM and WRON)	Wastewater treatment, pollution reduction: path defined by Member States; WISE monitoring system
Equity/ participation priorities	Participation of all stakeholders in water-related decision making, including: women, youth, indigenous people and local communities	Consultation required for Water Act. Participation of stakeholders in state and local level water-related policy making, planning and implementation: inclusion of indigenous interests and rights	Information, consultation and involvement of the public, including users – no specifications on minority group interests; Aarhus Convention
Governance scale priority	River basin	Catchment, basin	River basin
Significant differences	Focus on water as a life-support mechanism; funding sources and implementation mechanisms unclear	Focus on need to adapt, water market approaches, small population	Fixed vision of water issues and solution path, markedly top-down regulatory approach, reliance on science and economics

with less focus on participation of all stakeholders (especially local users), owing to the extensively defined roles for scientists and economists in collecting and analysing information (Steyaert and Ollivier, 2007). Some of the differences and similarities between the systems are outlined in Table 2.1.

From Table 2.1, it can be seen that each of these three governance systems has multiple objectives to reply to multiple challenges. The European position is the most fixed and spans the least number of issues related to different contextual elements of the complexity of water problem situations. It has a particularly well-defined legal base, is clearly anchored in an understanding of sustainability that is founded on the need to have healthy ecosystems, and is not as open to potential cultural, social and political influences as the two other systems. The international and Australian systems are aiming for more socially inclusive outcomes and currently rely on more local decision making processes from those on the ground with the resources. However, the Australian system is also set on the path of further introducing water markets as a means of improving the efficient and effective use of available water resources. It remains to be seen which approach, whether legally binding and directive, market-based or adaptive locally defined management systems, will achieve its goals over the next few years and further into the future.

2.2 CAN WE LEARN FROM THE PAST?

Human beings, who are almost unique in having the ability to learn from the experience of others, are also remarkable for their apparent disinclination to do so.
Douglas Adams (Adams and Carwardine, 1990)

Compared with any previous time on Earth, we have the most information and knowledge accessible to help us to make informed decisions. However, at the same time we are faced with the most uncertain, complex and conflict-ridden problem situations that have ever been experienced. As both planning and management are basically decision making processes (see Section A.2.2 for a review of planning and management mechanisms), today's and tomorrow's major challenge is about how to aid decision making processes to ensure better management outcomes.

Aiding appropriate decision making processes, and the most appropriate decision paths to pursue for water management and planning, are anything but evident, and are likely to vary widely based on contextual elements of the problem situations. However, as history has shown us (see Section A.2.1 for a short history of water management), some decision pathways are likely to be unsustainable. Just three seemingly unsustainable pathways are presented here: imposing purely 'scientifically' or 'technologically' based solutions without including affected communities in the decision making processes of choosing such solutions; 'mega-engineering' projects; and reactionary crisis management approaches. Alternative water management pathways that may hold more promise will briefly be mentioned, some of which, such as adaptive and risk management approaches, will be further investigated in later chapters of this book.

2.2.1 Technological imposition: community rejection

Technology alone and a traditional engineering-driven approach are very unlikely to solve our current water problems. Many authors and institutions have stated the need for a democratisation of decision making processes. Even if there are seemingly great technical solutions for community water problems, they may easily be rejected or their implementation fail for a variety of reasons when the affected communities and local experts have not taken part in the decision processes (Fischer, 2000; Gleick *et al.*, 2006). These are not purely hypothetical and empty conjectures, as there is mounting evidence against the above approaches around the world. For example, there have been countless failures of such technical approaches to sanitation and water supply system projects funded by external agencies in developing countries, as well as failures of propositions to implement more ecologically sustainable water solutions in developed countries, such as water reuse systems; just a few of which will be outlined here.

In many countries around the world, an all too common situation is for an aid organisation to see a water supply or sanitation issue, and with very little preliminary analysis or 'problem formulation' with the local communities or experts, decide that they have the perfect solution and that they will fund its implementation. Although the funding is often a cause for much hope in the local communities, the realities of the implemented 'solutions' are sometimes less welcome than expected. For example, a recent groundwater pumping system installed for the community of the small Pacific island of Chuuk ended up damaging the integrity of parts of the island's freshwater lens due to the poor siting, design and installation of pumps (White *et al.*, 2008; personal communication). The majority of the system quickly fell into disrepair because of a variety of problems, leaving the island's population with limited access to drinking water supplies. Had the aid agency involved simply consulted well-known island groundwater experts before implementing the design, the now more severe drinking water crisis and waste of money could easily have been averted (Falkland, 2007).

Major problems of a similar sort on a much larger scale have occurred with groundwater pumping in Bangladesh and India. Surface water of poor quality in both countries has long been a major source of disease, so in the 1970s UNICEF, with aid from the World Bank, embarked on a massive programme of bored tube wells with hand pumps to supply new 'clean' groundwater for communities in Bangladesh and the West Bengal area of India (WHO, 2000). Many communities were originally against the idea of drinking the new water source, calling it 'the devil's water'; yet the aid agencies apparently took little notice and did not try to discover why the communities named the groundwater in that way. Unfortunately, had the agencies probed further into this community concern, or tested the water quality before encouraging the communities to drink it, 'the largest mass poisoning of populations in history' (Pearce, 2004), caused by naturally occurring arsenic in the groundwater, might have been avoided. The dangerous levels of arsenic, a slow acting poison, in many of the bored wells' water are now thought to have affected 35 million people (Bearak, 2002). The discovery of elevated arsenic levels was only recently recognised by the world aid agencies originally involved in the well drilling programme, and progress on their behalf towards testing the potentially contaminated wells and developing solutions for the affected populations has been excruciatingly slow (Earthbeat, 2004).

A similar problem resulting from excess levels of naturally occurring fluoride in groundwater tube wells has also induced devastating health effects, such as skeletal fluorosis in communities in India and millions of people in up to 25 other countries, including China, Chile and Ethiopia. Despite high levels of fluoride being discovered 60 years ago in India, aid agencies and contractors sinking community groundwater wells lacked the vigilance to test the well water before declaring it clean and safe (Pearce, 2004). It is therefore evident that a number of today's water-related problems and health crises are known to have been aggravated by water managers or outside 'aid' organisations who have failed to adequately plan and work with local communities and a range of experts in a participatory approach to manage local water issues.

The lack of the use of a participatory approach in questions over the implementation of more sustainable water technologies has also had its ramifications in a number of developed countries. One such example was the public outcry that resulted when residents were forced to use recycled water systems that they were not aware of in their homes in Adelaide, in Australia (Marks, 2004). Many more examples of similar difficulties that can occur due to a lack of early inclusion of stakeholders in water reuse systems are outlined in GHD (2007). Water managers, scientists and policy makers in these situations have typically failed to take into account the underlying perception differences and other conflicts between stakeholders early in the decision making processes:

> *The various parties to water conflicts often have differing perceptions of legal rights, the technical nature of the problem, the cost of solving it, and the allocation of costs among stakeholders.*
>
> (Wolf *et al.*, 2005)

The types of scientific risk analysis that are undertaken (or may be not even undertaken!) by experts before suggesting such technologies tend to hide many value judgements, such as what are 'acceptable' levels of danger to communities, and this is just another reason for the failure of technocratic management approaches in many countries (Beck, 1992). In tomorrow's world, such technical approaches to impositions of seemingly 'good' solutions are likely to fail unless they are accompanied by certain forms of participation of affected people and stakeholders (Fischer, 2000).

2.2.2 'Mega-engineering' projects: ecological time-bombs

Examples of the extreme negative impacts of past 'mega-engineering' water projects that can be seen today are numerous. As the history of failed hydraulic civilisations appears to demonstrate, if nature is controlled for too long without care, ecological damage capable of removing the necessary environmental services required for human survival is likely to ensue (see Appendix A for further discussion). Perhaps one of the areas that has suffered the most horrendous impacts in the twentieth century and is now classed as an 'ecological disaster zone', which has also led to serious human health problems for the populations in the region, can be seen in the water basin of the Aral Sea (Greenberg, 2006).

Since the late 1950s when Soviet planners started to implement their plan for the region to become a major cotton or 'white gold' producer, the Aral Sea suffered a reduction of 74% of its surface area and 90% of its volume, as well as a tenfold increase in salinity (Micklin, 2007). The construction of the world's longest irrigation canal, the Karakum canal, which exports water from the Amu Dar'ya tributary to the adjacent Caspian Sea basin (outside the Aral Sea basin), is thought to be the greatest reason for the demise of the Aral Sea (Pearce, 2004). The irrigation water from this canal is predominantly used in an inefficient fashion by Turkmenistan and Uzbekistan, leaving the populations of the Karakalpakstan Republic and Kazakhstan that border the Aral Sea to suffer from the consequences of the retreating sea (Whish-Wilson, 2002). Some of the ancient fishing communities living in these regions are now tens of kilometres from the sea's edge and suffer from salt and toxic chemical dust storms coming from the parched sea bed, lack of drinking water and sanitation, a changed climate, salinisation and desertification of agricultural lands, as well as a lack of an economic livelihood (Micklin, 2007). Years of pesticides and toxic chemicals used on the cotton fields, lack of drainage channels from the irrigated fields, and even previous Soviet biological weapons testing on an island in the Aral Sea, are thought to add to the local population's health problems; the majority are ill with salt poisoning (Whish-Wilson, 2002). It is estimated that 97% of the female population in Karakalpakstan are clinically anaemic. The Republic also has the highest rate of oesophagus cancer in the world, with other rates of cancer and diseases crippling the population (Pearce, 2004). Since the fall of the former Soviet Union, many initiatives to help save the Aral Sea have been implemented with the aid of the international community, and although some small successes have occurred in the small northern Aral Sea, it is unlikely in the near future that much of the damage in the southern sea can be undone if irrigated agriculture in the region is to continue (Micklin, 2007). Unfortunately, the destruction caused by 'mega-engineering' projects cannot often be easily undone. However, the question remains as to whether lessons can be learnt from such mistakes.

There is still some continued enthusiasm for 'mega-engineering' projects, such as cross-continental water transfers in many countries, including: Libya's 'Great Man-Made River', which pumps water from the Nubian Sandstone Aquifer system below the Sahara desert and is planned, when finished, to supply 6.5 million m^3 per day (GRMA, 2006); China's planned 'north–south diversion', designed to bring river water from the Yangtze (equivalent in volume to the flow of the Yellow River) to the parched northern agricultural areas (Pearce, 2004); India's 64 km Kalpasar dam, as part of the 'Gulf of Khambhat Development Project' to create an enormous freshwater lake and harness tidal power (Saha and Alex, 2004); and the recent Chinese 'Three Gorges Dam', the world's largest dam, constructed to tame the Yangtze River (CTGP, 2008). Perhaps the lessons have not yet been learnt.

2.2.3 Reactionary management: from crisis to crisis

Another seemingly unsustainable decision pathway for water planning and management is reactionary crisis management. Although some crises may be true surprises to which reactionary responses are necessary, many crises related to water management are at least partially foreseeable and may be planned for proactively to mitigate their effects, avoid them entirely, or transform them into opportunities. Some of the largest foreseeable crises of our times in the water sector include floods and droughts. For example, the massive New Orleans flooding disaster caused by Hurricane Katrina had been foreseen by scientists for years (Travis, 2005). Despite numerous programmes for building world-leading structural flood defences to pre-empt flooding of the low-lying city region (80% below sea-level), the 1965 Hurricane Betsy still killed 75 people and created over one billion US dollars of flood-related damage (Burby *et al.*, 1999; US Government, 2006). However, in light of the obvious inadequacies of the city's flood defences to deal with such hurricanes, the policy makers of the time appear not to have sufficiently considered (or to have ignored) the principal causes of the high damage bill, one of which was allowing large populations to live in a highly vulnerable area, and reacted by deciding to increase the structural flood defences. Naturally, this gave people a greater sense of comfort, thinking that they were protected, but many scientists and engineers realised the heightened risks that hurricanes larger than the parameters designed for could pose to the city. In 1997, it was predicted that a level 5 hurricane hitting New Orleans could potentially kill more than 25 000 people and create 30 billion US dollars of damage (Burby *et al.*, 1999). Modelling in 2004 showed that even a category 3 hurricane's storm surge would be likely to cause overtopping of New Orlean's levees (Travis, 2005). Policy makers had therefore been warned, but attempts for further preventative planning, the development of evacuation and crisis management plans and community education were not sufficiently pursued and carried out in time for the next major hurricane-induced flooding. As a result, the foreseeable occurred in 2005, with a category 4 hurricane (Travis, 2005) causing over 90 billion US dollars of damage, at least 3300 deaths and the displacement of about a million people (US Government, 2006; Rivas, 2007). The question now remains as to whether the USA will become more proactive in its disaster prevention programmes, or whether the next foreseeable 'crisis' occurs.

A country with potentially similar issues of flooding and elevated risks due to structural protection is the Netherlands. Despite developing and maintaining evacuation programmes and considering potential policy options to 'accommodate' flood waters (i.e. selective flooding of the country), the Dutch government and citizens also must ask themselves whether they are truly ready for the predicted sea level rises, storm surges and inland river floods that will put their country and people at risk in the twenty-first century (Wesselink, 2007).

The same questions surround foreseeable crises resulting from climatic variability and climate change in other regions of the world. In Australia from late 2007, some farmers have been simultaneously receiving both flood and drought disaster subsidies (Ryan, 2007). This particularly absurd situation underlines current Australian policy flaws, and demonstrates the need for a rethink of the current reactionary approach to floods and droughts. As has been outlined in Section 2.1.2, extreme climatic variability in Australia is 'normal'. It is becoming increasingly important to embrace and learn to live with the true 'boom and bust' nature of the Australian environment, as well as any more permanent climatic shifts, as discussed briefly in Appendix B (Section B.3.2). This may mean making difficult policy decisions, including redefining the role of agriculture and working with communities to change old financial habits, in order to become more proactive in planning for the next flood and drought periods. For many years, Australian politicians have avoided such difficult questions; yet with the recent ravaging droughts, floods and water shortages the tide finally appears to be turning, as witnessed by articles on the subject in the national press (Walker, 2008). Whether real change away from the current disaster relief to treat flood and drought victims will occur soon is yet to be seen, although some positive signs of preparedness and effective flood-risk mitigation and capacity to mobilise the population for massive clean-ups were apparent in the devastating 2010–2011 floods in Queensland, Australia.

In the face of natural or human-induced disasters, crisis management of the reactive kind after the event has been the norm in most areas of the world. However, it is increasingly realised that proactive and forward thinking 'risk management' approaches could prove much more effective in preventing and mitigating effects of potential disasters or unwanted situations (Gleick, 2006). This is why some authors are also wary of 'adaptive' management approaches, as they feel the terminology will lead managers to wait and react to changes, such as the impacts of climate change, rather than act proactively in anticipation of impacts (Gleick, 1998). Other authors would contend that adaptive management is supposed to be forward thinking and proactive, to 'adapt' to the current contexts and adapt forward trajectories during implementation to achieve desired outcomes (Pahl-Wostl *et al.*, 2008). It is interesting to note that the word *crisis* stems from the Greek *krisis*, which signifies 'judgement', 'choice' or a time for decisive action, rather than referring specifically to 'disaster' (Priscoli, 1998). Like all other water planning and management decisions, those taken in times of crisis, and before and after them, give us the possibility of avoiding a repetition of history's past mistakes, and hence avoid the next foreseeable disasters. Whether we are able to think of foreseeable disasters as risks or still treat them as crises when they occur, we should see these events and the times leading up to them as opportunities to make decisions for action that may be able to help us to build a more sustainable future in an uncertain world.

2.2.4 Looking to the future

Considering these examples of seemingly unsustainable decision making and management pathways – imposing purely 'scientifically' or 'technologically' based solutions without including the affected communities in the processes of choosing such solutions; 'mega-engineering' projects; and reactionary crisis management approaches – it must now be asked what the plausible or possible alternatives could be, and how much better suited they are to the problems of the twenty-first century world.

It has been suggested that the single most important goal for a new approach to water planning and management should be to focus on increasing the productive use of water (Gleick, 1998). This is usually cited as the goal of 'demand-side management', where improving the efficiency of water use and water sharing between different uses is an aim that will require changes and innovation in technology, economic instruments and institutions (Vickers, 1991). However, just focusing on the 'productive use' of water limits all the other important roles that water has in this world and ignores the other complexities of the current water scarcities, especially some of the social ones. Increased conflict over management and planning values and objectives, lack of resource control by management authorities and democratic motives in western countries have been partially responsible for driving a range of participatory, interactive and collaborative planning processes (Forester, 1989; van Rooy *et al.*, 1998; Forester, 1999).

Proposals include adaptive, 'co-adaptive' or transition management approaches (Holling, 1978; Berkes and Folke, 1998; Levin, 1998; Cortner and Moote, 1999; Lee, 1999; Gunderson and Holling, 2002; Olsson *et al.*, 2004; van der Brugge *et al.*, 2005; Pahl-Wostl *et al.*, 2008), which can be considered close to, or part of, the Integrated Water Resources Management approach (Global Water Partnership, 2000; Pahl-Wostl, 2007). Most of these approaches explicitly include public participation or involve stakeholders through the management processes, although the practice often remains far from the theory (Ker Rault, 2008). Combined with the need to consider a range of human values and objectives, increasing concern about uncertainties and 'risks', as well as a lack of capacity to pre-empt, mitigate and adapt to their impacts, has driven the development of risk management, asset or value-based management approaches (Keeney, 1992; McDaniels *et al.*, 1999; Stirling, 1999; Slovic, 2000; Jaeger *et al.*, 2001; Klinke and Renn, 2002; Standards Australia, 2004, 2006; Wild River and Healy, 2006; Vance, 2007). Other approaches include those that focus on 'leadership' and have come from the realisation that management is predominantly not about managing objects, but rather about managing people (Adair, 1983), leading transformational change (Senge, 1990; Bass and Avolio, 1994) and learning (De Geus, 1988; Brews and Hunt, 1999). Social learning is now a commonly cited objective of water management (Ison *et al.*, 2004; Pahl-Wostl and Hare, 2004; HarmoniCOP, 2005b) but leadership appears to have received less frequent attention except when related to the need for local champions to aid the successful implementation of water development projects (Barnaud, 2008) and general mentions of it as a driver for adaptive management (Olsson *et al.*, 2006). Management and planning in these cases are considered to be 'process-orientated' rather than linear and 'goal-orientated' and place a large emphasis on the creation of knowledge, innovation, creativity and social or organisational learning (Nonaka and Takeuchi, 1995).

2.3 WATER MANAGEMENT COMPLEXITY

Perhaps the only certainty for anyone looking ahead is that the future is uncertain, unpredictable and complex.

Peter Gleick (1998)

Water planning and management around the world have many challenges to overcome, as has been seen from Sections 2.1 and 2.2 of this chapter. Furthermore, socially based disputes over water management in recent years have resulted from differences in values and viewpoints, interests on how water should be used, and power struggles over scarce water resources. Questions of whether water is of sufficient quality or if it is equitably distributed between people and geographical areas can spark highly value-charged debates (Delli Priscoli and Wolf, 2009). Protests and social activism over water management projects, such as dam construction, have also become more widespread (Hutton and Connors, 1999). Information transfer now allows local issues to become the international focus of protests, for example the human rights issues of the displacement of the estimated 1.3–1.9 million people for the Three Gorges Dam on China's Yangtze River (Gleick *et al.*, 2006).

The recent creation of the Internet and the realisation that all human activities can have global impacts through economic factors, such as trade, and societal changes at physical, cultural, environmental or individual human levels, for instance through changing beliefs, values, views, relations and practices, have also made water planning and management more complex. This is because not only the technical, economic and environmental factors have to be considered, but the social issues as well. This need has been further emphasised by the inclusion of the 'sustainability' or 'integrated water resources management' concepts in policy around the world. Contemporary water management is often characterised by a process of deciding at multiple levels of governance how water should or can be used and shared between a variety of stakeholders and the environment under conditions of major uncertainty. Such uncertainties and areas of potential rapid change include climatic conditions and natural hazards, such as floods, droughts, volcanic eruptions, disease outbreaks, earthquakes, tsunamis, cyclones, technology and scientific innovation, political regimes and priorities, and economic climate, as well as human behaviour and cultural imperatives.

Throughout history, a variety of planning and management approaches were developed and used to cope with the problem situations and water scarcities of the times, some of them reaching their desired outcomes and many more creating more issues that required evolutions in management strategies to deal with new complexities. Today, like water managers in previous times, there must be a focus on aiding the development of new strategies for managing new issues and not falling into the trap of repeating avoidable mistakes from history. Our specific challenge is to aid the management of water in an increasingly globalised, populated, environmentally degraded and inequitable world, with the increasingly affluent and poor; a world with unprecedented levels of complexity, uncertainty and conflict. This is made a more difficult task because water planning and management are now highly distributed activities occurring at a multitude of spatial, temporal and institutional scales. Each local region around the world has a variety of different water-related issues that it has to manage, and at each larger scale there are numerous planning and management responsibilities that attempt to oversee the coherence of these local efforts to ensure that more sustainable overall directions are pursued at that level. This has resulted in many layers of water planning and management, which are all interrelated and attempt to deal with different groups of stakeholder and their issues, needs, values, interests, representations, resources and actions in dynamic and messy situations; often with limited success.

2.3.1 Research needs and opportunities

Considering these issues of increasing water management complexity, *one of the most pressing needs is to develop and implement improved methods of aiding decision making processes for water planning and management.* These decision-aiding methods must be developed to better plan and manage water in:

- Multi-stakeholder and inter-organisational settings across spatial and administrative scales;
- Contexts exhibiting high levels of uncertainty, complexity and conflict; and
- A reflective and reasoned manner so as to learn from the past and adapt proactively in the face of future challenges.

In view of these needs, there is an opportunity to harness inspiration for new thinking and practice from alternative approaches for use in these decision-aiding processes and to improve certain aspects of the current plans and objectives of current water programmes around the world, including at an international, Australian and European level.

In particular, at an international level there is a need to move from rhetoric and design of the vision to implementation of specific actions to improve the world and to fulfil the communities' needs. Care should be taken to encourage reflective and critical practitioners and researchers in these pursuits, who will not fall into the traps of ignoring the context, local people and experts.

In Australia, there is still the need for more vision and reflection on how to live with a variety of extreme climatic and environmental uncertainties before implementing 'quick-fix' solutions that may limit adaptive capacity in later years. There is also an urgent need for more coordination between distributed management authorities over a variety of interrelated scales and for their decision making processes to be aided in a reflective and critical manner to drive them in more sustainable directions.

Taking the historical analyses of this chapter into account, in Europe there appears to be a risk of overly emphasising a technocratic approach to planning of the EU Water Framework Directive, which could lead to larger conflicts and problems in the future, although the new Federal approach to water policy in Australia could also hold the same risks. There may therefore be a need to investigate and embrace the use of more inclusive participatory approaches in the creation of basin management plans (from the initial phases of their conception) in both Europe and Australia to avoid potential future discontent by a range of stakeholders and the ensuing problems.

In light of the general need and research opportunity to develop and implement improved methods of aiding decision making processes for water planning and management, the following questions require further investigation:

- How are multi-stakeholder and inter-organisational decision making processes currently aided and how could these practices be improved?
- What decision-aiding methods are currently being used in water management contexts exhibiting high levels of uncertainty, complexity and conflict and to what extent do they appear to be successful? and
- How and to what extent can decision-aiding processes be carried out in a critically reflective manner so as to learn from the past and adapt proactively in the face of future challenges?

2.3.2 Conclusions

This chapter has presented a critical review of past and current water governance systems, their management priorities and strategies to examine whether water management has become increasingly complex, in response to the first objective of this book. This was carried out by examining the principal challenges facing water planners and managers in the twenty-first century, the current forms of governance at the Australian, European and international level used to manage these challenges and what can be learnt from history that may help us to manage today's and tomorrow's problems in the twenty-first century. It was found through this review that water management has indeed become more complex, in particular through the increasing number of interconnections between people aided by technology, a multiplicity of legislative levels and requirements, the dispersion of power and resources for water management and the need to address a range of social and environmental issues, as well as technical and economic ones, from multiple stakeholders' points of view. All of these factors affecting water management were not as predominant in past times. As a reflection on the analyses, this chapter outlined a number of future needs and opportunities, including the '*need to develop and implement improved methods of aiding decision making processes for water planning and management*', in particular for inter-organisational decision-aiding, which will be examined in the following chapter of this book. Confirmation for one of the underlying assumptions of this research – *that the increasing complexity of water-related problems has contributed to the need for improved inter-organisational decision-aiding for water planning and management* – has thus also been provided.

3 Decision-aiding for water management and planning

It is change, continuing change, inevitable change that is the dominant factor in society today. No sensible decision can be made any longer without taking into account not only the world as it is, but the world as it will be…

ISAAC ASIMOV (1978)

In Chapter 2, it was suggested that there was a '*need to develop and implement improved methods of aiding decision making processes for water planning and management*', in particular for inter-organisational decision-aiding for water planning and management for the twenty-first century. To meet this need, it is first necessary to understand the concept of 'decision-aiding', the methods that may be used to carry it out and their relation to the practice of water planning and management in a range of problem contexts. To assist this understanding and to treat the research questions outlined at the end of Chapter 2, this chapter will firstly give a theoretical introduction to the concept of 'decision-aiding' and introduce decision-aiding models from the engineering, policy and operational research domains relative to the inter-organisational water management context. The second part of the chapter will look at how such models might be operationalised in decision-aiding practice. The design of participatory structures and methods currently used in participatory modelling will be critically analysed and an integrated participatory modelling methodology for inter-organisational decision-aiding is proposed.

3.1 DECISION-AIDING AND ITS ROLE IN WATER MANAGEMENT

Be careful what you water your dreams with. Water them with worry and fear and you will produce weeds that choke the life from your dream. Water them with optimism and solutions and you will cultivate success. Always be on the lookout for ways to turn a problem into an opportunity for success. Always be on the lookout for ways to nurture your dream.

Lao Tzu (600 BC–531 BC)

3.1.1 What is decision-aiding?

Decision-aiding is common in everyday life where people help others to formulate their problems and make decisions. It has been studied in a number of disciplines, including operational research or management science, law and psychotherapy (Capurso and Tsoukiàs, 2003; Tsoukiàs, 2007). In this book, it is the operational research (OR) or management science vision of decision-aiding that will be predominantly examined, along with some typical engineering and policy visions, as they are most closely linked to the operational aspects of water management. Operational research is the discipline:

Concerned with scientifically deciding how to best design and operate man–machine systems, usually under conditions requiring the allocation of scarce resources.

(ORSA, 1977 cited in Müller-Merbach (2002))

In recent years, the focus of OR on just 'man–machine' or socio-technical systems has been broadened to incorporate other complex systems (EURO, 2008), such as socio-environmental systems and 'messes' (Ackoff, 1979). This is because of an increasing need to aid decision making processes for some of the most important challenges in the rapidly changing, uncertain and interconnected world outlined in Chapters 1 and 2, which are not only related to 'man–machine' systems. As described by Lachapelle *et al.* (2003):

Wicked problems and messy situations are typified by multiple and competing goals, little scientific agreement on cause–effect relationships, limited time and resources, lack of information, and structural inequities in access to information and the distribution of political power.

These conditions, which are further outlined in Box 3.1, have led to OR undergoing a paradigm shift (Kuhn, 1962) around the 1970s due to observable increases in messy problem situations and the inadequacy of 'hard' techniques of problem solving, planning and management to cope with them (Ackoff, 1979; Checkland, 1981). This mismatch between problem situations and relevant decision-aiding tools slowly led to the emergence

Box 3.1 Messes defined

Messes are dynamic situations that consist of complex systems of interacting and changing problems (Ackoff, 1979). The complexity of a problem situation is largely driven by the number of uncertainties, the number of interrelations and the number and level of conflicts present.

Messes are also referred to as:

- 'Practical problems' as opposed to 'technical problems' (Ravetz, 1971);
- 'Wicked' versus 'tame' problems (Rittel and Webber, 1973 Ref 714);
- 'Ill-structured' versus 'well-structured' problems (Simon, 1973);
- 'Soft' versus 'hard' thinking that can be applied in situations when the 'problem formulation' is either contested or uncontested (Checkland, 1978; 1985);
- 'Swampy lowland' compared with 'high-ground' problems (Schon, 1987); and
- 'Unstructured' versus 'structured' problems (Kolkman *et al.*, 2005).

Traditional methods of 'problem solving' are likely to be ineffective to manage messes, owing to the:

- Systemic nature of messes and the fact that the sum of all problem solutions is unlikely to equal the 'disentangling' of a mess;
- Unknown interactions of the 'problem solutions' with the rest of the mess; and
- Incapacity to extract a 'problem', especially one that will remain constant over time, and thus to find solutions that are relevant to the situation in the long-term – a one-off or quick-fix solution is not likely to fix the problem (or treat the rest of the mess) for long!

of 'soft' branches of OR, as outlined by Rosenhead and Mingers (2001b), Matthews (2004) and Kirby (2007).

From an OR viewpoint, decision-aiding typically refers to the process where a 'decision analyst' aids a decision maker or 'client' to formulate and analyse his or her 'decision problem' in a structured way before a decision is made. This differentiates it from general decision making processes that may be carried out without the aid of an 'analyst' (Tsoukiàs, 2007). This need for decision-aiding becomes particularly important when the place of unelected experts, the analysts, comes under scrutiny and when the legitimisation and understanding of decision problems and the ensuing final recommendations are required by elected officials or paid and legally responsible managers. The majority of OR decision-aiding research has focused on either one-to-one (analyst–client) or intra-organisational group decision-aiding, rather than on the inter-organisational and multi-stakeholder group decision-aiding situations that are common in the water sector. Research in decision-aiding, where the researcher acts as the 'analyst' with a client (or clients), tends to be carried out as 'intervention research' (Hatchuel and Molet, 1986; Hatchuel, 1994; Berry, 1995; Checkland and Holwell, 1998; Flood, 1998; Avenier *et al.*, 1999; David, 2000; Midgley, 2000). In this approach, theory is explicitly used to intervene to create collective action from which new insights can be drawn to adjust the theory and intervention 'en route'. Intervention in this research approach is considered '*a constitutive mechanism by which a conscious attempt is made to modify organisational phenomena according to some pre-established concepts or models*' (Hatchuel and Molet, 1986), or more simply by Midgley (2000) as '*purposeful action by a human agent to create change*'. The change may involve just mutual understanding and co-construction of problem formulations and management alternatives between the analyst and client, or be extended in the inter-organisational case to broad scale collective action. In all cases, if the final recommendations developed with the analyst's aid are implemented, further modification and system changes are likely to ensue. The theoretical bases of the intervention research approach are further outlined in Appendix B and will be reintroduced in Section 5.1 of this book, where the research protocols are developed.

One focus of OR, known as 'problem structuring', is perhaps the best exception to simple analyst–client and intra-organisational decision-aiding, with such frameworks as the 'strategic choice approach' (Friend and Hickling, 1987) and the 'soft-systems methodology' (Checkland, 1981) being used for complex and uncertain decision-aiding in the inter-organisational context. Such frameworks emphasise the importance of the problem identification or formulation phases of a decision process when dealing with 'unstructured' or 'messy' problems; phases that are typically taken as 'fixed' or 'given' in traditional decision-aiding and technical management approaches for 'structured' decision problems (Rosenhead and Mingers, 2001b).

This brief introduction to decision-aiding will now be broadened to examine the practical applications and historical trends of its use in water planning and management.

3.1.2 Decision-aiding in water planning and management

Decision-aiding in water planning and management has taken a number of forms through history, with the processes used and objectives of the interventions for 'aiding' being mostly based on the predominant rationalities and value systems of the societies in which they were performed. Until the eighteenth or nineteenth centuries, decision-aiding in water planning and management had been largely an 'art' or 'craft', with very limited theoretical scientific knowledge inputs (Dandy *et al.*, 2007). Early 'decision-aiding' in the water sector was also performed by philosophers and those studying the rules of law for counselling decision makers on how water should be allocated (Cech, 2005). Even in large, centralised civilisations with highly developed mathematical and mechanics knowledge, engineers in the role of 'decision-aiders' tended to rely upon tacit understanding and experience (Nonaka and Takeuchi, 1995), as well as empirical 'rules of thumb' to make their recommendations and inform their superiors. With the introduction of mathematics, mechanics and other sciences into the engineering curricula of the 'Grandes Ecoles' in France and others around the world, more scientific principles for 'rational' design, based on ideal economically rational behaviour, and implementation were developed and used by the engineers to aid civil service choices on water infrastructure systems (Florman, 1976).

Management sciences, operational research and a number of other similar scientific disciplines were also developed alongside, or as sub-strands of, the engineering sciences. As a result of these close linkages, water engineering, planning and management practices have closely followed the theoretical and practical trends of these disciplines and have been one of the focal areas of their application (see the engineering textbook, *Planning and Design of Engineering Systems: Revised Edition* (Dandy *et al.*, 2007), for evidence). Hence, decision-aiding in the water sector focused on using methods developed in these scientific fields, such as hard-systems theory, optimisation and statistical models, from around the mid-twentieth century.

Decision-aiding for planning processes then started to adopt the concept of Simon's (1954; 1977) 'bounded rationality' as the basis of a new repertoire of methods, along with the traditional 'economic rationality' models of human behaviour. Other statistical methods based on data mining techniques were also developed and used to avoid the necessity of initially assuming rationality models. Just some of the wide range of decision-aiding methods used through this period were: linear programming for reservoir design problems under multiple constraints; graph theory representations of flow networks for hydrologic and hydraulic modelling, and water allocation management; various forms of meta-heuristics for optimisation, such as genetic algorithms (Goldberg, 1989; Simpson *et al.*, 1993) for pipe network systems; and Monte-Carlo simulation and Markov-chain modelling (Meyn and Tweedie, 1993; Stewart, 1994; Kuczera and Parent, 1998; Rousseau *et al.*, 2001), as well as a range of empirical and statistical models, such as neural networks (Kohonen, 1988; Daniell, 1991) and Bayesian or probabilistic methods (Batchelor and Cain, 1999; Jensen, 2001; Ticehurst *et al.*, 2007) for rainfall runoff, water quality and other modelling applications.

With the rise of democratic movements and increasing awareness of environmental issues, decision-aiding in the water sector also started to undergo a paradigm shift. The reliance on only technical modelling methods, which were commonly carried out by water engineering specialists to inform the decisions of managers, also began to be rethought. As with the revolutions that were taking place in the domains of OR and management sciences about the importance of problem formulation and the search for new methods applicable to problem solving for messy, inter-organisational systems (Ackoff, 1979), water researchers and managers also started to examine new methods of coping with these situations (Lord *et al.*, 1979).

Mirroring the tendencies in OR, decision-aiding methods used in water management included: problem structuring methods, such as Checkland's (1981) soft-systems methodology, to aid the collective representation of complex systems; multi-criteria decision analysis (MCDA) methods to take multiple actors' preferences or 'utilities' into account (Saaty, 1980; Brans and Vincke, 1985; Roy, 1985); and multi-objective analysis methods, such as the Delphi method and the nominal group technique, to take a range of different human values into account in problem formulation through to recommending management alternatives (Goicoechea *et al.*, 1982; Fleming, 1999). Decision-aiders in water management also branched out further to integrate the disciplinary perspectives they believed they were lacking, such as sociological analyses to inform the decision-aiding processes. For example, large-scale consultation efforts through questionnaires and interviews were used to determine citizens' values, beliefs and actions related to water use and management for use in planning processes (e.g. CSIRO ASSERT, 1992; Marks, 2004).

Another of the water sector's responses to shifting needs was to use methods of participatory modelling as a decision-aiding process, also known as 'shared vision modelling' (Palmer *et al.*, 1993), 'group model building' (Vennix, 1996) or 'mediated modelling' (van den Belt, 2004), where the analyst takes the role of the facilitator or modeller attempting to understand and synthesise collective knowledge. Some of these processes have the potential to move closer towards being processes of 'rationalisation' in the sense of Habermas and his communicative action theory (Habermas, 1984). Such processes of co-construction theoretically occur in an 'ideal speech situation' where the confrontation of different stakeholders' rationalities through deliberative discourse and interaction can occur as a means of coming to commonly legitimised decisions (Habermas, 1996). To what extent this model of communicative action can be validated in practice through these processes and how participatory modelling can best be organised as a collective decision-aiding process is still in need of investigation.

The challenges of decision-aiding in the current and future world water sector are therefore in need of further research, especially on how collective decisions can be better aided. For example, one challenge is that it is increasingly difficult to legitimate the use of some OR decision-aiding tools, such as optimisation or hydrological models, on a purely normative basis, as was traditionally possible in technocratic societies where the place of the 'expert' was not challenged (Fischer, 1990). This is largely due to the realisation that there are multiple human rationalities, which are difficult to take into account in decision-aiding processes without adapted consultative or participatory methods. These processes would ideally include stakeholder communities, encompassing citizens or 'the public', officials or decision makers (policy makers and managers) and experts (Thomas, 2004). If the use of such OR tools or models can be constructively legitimised through such participatory modelling processes – in other words, the stakeholders take ownership of the problem, its formulation, the models developed and used, and the recommendations – these models may still prove valuable in the quest to manage and find sufficing and collectively legitimated solutions to complex water management challenges.

3.2 DECISION-AIDING MODELS AND APPROACHES

Decision-aiding models and methods are used with the goal of informing and improving the choice of actions. Many of the professions, such as law, medicine and engineering, have created models, methods and approaches over the years to aid their decisions-making practices and consequent actions. Some are more theoretically founded than others, having been developed and studied as part of scientific enquiry, whereas others have been built largely on experience over time or 'rules of thumb'. The decision-aiding processes used in the related professions of engineering, policy (linked to law) and operational research will be briefly examined and compared here, as well as their applicability to inter-organisational decision-aiding contexts.

3.2.1 Engineering approaches to decision-aiding

The word *engineer* appears to be derived from the Latin *ingenium*, which can be translated as 'mental power' (Lienhard, 2000). There are a myriad of definitions for engineering; yet they tend to have a number of similarities, including that the goal of engineering is to solve problems for the benefit of humanity. For example, the Institution of Engineers Australia (IEA, 2000) states that:

> *Engineering is a creative process of synthesising and implementing the knowledge and experience of humanity to enhance the welfare, health and safety of all members of the community, with due regard to the environment in which they live and the sustainability of the resources employed.*

Engineering as a process is applicable not only to sectors of the material world of resources such as civil and mechanical engineering (Dandy *et al.*, 2007) but also to the immaterial world through processes such as knowledge and decision engineering (March, 1978). Engineering can also be defined as a problem-solving process. The decision-aiding required to carry out this process for a client can be thought of as the process used to define: the 'current state' of a problem situation; the 'goal state' for the problem situation; and a set of procedures to progress from the current state to the goal state. These basic principles of the engineering process are outlined in Figure 3.1.

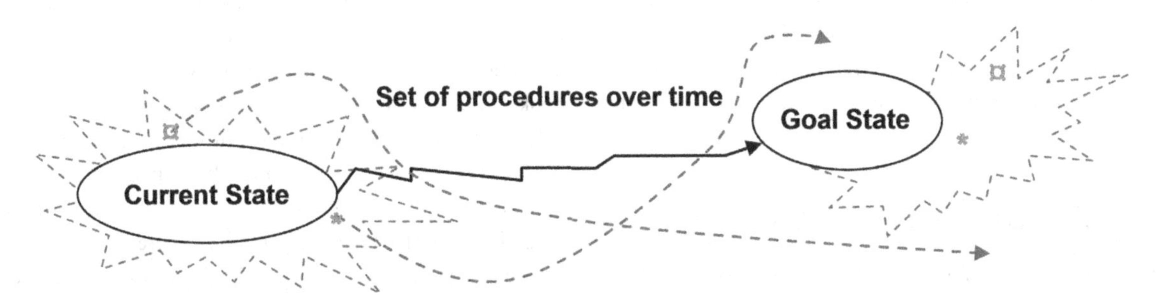

Figure 3.1 A typical engineering process. The black lines represent a 'closed' problem where it is relatively easy to define the problem (current state) and objectives (goal state), and the grey dotted lines represent an 'open' problem (a mess), where all phases of the decision-aiding (or engineering) process are likely to be difficult to define, owing to a multiplicity of views on what the current state represents and what the desired goal state should be. In such a situation, reaching the collectively defined goal state range is not assured, as this state is also likely to be continuously reconstructed through time through a process of rationalisation. Procedures of continuous innovation will therefore also be required to travel towards the moving target or broad 'vision'.

Like decision-aiding, which has taken many forms to adapt to the different challenges and value systems of society, engineers and the processes of engineering have also adapted to the changing needs of society, aiding decision makers and the communities they serve or represent to achieve their goals. The 'engineering' that many people first think of is still based on somewhat traditional value systems, where 'to engineer' is to 'control' nature and optimise systems (László, 2006). This concept is based on the idea of using rational theories of choice to determine optimal or at least 'sufficing' solutions to problems (March, 1978). However, like all other disciplines with their paradigmatic shifts starting from around the beginning of the 1970s, the engineering profession and its underlying principles have also been in constant evolution since this time. Today, engineering processes must take into account more environmental and social concerns than in the past and allow diverse groups of people to work together to manage and navigate through messes. Advancing the cause of 'sustainability' is now one of the central principles of engineering in a number of countries, such as Australia and the UK. This has incited a mass movement away from 'controlling' nature in a number of engineering disciplines towards attempting to work in alignment with natural processes and has led to initiatives such as 'water-sensitive urban design' (Wong and Eadie, 2000), 'low-impact development' (Coffman *et al.*, 1998), 'cradle-to-cradle product design' (Hargroves and Smith, 2005) and the environmental and systems branches of engineering. A number of national engineering accreditation bodies also have strict ethics codes to which its engineering professionals must adhere (IEA, 2000). Moreover, to treat increasing numbers of 'mess management' situations, it is increasingly recognised that engineers must not only act as 'problem solvers', but also as 'problem framers' (Donnelly and Boyle, 2006). The challenge for engineers in these messy situations is to aid the collective design of a desirable future (the goal or vision); determine how far away from it we are (the gap to the current state); aid the invention processes required to close the gap (the procedures), and evaluate how much progress has been made. This will require a re-examination of the use of rational theories of choice and of what types of rationality or processes of rationalisation are appropriate for these new forms of engineering decision-aiding.

3.2.2 Policy approaches to decision-aiding

> *To be rational in any sphere, to display good judgement in it, is to apply those methods which have turned out to work best in it. What is therefore rational in a scientist is therefore often Utopian in a historian or a politician (that is, it systematically fails to obtain the desired result), and vice versa.*
>
> Berlin (1996) in Althaus *et al.* (2007)

A broad range of guidelines exist on how decision making through the policy processes can be aided, many of which have their roots in political, policy and administrative management theory; or are based upon practitioners' experience in policy making. Various models of the policy process exist, which have led to different decision-aiding guidelines being developed to navigate through them. Many take a phase or step-wise approach, similar to the decision processes outlined in Section B.2, such as the 1951 policy process sequence of Harold Laswell; 'intelligence, recommendation, prescription, invocation, application, appraisal and termination' (Althaus *et al.*, 2007). Questions or a series of activities to be carried out for each phase then form the decision-aiding guidelines. Other basic models of the policy process that have informed such guideline proposals include the garbage can model (Cohen *et al.*, 1972), the 'funnel of causality' (Simeon, 1976), risk or uncertainty management approaches (e.g. Perrow, 1984; Boin *et al.*, 2005), the 'advocacy coalition framework' (Sabatier and Jenkins-Smith, 1993) and systems approaches (e.g. Colebatch, 2006). By way of example, a set of methodological guidelines for aiding a policy decision process for resource and environmental management, that is based on a hybrid understanding of a number of these models, is shown in Figure 3.2.

It can be seen that, although feedback between stages is possible, these decision-aiding guidelines suggest that a sequential and cyclic phase-type decision process should be used to aid the stakeholders and policy decision makers to formulate, implement and review the policy. Analysis of social goals, risks, system properties and other steps just in the framing phase helps to show the hybrid nature of these guidelines relative to existing policy process models. Although there is careful outlining of a number of activities that should be carried out to aid the decision making process, specific methods to be used are not dictated and thus they still need to be chosen by the analysts aiding this process. An alternative in-depth set of guidelines to aid progress and decision making through a typical eight-step policy cycle in Australian policy contexts, 'identifying issues, policy analysis, policy instruments, consultation, coordination, decision, implementation, evaluation', is given in Althaus *et al.* (2007). This book presents a broader view of the policy cycle that is applicable to any domain and also goes further in specifying particular methods or instruments that may be applicable in each stage. Nevertheless, the complexities of real policy processes that are unlikely to conform entirely to the ordered cycle leave many choices for the analyst and the need for his or her good sense and inspiration.

3.2.3 Operational research decision-aiding models

Decision-aiding, as studied in operational research, is strongly linked to the management sciences and so has a number of clear

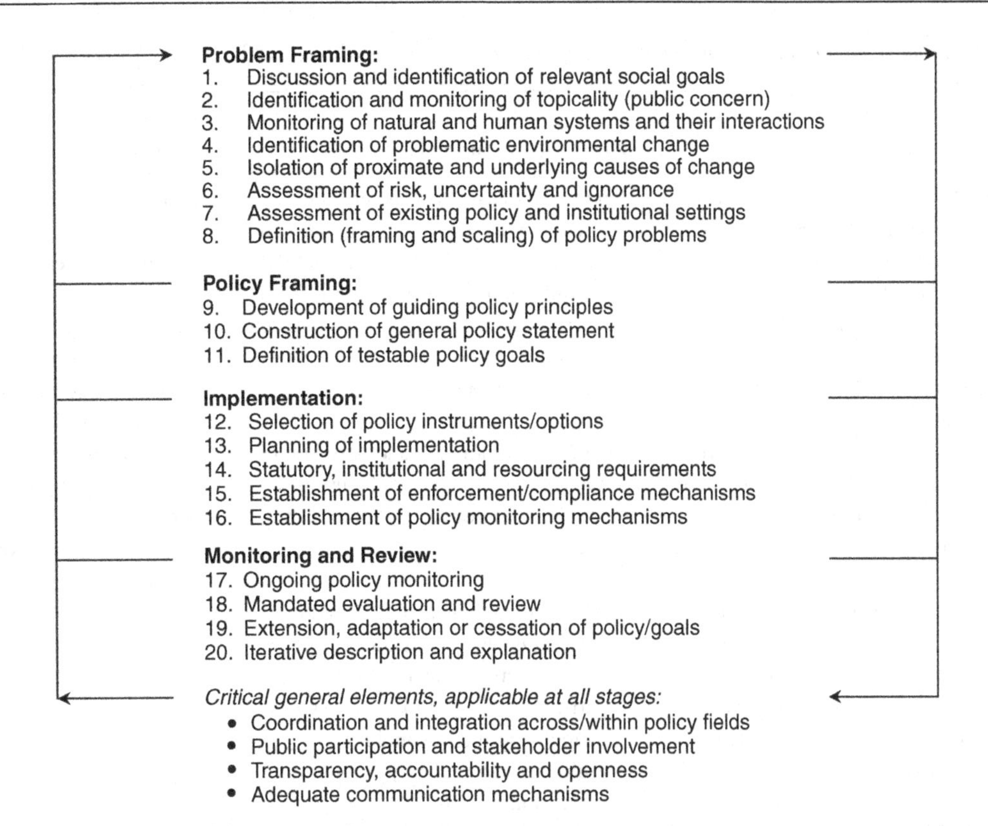

Figure 3.2 Decision-aiding process framework for resource and environmental management policy creation. Source: Dovers (1995; 2002; 2005).

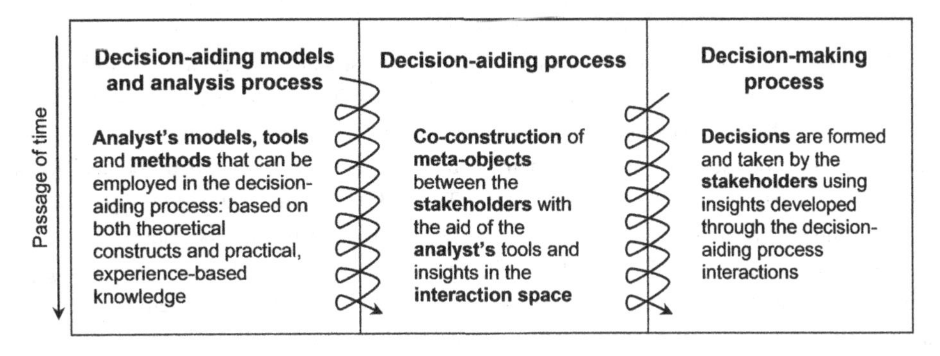

Figure 3.3 Interactions of the decision-aiding process.

links to the policy and engineering decision-aiding models. However, operational research typically has a stronger focus on abstract and mathematical formulation of its models. Many researchers claim to have developed models for aiding decision making, although many of these refer more to the 'evaluation methods and models' part of a decision-aiding process, for example the 'decision-aiding models' described in Gass (1983) and Appendix B, than to the whole process. As previously alluded to, decision-aiding processes and models are typically employed by the 'analyst' relative to a certain conception of how a decision making process is carried out, or rather how it should be carried out to obtain better informed decisions. Despite the theoretical validity issues, phase-type decision processes, such as Simon's 'intelligence', 'design', 'choice', 'implementation' and 'monitoring' phases (Simon, 1977), are still used as the conceptual basis for most decision-aiding models. A comparative description of decision making models has been provided in Section B.2. The relationship between these models, the decision making processes and the decision-aiding processes is represented in Figure 3.3.

In Figure 3.3, the continuous interaction that occurs in a bi-directional manner between the analyst's models and analysis and the decision-aiding process and between that process and the stakeholders' (individual and potentially collective) decision making process is shown by the spiral arrows. The analysts will use their insights to facilitate the decision-aiding process and create new insights for themselves through their participation in the 'interaction space' and co-constructions of 'meta-objects'

Table 3.1 *Classification of decision-aiding approaches. Adapted from: Dias and Tsoukiàs (2003) and Tsoukiàs (2007)*

Approach type	Rationality model derivation	Attribution of meaning	Interpretation of results for the decision maker(s)
Normative	From a priori established norms (i.e. ethical, legal, religious); rationality is exogenous and based on ideal economic behaviour	Norms are considered universal and necessary for rational behaviour. Any deviation from them by stakeholders is considered a mistake or shortcoming	Deviations need to be 'corrected' to aid the stakeholders to learn to decide in a rational manner
Descriptive	From observation of stakeholders' decision making; rationality is exogenous and based on empirical behaviour models	Patterns of process to quality of outcomes are noted. Models are generalisable for similar problem situations.	Derived 'laws' for the problem situation can be transferred to stakeholders to allow them to follow previous successful decision making processes
Prescriptive	Discovered or unveiled from answers to preference or value-based questions; rationality is endogenous and coherent with the decision situation	Models are contingent and based on the stakeholders' context and current responses (however incomplete they may be)	Improvements that could be made to the decision making process are 'prescribed' by the analyst based on his or her 'expert' knowledge and experience
Constructive	Built with the stakeholders for their particular context based on answers to preference or value-based questions; process of 'rationalisation' and collective learning that is coherent with the decision process	Models are co-constructed (based on consensus) and contingent on the stakeholders' context; they pay particular attention to structuring and formulating the stakeholders' problems, rather than just building evaluation models	The model-building process is aimed at directly allowing the stakeholders to make decisions for themselves

(these two concepts are further studied in Section 3.2.4). Similarly, the stakeholders will provide their knowledge, representations, beliefs and preferences to the decision-aiding process when required and, hopefully, receive useful insights for their decision making in return from the decision-aiding process.

Of all the decision-aiding models available in the literature, Dias and Tsoukiàs (2003) and Tsoukiàs (2007) classify the approaches into four main types: normative, descriptive, prescriptive and constructive. Each of these approaches can be differentiated according to the manner in which the 'rationality model' of the stakeholders is obtained, given meaning, and the results of the decision-aiding process are interpreted (Dias and Tsoukiàs, 2003). *Rationality* in this case is referred to as a process (rather than referring to the final decision) and is assumed to present some coherence, although not necessarily of the 'economically rational' kind (Tsoukiàs, 2007). Rationality is also considered to be 'bounded' in the sense of Simon (1954) and subjectively defined based on the decision maker's local context (i.e. time, space and cognitive capacity). This implies that each individual decision maker will possess his or her own unique rationality and that an overall 'group rationality' is unlikely to be an aggregation of the individual rationalities. The four approaches are outlined in Table 3.1.

It is noted that these descriptions refer to the 'processes' and not the actual methods or models used for decision-aiding. The analyst can thus decide which kind of stance to take in the decision-aiding situations, and which range of models and analysis tools to use in relation to this stance. The similarity between the underlying philosophies of the decision-aiding approaches in Table 3.1 and the evaluation types outlined in Appendix B (Table B.4) can also be observed. If this classification is used to look at the policy decision-aiding guidelines outlined in Figure 3.2, considering the emphasis on the problem framing section, the approach would most probably be carried out in a constructive or perhaps prescriptive manner, owing to the explicit prescription to include public participation and stakeholder involvement in all stages. The final approach taken would still depend on the position taken by the analyst.

In operational research, decision-aiding (process) models do not just include guidelines, but also models based on formal and abstract language. For example, the model of Tsoukiàs (2007) defines that a number of cognitive artefacts (meta-objects) are to be generated through the phases of the decision process phases. Each cognitive artefact is then constituted of a number of sets of elements, which can be represented using set theory. The decision-aiding process model producing these artefacts is represented diagrammatically in Figure 3.4. There may be iteration between the building of these cognitive artefacts and feedback to previous phases to update them, which is represented by the dotted lines in Figure 3.4.

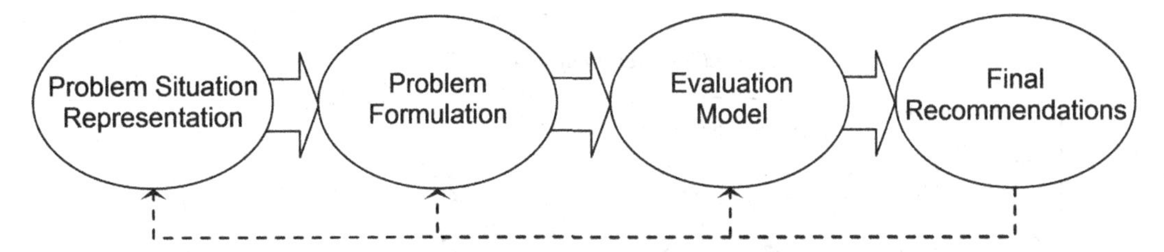

Figure 3.4 Four-stage decision-aiding process model (Daniell *et al.*, 2010a). Derived from: Tsoukiàs (2007).

It is suggested that using 'formal' and 'abstract' models as a basis for decision-aiding can provide a number of benefits over other types of decision-aiding such as guidelines in certain problem situations, including that they can be used to reduce the ambiguity that is structured in human communication and that they are generalised theoretical constructs that are independent of the decision-aiding context (Tsoukiàs, 2007). Such decision-aiding models could be of particular value for an analyst in an intervention research context, as the generalisable theoretical perspective could be adapted for use in many situations. Specific methods of co-constructing the 'meta-objects' chosen, as required to spawn new insights, could also add to the model's scientific validation and create 'actionable' knowledge (Hatchuel and Molet, 1986; Roy, 1993). However, one of the issues of this particular decision-aiding process model is that it was originally developed for a simple analyst–client relationship, rather than for the intra- or inter-organisational context. To determine what adaptations of the model might be required for it to be used in the inter-organisational context, the next section examines the specificities of the context. The model's potential adaptation for use in the inter-organisational intervention research context will then be outlined in Section 5.1.3.

Other approaches stemming from the 'soft' branch of operational research could also be considered as decision-aiding models that are more similar to the guideline types. Just two of these approaches, the 'strategic choice approach' (Friend, 2001) and 'journey making' (Eden and Ackermann, 1998), are presented in Appendix C (Section C.3.3). Both of these approaches are more highly specified in terms of the methods that should or could be used in each part of the decision-aiding process, although their underlying principles may be of use in many situations. In particular, the strategic choice approach has been developed and widely used in the inter-organisational context and stresses the need for flexibility to oscillate or switch freely between different modes (shaping, designing, comparing, choosing) of thinking which occur during a decision-aiding process.

3.2.4 Context of inter-organisational decision-aiding

Decision-aiding for 'messy' problems in the inter-organisational context focuses on providing a 'decision analyst' or analysis steering group with methodological aids that allow them to facilitate a group in a transparent manner to structure and exchange views on issues ranging from identification of the problem and objectives to final recommendations or 'choices'. This process can be considered to occur in an 'interaction space' (Ostanello and Tsoukiàs, 1993) or a 'collaborative space' (Digenti, 1999), where the collective construction of the participants' representations of the problem can be regarded as a 'model', 'meta-object' (Tsoukiàs, 2007) or 'intermediary object' (Vinck and Jeantet, 1995), which can form the basis for further collective discussion and decision making. The 'models' considered here are thought of as '*instruments that aid the production of knowledge*' (Poussin, 1987). Interactions between the various process participants are governed by rules that may only exist within the 'interaction space'. Related to this description, Mazri (2007) defines an *interaction space* as: '*A formal or informal structure that is governed by a number of rules and is destined to provide a field of interaction to a finite set of actors.*'

Ostanello and Tsoukiàs (1993) provide a descriptive model of an inter-organisational decision process, which outlines a number of characteristic 'states' based on the ideas of MacKenzie's (1986) 'process laws', which could be observed in the evolution of the interaction space. In this case, the process state is defined in terms of: the participating 'actors': 'objects' (the concerns or stakes of the actors); 'resources' that the actors are willing to commit to their objects of interest; and the 'relations' between these elements (Ostanello and Tsoukiàs, 1993). Such a model, even if simplified to just a few theoretically important variables, can provide a useful basis for understanding inter-organisational decision dynamics and therefore provides a conceptual framework for the development of a decision-aiding approach to guide the evolution of the interaction space in a favourable direction.

In inter-organisational groups, a number of other context specificities should be taken into account by analysts developing decision-aiding approaches. For example, unlike in groups that share the same organisational background and accountability structures, there will be outside factors, interests or rules that will affect the ability of each participant to agree on decisions. Participants may only have limited power to enter into commitments on behalf of their organisations, making such a working group 'multi-accountable', unlike a traditional 'team-like' group (Friend, 1993). In this context, it is likely that the 'interaction

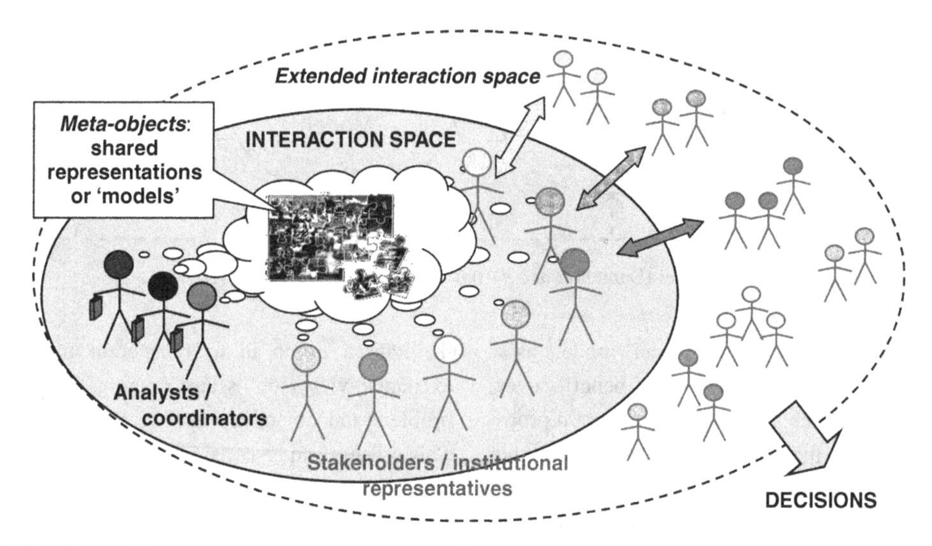

Figure 3.5 Decision-aiding for 'multi-accountable' groups in a stretched interaction space. Source: Daniell *et al.* (2010a).

space' of the decision processes will not be limited to just a working group which meets, but rather expand to include the external interactions and negotiations that are likely to occur between organisations at different managerial and government administration levels. A graphical representation of this expanded interaction space for inter-organisational, 'multi-accountable' groups is given in Figure 3.5.

It is noted that the *organisation* in the 'inter-organisational' groups is considered here in the broadest possible manner as a group, association, business, institution or government or any ensemble of two or more people who share at least one common characteristic, interest, vision or goal. This can include individual citizens, as they can be considered as representatives of their country, region, sex, religion, ethnicity, age group, etc. Each stakeholder group in Figure 3.5 is represented by a different pattern of shading; relationships between individuals in the group are represented by the double-headed arrows.

The decision-aiding processes used for multi-accountable groups, outlined in Figure 3.5, must be applicable to these more complex environments; in particular where those participating in the 'core' interaction space with the analysts rely upon others from the 'outer' interaction space to make the final decisions. It is apparent from Alexander's (1993) theory of 'inter-organisational coordination' that the equivalent of the 'core' interaction space is an 'inter-organisational group' and the total 'outer' interaction space constitutes the whole 'inter-organisational network'.

In some cases, a number of analysts or coordinators might be marshalling the different skills and knowledge required for working with a diverse group of stakeholders, as represented by the 'coordination structures' of Alexander's (1993) framework. Akkermans (2001) suggested that in this type of situation, traditional organisational theory from the management sciences, focusing on command and centralised hierarchical control, is of limited value because of the lack of a formal single locus of authority. Rather, the networks' decisions and collective actions are driven by the power and influence of the individual stakeholders and their supporting organisations and by communication, persuasion and consensus building (Akkermans, 2001).

A range of other names for 'multi-accountable' groups and bodies or corresponding research equivalent to these terms can be found in the literature. This includes work on: 'multi-organisations' (Stringer, 1967), 'inter-organisational networks' (Benson, 1975), 'inter-institutional networks' (Wenk Jr. and Kuehn, 1977), 'inter-organisational fields' (Warren, 1967), 'organisation-sets' (Aldrich and Whetten, 1981), 'implementation structures' (Hjern and Porter, 1981), 'policy-issue networks' (Kirst, 1984), 'multi-actor implementations' (O'Toole Jr., 1986), 'multi-party' or 'collaborative' alliances (Gray, 1989; Huxham, 1996), 'multi-organisational fields' (Friend, 1993), the 'inter-organisational domain' (Hardy and Phillips, 1998) and 'multi-stakeholder' collaborations, partnerships or groups (Poncelet, 2001). In some cases, other types of group, such as communities of practice (Wenger, 1998), could also refer to 'multi-accountable groups'. This multiplicity of terms is symptomatic of both the wide variety of independent disciplines relevant in this area, and the challenges faced in researching it.

3.2.5 Critique of decision-aiding models for the inter-organisational context

Each of the engineering, policy and organisational research approaches to decision-aiding discussed in this chapter has some common elements. They all have strong links to management and administrative sciences and are based on assumptions of evolving rationality models of human behaviour. Most started with standardised processes based on assumptions of economically rational behaviour and have evolved to develop decision-aiding guidelines or models that can be used in complex environments by analysts. These approaches assume that a variety of rationalities exist or aid participatory processes of rationalisation in the sense of Habermas (1984; 1996). Many have also adopted phase-type decision making process models to underlie their decision-aiding processes. The strategic choice approach (Friend, 2001) is an example of a process that adopts a more chaotic and iterative view of the decision-aiding process.

The focus on goals, the content of the decision-aiding process and the current situation are all variable in the aforementioned approaches. Engineering approaches to decision-aiding are typically highly purposeful where goal or vision setting is of high importance. This 'engineering decision-aiding' often works at the operational action end of the decision-aiding spectrum. Policy approaches, often trying to create 'social goals', take a broader approach and consider other factors, such as topicality and risk assessment. The creation of such goals appears important when general policy coherence in the wider political sphere is required. Operational research models for decision-aiding appear to be adaptable to multiple levels of analysis. The 'journey-making' (Eden and Ackermann, 1998) model stresses that it is structured as a 'conscious and purposeful activity', which could be used at either operational or political levels of decision making. The strategic choice approach (Friend, 2001) and the Tsoukiàs (2007) decision-aiding models appear less 'purposeful' but could be also used in this manner.

The key themes elicited from the decision-aiding models described here can also easily be compared and reinterpreted using Ackoff's description and operational principles of 'interactive planning' for organisations from the management sciences (Ackoff, 1999), but must be extrapolated to the inter-organisational case. Ackoff considers interactive planning to consist of '*the design of a desirable future and the selection or invention of ways of bringing it about as closely as possible*'. This closely resembles the principles underlying the engineering approach and some of the operational research approaches. The objective is to 'idealise' the future and determine an appropriate planning process that can be used to reach it. The focus is on increasing adaptivity and the ability to control or influence a change or its effects, as well as to respond rapidly and effectively to uncontrollable or unforeseen changes (Ackoff, 1999). This principle is closer to the broader method taken by policy approaches to decision-aiding. To carry out such planning, Ackoff suggests three 'operating principles':

- *The participative principle*. Undertaking the 'process' is the most important benefit of planning; all stakeholders should be involved in planning for their own collective future. The role of the planner is thus to 'facilitate' this interactive planning process as best as possible.
- *The principle of continuity*. There should be continuous monitoring of the assumptions that underlie the planned trajectories, progress performance and planned expectations. If any assumptions are disproved or deviations from the planned actions noted, then re-adaptation, correction and improvement of the planned and implemented actions are required. Designing flexibility into systems and implementing actions in a relatively controlled experimental manner is considered an integral element of interactive planning.
- *The holistic principle*: with two sub-principles – *coordination* and *integration*. Both imply a need to simultaneously and interdependently plan the trajectories of different parts of the system, effectively concentrating on the breadth and interactions of planning (rather than the depth and specificity of individual actions), as well as the vertical integration of interdependent and synchronous planning at all decision making levels; in other words, integration of top-down/bottom-up planning to make it 'all-over-at-once'.

These principles are to be used through Ackoff's 'five phases of interactive planning', outlined in Table 3.2, which may be carried out in sequence, or may interact or have feedback between them, or be carried out in a more simultaneous manner indefinitely (Ackoff, 1999).

It can be seen from this description of interactive planning that there are a number of features in common with both the elements of the decision-aiding model proposed by Tsoukiàs (2007) and typical engineering processes, where the focus is much more strongly on bridging the 'gap' and inventing or designing the means to implement and achieve this bridge. The main potential difference is in the capacity to continue the process through adaptive cycles, owing to the probable short intervention of the decision analyst or consulting engineer rather than a full-time planner. The operating principles closely resemble the challenges of the policy process and decision-aiding in the inter-organisational case, and seem consistent with a constructive approach to decision-aiding. However, there is less emphasis on the monitoring and evaluation that might be necessary to complete the planning cycle. The challenges of operationalising decision-aiding models and approaches will be further examined in the next section.

3.3 OPERATIONAL USE OF DECISION-AIDING MODELS

Inter-organisational decision-aiding approaches require the use of methods to negotiate through the uncertainties, complexities and conflict likely to arise in the messy problem situations that have been previously discussed in this book. In Sections 2.2.4, 3.1.2 and 3.2.5, a number of methodological approaches to management have been suggested that may be appropriate in this 'mess management' context, such as adaptive management, risk

Table 3.2 *The five phases of interactive planning. Adapted from: Ackoff (1999)*

Interactive planning phase	Objective	Element elicitation	Principal output
Formulation of the mess	Determine problems and opportunities	Problems, opportunities and their interactions; constraints and obstructions to effective management	Reference scenario
Ends planning	Determine an idealised re-design and the gaps to be closed through the planning process	Extract goals, objectives and management ideals that are part of the system's idealised re-design	Gap identification between idealised re-design and reference scenario
Means planning	Choose appropriate mechanisms to close the gaps	Invent and select means: alternatives for policies, programmes, projects, practices and actions	Set of means that can be used to close the gap
Resource planning	Determine resources required to allow chosen means to be implemented	Define and classify resource needs by type, quantity and timing, and state whether they are available or how they will be generated or acquired	Set of required resources for chosen means
Implementation and control	Determine responsibilities and schedules for means implementation	Define who is responsible for which means, when they are to be implemented and how they are to be monitored to ensure performance expectations are being reached	Schedule and responsibilities for plan implementation

management and Ackoff's (1999) interactive planning. However, exactly what types of method or process could be used to improve decision-aiding processes for mess management have yet to be specifically analysed here. A number of key themes emerge for dealing with uncertainty, complexity and conflict in practical, decision-aiding approaches, as previously touched on.

As part of operationalising the 'participative principle' outlined by Ackoff (1999), there is likely to be a strong reliance on the analyst to aid the creation of a 'representation' or 'model' of important elements in the problem system that can then be used to promote sufficient understanding to come to more informed decisions, as visualised in Figure 3.5. When conflict and disagreements arise, emphasis must be placed on how to accommodate the participation of conflicting parties and associated decision makers, to negotiate collectively through the phases or iterative elements of the decision-aiding process. Dealing with conflict thus makes the *process of interaction and participation* a necessity for mess management. It is suggested, therefore, that 'participatory modelling' could be used in this context where the representation development and understanding, the modelling, can take place in a participatory manner through the whole decision-aiding process. A specific attempt at defining 'participatory modelling', which is based on the active involvement of different participants through phases of a typical decision process, is given in Appendix C. Exactly how such a process should be designed and conducted is still under fierce debate (e.g. Zagonel, 2002).

Before considering potential methods and how to conduct a participatory decision-aiding process, some issues of process initiation and design from both the water management and operational research domains will be outlined. When commencing an inter-organisational decision-aiding exercise, it is probable that there will be some preliminary interaction between the 'decision analyst' and one or more of the stakeholders. During this preliminary interaction, an agreement may be made to help these stakeholders structure and manage a particular issue under certain rules of engagement (Avenier *et al.*, 1999). Once an agreement has been created, a general process design, including the 'participatory structure' under which the exchanges in the interaction space are to take place (Mazri, 2007), needs to be developed. In this process, the decision analyst needs to be legitimated in his or her capacity to engage the participants (Creighton, 2005). Stakeholders from different organisations who are considered to have interests in the issue can then be invited to participate in the decision-aiding process through the construction of the meta-objects or 'models' outlined in the previous section. Two methods for designing such participatory structures will be discussed next.

3.3.1 Determining participants and process structure

Until recently, research on 'participatory structures' for organising the participation of different groups of actors from the policy, public and science arenas in decision-aiding processes has principally focused on how to choose between or evaluate a number of common structures (Rowe and Frewer, 2000; Beierle and Cayford, 2002; Lynham *et al.*, 2007; Mazri, 2007; Bayley and French, 2008; Daniell, 2011). These include public meetings, citizens' juries, consensus conferences, focus groups, Samoan circles, open-space meetings, future searches, online conferences, Delphi, interviews, questionnaires and many more (OCDE, 2001; Elliott *et al.*, 2005; HarmoniCOP, 2005a). Rather

Phase 1: Decision Analysis

- Who needs to be involved in the decision analysis?
- Who is the decision maker?
- What are the decisions being made or the problems being addressed?
- What are the stages in the decision making process and what is the schedule for each stage?
- What institutional constraints or special circumstances could influence the public participation process?
- Is public participation needed and if so, what level of participation is needed?

↓

Phase 2: Process Planning

- Who needs to be on the planning team?
- What are the issues and who are the public or stakeholders for these decisions?
- What are the decisions being made or the problems being addressed?
- What are the potential levels of controversy and how can they be planned for?
- For each step in the decision making process: what is to be accomplished with the public and what are the participation objectives?
- What does the public need to know to participate effectively and what needs to be learnt from the public?
- What special circumstances affect the selection of public participation techniques?
- Which public participation techniques are appropriate?
- What shall be in the public participation plan?

↓

Phase 3: Implementation Planning

- Has approval for the public participation plan been given by the relevant authorities or funding bodies?
- How much time is needed for the plan implementation and where will the activities take place?
- What meeting facilities and technologies will be used, and who will organise them (logistical preparation)?
- Who is to participate in each meeting or activity and what are their roles (i.e. publicists, facilitators, spokespeople, technical experts, report writers)?
- What is the meeting agenda or activity outline?

Figure 3.6 Stages of public participation programme design. Adapted from: Creighton (2005).

than determining how best to fit the available structures to the context, it has been emphasised that there is a need for increased theoretical and practical understanding of how and what types of participatory structure should be conceived or designed to best deal with specific problem constraints (Rowe and Frewer, 2000; Wiedemann *et al.*, 2000; Mazri, 2007). Many years ago, Ackoff had similarly outlined this issue in his critique of the evolution of operational research (OR) practice. He thought OR used to be '*dictated by the nature of the problem situations it faced. Now the nature of the situations it faces is dictated by the techniques it has at its command*' (Ackoff, 1979).

Recently, the process of participatory structure design has been examined in more detail, including by Creighton (2005) and Mazri (2007). From a practitioner's point of view, Creighton (2005) lays out a detailed guide of process steps that may be followed to design and plan for a public participation programme, an abbreviated representation of which is outlined in Figure 3.6.

For many of the specific questions related to each phase of programme design, Creighton (2005) has also developed a number of simple checklist or matrix analysis tools based on his experience in the field, to aid other practitioners. The practical insights on suggested processes that precede implementation of public participation provide a useful basis for reflecting on the needs of implementing decision-aiding processes, such as participatory modelling. Many of the points, especially in the decision analysis and process planning stages outlined by Creighton (2005) still require further research or deeper theoretical insight. It is noted that there is no explicit mention of monitoring or evaluation in the programme design, although Creighton (2005) does include pointers for evaluating public participation late in his handbook. One of the issues with using such an outline for participatory process design and comparing it with current research is in understanding exactly which activities are carried out in which phases. For example, 'stakeholder analysis' (Freeman, 1984; Grimble and Wellard, 1997;

Mitchell *et al.*, 1997; Ramirez, 1999; Brugha and Varvasovszky, 2000) may occur in a number of rounds through a complete participatory structure design and participatory process implementation: in the decision analysis phase (most commonly performed informally by the project initiators); twice in the process planning phase (to determine planning team members and then to determine broader stakeholder groups who should participate), with the second part being where more complex 'stakeholder analysis' techniques may be used; in the implementation planning phase (to determine specific invitees or participants in each activity); and then potentially in the actual decision-aiding process (for example 'actors' are to be defined in the construction of the 'problem situation' with participants – refer to Figure 3.4 and Section B.3). This type of complexity for organisers and implementers of participatory processes is rarely recognised and thus appears to be an interesting topic that merits further investigation. Several other methodological guides with suggested phases similar to Creighton (2005) include those given in Bertrand and Martel (2002), on their participatory multi-criteria approach, Beierle and Cayford (2002), for environmental decision making, and Wiedemann *et al.* (2000) and Stern and Fineberg (1996), specifically designed for risk communication and management programmes. Kelly (2001) also provides a comprehensive description and descriptive model of elements required for design of community participation activities in rangeland management.

A number of elements related to this research area, especially methodological theory that can be employed as an equivalent to parts of Creighton's (2005) decision analysis stage and process planning stages, have recently been proposed by Mazri (2007) in his PhD thesis on structuring public decision processes in participatory contexts in the field of decision-aiding for risk management. His descriptive model of the participatory structure design process is shown in Figure 3.7.

Of particular interest in this more formalised model is the iterative nature of the process and the search for continual improvement and social learning as the process progresses. Mazri (2007) notes that there may also be other types of feedback in between different steps of the process, and not only around the represented loop, which would help encourage further reflexivity of, and social learning between, the actors. For the characterisations, the 'intrinsic' refers to basic elements that the actor uses to build a world view including 'stakes', which may be *concerns or interests* in a wide range of contextual elements (i.e. social, environmental – refer to Section B.3 for a more complete list), and 'resources' belonging to the actor. These resources incorporate: *a set of knowledge bodies* (scientific, practical, contextual, etc.) that can be employed to understand and study the problem situation; *value systems*, considered legitimate for the problem situation under study, that can be used to influence the outcomes of a participatory process; and *attributes that confer powers of influence*, such as judicial attributes, including legal responsibility, economic attributes, including financial resources, and social attributes, including respect, charisma and confidence (Mazri, 2007). 'Extrinsic' characteristics are those elements that can be used to describe the representation of the problem situation in which the actor is evolving, in this case relative to the creation of a participatory structure, and therefore: who is relevant to the problem context that the actor thinks should be involved in the creation of the participatory structure and why; what are the most critical elements of interest in this context that should become objects of debate considered by future participants in a participatory process; and for which objectives, such as a legitimate decision making forum, conflict resolution or an advisory process, the participatory structure should be designed (Mazri, 2007).

Another point in Figure 3.7 that requires further explanation is the definition of the principles of 'competence' and 'equity' referred to in Step 2. Firstly, the principle of 'competence' of the actors stems from the work of Habermas (1984), where it is used to describe a set of possibilities and talents of an actor (i.e. cognitive, linguistic, pragmatic and interactive competencies), which is also much the same as the set of deliberative rules suggested by Webler (1995) and Mazri (2007). Similarly, the principle of equity is based on the work of Habermas (1984) and that reformulated by Webler (1995), and is to be enacted when an 'ideal speech situation' is maintained in which the set of participating actors have equal chances to: formulate and explain their declarations; present and defend their positions relative to the four validity constraints of comprehensibility, truth in the scientific sense, normative rightness and sincerity (truthfulness); contest validity claims of other participants; and influence the final modes of validation or decision making rules and hence selection of final recommendations (Webler, 1995; Mazri, 2007).

The last point of note relates to the creation of the organisational model in Step 4, where the dependence relations between each two objects of interest in the participatory structure are developed. Mazri (2007) defines three types of dependence relation between the objects: dependence (uni-directional relation); interdependence (i.e. bi-directional relation); and independence (no relation). From these dependencies, a logical series of objects to be treated in the participatory structure (i.e. which objects to treat when through time) can be developed, similar to a work flow diagram, where the critical path can then be established to aid additional implementation planning. An example diagram is shown in Figure 3.8 where: *participants to deliberate* include those in the *implication* and *consultation* categories; and *participants informed* include those in the *information and listening* and *information* categories.

As such a model is 'just a skeleton' (Mazri, 2007), further study and selection of specific methods to be used are required, as well as more detailed logistical planning (such as the phase

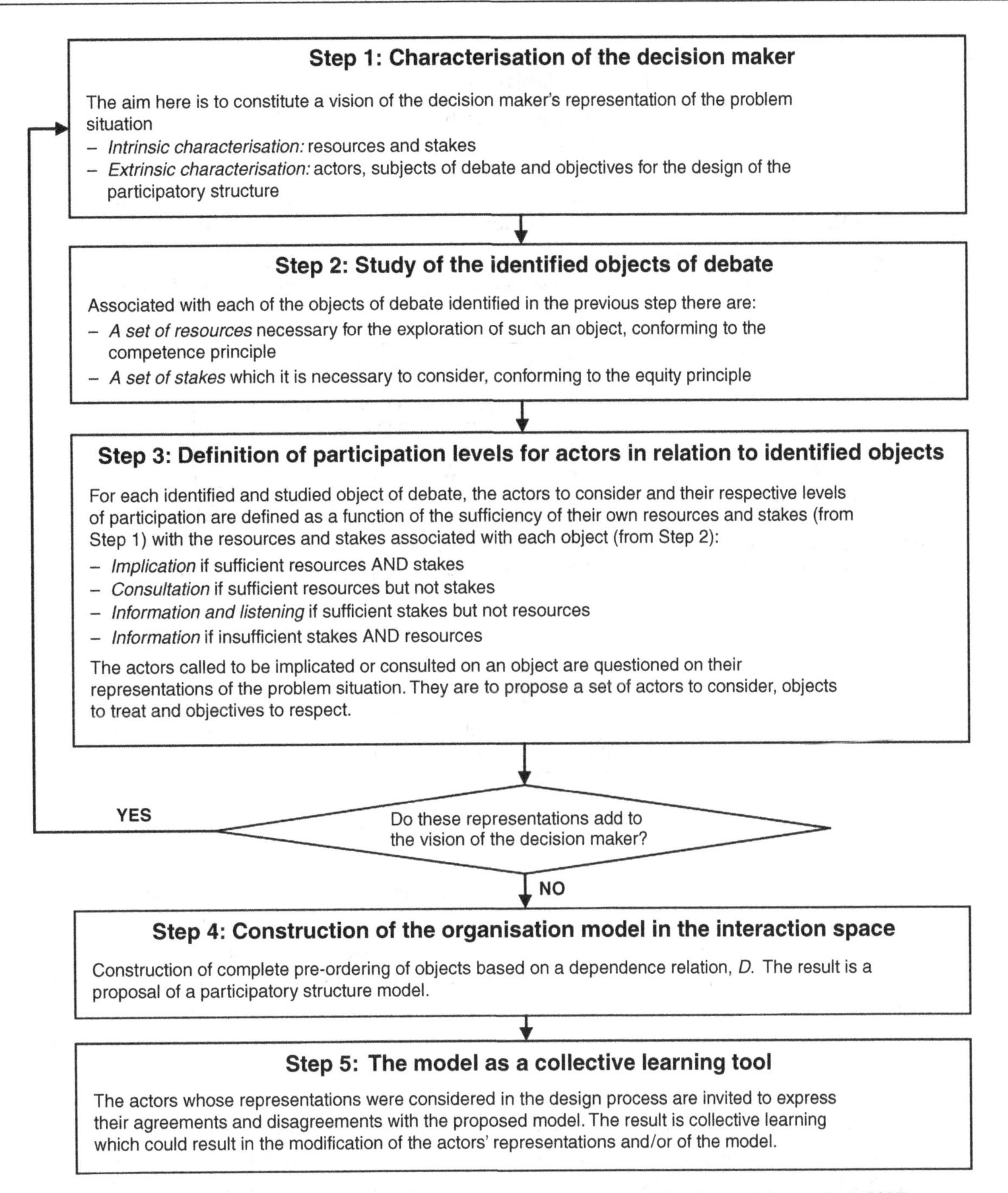

Figure 3.7 Descriptive model of the participatory structure design process (Daniell *et al.*, 2010a). Adapted from: Mazri (2007).

guidelines presented in Figure 3.6). Validity of the model is based on adherence to the ideals of competence, equity, efficiency (the structure permits efficient use of resources) and legitimacy (the structure is accepted and legitimised in the context of the decision-aiding that it has been designed for) (Mazri, 2007).

Recent work by von Korff *et al.* (2010) has attempted to further this area of participatory structure or participation plan design by comparing the suggested phases and design steps of Creighton (2005), Mazri (2007), Stern and Fineberg (1996), Beierle and Cayford (2002) and d'Aquino (2007; 2008). They find that these works contain many similar and complementary elements. This leads them to suggest that a new, more inclusive, guide for aiding the design of participatory process could be created and systematically tested and evaluated. Such a guide would contain the characteristics outlined in Figure 3.9.

The proposal shows convergence towards a number of commonly required activities, despite the different normative

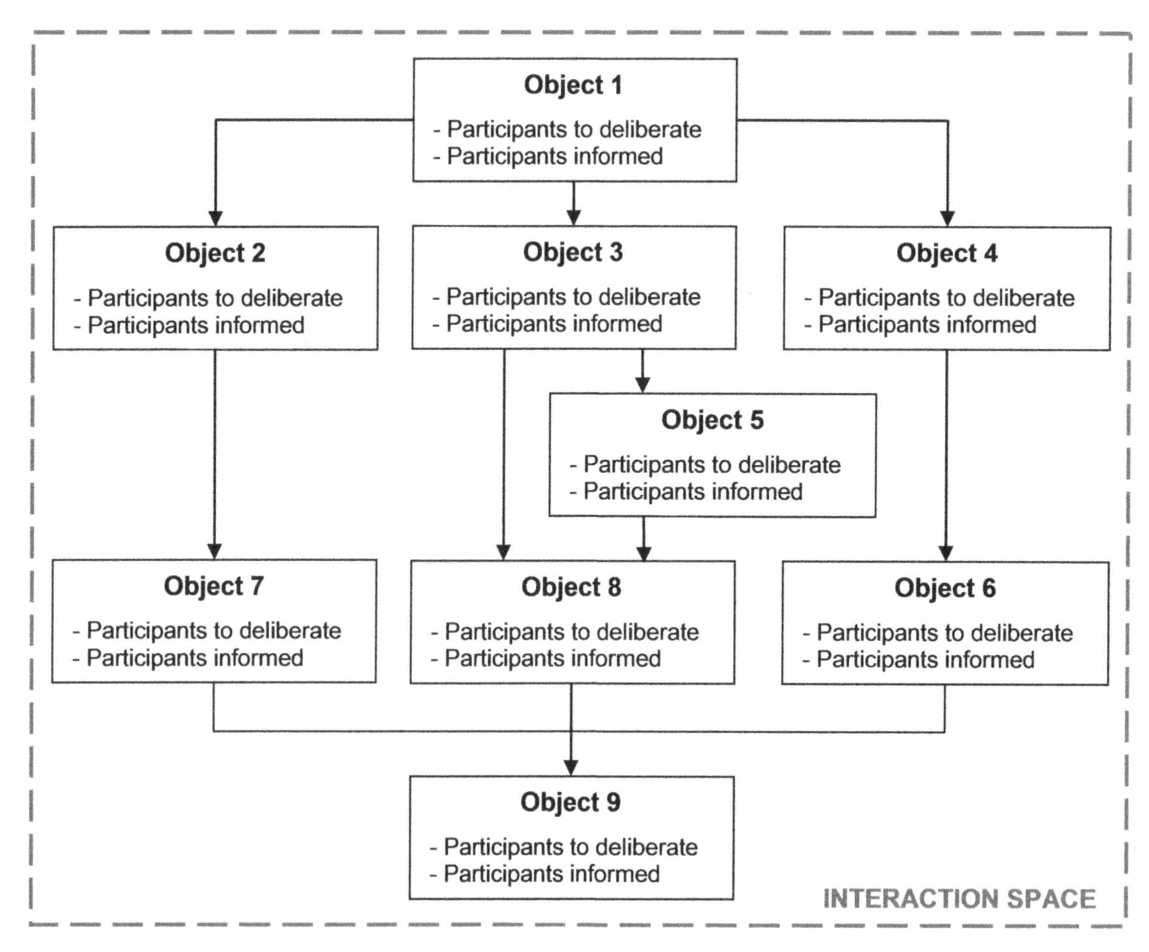

Figure 3.8 Example organisation model of a participatory structure. Adapted from: Mazri (2007).

positions and backgrounds of the authors of the analysed works. However, it also leaves a number of issues untreated, including what could be expected if this design process was itself participatory, as Mazri (2007) suggests, with multiple designers competing to use different tools based on their own values and objectives, and to what extent it is actually possible to suggest steps and tools for certain contexts.

Related to these issues, some of the methods employed in typical participatory decision-aiding exercises once the structure has been defined will be briefly noted and analysed in the following section, with particular attention to those currently used in the water sector. Further issues associated with choosing tools for certain contexts and the issues of working in an 'analyst team' where there are multiple designers will be examined in Chapter 4.

3.3.2 Determining methods for inter-organisational decision-aiding

If the formal Tsoukiàs (2007) decision-aiding process model is considered as the four 'meta-objects' to be constructed (i.e. the problem situation, problem formulation, evaluation model and final recommendations), it may be possible to determine which kinds of existing method could be used for the collective construction and legitimisation of these meta-objects in a messy inter-organisational context. A large number of those previously mentioned in this chapter and in Section B.3 concentrate on the construction and use of the evaluation model of the decision-aiding process to aid the production of final recommendations. These include the statistical, probabilistic and multi-criteria or multi-objective decision analysis methods briefly mentioned in Section 3.1.2, as well as many other types of simulation model based on causal or behavioural rule models. How some of these can be developed, and not merely used, in a collective setting, especially with stakeholders, scientific experts and decision makers present, is far from evident. A reasonable number of methods can also be used to concentrate on developing the problem situation and problem formulation stages of the decision-aiding process in inter-organisational settings, for example cognitive mapping and problem structuring methods (Simon, 1973). If issues arising from the collective building of evaluation models can be further investigated, there appears to be a reasonable potential to combine and adapt methods through a complete decision-aiding

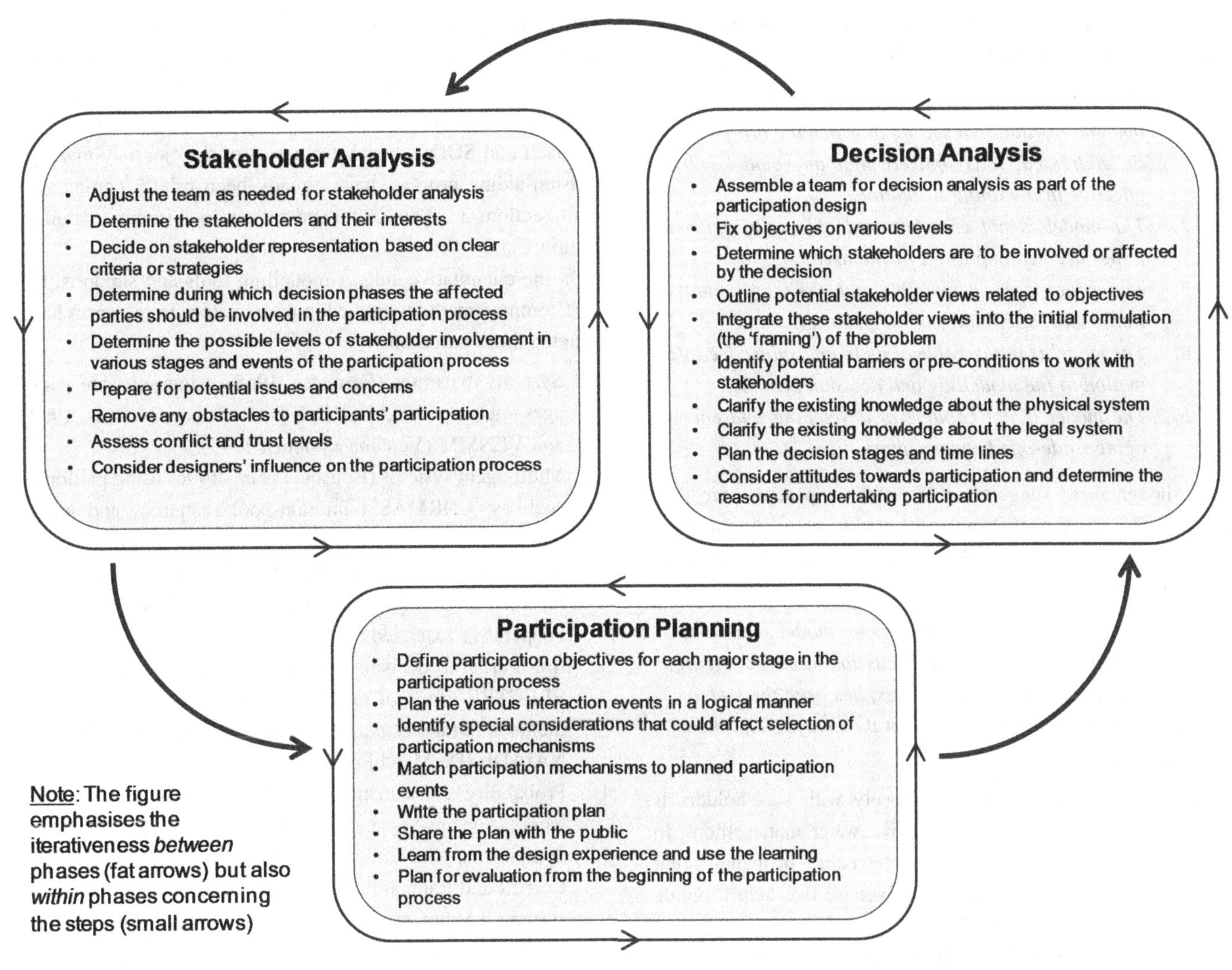

Figure 3.9 Suggested characteristics for an integrated participation design guide. Source: von Korff *et al.* (2010).

process. Apart from a few specific examples of practices, such as some uses of the strategic choice approach (Friend and Hickling, 1987), decision conferencing (Quaddus and Siddique, 2001; Rosenhead and Mingers, 2001a) and reasoning maps (Montibeller *et al.*, 2001), and theoretical articles on 'multimethod' or 'multimethodology' practice (Midgley, 2000; Mingers, 2001; Matthews, 2004), which will be treated in the next chapter, methods that can be used effectively to bridge the problem formulation to evaluation model stages still appear relatively rare and in need of further investigation (Watson and Buede, 1987; Forrester, 1992b).

One broadly classified exception to this observation is a variety of 'participatory modelling' methods (Hare *et al.*, 2003; Voinov and Bousquet, 2010). As models are built in close collaboration with a range of stakeholders, sometimes in inter-organisational contexts, and often right from the problem situation stage, the next section will introduce and briefly analyse some of these methods.

3.3.3 A critique of participatory modelling methods for inter-organisational decision-aiding

Participatory modelling methods have already been used in a variety of water management and planning applications around the world at several different management scales. These are now being mooted as some of the 'best practice' methods that can be used to satisfy international water and sustainability directives. Even if participatory modelling is only just becoming 'in vogue', its potential for helping to solve complex environmental problems was originally mooted by C. S. Holling as an aid in his 'adaptive management' paradigm in the 1970s (Holling, 1978), suggesting that in order to combine the requirements of various decision makers and stakeholders with water resources engineering analyses, computers could become a useful tool (Werick, 2000). However, there are many reasons why the use of computer models in water resources management and planning applications can often be considered as less than totally successful.

Loucks *et al.* (1985) outlined a number of traps that water resources modellers commonly fall into:

1. *Unless model builders are familiar with both the problem and institutional setting in which the problem is to be addressed, it is unlikely that any model will be effective in obtaining a solution.*
2. *The model is not complete and able to examine all issues deemed important by the user.*
3. *The model is not compatible with the (conceptualisation) model users have of the problem.*
4. *The model is not capable of including subjective information in the modelling and decision process.*
5. *The model is not capable of developing and encouraging trade-offs between alternatives.*

Even though these suggestions of pitfalls appeared more than 20 years ago, it seems that most still plague many current water resources management models. Loucks *et al.* (1985) then went on to say that:

> *Perhaps one of the biggest reasons for model solution rejection, even as a basis for discussion in the managing, planning or policy-making process, has been the lack of adequate communication between the analysts and their clients.*

Undertaking modelling activities closely with stakeholders is therefore highly important for effective water management. In particular, it should ideally be the water policy or management objectives underlying the modelling exercise that help to guide decisions on which stakeholders need to be included and what specific methods or tools might be appropriate. Participatory modelling exercises can theoretically include anyone from the general public, policy makers and managers, as well as scientific experts or technical specialists, as touched upon in Section 3.1.2 and expanded upon in Appendix C.

In most cases, participatory modelling varies from other collective problem solving or conflict resolution methods in that the process calls upon 'modelling tools' or supports to focus participant energy, and to shape contributions into compatible formats, in order to create a common model or 'meta-object', as introduced in Sections 3.2.3 and 3.2.4. These modelling 'tools' may be qualitative or quantitative methods with supports. For example, qualitative methods such as cognitive mapping (Axelrod, 1976) can either be used in a basic form, which may require only writing equipment as a support, or in other forms, for example computer software packages such as DecisionExplorer® (Ackermann and Eden, 2001) and DANA (Bots *et al.*, 1999) that allow more complex forms of analysis between participants' individual viewpoints. Other qualitative methods for participatory modelling fall into what are termed 'problem structuring methods' (Rosenhead and Mingers, 2001b), including such methods as the soft systems methodology (Checkland, 1981), the strategic choice approach (Friend and Hickling, 1987; Friend, 2001) and SODA (Eden, 1989; Eden and Ackermann, 2001). The bases of the strategic choice approach and SODA, which form part of the 'journey-making' decision-aiding process, have already been briefly commented on in Sections 3.2.3 and 3.2.5 and are outlined in more detail in Section C.3.3.

On the quantitative side of modelling tools and supports, the most common methods applied in participatory frameworks have included those based on:

1. Systems dynamics (Forrester, 1968) using software packages such as STELLA (High Performance Systems, 1992) and VENSIM (Ventana Systems, 1995);
2. Multi-agent systems (Bousquet *et al.*, 1996) using platforms such as: CORMAS (*co*mmon-pool *r*esources and *m*ulti-*a*gents *s*ystems); REPAST (*re*cursive *p*orous *a*gent *s*imulation *t*oolkit); and the DIAS/FACET (the *d*ynamic *i*nformation *a*rchitecture *s*ystem/*f*ramework for *a*ddressing *c*ooperative *e*xtended *t*ransactions) (ECAABC, 2004);
3. Multi-criteria decision analysis (MCDA), such as: the ELECTRE group of methods (Roy, 1985); PROMETHEE methods (Brans *et al.*, 1986); SMCE (Munda, 2004); and NAIADE (De Marchi *et al.*, 2000);
4. Probability and statistical methods, such as Bayesian networks (Soncini-Sessa *et al.*, 2002);
5. Spatial mapping and visualisation tools (Maurel, 2001; Corbett and Rambaldi, 2009) and geographical information systems (Abbot *et al.*, 1998), also known as 'community mapping' or 'public participation GIS' (Craig *et al.*, 2002; Nicholson-Cole, 2005; Shaw *et al.*, 2009; Cutts *et al.*, 2011), which can be used with the help of computer platforms such as MAPTALK™ (Wien *et al.*, 2003).

In-depth investigation and description of just the first two quantitative methods, systems dynamics and multi-agent systems, as applied in participatory settings through the 'shared vision modelling' (Palmer *et al.*, 1993; Werick and Whipple Jr., 1994; Stephenson, 2003), 'mediated modelling' (van den Belt, 2004) and 'companion modelling' (Bousquet *et al.*, 2002; D'Aquino *et al.*, 2002a; Barreteau, 2003a, 2003b) participatory approaches, are available in Sections C.3.1 and C.3.2. Similarly, the problem structuring methods, the strategic choice approach (Friend and Hickling, 1987; Friend, 2001) and SODA (Eden, 1989; Eden and Ackermann, 2001) are outlined in Section C.3.3, as previously mentioned. The next section will focus on a brief critical comparison of these methods as an introduction to the issues and opportunities that different participatory modelling methods could offer for inter-organisational decision-aiding. A potential integrated methodology using these methods as a basis will then be proposed.

3.3.4 Participatory modelling methods: comparison, issues and opportunities

Most currently used participatory modelling methods have both strengths and weaknesses when used in water sector applications.

System dynamics (1961), as used in the 'shared vision modelling' and 'mediated modelling' approaches, has the often-cited drawback of not being able to include behavioural or decision making information in its simulations, since all relationships must be hard programmed with quantitative measures. It has sometimes been successfully used in planning processes for larger-scale problems and collective decision-aiding, such as in the US National Drought Study (Palmer *et al.*, 1993; Werick and Whipple Jr., 1994), in cases where political timing and conflicts were not overly limiting.

Multi-agent systems, as used in the 'companion modelling approach', have the ability to include stakeholder actions and decisions as part of models (Scholl, 2001), which can be explored and validated by the use of role-playing games (D'Aquino *et al.*, 2002a); a process typically developed for building social capacity to aid in future adaptive management (Bousquet *et al.*, 2002). However, they have the drawback of requiring extensive information on individual agent behaviour to complete the models, and this has so far limited their application to predominantly small-scale water management and planning processes. One exception is the groundwater management process in Tarawa in the Pacific Islands, which could be considered 'inter-organisational' although it was carried out with only a small group of participants (Dray *et al.*, 2005).

Both the shared vision modelling using systems dynamics models and the companion modelling approaches using multi-agent systems models also require a reasonable amount of time to complete and validate the computer models. This can prove problematic when a decision needs to be made urgently. Many problem-structuring methods are able to overcome these sorts of time constraints, owing to their focus on using more qualitative tools in 'real-time' with participants (Rosenhead and Mingers, 2001b). They also present formalised methods that may be used in the construction of collective views on the 'problem situation' and 'problem formulation' meta-objects of the decision-aiding process. When such methods are used as 'evaluation models', some of the methods have been criticised for a number of reasons, including their inability to allow participants to analyse feedback mechanisms correctly (Forrester, 1992b).

Participatory modelling methods may be used in water resources management and planning applications for a variety of objectives. For example, systems dynamics in shared vision modelling is used for large-scale practitioner-run planning processes where operational or policy outcomes are required; multi-agent systems in the companion modelling approach are used for small-group system exploration, learning and social capacity building; and problem structuring methods are typically used to identify, structure and strategically manage problems in fast-changing, complex, uncertain and conflict-ridden environments. However, there appears to be no major reason why methods like these could not be integrated to meet other sets of dominant objectives, based on the needs of specific contexts, especially to better bridge the gap between 'problem formulation' and 'evaluation model' that is sometimes present in decision-aiding processes, as outlined in Section 3.3.2. An example of an integrated methodology that could be used to bridge this gap is outlined in the next section.

3.3.5 An example of an integrated participatory modelling methodology for inter-organisational decision-aiding

New methodologies could be proposed that attempt to take some of the best characteristics of each of the participatory modelling processes, as outlined in Section C.3. For example, the proposed methodology could pursue a roughly cyclic, adaptive and interactive planning process (Holling, 1978; Deming, 1986; Ackoff, 1999; Strömgren, 2003), which is similar to that pursued in the 'shared vision planning' methodology outlined in Section C.3.1. A few differences could be envisaged. In the shared vision planning and modelling process, rather than using 'collaborative negotiation' and 'deliberative decision processes' between the stakeholders (Stephenson, 2003) for problem identification and objective setting, or development of visions of the 'status quo' and 'the future', a number of problem structuring methods could be used. This could be particularly valuable for focusing the next stages of modelling building on the most important problem areas. Cognitive mapping, such as elements of SODA, or the strategic choice approach, would appear well suited to focusing and shaping stakeholder discussions and aiding perspective sharing and common vision construction for these preliminary phases of the modelling process.

One of the drawbacks of many current multi-agent systems approaches is that they take a long time, and it is rather difficult to include behavioural information explicitly in the individual agents that can be validated by observing real stakeholder behaviour in role-playing games. It is, however, extremely important to be able to take into account stakeholder behavioural and human decision effects on the systems when scenario analyses or management options are assessed with the aid of models. There are difficulties with this aspect in shared vision planning and modelling methodology. For this reason, it is proposed that, instead of physically hardwiring stakeholder behaviours into models during a participatory modelling process, participants or stakeholders in the participatory modelling process could 'play out' their roles or decisions under a certain number of model scenarios. Such a process should encourage learning and the possibility for stakeholders to gain an understanding of the potential impact of their actions and decisions, or those of others, on the modelled 'micro-world system', which should, in turn, inform the final decision making or plan formation and implementation processes. An explorative role analysis or 'role-playing game' could therefore be utilised to examine and use the model in a traditional adaptive management cycle or planning process and would include a component of continual monitoring and evaluation of the process and results related to objectives. In this process, a range of modelling techniques could be used as a base, from the qualitative, semi-qualitative and quantitative, such as multi-agent modelling or Bayesian belief networks, or the entirely quantitative, such as systems dynamics models, provided the model and its interface can take into account the actions or decisions of the stakeholders, so that the potential system impacts can be examined.

For such an integrated methodology, the most important properties of the methods used should ensure that all cases are efficient and effective at achieving their goals, remain open in nature and explicitly allow the participation of multiple stakeholders with the interaction and exchange of different viewpoints and perspectives. Evaluation is claimed to be one of the principle stages in the planning and decision process theory. However, the evaluation of processes and results of participatory modelling experiences remains an underdeveloped practice (Jones *et al.*, 2009). An integrated methodology should therefore also include a continuous and participatory monitoring and evaluation programme. Evaluation theory and a proposal for an evaluation protocol to contribute to this integrated participatory modelling process will be outlined in Chapter 5.

Such an integrated process could better contribute to the largest range of objectives of participatory modelling, outlined in Appendix C, such as helping to examine the 'real' underlying problems, increasing trust, appropriation and understanding of the models created, leading to greater individual and social learning, and producing richer and more realistic action plans. However, to what extent these objectives can be met in a range of contexts still requires further investigation, testing and validation. One of the potential issues with this kind of integrated methodology or 'multimethodology' (Mingers, 2001) is that the various components of the methodologies that could be chosen often possess quite different underlying philosophies and have been developed with different epistemological backgrounds and theoretical bases. Finding designers and implementers with sufficient knowledge and skills to understand and embark on such integration may prove another limiting factor to the feasibility and adoption of such approaches (Mingers, 2001) and seems to imply that a team approach is necessary. Understanding how a group or 'project team' of such designers, implementers or analysts can work together to 'co-engineer' or organise such processes to reach the desired objectives of a participatory process through the project's collectively driven initiation, design and implementation is therefore fundamentally important. These constraints can thus create a number of challenges for deciding what methodological elements to use and how to choose various methods in certain contexts.

Potential means of understanding and overcoming such challenges more effectively will be addressed in Chapter 4. The theoretical background behind the concept of mixing and matching, or creatively combining and designing new methods for context specificities, will be critically reviewed, as well as the concept of 'co-engineering' participatory modelling processes to help build knowledge on these current gaps in understanding.

3.4 CONCLUSIONS

This chapter has focused on the second of the objectives of this book: *to critically review decision-aiding theory and methods, including participatory modelling, and the way in which they could be used to improve water planning and management.* To achieve this objective, a number of domains related to decision-aiding have been reviewed and analysed, and answers provided to two of the questions posed at the end of Chapter 2:

1. How are multi-stakeholder and inter-organisational decision making processes currently aided and how could these practices be improved?
2. What decision-aiding methods are currently being used in water management contexts exhibiting high levels of uncertainty, complexity and conflict and to what extent do they appear to be successful?

The concept of decision-aiding was discussed in relation to its use in the water sector. The theoretical background of decision-aiding processes and models was then analysed relative to engineering practice, policy approaches and operational research decision-aiding approaches. This analysis revealed that decision-aiding processes typically follow similar processes, linked to underlying organisational management and decision making theories; yet their level of application to operational management or policy problems varies. The processes' underlying ideologies also vary greatly, depending on the analyst's conception of rationality. For inter-organisational decision-aiding processes in messy situations, constructive approaches to decision-aiding that work through a process of rationalisation could prove most worthwhile.

The operationalisation of decision-aiding models was studied, by presenting theory and practical methods on how participatory structures for decision-aiding processes can be designed. A range of methods for implementing these structures was then examined. Participatory modelling methods appear to be appropriate for bridging the gap between 'problem formulation' and 'evaluation model', which can sometimes occur in decision-aiding processes. A small selection of current and potential participatory modelling methods was then compared for inter-organisational decision-aiding in water planning and management. This led to a proposal for integrated participatory modelling specifically tailored to inter-organisational decision-aiding.

From the analysis presented in this chapter, a number of key areas warranting further research have been identified:

- How the choice and concept of combining and designing new methods for context specificities can be undertaken in the development of participatory modelling methodologies;
- The '*co-engineering*' concept as an evolving role for decision analysts' work to aid mess management in inter-organisational and multi-accountable group situations; and
- *How participatory modelling processes for inter-organisational decision-aiding can be evaluated and compared*, for example, by examining the efficacy, efficiency and other outcomes and metrics of the methods and process used, and these interventions' results on improving water planning and management.

The first two of these areas will be further examined in the next chapter, and the final area will be investigated in Chapter 5.

4 Co-engineering participatory modelling processes

It is evident that in the area of perception we have reached a realization analogous to the Heisenberg uncertainty principle. We do not observe the physical world. We participate with it. Our senses are not separate from what is 'out there', but are intimately involved in a highly complex feedback process whose final result is to actually create what is 'out there'.

TALBOT *(1993)*

The analysis of inter-organisational decision-aiding and its operationalisation for the water sector outlined in Chapter 3 identified the need to understand and be able to take a multi-modal contextual approach to method choice for participatory modelling interventions, as well as to understand better the roles of multiple analysts in project teams who design and implement broad-scale participatory processes. This chapter will aim to meet the first need by critically reviewing a number of current approaches designed to aid the choice, combination or creation of methods in participatory interventions and identifying any gaps in knowledge. The current gap in understanding of the co-engineering of participatory modelling processes for decision-aiding will then be analysed. A definition of the co-engineering process will be offered, followed by a critical, trans-disciplinary review of literature to build insights into the concept. From these insights, a research agenda on the co-engineering of participatory modelling processes will be drawn up.

4.1 PARTICIPATORY MULTI-MODAL CONTEXTUAL APPROACHES TO INTERVENTION DESIGN AND IMPLEMENTATION

There are a number of theoretical and practical questions concerning inter-organisational interventions, some of which have already been touched on in this book, such as:

- Are certain methods more appropriate for certain contexts and, if so, do systems for aiding choice of methods exist?
- How can underlying differences in philosophical, epistemological and methodological assumptions of methods be accommodated when choosing, mixing or developing decision-aiding methods?
- Who should take part in the multi-stakeholder groups?
- Who should design and implement decision-aiding interventions?

These questions will be discussed through this chapter, together with a critique of some recent work on pluralist and contextual approaches to decision-aiding interventions, followed by the introduction of the concept of 'co-engineering' to describe the idea of participatory design and implementation by teams for decision-aiding approaches.

4.1.1 Introducing pluralist and contextual approaches to decision-aiding interventions

Although the exploration, theorisation and use of multimethodology design or combinatorial method choice for decision-aiding interventions is not a recent pursuit (e.g. Jackson and Keys, 1984; Flood and Jackson, 1991b; Midgley, 1997b; Mingers and Gill, 1997), there are still relatively few authors actively interested in the topic. The largest and easily identifiable bodies of literature on the topic include the work emanating from a group of researchers from the University of Hull's Centre for Systems Studies and their associates, including Mike Jackson, Paul Keys, Robert Flood, Gerard Midgley, Wendy Gregory and Norma Romm. Equally well known is the work of John Mingers, John Brocklesby, Tom Omerod, Ann Taket, Leroy White and their collaborators. Related work also stems from the work of Werner Ulrich and other researchers focused on the idea of 'boundary critique' as the foundation for systemic interventions.

From these authors' bodies of work, most of which have been related to the 'soft-systems' movement in operational research, the clearest and most well-known propositions for multimethodology design or combinatorial method choice include the:

- System of systems methodologies (Jackson and Keys, 1984; Jackson, 1988; 1990) and total systems intervention (Flood and Jackson, 1991a, 1991b; Flood, 1995; Jackson, 2003);

- Multimethodology process (Mingers and Gill, 1997; Mingers, 2001, 2003); and
- Creative design of methods, based on boundary critique (Midgley, 1997b, 1997a, 2000).

The common underlying assumption in these approaches is that there is no individual approach, methodology or method that can be used to appreciate the complexity of social reality (Yu, 2004). The underlying principles and theories of how methods or methodologies can or should be chosen differ in each of the combined method design or choice propositions.

4.1.2 System of systems methodologies (SOSM) and total systems intervention (TSI)

In 1984, Jackson and Keys proposed that the existing range of systems methodologies or methods could be used together in a complementary fashion to help deal with different problem contexts. The problem contexts were classified according to the perception of the system type in which the problem situation occurred, as either *mechanical* or *systemic*, based on the work of Ackoff (1974), and to the relation between the intervention's participants or stakeholders, as either *unitary* or *pluralist*. The relation was considered unitary if the stakeholders were perceived to be in agreement over the problem situation or goals of the intervention, or pluralist if there was disagreement. A third category of relations, *coercive*, was also suggested and included in Jackson's later works (e.g. Jackson, 1988; 1990). In the coercive case, power differences between participants exist that may prevent latent disagreements from surfacing. Matthews (2004) pointed out that these three categories show close correlation to Habermas' (1972) 'knowledge-constitutive human interests' and their corresponding 'goals' and 'characteristic sciences', in which the:

- *Unitary* category can present an underlying 'technical human interest', where the empirical and analytical sciences may be used to '*achieve technical control over objectified processes*';
- *Pluralist* category can present an underlying 'communicative human interest', where the historical and interpretive sciences may be used to '*preserve and extend consensus towards action*'; and
- *Coercive* category can present an underlying 'emancipatory human interest', where the critical sciences may be used to '*dissolve the apparently objective, but in principle alterable, relationships of dependency*'.

The various systems approaches and some of the common methodologies from the operational research domain were mapped into each of the problem contexts, to create the 'system of systems methodologies', SOSM, outlined in Table 4.1. The 'total systems intervention', TSI, then forms a complete approach to problem solving for messes by 'creatively *surfacing* issues' and examining a number of possible system metaphors that can be used to help 'choose' relevant methodologies using the SOSM (Flood and Jackson, 1991b), as shown in Table 4.1, before 'implementing' them and repeating the phases in an iterative fashion.

Both SOSM and TSI aim to transcend the underlying ontological, epistemological and paradigmatic assumptions of these different systems approaches and methodologies by taking a 'complementarist' stance and developing a 'meta-methodological' viewpoint (Flood and Jackson, 1991b). However, they have been criticised by a number of authors, including Taket (1992), Tsoukas (1993), Flood (1995), Midgley (2000) and Matthews (2004). The most important criticism from these authors is that the SOSM and TSI are largely based on Habermas' (1972) theory of knowledge-constitutive interests, which itself has a strong epistemological–sociological foundation. Therefore, the SOSM objective of 'transcending paradigms' is not achievable as it '*sets itself up as the ruler and judge of which paradigm is appropriate for which context*' (Matthews, 2004). Rather than being 'complementarist', the approach becomes 'imperialist' in the sense defined by Flood and Jackson (1991b; see also Gregory, 1996; Matthews, 2004). Although more recent versions of the TSI (Flood, 1995; Jackson, 2000) have been adapted that have omitted SOSM entirely, focusing instead on critical systems thinking and practice, with these criticisms in mind and the important progress that has been made, the type of 'complementarist' pluralism still advocated in the TSI remains highly contentious. Some of these critiques and different concepts of pluralism will be discussed later in this section.

4.1.3 Multimethodology

Mingers' (2001, 2003) 'multimethodology', or framework to inform 'mixing and matching methods' during an intervention, bears a number of similarities to SOSM and TSI in its objectives of informing method mixing and choice. However, it attempts to soften the classification of methods by analysing the roles that the methods could play in different phases of an intervention – *appreciation, analysis, assessment* and *action* – in relation to Habermas' (1984) 'three worlds': *social, personal* and *material*. Each method is then mapped across a tabular framework (shown in Table 4.2) with its corresponding capacity to address the specific dimensions in each phase of the process.

The mapping of methods then forms just one tool, like the SOSM in the TSI, to aid critical reflection on the design context of the intervention's multimethodology, which is represented in Figure 4.1

Critical reflection must be undertaken on the nature of the relations between the different systems in the intervention context, given as A, B and C in Figure 4.1. These include the nature of the commitments, power and values of agents who are

Table 4.1 *The system of systems methodologies and corresponding system metaphors as used in the total systems intervention. Adapted from: Jackson and Keys (1984), Flood and Jackson (1991b), Midgley (2000), Matthews (2004)*

System characterisation	Relationship of participants		
	Unitary	Pluralist	Coercive
Mechanical or simple	**Hard systems thinking** *Machine metaphor* *Team metaphor* Operations research Systems analysis Systems engineering	**Soft operations research** *Machine metaphor* *Coalition metaphor* *Culture metaphor* Social systems design (Churchman, 1968; 1979) Strategic assumptions and surfacing testing (Mason and Mitroff, 1981) Cognitive mapping (i.e. SODA as outlined in Section C.3.3)	**Critical systems heuristics** (Ulrich, 1983; 1991) *Machine metaphor* *Organism metaphor* *Prison metaphor*
Systemic or complex	**Organisational cybernetics** *Organism metaphor* *Brain metaphor* *Team metaphor* Viable systems modelling (Beer, 1984) Contingency theory Socio-technical systems theory	**Soft systems thinking** *Organism metaphor* *Coalition metaphor* *Culture metaphor* Soft systems methodology (Checkland, 1981) Interactive planning (Ackoff (1999) as outlined in Section 3.2.5)	?

Table 4.2 *Framework for mapping methodologies. Source: Mingers (2001)*

	Appreciation of	Analysis of	Assessment of	Action to
Social World	Roles, norms, social practices, culture and power relations	Underlying social structures: distortions, conflicts, interests	Ways of changing existing structures, practices and culture	Generate enlightenment of the social situation and empowerment
Personal World	Individual beliefs, meanings, values and emotions	Differing Weltanschauungen (world views, perceptions) and personal rationalities	Alternative conceptualisations and constructions	Generate understanding, personal learning and accommodation of views
Material World	Material and physical processes and arrangements	Underlying causal structures	Alternative physical and structural arrangements	Select and implement best alternatives

involved in the intervention, relative to the problem context system, or their knowledge and competence to choose and implement particular methodologies suited to the problem content system (Mingers, 2003). A useful list of 'critical questions' for reflection on these relations is given in Mingers (2001). Further critical reflection on the underlying philosophical assumptions of methods and whether they may be suited to a particular intervention context may also be found in Mingers' (2003) framework for characterising operational research and management science methods, which characterises each method in terms of its ontology (what is modelled), epistemology (how modelling takes place) and axiology (why modelling takes place). This typology and Mingers' analyses of many of the methods noted in Table 4.1 are supposed to aid practitioners to choose and combine methods for their interventions.

Although the multimethodology approach adopts a broader and generally more critical analysis of the potential usefulness of methods for different purposes, intervention phases and 'worlds' of reality, it remains, like TSI, a 'meta-methodology' for method choice and shares some of the same criticisms. The frameworks aiding method choice could diminish the amount of critical reflection that analysts engage in in their method selection, as fundamental differences in the methods and their assumptions are less likely to be debated when they are already laid out clearly in a nicely organised table!

4.1.4 Which pluralism?

Gregory (1996) criticises the type of pluralism that underpins TSI, and to a lesser extent multimethodology, which tries to minimise

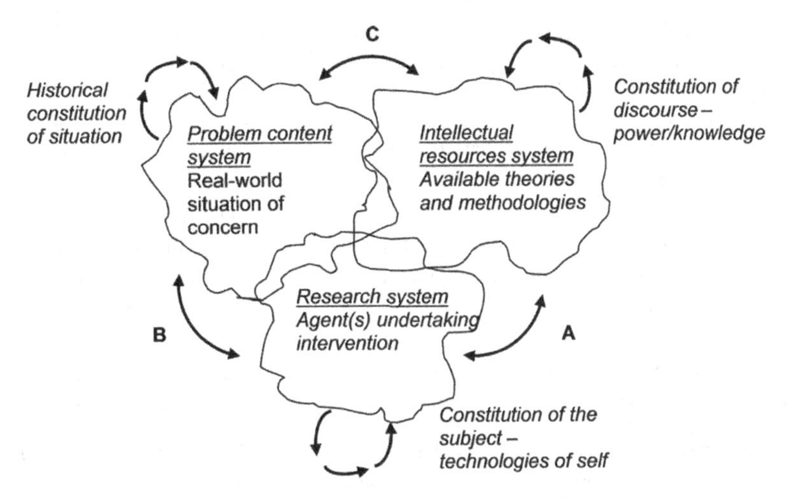

Figure 4.1 Context of multimethodology interventions. Source: Mingers (2003). A, B and C are the relations between different systems.

tensions and ignore radical differences. She instead considers that discord and diversity are the most fundamental parts of pluralism, and that learning to *'critically appreciate other viewpoints'* can lead to *'transformation through understanding of self and others'*. She goes on to define this type of pluralism as 'discordant pluralism', which provides a solid basis for critical systems thinking and possesses the following characteristics:

1. It has a *'local, contingent and historically situated nature'*;
2. It *'promotes communication with other radically different and alien perspectives'*; and
3. It uses the *'insights gained through such communication to provide for ethical decision making'* by the *'juxtapositioning of oppositional viewpoints within a constellation which supports both one perspective and the other'*.

Such pluralism, rather than conciliating alien paradigms and failing to recognise fundamental differences, instead allows for paradigmatic incompatibilities by legitimating '*both* similarities *and* differences', in order to promote continuous debate, gain a greater critical appreciation and encourage transformation. Although theoretically interesting, how to operationalise such pluralism through the use and creation of useful decision-aiding methods is unclear. What may be used as a guide by analysts is that, in participatory modelling processes, seeking consensus or to 'win' an argument over another perspective is not preferred. Different perspectives should rather be critically appreciated for themselves and management solutions sought that maintain or allow the transformation of one or both desired perspectives, i.e. the 'win–win' equivalent of negotiations or 'reframing of problems', where transformative learning may occur (Leach and Wallwork, 2003).

Based on similar critiques to Gregory (1996) and drawing on post-modern and post-structuralist perspectives, including the work of Foucault on 'power–knowledge' relations (Foucault, 1977; 1980; 1982), Taket and White (1993, 1996; White and Taket, 1998) have also suggested another possible type of pluralism for consideration in interventions; 'pragmatic pluralism'. This type of pluralism concentrates on the need for pluralism in a number of areas and aims to promote a *'process of critically reflective practice'*. From the synthesis of White and Taket's work in Matthews (2004), this includes encouraging pluralism in the:

1. *Facilitation process* – which can be achieved by *'deconstructing the traditional role assumed by systems (or OR) practitioners'*;
2. *Modes of representation employed* – which can be achieved by *'deconstructing the traditional claims to objectivity suggested by certain types of representation'*;
3. *Use of specific methods or techniques* – which can be achieved by resisting the *'will to methodology'* and thus rejecting *'the development of any overarching meta-methodology that aims to direct practitioners in methodological choice'*; and
4. *Nature of the client* – which can be achieved by *'acknowledging and respecting the views of a wide range of stakeholders in the intervention'*.

Although deconstructing and critiquing a priori assumptions and creatively designing and adapting new methods to each local situation with stakeholders is by no means a simple endeavour (Matthews, 2004), practices and theoretical insights that follow the basis of this type of pluralism do appear to be emerging. Midgley's 'creative design of methods' (Midgley, 1997a; 1997b; 2000) is just one such example of intervention design and practice that considers elements of White and Taket's 'critically reflective practice' and Gregory's 'discordant pluralism'.

4.1.5 Creative design of methods

Midgley (1997a, 1997b, 2000) considers that simple methodology choice and mixing pre-designed methods, as advocated

Table 4.3 *Ulrich's 12 critical heuristic boundary questions to aid analysis of normative premises. Source: Matthews (2004)*

Systems categories		'Ought' mode	'Is' mode
Social system to be bounded – Involved	Sources of motivation	Who ought to be the *client* (beneficiary of the system to be designed or improved)?	Who is the actual *client* of the system design (i.e. who belongs to the group of those whose purposes, interests and values are served)?
		What ought to be the *purpose* of the system (i.e. what goal states ought the system to be able to achieve so as to serve the client)?	What is the actual *purpose* of the system (as being measured not in terms of the declared intentions of the involved but in terms of the actual consequences)?
		What ought to be the system's *measure of success* (or improvement)?	What, judged by the design's consequences, is its built-in *measure of success*?
	Sources of control	Who ought to be the *decision maker* (i.e. have the power to change the system's measure of improvement)?	Who is actually the *decision maker* (i.e. who can actually change the measure of success)?
		What *components* (resources and constraints) of the system ought to be controlled by the decision maker?	What *components* (resources and constraints) of the system are actually controlled by the decision maker?
		What resources and constraints should not be controlled by the decision maker (i.e. what ought to be his/her decision *environment*)?	What resources and constraints are not controlled by the decision maker (i.e. what is his/her decision *environment*)?
	Sources of expertise	Who ought to be involved as the *designer/planner* of the system?	Who is actually involved as the *designer/planner* of the system?
		What kind of *expertise* ought to flow into the design of the system (i.e. who ought to be considered an expert and what should his/her role be)?	What kind of *expertise* is flowing into the design of the system (i.e. who is involved as an expert and what role does he or she play)?
		Where ought the involved seek the *guarantee* that their planning (designs) will be successful (judged by the system's measure of success)?	Where do the involved seek the *guarantee* that their planning (designs) will be successful (judged by the system's measure of success)?
Social system to be bounded – Affected	Sources of legitimation	Who ought to belong to the *witnesses* (who represent the concerns of the citizens that will or might be affected by the design of the system)? That is to say, who among the affected ought to be involved?	Who, if anyone, belongs to the *witnesses* (who represent the concerns of the citizens that will or might be affected by the design of the system)? That is to say, who among the affected is involved?
		To what degree and in what way ought the affected be given a chance of *emancipation* from the premises and promises of the involved? Ought the affected be considered as clients?	To what degree and in what way are the affected being given a chance of *emancipation* from the premises and promises of the involved? Are the affected being considered as clients?
		What *world view(s)* ought to underlie the system's design? Should the world view(s) of the affected be incorporated?	What *world view(s)* are underlying the system's design? Are the world view(s) of the affected being incorporated?

in the two approaches just mentioned, is commonly insufficient for many intervention contexts. Rather, he considers that most intervention situations are sufficiently complex to warrant the use of a variety of methods that may often have to be designed from scratch as the intervention progresses with other stakeholders or experts, rather than choosing a collection of 'off-the-shelf' methodologies at the beginning of an intervention (Midgley, 2000).

For the 'creative design of methods', a number of elements that are particularly important are the aspect of times or 'moments' of enquiry through the process, as well as the 'multi-layered' aspects of local context, which may require the use of different methods during the intervention (Midgley, 2000). The approach considers that intuition, as well as critical reflection and debate over 'boundary judgements', can be used to enhance creation and choice of appropriate methods or 'synergy of methods', as well as the learning of interveners. Although drawing upon and extending the work of Ulrich (1983, 1991) for the types of questions that can be used for boundary critique and aiding the choice of methods (see Table 4.3), Midgley disagrees with Ulrich's idea of the need to '*transfer responsibility for ethical decision making wholly to participative stakeholder groups*'. Midgley's rejection of the idea of complete responsibility transfer to stakeholders underlies the development of his own related theory of 'marginalisation' (Midgley, 2000) of specific groups when certain sets of stakeholders take hold of decision making power.

Midgley strongly believes that the intervener can play a pivotal role in working with and resolving tensions between stakeholder groups, which in some cases may require bringing the interests of un-represented groups, such as 'non-humans' (Barbier and Trepos, 2007) or repressed minorities, into the debate and boundary critiquing process or holding confidential talks with individuals or stakeholder groups throughout the different stages of the intervention. It will also be largely the role of the intervener to assess and reflect on the *purposes, principles, theory, ideology* and *past practice* of a variety of methods as part of the creative development and design of methods for the intervention. Prior experience, intuition during the intervention and reflection on methods will all drive intervener learning.

Some questions that should be reflected upon by the intervener and other associated participants include:

- *Boundary questions*, which lead to the design of methods for defining issues;
- *Issue-related questions*, which lead to the design of methods for addressing the issues already defined; and
- *Knowledge-related questions*, which enable explorations of the intervener's and participants' intellectual resources (Midgley, 2000).

Midgley's 'creative design of methods' for intervention is strongly based on a 'process philosophy' (Midgley, 2000), which differentiates it from TSI and multimethodology. He also advocates a 'fragmentary whole' version of methodological pluralism and philosophical pluralism, which he names 'critical pluralism'. This type of pluralism aims to maintain the creative tension between both differences and similarities of viewpoints as advocated by Gregory (1996), potentially approaching the yin–yang type of creative tension found in some of the eastern philosophies (see Zhu, 1998; 2010). Combined with these pluralist perspectives must be reflections on the linking of practice and theory before, throughout and following interventions. This includes reflecting on the gaps between the intervener's perception of his or her own 'espoused theory and methodology' and 'theories and methodologies in use' to maximise the depth of learning (Argyris and Schön, 1996).

Although Midgley's 'creative design of methods' presents many interesting and well-argued propositions for contextually based intervention design and implementation, it could potentially be criticised for over-emphasising the role of 'the intervener' as a prime decision maker or shaper of the intervention. The ethics related to taking on such a position are debatable. Why should the intervener have the right to judge what should be brought into debates and how is this legitimacy constructed? What are the risks of power being misused by interveners in such situations and how might these risks be monitored? Another issue that is not raised in Midgley's work to a great degree is in the domain of inter-organisational processes, such as those often pursued in water planning and management. In that context, it is often common to have more than one 'intervener'. Would the underlying process design be any different when a group of diverse 'interveners' with different objectives and underlying value-systems are required to work together? What extra issues would need to be investigated in this case?

This object of reflection is currently under-studied, especially in the operational research literature. The next section proposes the type of process that might occur and examines the key knowledge gaps where further investigation is required.

4.2 TO ENGINEER OR CO-ENGINEER?

As emphasised at the end of Section 4.1.5, much current research highlights the role of the analyst, modeller or process 'engineer' as the prime decision maker and designer of the methods or modelling techniques to be used in the participatory interventions. As was outlined in Section 3.3.5, the common situation of intervention in water planning and management is one of a project team who must work together to collectively initiate, design, implement and manage participatory decision-aiding processes in an inter-organisational context, in an attempt to achieve a broad range of goals. The concept of 'co-engineering' is therefore introduced to describe this collective process and a literature review is provided on the topic.

4.2.1 Defining co-engineering

The proposition made in this book of what constitutes the co-engineering of participatory processes is strongly linked to intentional or purposeful action (Checkland, 1981; Midgley, 2000). The term *engineering* typically refers to the process of formulating goals and then attempting to design and manage systems to help attain them, as outlined in Section 3.2.1. Of interest here is determining how forms of collective action to support water management are themselves engineered by internal or external agents to the management system. This co-engineering is a form of meta-level engineering and organisational decision making process that defines the rules and processes for collective choice in water management planning, as outlined in Figure 1.1.

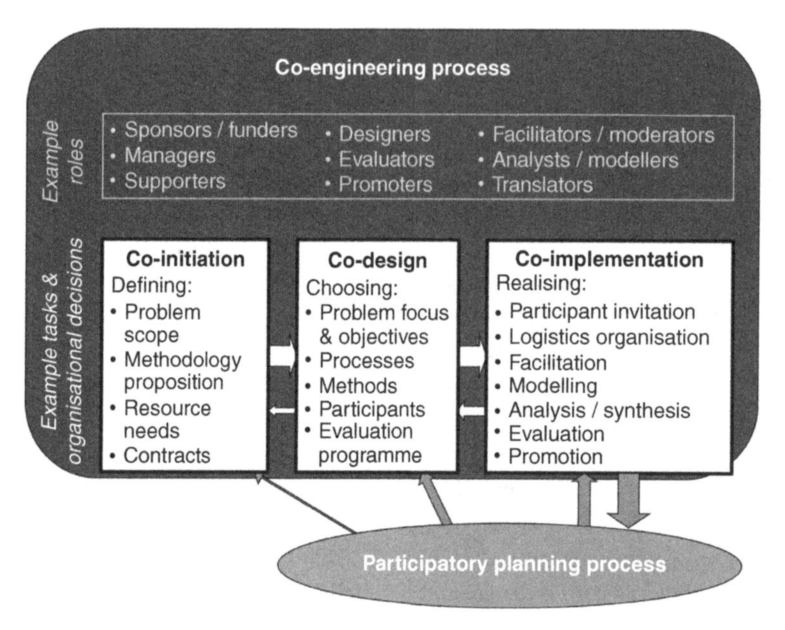

Figure 4.2 Typical co-engineering process roles and tasks.

It is co-engineering, rather than engineering, which is here assumed pertinent. The prefix 'co-' means 'together' or 'with'. Co-engineering can therefore be defined as a process of engineering performed 'together' or 'with' others. For example, the form of engineering could be collaborative, concurrent, cooperative, coordinated or even conflict-ridden. It is this collective action, the nature of the social processes associated with engineering practice (Bucciarelli, 1994), and the way they influence participatory planning processes, which are of particular interest. This proposal of the co-engineering concept is equivalent to attempts to promote the collaborative aspects of adaptive management of socio-ecological systems by re-labelling it adaptive co-management (www.resalliance.org/2448.php). Leaving aside the spatial and temporal dimensions, Bouwen and Thallieu (2004) recognised the importance of this relational dimension and the need to focus on studying the quality of relational practices, rather than only concentrating on the engineering-task-orientated problem solving aspects that take place in multi-party collaborations. The relatively open concept of 'co-engineering' presented here therefore provides broad scope for investigations of these processes by encouraging the need for relational, operational, temporal and spatial analysis of the concept.

The co-engineering process is more specifically defined as the succession of a co-initiation, co-design and co-implementation phase. These phases are similar to those used in the participatory modelling processes for decision-aiding, which can be considered to take the form of a number of facilitated or collectively managed stages of situating and formulating problems or issues of interest through to evaluation of management alternatives, choice and implementation, as outlined in Chapter 3. The three phases of the co-engineering process are more likely to overlap with, and iterate between, one another, than to be carried out sequentially. A co-engineered participatory process differs from an engineered one by the presence of a project team working collectively, such as modellers, facilitators and other project managers who have some shared decision making power over the objectives, design, choice of methods and implementation. Typical roles and tasks of the members of the co-engineering process are outlined in Figure 4.2.

In Figure 4.2, each phase of the co-engineering process – co-initiation, co-design and co-implementation – can be seen to lead on to the next and also to the realisation of the participatory planning process. The first co-engineering phase, the co-initiation, involves the definition of certain aspects of the participatory planning process, such as the problem scope, a methodology proposition and resource needs. It is also typically the phase when contractual arrangements between participatory process organisers – the co-engineers – will be first formalised. The next phase, the co-design, then focuses on the choosing of a problem focus and objectives, processes and schedules, methods to be used, participants and a programme of evaluation. Ideally, prior to the commencement of the co-initiation phase, a 'participation plan' for the intervention should result from co-design outlining these organisational decisions (von Korff *et al.*, 2010). Then the co-implementation of the plan can commence, which as well as inviting participants and arranging the logistical aspects of organising the participatory planning process, also includes the co-engineers' interactions with the participants through the participatory modelling decision-aiding process. This will normally take the form of meetings or workshops and is aided by a variety of techniques or tools, such as facilitated discussions, cognitive mapping, brainstorming, story-telling, visioning, model building, role-playing games, ranking or voting methods and evaluation mechanisms. Feedback and iterations between co-engineering phases and during the participatory planning process are possible. Typical magnitudes of influence of processes on each other are represented by varying arrow widths in Figure 4.2. The co-engineering process ends when the project team disperses, which often occurs when the operational management phase commences after an agreed-upon action plan has been formulated.

4.2.2 Review of co-engineering participatory water management processes

In this section, the co-engineering phases will be worked through backwards, to highlight available literature and insights relevant to the co-implementation, co-design and co-initiation phases. As far as is known, the term *co-engineering* has never been applied to multi-stakeholder or participatory water planning and management processes. However, the term can be found in other domains, such as information technology or manufacturing sectors, but often as a contraction of 'concurrent engineering' with a focus on process timing, rather than on 'collective' work practices with a focus on relations. Insights related to the co-engineering scale of analysis have been found scattered through an extraordinarily diverse range of literature. One of the current difficulties in advancing knowledge on research areas that require transdisciplinary forms of analysis, relates to difficulties in understanding or critically appreciating the ideas behind a range of disciplinary vocabularies and thus locating relevant research (Bammer, 2005). This review attempts to bring together a number of relevant research strands to build an integrated picture of existing knowledge on co-engineering of participatory water management processes.

CO-IMPLEMENTATION

The largest bodies of literature related to co-engineering participatory water management processes tend to focus on the (co-) implementation phase of participatory water management processes. Most papers on participatory modelling and its variants in water management, such as Palmer *et al.* (1993), Hare *et al.* (2003), Pahl-Wostl and Hare (2004) and Dray *et al.* (2005), fit into this category; yet they rarely focus on the relational dimension of project teams during the implementation process. One quality exception stems from systems dynamics literature, where the roles and interactions that appear to be beneficial in the co-implementation phase for group model building using system dynamics models have been outlined (Richardson *et al.*, 1992; Andersen and Richardson, 1997; Luna-Reyes *et al.*, 2006). However, this literature is based on the use of one type of modelling method. Issues of how method choice occurs in project teams are therefore outside the scope of these works.

Broader views on the topic are found in adaptive co-management literature (Berkes and Folke, 1998; Olsson *et al.*, 2004) or in multi-stakeholder platform literature (e.g. Fayesse, 2006). These views provide some relevant insights for co-engineering systems. Some of the most relevant insights on project team roles, which even touch on some aspects of design and initiation, as well as the co-implementation process, are present in Levrel and Bouamrane (2008), Kelly (2001) and Bots *et al.* (2011). For example, Levrel and Bouamrane's (2008) article on the co-construction of interaction indicators for biosphere reserves in West Africa offers some insights on the roles of mediators, experts and reserve managers through the participatory process and the associated issues of power relations, foreign language use, participatory literacy levels and trust between different stakeholders and the organising team. Some design principles, or methodological reasoning and assumptions, are also noted, but not how the co-design phase took place or how any possible agreement on these design principles was achieved. One perspective on both the operational and relational aspects of the co-implementation part of the process is outlined by Bouwen and Taillieu (2004) in their paper on multi-party collaborations in the natural resources sector.

CO-DESIGN

Recent research on the design phase, as has already been mentioned in Section 3.3.1, often focuses on ways to select and evaluate participatory methods and tools in a given context (Rowe and Frewer, 2000; Beierle and Cayford, 2002; Lynham *et al.*, 2007; Mazri, 2007; Bayley and French, 2008). However, rather than determining how to best fit the available approaches to the context, some researchers recognise the need for increased theoretical and practical understanding on how and what types of participatory structures could be conceived or designed to best deal with specific contextual problem constraints.

This challenge has been addressed in part through the operational research literature, including the meta-design framework laid out in the *system of systems methodologies* (Flood and Jackson, 1991b) and subsequent approaches, such as *multimethodology* (Mingers, 2001; 2003) or the *creative design of methods* (Midgley, 1997a; 1997b; 2000), as outlined in Section 4.1. There has also been considerable attention to the question of whom to include in participatory processes. This has been treated both operationally and philosophically in the copious literature on stakeholder analysis (e.g. Freeman, 1984; Grimble and Wellard, 1997; Mitchell *et al.*, 1997; Ramirez, 1999; Brugha and Varvasovszky, 2000) and participation and democracy theory (e.g. Arnstein, 1969; Pateman, 1970; Fischer,

1990; Mostert, 2003b), which is briefly reviewed in Section C.1.1. However, the majority of this research still neglects the relational aspects of the design process. Further information on the design phase, including issues of process planning, stakeholder analysis and decision analysis (Creighton, 2005) is covered in von Korff *et al.* (2010), but again without emphasis on the collective aspects.

There has also been increasing interest in the co-design component of participatory intervention processes since the term *co-design* was introduced into the systems movement by C. West Churchman (1968), based on his understanding of Kant's (1781) philosophy and the a priori content of all knowledge (McIntyre-Mills, 2006). Researchers such as Ulrich (1983, 1991) and Midgley (2000), whose work was mentioned in Section 4.1.5, have further followed this line of enquiry into the co-design process, placing an emphasis on uncovering normative premises and making explicit the boundary judgements of various stakeholders in design processes, where boundary judgements '*define what constitutes "content" in any particular process*' and lead to '*distinctions of what exists*' and the concepts of inclusion and exclusion (Midgley, 2000). Despite the fact that their co-design processes often refer to the co-implementation phase in the definitions taken here, or the arena of collective choice (see Figure 1.1), some of their insights on the relational aspects may still be applicable to the design of participatory modelling systems. Other research into the interactions between cooperation and design can be found in a number of domains, including the management sciences (e.g. Nonaka and Takeuchi, 1995; De Terssac and Friedberg, 2002; Bouzon, 2004; Fuchs, 2004; Hatchuel, 2004; Kolfschoten *et al.*, 2004; Kazakçi and Tsoukiàs, 2005), ergonomics (e.g. Gaillard and Lamonde, 2001), sociology of work and science (e.g. Vinck, 1999; Callon *et al.*, 2001) and policy and institutional analyses (e.g. Ostrom, 1990; 1996; Edelenbos, 1999; Enserink and Monnikhof, 2003; Bots, 2007), which could inform our analyses.

CO-INITIATION

Research that focuses on the co-initiation stage of participatory processes is most commonly found in public administration, policy or development studies. For example, some articles linked to change and development studies in a range of domains, such as co-management, urban planning and education programme development, have started to show some of the roles that development workers, governments, researchers, NGOs and other institutional actors play in setting up and influencing participatory processes (e.g. Sundar, 2000; Watson, 2000; McKinnon, 2007; Helfgott, 2008). Nevertheless, there appears to be relatively little research specifically linked to different types of co-initiation structures that are used to set up and aid participatory water management processes.

THE WHOLE CO-ENGINEERING PROCESS

A handbook for co-management interventions developed by Borrini-Feyerabend *et al.* (2000) takes a more operational approach to outlining the phases required. Their operational phases – entitled 'a point of departure', 'organising for the partnership' and 'negotiating plans and agreements' – provide an example of typical co-engineering process phases, although the relational issues and interactions required or expected between project team members are not a major focus. Similarly, a number of other reports on developing processes and tools to support social learning in water management (e.g. Ison *et al.*, 2004; HarmoniCOP, 2005a), reviews of participatory environmental policy practice (e.g. Holmes and Scoones, 2000) and guides on building broad-scale public participation programmes (e.g. Leeuwis, 2000; Wiedemann *et al.*, 2000; Bertrand and Martel, 2002; Creighton, 2005; CEAA, 2008) outline phases, issues or questions to be considered in participatory process engineering. However, most still lack insights or questions related to the management of the relational aspects of the project teams and stakeholders involved in co-engineering, as well as how potentially diverse people can effectively work together. This is an issue that Syme and Sadler (1994) and other management and engineering-design literature (e.g. Bucciarelli, 1994; Katzenbach and Smith, 2002; Dandy *et al.*, 2007; Page, 2007) note to be significant. The study of relational issues and team-member diversity is largely neglected; however, it can be significant to the group's collective performance and achievement of individual and mutual objectives (Hong and Page, 2004). Further discussions on the tensions among individual and collective interests, in particular in inter-organisational settings, can be found in the collaboration literature (e.g. Huxham, 1996; Thomson and Perry, 2006), along with some examples of how collaborations are convened or co-engineered. Other relevant literature on relational aspects of project teams is available in the fields of policy development and strategy building (e.g. PMSU, 2004), organisational and engineering management (e.g. Dandy *et al.*, 2007), negotiation and conflict management (e.g. Thomas, 1976; Fisher and Ury, 1981; Leeuwis, 2000; Lewicki *et al.*, 2001; Leach and Wallwork, 2003; Rinaudo and Garin, 2003), or leadership, teamwork, organisational or social psychology literature (e.g. Senge, 1990; Bass and Avolio, 1994; Schein, 1999; Katzenbach and Smith, 2002; Stewart, 2008), where there is a much stronger tradition of using negotiation and team building theory linked to appreciating personality and skill differences required for effective relational and operational management. Systematic evaluation of co-engineering processes also appears rare, although elements of qualitative description of such processes are present in a few articles, some of which do not focus specifically on participatory water management processes (e.g. Syme and Sadler, 1994; Berry, 1995; Midgley, 2000; Creighton, 2005; Eden *et al.*, 2009).

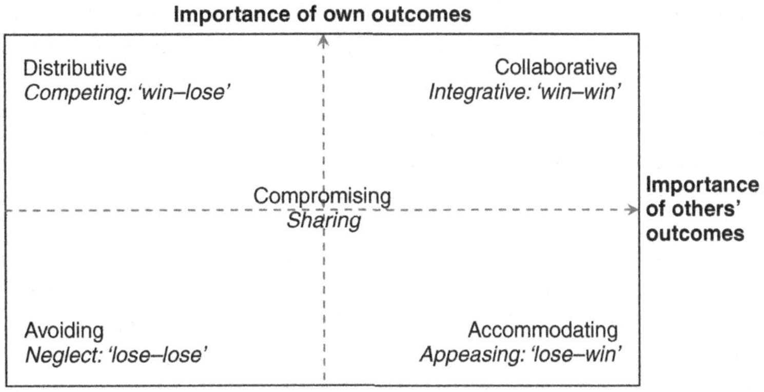

Importance of own outcomes

Distributive
Competing: 'win–lose'

Collaborative
Integrative: 'win–win'

Compromising
Sharing

Importance of others' outcomes

Avoiding
Neglect: 'lose–lose'

Accommodating
Appeasing: 'lose–win'

Figure 4.3 Negotiation modes. Adapted from Leach and Wallwork (2003), Thomas (1976), Fisher and Ury (1981), Lewicki *et al.* (2001).

4.2.3 Salient concepts of potential importance for co-engineering

This review of available literature relevant to co-engineering demonstrates that there are quite a number of potentially useful concepts dispersed over a number of domains. Previously in this book, it has predominantly been the operational types of concepts relevant to the investigation of co-engineering that have been outlined, such as what types of decision-aiding process methods and process models are available to guide participatory processes (Chapter 3) and how contextually adapted methods for participatory process organisation could be chosen or designed (Section 4.1). As in many of the sources reviewed previously, the critical reviews in this book have not yet focused on understanding the relational aspects of organisation of participatory processes that are likely to be needed for appreciating the co-engineering process more fully. This section will therefore briefly present a few concepts on the relational aspects of project teams that were found through the literature review and could be remobilised for investigating co-engineering interventions in the second part of this book.

NEGOTIATION AND CONFLICT ANALYSIS

One of the key themes that emerged on relational aspects through the co-engineering review related to conflict and how this is managed or resolved through negotiation. Huge bodies of literature exist on conflict and negotiation in different contexts, including in the water domain (Delli Priscoli and Wolf, 2009). Some of this work on conflict and negotiation has been outlined in Sections B.3.4 and B.3.5, as part of the investigation of potential decision-aiding approaches for managing conflict. One of the key strands of this body of literature relates to modes of negotiation and the linked manners in which conflicts are managed, which could be used as a tool for understanding and interpreting key events that occur through a co-engineering process.

Typical negotiation modes are derived from Thomas (1976), Fisher and Ury (1981) and Lewicki *et al.* (2001) and include the collaborative (integrative), distributive (competing), compromising (sharing), accommodating (appeasing) and avoiding (neglect) modes, which are based on the interaction of the level importance and energy that negotiation participants place in their own and others' outcomes (Leach and Wallwork, 2003), as visualised in Figure 4.3.

Such a scheme could be used descriptively as an interpretation scheme for negotiations that occur as part of the co-engineering process. It could also be used normatively to drive personal behaviour as an intervener in the co-engineering of participatory modelling processes, to aim for collaborative behaviour amongst the co-engineers and integrative 'win–win' negotiation outcomes, as well as to enable increased collective learning (Leeuwis, 2000).

TEAMWORK AND LEADERSHIP STYLES

The co-engineering review also highlighted that bodies of knowledge on team building and leadership, which are commonly used in organisational management practice, may also be appropriate for interpreting and understanding co-engineering relational dynamics. Group work and leadership are often interlinked concepts. The two schemes shown in this section to aid interpretation of the co-engineering interventions in Part II of this book have been chosen because they have different scales of focus (i.e. the group and the individual group member or leader), which could add richness to future analyses and focus analysis on a range of group and personal attributes. They also stem from academically peer-reviewed and heavily used sources, which appear to be safe choices for preliminary investigative work in this area.

The first scheme of Katzenbach and Smith (2002), depicted in Table 4.4, attempts to specify fundamental differences between different types of groups and performance units or 'teams' based on a number of attributes, such as the type of purpose that is defined, leadership structure, types of goals, role distributions and accountability structure.

Analysing the contextual needs for the participatory modelling process and the match or mismatch of the co-engineers' relational dynamics according to this scheme may help to

Table 4.4 *Effective group fundamentals versus the single leader or team disciplines. Adapted from: Katzenbach and Smith (2002)*

Attribute of interest	Effective group	Single-leader discipline: performance unit	Real-team discipline: performance unit
Purpose definition	Clearly understood charter or purpose (not necessarily related to performance)	Strong performance charter and purpose comprised mostly of individual contributions	Compelling performance challenge comprised of many collective work products
Leadership structure	Hierarchical leader promotes open communication and coordination	Focused, single leader applies relevant experience and know-how to create performance focus	Leadership role shifted between or shared among members to reflect and exploit performance potential
Goal types	Individual goals seldom add up to a clear performance purpose for the group	Individual outcome-based goals and work products that add up to the performance purpose	Outcome-based goals include both individual and collective work products (the latter predominates)
Role distribution	Clear roles and areas of responsibility remain constant through the group effort	Stable roles and contributions reflect talents and skills of members	Shifting roles and contributions to match varying performance tasks, as well as exploiting and developing members' skills and talents
Accountability structure	Accountability is understood, but consequence management principles seldom prevail	Individual accountability enforced primarily by leader; consequence management usually prevails	Both individual and mutual accountability, largely peer and self-enforced. However, only the team can 'fail'.

build insights into how co-engineering processes could potentially be improved.

To complement the group-level scheme, an individual-level scheme looking at leadership and delegation capacities to others in the group is shown in Table 4.5.

This scheme may help to demonstrate how co-engineering performance is linked to leadership types and certain behaviours exhibited by co-engineers through the process, which could be seen to help or hinder collective work and performance objectives.

A number of further schemes that could be used to aid evaluation of co-engineering processes and their impacts on participatory modelling processes will be detailed in the next chapter, where the framework for intervention research is to be developed, addressing another of the gaps highlighted in the co-engineering literature review.

4.3 CONCLUDING REMARKS ON THE STUDY OF CO-ENGINEERING PARTICIPATORY MODELLING PROCESSES

It has been shown in this chapter that there is a broad base of potentially relevant elements available in the literature from a large variety of disciplines, which could be used to study co-engineering processes of participatory water management projects. It was shown that some existing theory and methodology on how multi-modal contextual method choice or design could be drawn upon for designing contextually relevant participatory modelling methodologies for practical interventions. Some of the methodology-making recipes will require considerable critical thought in their application to avoid potential ethical concerns related to the position of the interveners. For situations where there are multiple interveners, the concept of co-engineering was proposed and a trans-disciplinary review was carried out to examine what existing research was apposite to the topic. This work fulfilled the third objective of this book: *to develop a definition of, and critically review, the concept of co-engineering as it relates to the organisation of participatory modelling processes for water management.* Co-engineering was defined as having three principle stages – co-initiation, co-design and co-implementation – in all of which a number of people work collectively to realise a participatory modelling process for water planning and management. A trans-disciplinary review of literature relevant to these three phases and the overall co-engineering process then demonstrated that a broad range of potentially relevant literature was available for studying their different aspects, but that little work had previously focused on the equivalent of the whole co-engineering process for participatory modelling processes. The literature review also found no evidence to refute the assumption that *participatory modelling processes used for inter-organisational decision-aiding in complex water management contexts are co-engineered*, but a number of practical cases that supported it. Knowledge gaps on the combination of operational and relational aspects of co-engineering processes and their systematic evaluation were identified.

Consequently, this research now aims to:

- *Investigate some of these relational and operational mechanisms through whole co-engineering processes* and to draw together insights from this diverse literature to understand how project

Table 4.5 *Leadership models to delegation. Source: Kuhnert (1994)*

	Model I: the transactional operator	Model II: the team player	Model III: the transformational 'self-defining' leader
Major attributes	Operates out of own needs and agenda 'Manipulates' others and situations Seeks concrete evidence of success	Very sensitive to how he or she is viewed or experienced internally by others Self-definition derives in part from how he or she is experienced by others Lives in a world of inter-personal roles and connections	Concerned about values, ethics, standards and long-term goals Self-contained and self-defining
View of others	Others seen as facilitators or obstacles to meeting own goals Others seek own payoffs and can be manipulated with that knowledge	Thinks others define themselves by how she or he experiences them, so feels responsible for others' self-esteem	Able to grant others autonomy and individuality Concerned about others without feeling responsible for their self-esteem
Leadership philosophy	Play by my rules and I will get you what you want	Show associates consideration and respect and they will follow you anywhere The 'unit' and team morale are paramount	Articulates clear long-term standards and goals Bases decisions on broad view of the situation, not just immediate factors
Follower philosopher	Let me know what you want and I will get it for you (if you take care of my needs)	I will do what it takes to earn your respect, but in return you must let me know how you feel about me	Give me autonomy to pursue broad organisational goals Do not ask me to compromise my own values or standards of self-respect, unless it is for the good of the group or organisation
Major blind spots in delegation	Cannot suspend agenda or coordinate agenda with others Cannot think of others as thinking about him/her; lack of trust Does not understand that some people forego immediate payoffs to maintain a relationship of mutual trust or respect	Unable to define self, independent of others' view or independent of role expectations Unable to make difficult decisions that entail a loss of respect	Can be too self-contained and reluctant to delegate May become isolated in leadership role

team member interactions, conflicts and collective choices shape the intervention process and the outcomes of participatory modelling processes designed for inter-organisational decision-aiding for water planning and management.

- *Determine how participatory modelling processes for inter-organisational decision-aiding can be evaluated and compared*, for example, by examining the efficacy, efficiency and other outcomes and metrics of the methods and process used, and these interventions' results in improving water planning and management.

The remainder of this book will concentrate on these practical research challenges. In particular, Chapter 5 will focus on developing an evaluation protocol for investigating the co-engineering of participatory modelling processes and an adapted inter-organisational decision-aiding process model in water planning and management. Part II of the book will then present the intervention research programme where the model is introduced and the evaluation protocol used through interventions in Australia and Bulgaria to produce a range of insights, followed by further discussion.

5 Intervention research and participatory process evaluation

It is through methodology, which sweeps in philosophical reflection, that we can better understand how methods of intervention can be used to create and sustain valued personal, social and ecological change.

GERARD MIDGLEY (2000)

This chapter firstly describes the 'participatory intervention research process' and research protocols that can be used to *investigate the impacts of co-engineering on participatory modelling processes for inter-organisational decision-aiding in water planning and management.* The principal objects of interest within this process for the purposes of this book, the 'co-engineering process' and the internal 'participatory modelling process', are then delimited and the choice of setting these boundaries discussed. An adaptation of the Tsoukiàs (2007) decision-aiding process model to the inter-organisational context is proposed as the basis for constructing contextualised participatory modelling methodologies. This is followed by development of evaluation protocols and methods that could be used to monitor and develop further insights on the co-engineering of participatory modelling processes for inter-organisational decision-aiding for water planning and management. Finally, considerations of the validation and legitimisation of research insights created through an intervention research posture will be outlined.

5.1 PARTICIPATORY INTERVENTION RESEARCH PROCESS DESCRIPTION

The research reviewed in Chapters 2, 3 and 4, and the intervention-based case research to be presented and discussed in the next part of this book, have been informed through the process that will be termed 'participatory intervention research', as presented in Figure 5.1.

The research approach is based on the 'intervention research' stance (Hatchuel and Molet, 1986; Hatchuel, 1994; Berry, 1995; Flood, 1998; Avenier *et al.*, 1999; David, 2000; Midgley, 2000) outlined in Section 3.1.1, where 'intervention' is considered as '*a constitutive mechanism by which a conscious attempt is made to modify organisational phenomena according to some pre-established concepts or models*' (Hatchuel and Molet, 1986) or as purposeful action to create change. The 'concepts or pre-established models' to be introduced as part of the interventions outlined in this book are participatory modelling methodologies, developed with the elements of the inter-organisational decision-aiding process model based on Tsoukiàs (2007), and used in different contexts with a variety of objectives. Collective action around these interventions will then be formed as a process of collectively constructing knowledge and relations through which the development of new insights and transformation of action may occur (Hatchuel, 2000). Research based on these kinds of intervention therefore aims to explore how collective action determines, considers or influences two principal types of fundamental relationships. As explained by Hatchuel (2005), these include all 'differentiations' that can be used to describe:

- *Subject–subject* and *subject–collective* relationships, such as power, trust, hierarchy and ethnicity or other social groupings, which can be thought of as 'operators of relation';
- *Subject–object* relationships through rationalisations and codifications (i.e. language, representations, memories, medias), such as emotions, senses, symbols or gestures, which can be thought of as 'operators of knowledge'.

The aim of intervention research is to create 'actionable knowledge' (Hatchuel, 2005; David and Hatchuel, 2007) by understanding the roles of relations and knowledge of models of collective action that could in turn lead to new theories and transformation of current forms of collective action, as represented by the 'reconstruction of theory' in Figure 5.1. In the case of the research presented in this book, it is actionable knowledge of the interaction of co-engineering and participatory modelling processes, shown as the bold lines in Figure 5.1, that take place with the main purpose of collective or inter-organisational decision-aiding, which is of interest. The delimitation of these research objects, and the methodologies and evaluation protocols to be introduced in the interventions, as represented in Figure 5.1, will be further discussed in the next

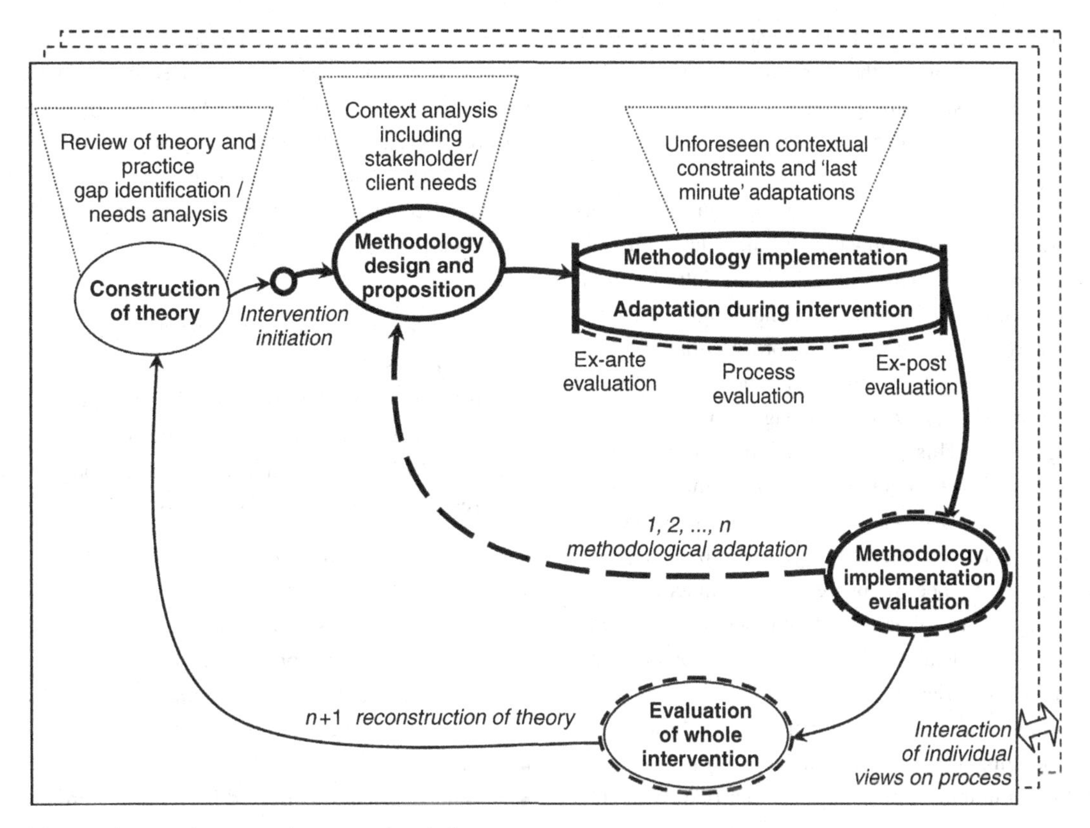

Figure 5.1 Participatory intervention research process description.

sections. Additional explanation of the intervention research approach is given in Section B.1.

5.1.1 Principal research object boundary setting

Many aspects of the initialisation and design processes related to participatory modelling processes are glossed over in the literature. It appears important to question how, why and to what extent they affect the process, content and outcomes of participatory modelling exercises for collective decision-aiding in water governance contexts. The principal interest in this book is the *co-engineering process* that includes and drives participatory modelling process initiation, design and implementation. A secondary and important research object contained within this is that of the *participatory modelling process* implementation, as mapped onto the inter-organisational decision-aiding process model adapted from Tsoukiàs (2007).

This choice of boundaries, outlined as the research focus area in Figure 1.1, limits the objects of research to those actions carried out by the organising team relative to the participatory modelling process and to the activities of the participatory modelling intervention. This means that the implementation of decisions that are taken during the final phase of participatory modelling processes for collective decision-aiding, focused on aspects of water governance and the processes on improving the external water-related systems, are not the key focus of the rest of this book. This choice was made as the research interventions discussed in this book are relatively recent and doing justice to this third research object is not possible at this stage, owing to the long timeframes of observation typically required to discern the effects of governance actions on complex socio-ecological systems.

Having identified the principal boundaries and research objects of interest, the next section presents a description of the types of methodology that will be introduced in the interventions. The co-construction of specific methodologies based on pre-existing theory for each intervention will be examined as a central theme of the co-engineering process. The effects and changes in collective action and knowledge construction that the introduction of these methodologies induces through the participatory modelling process will then be studied. This will aid their validation or the creation of new insights, theories or models of collective action requiring further testing.

5.1.2 Developing methodologies and models for the interventions

Chapter 3 concluded that an inter-organisational decision-aiding process model could provide a useful theoretical basis on which to choose or build adapted content-based modelling methods. It was then suggested that 'participatory modelling', where a broad range of stakeholders from different organisations are involved in the construction of each phase of the decision-aiding process and in the construction of each of the four 'cognitive artefacts' of the Tsoukiàs (2007) decision-aiding model, could prove beneficial for a variety of reasons. These are presented as hypotheses in Appendix C.

Participatory modelling processes based on an inter-organisational decision-aiding model, outlined later in this section, will be the prime methodologies to be introduced in the research interventions. These initial methodologies will then be developed as part of the co-engineering process that will take into account further contextual constraints and preferences to construct an intervention 'methodology design and proposition', as shown in Figure 5.1. This process will follow a philosophy similar to Midgley's (2000) creative design of methods, as outlined in Section 4.1.5. How this collective action process occurs to create this 'contextualised' methodology will provide a range of insights on the 'co-engineering process'. The implementation of this methodology as the collective action process of participatory modelling will then lead to new insights, such as how and why collective practice varies from what was theorised. There is a variety of cognitive artefacts to be constructed through the participatory modelling process for collective decision-aiding. The methodology implementation may be carried out in a number of stages, with each stage evaluated separately during an intervention to allow informed en-route adaptations to the overall methodological design. This idea is 'adaptation during intervention' and the loop related to 'methodology implementation evaluation' in Figure 5.1.

An adaptation of the Tsoukiàs (2007) decision-aiding process model to the inter-organisational context is proposed as the basis for constructing contextualised participatory modelling methodologies in the following section.

5.1.3 Development of an inter-organisational decision-aiding process model

As most water planning and management problem situations are 'messy', the decision-aiding models used to inform management need to be adapted to the inter-organisational and multi-stakeholder contexts, as discussed in Section 3.1.1. This section will therefore describe how the Tsoukiàs (2007) decision-aiding model could be adapted to such a context. Some issues related to the practical application and validation of such a model will also be considered.

Tsoukiàs (2007) considers a decision-aiding process as a distributed cognitive process, where operationally a number of shared cognitive artefacts or 'meta-objects', as represented in Figure 3.4 and Figure 3.5, are produced as outcomes of 'deliverables' of the decision-aiding process, which takes place in the 'interaction space' represented in Figure 3.5. The elements of these four meta-objects for inter-organisational decision-aiding are outlined next. Most of the elements here are given in, or adapted from, the original model description in Tsoukiàs (2007) and were further specified in Mazri (2007). Some of the main differences relate to the multiplicity of participating or associated actors and the need to consider their perspectives to ensure the final acceptance of the recommendations and legitimacy of the process.

META-OBJECT 1: REPRESENTATION OF THE PROBLEM SITUATION

The first meta-object to be created in the process for decision-aiding is the definition of an 'inter-organisational network' (Benson, 1975) around a problem situation of interest. This requires answers to questions such as:

- Who has a problem or issue to resolve?
- What is the problem?
- Why is this considered a problem?
- Is this a problem for anyone else?
- Who has the resources to manage this problem?
- Who makes the final decision?
- Can the decision be influenced and by whom?
- Who else will be affected by the decision?

A representation of the problem situation, **P**, which is likely to evolve during the decision-aiding process, can be given as a triplet,

$$\mathbf{P} = \langle \mathbf{A}, \mathbf{O}, \mathbf{R} \rangle,$$

where:

- **A** is a set of 'actors', who are the participants and stakeholders: individuals or organisations associated with, interested in or affected by the decision process;
- **O** is a set of 'objects', such as concerns, interests or stakes, for each of the identifiable actors; and
- **R** is a set of 'resources', either physical or abstract factors, linked to the actors and objects. These resources may be either currently available or unavailable to the actors.

In the inter-organisational context (considering Figure 3.5), the set of actors, **A**, can be further specified to include:

- A subset of 'core participants', **C**, who interact in the 'interaction space' $\rightarrow \mathbf{C} \subseteq \mathbf{A}$;
- A subset of 'associated stakeholders', **K**, who may be either directly related to the core participants through organisational or personal affiliation, or unrelated to the core participants, where their stake in the problem situation may be known or unknown to core participants $\rightarrow \mathbf{K} \subseteq \mathbf{A}$;
- A subset of 'project team members', **T**, such as the 'analysts', who are responsible for facilitating, organising and

managing the decision-aiding process, including any required external analysis outside the interaction space. Members of this set may either also be 'core participants' or 'associated stakeholders' at any point in the decision-aiding process $\rightarrow \mathbf{T} \subseteq \mathbf{A}, \mathbf{A} = \mathbf{C} \cup \mathbf{K}$.

A range of methods, such as mapping exercises, individual reflection and collective discussion and analysis, may be used to elicit the set elements and relations between them within the problem situation, dependent on the context and the project team's capacities and preferences and the stakeholders' needs. The resulting meta-object is likely to perform a descriptive or explicative role, upon which the 'problem formulation' can be constructed (Tsoukiàs, 2007).

META-OBJECT 2: FORMULATION OF THE PROBLEM AND OBJECTIVES

The aim of the second stage of the decision-aiding process is to formalise which parts of the problem situation are to be focused on, and what decisions will need to be made at the end of the process. This means that communal objectives have to be decided on, resulting in a 'problem formulation', $\boldsymbol{\Gamma}$, which is given as the triplet

$$\boldsymbol{\Gamma} = \langle \boldsymbol{\Pi}, \boldsymbol{A}, \mathbf{V} \rangle,$$

where:

- $\boldsymbol{\Pi}$ is a set of 'problem statements' on areas identified in the problem situation that require decisions. In the case of developing water plans and management strategies in the inter-organisational context, it is likely that several problem areas will be addressed, each of which will have a specific 'problem statement'.
- $\boldsymbol{A}$ is a set of potential 'actions' that each actor or group of actors could undertake relative to the set of problem statements and within the defined problem situation;
- $\mathbf{V}$ is a set of potential 'points of view' with which each actor will observe, evaluate, analyse and compare the set of actions.

Within this triplet, it is clear that making explicit the set of problem statements is likely to be the most valuable activity in terms of limiting differences in interpretation that could negatively affect the attainment of collective goals. It is through this stage that the 'ambiguous' problem situation description can be developed into a 'formal' problem (Tsoukiàs, 2007). If effective problem structuring tools are selected, this stage should provide a good opportunity to encourage participants' understanding, learning and knowledge production, based on the analysis and integration of other actors' views. From the 'problem formulation', the evaluation model can then be constructed as the next phase of the process.

META-OBJECT 3: MODEL EXPLORATION AND OPTIONS EVALUATION

Traditional decision-aiding for water management and planning typically starts at this stage, where the problem formulation is taken as predominantly 'given'. Based on the previous stages of representing and formulating the problem and objectives, collective decisions need to be made on the elements of the evaluation model, $\boldsymbol{M}$, which is given by the following n-tuplet:

$$\boldsymbol{M} = \langle \boldsymbol{A}^*, \{\mathbf{D}, \mathbf{E}\}, \boldsymbol{H}, \boldsymbol{U}, \boldsymbol{F} \rangle,$$

where:

- $\boldsymbol{A}^*$ is the set of alternative actions to be evaluated as potential options for decisions. These alternatives or scenarios will help to dictate the relations and functions that must be considered in a model or models if there is more than one problem statement.
- $\mathbf{D}$ is a set of dimensions, attributes or indicators, under which the alternatives will be described or measured;
- $\mathbf{E}$ is a set of corresponding scales to each element of $\mathbf{D}$;
- $\boldsymbol{H}$ is a set of criteria against which the alternatives are evaluated, to take into account the actors' preferences;
- $\boldsymbol{U}$ is an uncertainty structure;
- $\boldsymbol{F}$ is a set of operators that allows the synthesis and manipulation of the above information to aid decision making.

Most commonly used modelling methods in water management and planning, whether qualitative or quantitative, could be described in terms of some of these elements. Such models should be subjected to a number of conceptual, logical, experimental and operational validation processes before the next stage of defining 'final recommendations' can be pursued (Tsoukiàs, 2007). By the end of this stage, the 'model' or 'models' required to explore and allow the option evaluation will have been constructed and used by the 'core participants' in the interaction space, and ideally approved by the 'associated stakeholders'.

META-OBJECT 4: FINAL RECOMMENDATIONS

The final stage of the process may take place after a number of feedback loops or iterations through the other stages, including formal or informal input from the 'associated stakeholders'. The purpose of this stage is to make choices about the final alternatives, decisions or a set of 'final recommendations', $\boldsymbol{\Phi}$, to respond to the set of 'problem statements' defined as 'meta-object 2'. When constructing and evaluating these final recommendations and the methods used to obtain them, a number of questions should be asked about their validity (Landry *et al.*, 1983), some of which may be aided by performing sensitivity or robustness analyses, and about their legitimacy (Landry *et al.*, 1996). If there are validity or legitimisation concerns, then the cognitive artefacts constructed in the previous stages of the decision-aiding process may need to be updated or revised

(Tsoukiàs, 2007). Furthermore, in the inter-organisational context, issues such as how these decisions are going to be published, distributed, legitimated by associated stakeholders, implemented and used should be considered, as well as the core participants' and others' views of the success of the process and its outcomes.

ISSUES OF MODEL APPLICATION

The model presented here is constituted of purely formal and abstract constructs. How it is to be used must be determined by the analysts taking part in the co-engineering process for constructing a participatory modelling methodology for collective decision-aiding. A decision-aiding process in an inter-organisational context and the use of a decision-aiding model are likely to be part of a larger planning and management process, the context of which must be taken into account by the analysts (refer to Appendix A for possible contextual elements to consider). Under these constraints, there is then a need for method choice or design to obtain the formal elements of the model and use them in a coherent manner that can be tested for their validity and legitimacy, an issue that is discussed at the end of Section C.1.2. In general, a variety of possible methods might be acceptable for allowing the process of modelling and exchanging views on certain elements of the meta-objects. Some of these were discussed in Section 3.1.2 and more are noted in Sections B.3 and C.3. How they may be chosen or others constructed was mentioned in Section 4.1. As this decision-aiding model is to be used as the base of a 'participatory modelling' exercise, shown in Figure C.10, the participation of relevant actors is required in the construction of each of the four meta-objects of the decision-aiding model.

To improve model applicability, continuous monitoring and evaluation can be carried out throughout the participatory modelling process as a part of each of the stages of its implementation, as well as after the final stage, as indicated by the evaluation stages shown in Figure 5.1. Monitoring and evaluation procedures can have a number of aims, including: determining whether objectives for the process are achieved; encouraging individual reflection by participants and analysts; early identification of process problems or inefficiencies so that adjustments can be made; identification of what the process has achieved; determination of whether the final recommendations adequately address the problem statements and have been implemented; and ascertainment of whether the methodology used was acceptable. Some further issues of validation and legitimatisation of the methodologies and their underlying decision-aiding model introduced through these interventions will be discussed in the next section. Depending on the specific aims of the interventions, the evaluation process may be participatory or externally audited and may use a range of methods, some of which are outlined in Section B.2.2.

As this evaluation and its methods will form one of the key aids to reflection through the intervention research process, Sections 5.3 to 5.5 will describe in greater detail the evaluation procedures and protocol development that will be used as part of this intervention research programme.

5.2 CONSIDERATIONS FOR VALIDATION AND LEGITIMISATION OF INSIGHTS CREATED THROUGH INTERVENTION RESEARCH

Hatchuel (2005) explains that research in the management sciences can in general be defined as the *'identification, criticism and invention of models of collective action'*, a type of research which *'makes sense only in human contexts where collective action is transformative and creative and where the definition of the "truth" or the "real" depends on models of action that determine a knowing process'*. Under this disciplinary context, David and Hatchuel (2007) then suggest that models of collective action, either created through theoretical insights or during intervention innovations, may be validated by introducing them or re-examining them operationally in *in-situ* management situations: in the 'inter-organisational' context for the research presented in this book.

Landry *et al.* (1996) consider that the value of the models of collective action created in the domain of operational research must be both 'legitimated' by the actors that are involved in the collective action, and 'validated' by academics or practitioners by reapplication of the models or scientific peer review, as outlined in Section C.1.2. Other models and scientific results developed in a pluralist manner throughout the intervention using different 'knowledge-generating systems' (Midgley, 2000) or 'models of action that determine a knowing process' (Hatchuel, 2005) need to be validated and legitimated differently, based on their own underlying philosophical assumptions and epistemological backgrounds.

To aid the scientific review process and validation of models of collective action, Hatchuel and Molet (1986) suggest that 'experimental reports', which adopt a critically reflective attitude to analysing interventions are indispensable. These reports by research analysts should give clear conceptualisations of the intervention process, including a presentation of the tools used, as well as precise descriptions of the problems and difficulties met by the researchers and the reactions of the other project team members and stakeholders involved in the process. For each of the intervention cases to be presented in this book, this idea of an experimental report outlining the co-engineering and participatory modelling processes will be used: each report will include the evaluation findings, developed through the evaluation procedures, in line with the general protocol to be outlined in Section 5.5 and a discussion of other insights gained. Further comparative discussion about intervention cases and issues of validation will then be outlined in Chapter 9.

5.3 EVALUATION PROCEDURES AND PROTOCOL DEVELOPMENT

Two principal research objects are under examination in this book: the co-engineering process; and the participatory modelling process linked to it. Some form of critical reflection over, or evaluation of, both is required in order to study and respond to the key research hypotheses and aims outlined in Sections 1.4 and 1.5. Data that can be manipulated to provide insights are needed for both processes.

As argued in Section B.2.2, many different approaches to evaluation may be taken, depending on the needs and objectives of the evaluation. Different types of evaluation will use different approaches for the construction of their evaluation protocols. For example, there may be pre-formed objectives on which progress must be measured so that decisions on relative criteria, indicators and data for measuring them may be sought. On the other hand, field data on effects noticed through an intervention may have to be analysed by *ex-post* construction of indicators, criteria and values from these data. These different approaches to the construction of evaluations are represented in Figure 5.2.

For the research presented in this book, a pluralist approach to evaluation is to be taken, in order to capture as much relevant information and develop as many different insights as possible through the intervention processes. Therefore, top-down approaches to evaluation are to be used to capture specific information on certain objectives or criteria considered important in the participatory modelling processes, while bottom-up approaches to evaluation will help to identify unknown points of interest.

Top-down approaches can involve such methods as closed-response questionnaires and structured interview questions or cognitive mapping techniques with a focus on particular objectives or values to be explored. Bottom-up approaches for gaining data generally have less planning for how they will be interpreted and could include open-response questionnaires and interviewing techniques, story-telling, role-playing games, group cognitive mapping, document analyses, and process and participant observation through audio and video recordings, to note and further analyse aspects of interest.

This plurality of evaluation approaches is thought to be of particular use for developing actionable knowledge through the process of intervention research, as one of the aims of such a research approach is to help create innovative forms of collective action, many of which may not be able to be pre-formulated by theory (Hatchuel, 2005). Employing only evaluation methods that support the essentially 'top-down' deductive 'hypothesis testing' research approach may lead to innovations being ignored if they occur outside hypotheses of interest. It is therefore considered that evaluation methods supporting mainly 'bottom-up' inductive, or 'cyclic' abductive research styles, should also be employed to help discover, when possible, the 'unknowable' (Levin-Rozalis, 2004).

5.4 EVALUATING CO-ENGINEERING AND PARTICIPATORY MODELLING PROCESSES

Structured methods and examples of evaluation of participatory modelling processes used for collective decision-aiding and their associated co-engineering processes remain rare (Andersen *et al.*, 1997), despite the recent surge in interest in evaluation of public participation programmes or participatory processes (Wiedemann and Femers, 1993; Syme and Sadler,

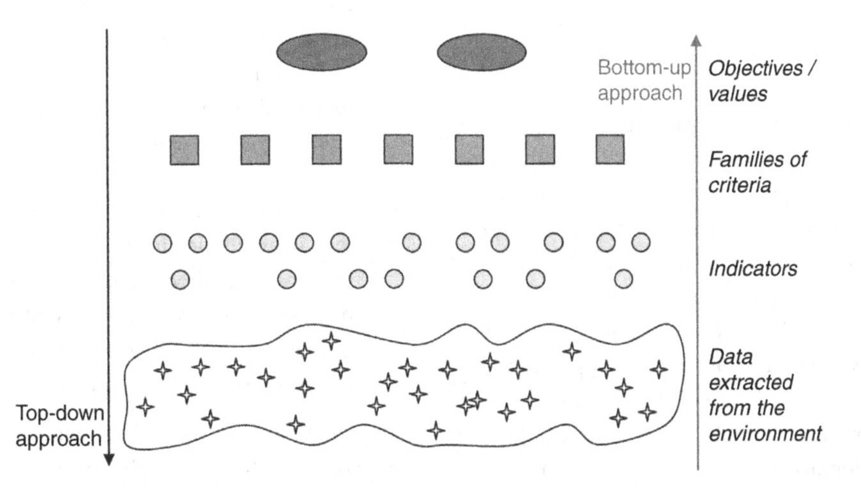

Figure 5.2 Different approaches to evaluation construction. Adapted from Damart (2003).

1994; Renn *et al.*, 1995; Webler, 1995; Webler *et al.*, 1995; Carnes *et al.*, 1996; Bellamy *et al.*, 2001; Connick and Innes, 2001; Marsh *et al.*, 2001; Webler *et al.*, 2001; Brinkerhoff, 2002; Colianese, 2002; Rowe and Frewer, 2004; Blackstock *et al.*, 2007), participatory research and evaluation processes (Guba and Lincoln, 1989; Webber and Ison, 1995; Estrella and Gaventa, 1998; Murray, 2000) and a certain number of methods or tools for structured participation (Rowe and Frewer, 2000; Beierle and Cayford, 2002; Maurel *et al.*, 2004; Creighton, 2005; HarmoniCOP, 2005b; Lynham *et al.*, 2007). A few comparative assessments of participatory modelling processes do exist, which include: process descriptions of companion modelling interventions (Bousquet *et al.*, 2002); process comparison based on observable external criteria, such as project objectives, number of participants or participant types (Hare *et al.*, 2003); a comparative case study evaluation based on in-depth evaluation of clients' opinions on group model building (Akkermans and Vennix, 1997); and a study of the use of computer models in four large integrated assessment projects, based on their normative function, substantive rationale, instrumental rationale and ability to instil mutual learning (Siebenhüner and Barth, 2005). Systematic evaluation of co-engineering processes appears rarer, as noted in Section 4.2.2, although elements of qualitative description of such processes is present in a few articles that do not focus on participatory modelling (i.e. Syme and Sadler, 1994; Berry, 1995; Midgley, 2000; Creighton, 2005).

Despite these limited investigations, more in-depth evaluation and analysis of participatory modelling processes for collective decision-aiding and their enveloping co-engineering processes, particularly for water planning and management, are required, to understand the extent to which implemented processes match their theorised forms and provide new insights. Formative (ongoing) evaluation could also help to: improve participatory process management; increase levels of process accountability and transparency; foster participant and project team member learning and reflection; and reduce process risks and uncertainties by uncovering potential problems earlier.

As there is very little theory concerning the co-engineering of participatory modelling processes in the literature, more inductive or abductive research methods may need to be employed to construct theory and to test further any hypotheses as the research programme progresses through the interventions. For the participatory modelling processes, an evaluation protocol that could be followed in each intervention, in order to study a number of existing hypotheses (see Section 3.3.5 and Appendix C), will be constructed here. Developing a number of key fields where each of the processes may be later compared may be beneficial for gaining new insights.

5.5 DEVELOPMENT OF AN INTERVENTION EVALUATION PROTOCOL

This book uses ideas from the evaluation literature cited previously, to suggest research on how co-engineering and participatory modelling processes may be described, analysed and evaluated. The evaluation protocol outlined in this section will need to be specifically adapted to the intervention context, needs and constraints of individual case studies. The design of the evaluation programme to be completed by participatory modelling process participants will be co-developed, or at least validated, by other project team members. The evaluation protocol is based on the contextual elements outlined in the work of Bellamy *et al.* (2001):

- Objectives or intent of the water planning and management intervention;
- Instrumental assumptions underpinning the implementation of the intervention;
- Objectives or intent of the evaluation itself;
- Feasibility and practicality of methods given the resources available; and
- Constraints of the implementation context.

The evaluation of the co-engineering and participatory processes can be used to maximise the number of insights by including a variety of forms of evaluation, as represented in Figure 5.1:

- *Ex-ante*: before the intervention or stages of implementation – focusing on *context*, objectives and theoretical and methodological assumptions;
- *Process evaluation*: formative monitoring through the process – focusing on intermediate outcomes, *process* dynamics and changes that have occurred since the *ex-ante* evaluation; and
- *Ex-post*: just after the intervention or stages of implementation – focusing on *outcomes*, the differences between the 'espoused theory and methodology' and the 'theory and methodology that was practically used' in the intervention and analyses of *why* differences or changes have occurred. Long-term *ex-post* evaluation is likely to give further valuable information, especially to the longevity and impacts of the intervention in the associated socio-environmental systems; yet direct causal inferences to the participatory process are likely to be difficult to validate and time is a major constraint for such forms of evaluation in the context of short research programmes.

The larger the variety of both qualitative and quantitative data sources available or sought under a variety of criteria fields during these three phases, the more likely it is that interesting insights may be obtained and evaluation results triangulated or validated (Blackstock *et al.*, 2007).

An evaluation could begin by attempting to analyse to what extent participatory processes carry out their assumed normative, substantive, instrumental and social learning functions (Siebenhüner and Barth, 2005). The development of criteria for evaluating the normative functions of participatory processes, such as building the legitimacy of the knowledge generation process through the inclusion of a broad range of stakeholders and institutions and choice of appropriate methods, appears to have received much more attention in the literature than the other functions. Marsh *et al.* (2001) provide ready-made questions corresponding to a variety of 'best practice' *acceptance* and *process* groups of criteria:

Representativeness Are participants broadly representative of the affected and managerially responsible actors?

Influence (impacts) Does the output of the participatory process have a genuine impact on policy and do the participants receive something positive from their involvement in the process?

Independence Is the participatory process designed and implemented in an independent and unbiased manner?

Transparency Is the process sufficiently transparent that both participating and external actors can see how the process is carried out and decisions made?

Early involvement (timeliness) Are participants involved as early as possible in the process and as soon as value judgements become salient?

Task definition Is the nature and task of the participatory process clearly defined?

Structured decision making Does the participatory process use or provide appropriate mechanisms for structuring and displaying the decision making process?

Resource accessibility Do the participants have access to the appropriate resources to enable them to successfully fulfil their task?

Cost effectiveness (cost–benefit) Is the process in some way cost effective?

Such criteria of what 'should' theoretically be aimed for are normative assumptions that need to be assessed for appropriateness and relevance in different contexts. For example, there may be certain situations, such as in repressive political regimes, where carrying out an independent and transparent process could carry large risks to project team members and participants, as well as jeopardise progress towards other objectives.

For the substantive, instrumental and social learning functions, a limited number of criteria and potential questions for evaluating participatory modelling processes exist in the literature. However, the development of an evaluation model, specifically for assessing collective action processes, could help to fill these gaps.

ENCORE (Ferrand, 2004a; 2004b; Ferrand and Daniell, 2006) is an observation model for management processes, named for an acronym of its key dimensions: *external*; *normative*; *cognitive*; *operational*; *relational* and *equity*. It was developed following experiences in the HarmoniCOP European project (e.g. Maurel, 2003) and through companion modelling exercises (Barreteau, 2003b), as outlined in Section C.3.2. Its philosophical underpinnings are strongly influenced by the concept of 'social learning', which has been increasingly adopted, especially in European projects (e.g. Ison *et al.*, 2004), for describing the advantages of adopting participatory approaches as a theoretical means of achieving more sustainable forms of development and water management. The achievement of these 'on the ground' development or management objectives is considered as part of the 'external' impacts component of ENCORE. On the aspects of social learning, the work of Webler *et al.* (1995) on a number of possible forms of cognitive and moral learning has informed the ENCORE model. However, the model then extends the view of social learning to focus also on the development of practices, human relations and social justice regimes (Ferrand and Daniell, 2006) and becomes consistent with the extended definition of social learning provided by the SLIM project (Ison *et al.*, 2004). ENCORE is also based on the assumption of an intentional deliberative model of an actor, who is equipped with a set of beliefs or representations of the world, normative preferences and a multiplex social network, all of which affect the actor's decision making process in the world, leading to observable behavioural effects (Ferrand and Daniell, 2006).

An example of what frame the ENCORE model could provide for evaluating the impacts of participatory modelling processes for collective decision-aiding in the domain of water planning and management is outlined in Table 5.1. 'Participants' indicates the 'core participants' who are involved in the participatory modelling process that takes place in the intervention's 'interaction space'.

These dimensions provide a useful basis for analysing the changes and impacts induced on the group of participants involved in the participatory modelling process and the external socio-ecological system. A range of potential evaluation methods that could be used to elicit information on each of these dimensions is outlined by Ferrand and Daniell (2006), who include common methods, such as individual questionnaires used throughout the process, group discussion and participant observation, to more innovative forms of evaluation, such as experimental tests, role-playing games and reflexive participatory modelling.

To judge to what extent operational goals have been met by using a participatory modelling process and its associated co-engineering process, other criteria may need to be added to the evaluation protocol. These could include those suggested by Checkland (1981) for analysing the outcomes of his soft systems methodology – *effectiveness*, *efficacy* and *efficiency* – gauged from both participant and design team perspectives, which could

Table 5.1 *Potential use of the ENCORE model for understanding impacts of participatory modelling processes. Adapted from Ferrand and Daniell (2006)*

Evaluation dimension	Potential impacts of participatory modelling processes	Example questions for framing evaluation
External: improvement in the water management situation	Direct and indirect impacts on the external context. These could be measured using multi-criteria evaluation techniques looking at sustainable development indices related to the range of contextual elements outlined in Section B.3	Have there been any measurable external effects of the participatory modelling process on the water management situation (i.e. on environmental, social, economic or infrastructure elements of the water-related management systems)?
Normative: values and preferences of the participants	Reconsideration of values and preferences, including those linked to cooperation, otherness, the common good and the meaning of 'long term'. Perception (cognitive function) of mutual dependencies and complex dependencies with the environment, which could lead to a revision of preferences linked to self, society and the environment	Have any of the participants' values or preferences changed as a result of the participatory modelling process?
Cognitive: representations and beliefs of the participants	Learning and comprehension related to the environment and others' views could aid individual and collective decision making to become more coherent with the more broadly assumed reality of environmental dynamics and constraints. Explored management options for water management are cognitively suitable, in that they are understood and eligible for consideration (according to the participants' worldviews)	Have the participants acquired a more global or integrated vision of the water management-related systems and the possible impacts of different actions? Have the participants integrated the role and diversity of others' perspectives into their own representations? Have the participants perceived a range of possible long- and short-term scenarios?
Operational: practices and actions of the participants	Changes in practices that affect water management and resource use	Have there been changes in the participants' practices or actions, especially behavioural changes that take into account new understanding of the impacts and constraints of these practices on water management and resource use that have been developed through the participatory modelling process?
Relational: social relationships between participants	Development of mutual understanding and social relations, such as confidence and trust in others	Has there been a change in social networks or the emergence of new coalitions or common interest groups, which could contribute to more coherent choices of options for water management?
Equity: social justice regime and distribution of resources between participants	Changes to the social justice regime between participants and an evolving distribution of resources throughout and following the participatory modelling process	Has there been a change in the distribution or an equitable usage of resources throughout the participatory modelling process and in accompanying water management practices?

equally be applied to the evaluation of the participatory modelling interventions. A final criterion to help gauge the scientific value of the intervention could be added: that of *innovation*, in other words, the extent to which new forms of collective action and knowledge were discovered through the intervention, or as a result of it.

These criteria could all be used to formulate a range of questions or areas of interest for analysis and evaluation of the co-engineering and participatory modelling processes in terms of: their contexts, including an analysis of the management system and objectives; their processes and content, including evolutions in the ENCORE dimensions, methods used and issues treated; and their outcomes, including impacts of the processes, levels of satisfaction and innovations, as outlined in the summary evaluation protocol in Table 5.2.

The protocol will be considered by assuming a need for methodological pluralism in each intervention and by employing a range of qualitative and quantitative methods, where applicable, within contextual constraints. The definition of this evaluation protocol ends the theoretical development of protocols required before the research interventions can be undertaken.

Table 5.2 *Summary evaluation protocol for the co-engineering and participatory modelling processes*

Evaluation object of interest	Possible leading questions for analyses of the co-engineering process
Co-initiation process: establishment of the intention to undertake an intervention *Project team views*	What is the context of the intervention (socio-political and physical environment system and understanding of boundaries)? Who is involved and what are their objectives, stakes, resources and roles for the intervention? Are there any potential divergences/commonalities in these objectives? What levels of complexity, uncertainty and conflict are perceived relative to the objectives and corresponding systems? How is the choice made on whom to involve in the design phase?
Co-design process: creation of the project team and a preliminary collectively agreed methodology for participatory modelling *Project team views*	Who is involved? What are their objectives, stakes, resources and roles? To what extent have these changed since initiation? Are there any potential divergences/commonalities in these objectives? To what extent does the design team work effectively together? What is the scope of the participatory modelling process design and how and why have changes occurred since intervention initiation (redefinition of boundaries)? How and why are participants chosen or come to participate, and who are invited? How and why is the participatory modelling process methodology and timetable decided? How and why are certain participatory modelling and evaluation methods chosen? Which objectives and criteria drive this process?
Co-implementation process: participatory modelling methodology is implemented by the project team and adapted as required *Project team views*	*Implementation for each stage or workshop* Who is involved in the implementation process and why? What are their objectives, stakes, resources and roles? Are there any potential divergences/commonalities in these objectives or conflicts? To what extent does the implementation team work effectively together? How and why are participants chosen or come to participate in the process stage, and who are chosen? How and why are the participatory modelling process methods and agenda decided? What is the nature of the implementation and which changes occurred and why did they differ from what was designed? Which elements of the inter-organisational decision-aiding model are considered, to what extent and why? How/why/on what/when/with whom/by whom are the evaluation procedures carried out? What are the implementers' and participants' perspectives on each other – conflicts/similar views? To what extent are process stage expectations met? What differences between the theory and the practice of the intervention are perceivable? What was surprising?
Participatory modelling process: actors participate in a collective decision-aiding process for water planning and management *Participant views*	*Evaluation of each stage of the participatory modelling process* What are the participants' views on context: socio-political and physical environment system and understanding of boundaries? What are the participants' objectives, issues, stakes, resources and roles? What are the participants' understandings of collective objectives and on whether the right representation of stakeholders in the process has been achieved? What are the participants' understandings of the relevant stage of the decision-aiding process and the participatory modelling methods used? Are the methods used and process results obtained satisfactory? To what extent are the objectives achieved? What learning has occurred and what other changes and external impacts have occurred as a result of the process (based on the ENCORE model dimensions)? Are any process, method or facilitation improvements possible? Do participants have any other general insights? Was anything surprising to them?
Overall intervention outcomes: impacts of co-engineering participatory modelling processes *Diverse views*	*Project team, participant, external stakeholder and external views* *Effectiveness:* To what extent was this approach the right one to take (legitimisation/validation)? *Efficacy:* To what extent did the means work to achieve their objectives? *Efficiency:* To what extent was there a minimum use of resources? *Innovations:* To what extent did new forms of collective action and knowledge result?

5.6 CHAPTER CONCLUSIONS: SUMMARY OF PROTOCOL DEVELOPMENT

In preparation for the research interventions to be undertaken in Part II of this book, this chapter has specified a formal decision-aiding model to underlie intervention cases and has developed an evaluation protocol for individual case and comparative analysis. This was in order to meet the fourth objective of this book: *to formulate an intervention research programme and evaluation protocol for investigating the impact of co-engineering on participatory modelling processes for inter-organisational decision-aiding in water planning and management.* More specifically, the work outlined in this chapter included:

- The definition of the participatory intervention research process;
- A delineation and discussion of the boundaries of the two research objects of interest: the *co-engineering process* and the *participatory modelling process* that it organises;
- The presentation and *expansion of Tsoukiàs' analyst–client decision-aiding process model* to the multi-stakeholder, inter-organisational or 'multi-accountable' group situation, so that it can be considered as a basis of elements on which to build participatory modelling processes for collective decision-aiding; and
- An outline and analysis of a number of existing bodies of literature on participatory process-related evaluation, which were carried out to aid the *development of an evaluation protocol* that is relevant for investigating co-engineering and participatory modelling processes and that supports pluralist use of evaluation techniques.

5.7 BOOK SUMMARY: PART I

This chapter concludes Part I of this book. So far, attempts have been made to outline the context required to understand the aim of the book, which was to *investigate co-engineering of participatory modelling processes for inter-organisational decision-aiding in water planning and management.* The first four objectives set out at the beginning of the book have been addressed and the first three underlying hypotheses confirmed. Before moving on to outline the lessons learnt through intervention research in Part II, to address the fifth objective of the book, a brief summary of the major issues and areas investigated to date will be provided.

Throughout Part I of this book, a broad range of areas under the general themes of water planning and management, decision-aiding processes and models, participatory modelling processes and method choice and design, as well as the concept of co-engineering both in terms of past and present theory and practice, have been examined. From this broad review and subsequent analyses, a number of key knowledge gaps and needs have been identified.

The review and analysis of water governance systems, challenges, mistakes and opportunities in Chapter 2 demonstrated that *one of the most pressing needs is to develop and implement improved methods of aiding decision making processes for water planning and management*, and in particular, that decision-aiding processes and the methods they use must be developed to better plan and manage water in:

- Multi-stakeholder and inter-organisational settings across spatial and administrative scales;
- Messy problem situations exhibiting high levels of uncertainty, complexity and conflict;

and using:

- A critically reflective manner, so as to learn from the past and adapt proactively in the face of future challenges.

From the investigation of past and current theory and practice of 'decision-aiding' processes related to the needs outlined in Chapter 2, Chapter 3 highlighted that there is relatively little theory developed on how inter-organisational or 'multi-accountable' groups can be aided by analysts, but the most promising insights came from researchers investigating 'participatory modelling', formal decision-aiding process models and intervention research practice. How decision-aiding models and methods can be operationalised in messy problem situations was also investigated, with participatory modelling appearing an appropriate means. The chapter then presented an example of integrated participatory modelling methodology after a brief critical review of three commonly used approaches, and summarised a number of key areas warranting further investigation, analysis or research, including:

- How the choice, concept of mixing and matching, or creatively combining and designing new methods for context specificities can be undertaken in the development of participatory modelling methodologies; and
- The utility of the 'co-engineering' concept as an evolving role for decision analysts' work to aid mess management in inter-organisational and multi-accountable group situations.

Chapter 4 then analysed and developed the ideas in the literature pertaining to these areas, in particular on how existing literature on multi-modal contextual method choice or design could be drawn upon for designing contextually relevant participatory modelling methodologies for practical interventions. The concept of 'co-engineering' was then defined and the limited trans-disciplinary review of existing research carried out on the topic. From these analyses, just two of the most important areas requiring further research were identified:

- The concept of *'co-engineering' participatory modelling processes for decision-aiding in inter-organisational and 'multi-accountable' groups*, in order to examine how and to what extent process design and implementation in teams can be carried out to achieve a variety of objectives;
- *How participatory modelling interventions for inter-organisational decision-aiding can be evaluated and compared*: for example, examining the efficacy, efficiency and other outcomes and metrics of the methods and process used, and these interventions' results on improving water planning and management.

Chapter 5 then looked into this last question, which resulted in the development of the evaluation protocol to be applied in co-engineering participatory modelling processes, along with the inter-organisational decision-aiding model, which will be introduced in the interaction space to underlie the contextually developed participatory modelling process research interventions, as summarised in Section 5.6.

Part II of this book will now describe how these theoretical concepts have been transferred into real-world water management contexts to drive the creation of actionable knowledge.

Part II
Learning through intervention

6 Intervention cases and lessons from the pilot trial

This first chapter of Part II describes the practical intervention cases used to create actionable knowledge through recent interventions of co-engineering participatory modelling processes for inter-organisational decision-aiding in water planning and management. This is specifically to begin to address the fifth objective of this book: *to outline the lessons learnt through individual and comparative intervention case analysis to determine to what extent co-engineering can critically impact on participatory modelling processes and their outcomes.* An overview of the lessons learnt from the pilot intervention case study, carried out in Montpellier, France, which were used to inform the next two interventions in Australia and Bulgaria, is provided. Information on the selection of the two case studies and a background to the cases, including data sources and interpretation schemes, are also given. Further work to fulfil this objective will be presented in Chapters 7, 8 and 9, as well as insights from extension case studies held in Germany, Algeria and the Republic of Kiribati.

Drawing on the reviews and conclusions in Part I, a number of research questions in need of investigation that could be resolved by practical interventions are highlighted. These specifically relate to the aim of this book, which is to *investigate the impact of co-engineering on participatory modelling processes for inter-organisational decision-aiding in water planning and management.*

6.1 INTERVENTION RESEARCH CASE QUESTIONS AND PROPOSITIONS

Key research questions of interest for the intervention research in different cases include:

- How does the co-engineering of participatory modelling processes for inter-organisational decision-aiding processes in water planning and management occur?
- How critical is co-engineering to participatory process outcomes?
- What factors contribute to successful co-engineering of participatory processes?

The two-intervention case study comparison in the following chapters aims to show why co-engineering processes are important. It also aims to show why and how vision and values helped to drive these cases to successful outcomes. The discussion is then further expanded with insights from other cases of co-engineering for participatory water management carried out by other researchers from around the world.

6.2 INTERVENTION RESEARCH CASE SELECTION

The selection of cases for intervention research has been performed in an iterative fashion using insights from, and adaptation of work in, the case studies to redefine key research themes, hypotheses and objectives. This iterative practice is typical of the learning that takes place through the process of many time-limited research programmes, such as those of PhD projects (Beaud, 2003). The pilot trial intervention in France and the two cases in Australia and Bulgaria were all chosen on pragmatic grounds. These included a high probability of finishing the interventions within the dedicated research programme timeframe, high levels of interest and operational support at ground level, and a central theme of water management.

The manner in which the final interventions were carried out is explored in this part of the book, to investigate the impact of co-engineering on participatory modelling processes for collective decision-aiding. It represents a hybrid form of intervention research (Hatchuel and Molet, 1986; David, 2000; Midgley, 2000), as presented in Section 5.1, and case study research (Yin, 2003) using more traditional social science methods. This is because some analyses were based on documentation studies and interviews, especially to elicit information at the beginning of the Bulgarian case study, as I only started to intervene in the co-engineering team of this case partway through the process. Other analyses using available data sources relevant to some of the evaluation protocol questions were also carried out *ex-post*. These data sources are outlined in Section 6.3.2.

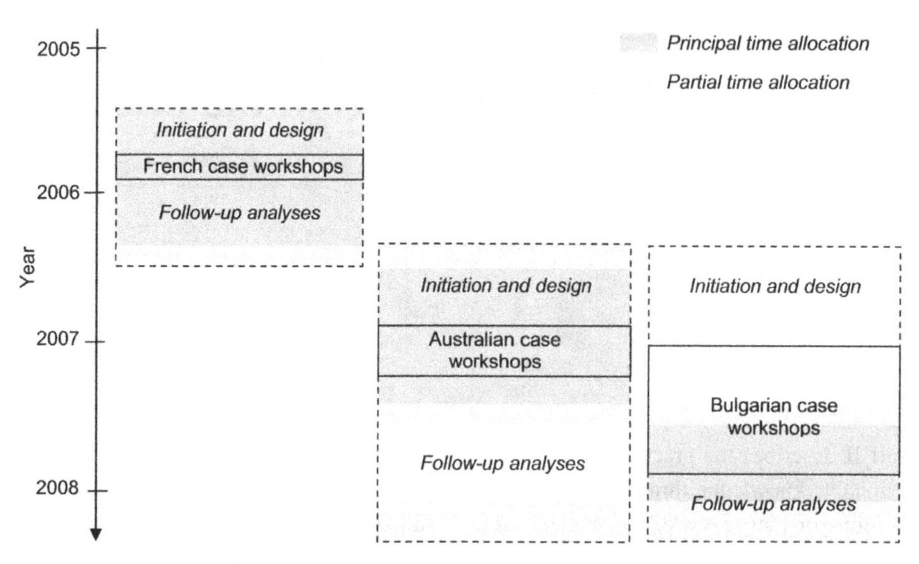

Figure 6.1 Research time allocation across cases.

In line with the case study research design outlined by Yin (2003), three large loops were taken through the participatory intervention research process, as depicted in Figure 5.1. These can be classified in case study research terms as:

- A pilot intervention case study trial, undertaken in Montpellier, France;
- A first intervention research case study, undertaken in the Lower Hawkesbury Estuary, Australia; and
- A second intervention research case study, undertaken in the Upper Iskar Basin, Bulgaria.

The time allocated to different activities in the interventions is shown in Figure 6.1.

Further information on the French pilot trial and the Australian and Bulgarian intervention cases is provided in the following parts of this section. Lessons learnt through the pilot trial will be discussed at the end of this chapter. The Australian and Bulgarian intervention cases will then be discussed in detail in Chapters 7 and 8 and compared in Chapter 9.

6.2.1 Pilot case definition

The Montpellier intervention was developed as a pilot trial of a participatory modelling process with a group of French students. It was co-engineered under a considerable number of constraints, including high time pressures, owing to student calendars, and resource constraints in the organising group. Although the pilot trial intervention context did not meet all the ideals of a 'real world' messy problem situation, or a participant group with representation from policy, expert and local stakeholder communities (refer to Section C.2.1), it was still of value, in particular as it allowed experimental testing of a theoretically developed participatory modelling methodology and evaluation protocol. This testing could also take place in a relatively risk-free environment of 'learning by doing', in which practical experience in designing and mixing methods, organising, facilitating, modelling and simultaneous evaluation of participatory processes could be obtained. Such spaces for experimenting with co-engineering participatory processes have since been formalised, as will be discussed in Section 9.5.2.

6.2.2 Intervention definition: embedded two-case study design

The two interventions in Australia and Bulgaria in this book targeted regions that presented a range of water management and planning issues that appeared amenable to the use of participatory modelling for aiding inter-organisational decision making processes. High levels of mutual interest and will to invest in building a collaborative project were present in the Hornsby Shire of Northern Sydney and in the Sofia region of Bulgaria, although a *final recommendation* on decisions was not an initial objective of the Bulgarian case. The two intervention cases also represent countries at markedly different stages and experience in broad stakeholder participation in water management. In Australia, a strong democratic tradition has existed for well over a century and there are more than 30 years' experience of very active community participation in water and land management. In Bulgaria, democratic traditions have been strengthening over the past 18 years but the country has very little experience with community participation in water planning and management. These similarities and differences in the two regions present a number of opportunities for comparative analysis of co-engineering processes for participatory water management.

A number of characteristics of the two intervention cases are highlighted in Table 6.1.

Table 6.1 *Comparative intervention case characteristics*

Characteristic	Australian process	Bulgarian process
Process type	Management-driven, supported by research	Research-driven to support management
Case location	Lower Hawkesbury Estuary, NSW, Australia	Upper Iskar Basin, Sofia region, Bulgaria
Principal shared objective	Create a regional estuary management plan	Test a multi-level participatory modelling process for joint flood and drought risk management
Principal project funder	Hornsby Shire Council (local government), NSW, Australia	European Union, as part of the AquaStress Integrated Project (Framework 6 Programme)
Principal co-engineering institutions	Local government water managers, university researchers, private environmental engineering consultants	University researchers, government institute researchers, private research consultants, stakeholder group
Participatory process stakeholder and administrative level inclusion	State government departmental representatives; local government councillors; managers, planners and scientists; national environmental NGO; catchment management authority representative; regional associations, industries and commerce; regional water agency managers; local residents	National ministers and departmental representatives; national NGOs, association representatives and water experts; water basin directorate representatives; regional mayors; regional water agency manager; municipal representatives; local residents
Participatory process (and average no. of participants at each workshop)	Three workshops over four months with 38 participants (average of 22 per workshop)	Two sets of interviews and 15 workshops over one year with approximately 135 participants (first ten workshops: average eight; last two workshops: average 26)

Details of these two cases are provided in Chapters 7 and 8, and case comparisons are discussed in Chapter 9.

6.2.3 Analysis boundaries

As outlined in Section 5.1.1, the units of analysis and their boundaries are around the participatory modelling process interaction space and the co-engineering process interaction space. The actions of individual actors relative to objects and resources considered in these interaction spaces, as well as relations between the two interaction spaces, add another pseudo-level of analysis to the case studies.

6.3 MODEL AND PROTOCOL USE IN THE INTERVENTIONS

In the intervention cases, participatory modelling methodologies were used as the key theoretical constructs or models around which co-engineer and participant action occurred. An evaluation protocol was also carefully followed through the interventions to obtain and interpret data on the questions of interest related to the cases. How each of these was used is described separately in the following sections.

6.3.1 Participatory modelling methodologies

Some of the 'concepts or pre-established models' introduced as part of the interventions in Australia and Bulgaria to generate actionable knowledge were participatory modelling methodologies. Each of these methodologies was developed and designed along the lines of the integrated participatory modelling methodology outlined in Section 3.3.5. Underlying this methodology were elements of the inter-organisational decision-aiding process model described in Section 5.1.3 that was adapted from Tsoukiàs (2007) and Ostanello and Tsoukiàs (1993). These included elements to be considered through the decision-aiding process – from defining the situation and formulating the problems requiring management to developing and using an evaluation model to assess potential management alternatives – before finally choosing and recommending the most desired courses of action. The idea is that there will be attempts made to ensure that these decision-aiding process stages will take place in the participatory planning process (Figure 1.1); yet how this is to occur will be dictated by the organisational and operational decisions of the design and implementation phases, which are defined at the level of the co-engineering process (Figure 1.1). In this book, the objectives of intervention research include developing an understanding of and demonstrating how aspects of these participatory modelling methodologies were debated, changed or re-designed and adapted for implementation by the co-engineering teams in the Australian and Bulgarian cases. An understanding of the impacts of this co-engineering on the participatory water planning processes is also sought.

6.3.2 Evaluation protocol use and data interpretation

The evaluation protocol, described in Section 5.5, was used to support the systematic collection of rich qualitative and quantitative multi-source data on the co-engineering and participatory water management processes for all cases. The participatory intervention research involved using a plurality of approaches to data collection and evaluation within the intervention cases to investigate the processes and their effects (Levin-Rozalis, 2004). For both intervention cases, a dedicated formal evaluation was conducted by an external observer, who attended some design meetings and all of the participatory planning workshops (Jones, 2007a; Vasileva, 2007; 2008; Jones *et al.*, 2009), but who was not a member of the project team responsible for the participatory process and its outcomes. These two evaluators were responsible for preparing and administering the participant questionnaires at the end of each participatory workshop, which had been developed in collaboration with some of the co-engineers. The questionnaires typically included between 10 and 25 open- and closed-answer questions (examples are available in Daniell (2008)), which allowed participant-perceived impacts of the co-engineering process to be elicited. For example, if a co-engineer had an objective of creating a workshop to enhance participant interaction and learning, questions probing participants on what they had learned during the workshop could help to gauge the efficacy of the participatory processes' co-engineering on this aspect. Evaluators also recorded observations on participants' and co-engineers' behaviours and relations to each other. Informal interviews were carried out as well, to further investigate observations of interest. Particular attention was paid to the phenomenon of discordance, i.e. participants manifesting different objectives or interests, or discomfort e.g. negative body language or verbal complaints, in each other's presence. For example, body language of a 'blocking or defensive' type (Pease and Pease, 2005) could signify conflicts or their impacts. Evaluators' impressions and interpretations could also be re-analysed by the other co-engineers *ex-post*, using video recordings.

For some elements of analysis related to the research questions outlined in Section 6.1, negotiation episodes will be taken as a focus, as they constitute specific co-engineering events where operational preferences and relations between co-engineers can undergo the most rapid changes. They are typically decision making episodes in which divergent or common objectives, interests and conflicts are also more evident. Decision impacts, or the results of the negotiations, can then be tracked onto the participatory planning processes. A focus on these episodes also allows analysis of underlying team dynamics to be informed, and use of the leadership and negotiation theories and frameworks to provide a structure for data interpretation, as outlined in Section 4.2.3. Focusing on negotiation episodes in participatory water planning and management processes is intended to contribute to the current debate on 'rethinking' participatory processes in natural resources management as ongoing negotiation and conflict resolution processes (e.g. Leeuwis, 2000; Leach and Wallwork, 2003). It is thought that it could be an equally valuable approach to study the co-engineering processes that organise them.

Data focusing on co-engineering negotiations and conflicts were obtained through en-route and *ex-post* analyses, personal reflections and reporting performed by a number of the participating co-engineers. For example, in my participation as a researcher in the two cases, I developed 'experimental reports' (Hatchuel and Molet, 1986) on the interventions, which involved making notes on co-engineering negotiations. This included keeping records of what was said and the co-engineers' personal feelings about the progression or diffusion of conflicts, including whether these conflicts led to learning about their own and others' objectives. If such learning was deemed to have occurred on two or more sides of a negotiation, this was interpreted as a move toward the collaborative mode of negotiation (see Figure 4.3). I also carried out in-depth semi-directive interviews with other co-engineers in both cases to delve further into the negotiations and conflicts of the co-engineering groups. These interviews were audio-recorded but only partially transcribed. They supplemented and allowed cross-examination of personal reports, observations and the process documentation analyses of both cases, i.e. of scoping documents, email correspondence, meeting minutes and notes, debriefing session records, project team-member interviews, participant and facilitator evaluation questionnaires, photos, reports, and some audio and video recordings of workshops.

As a result of the participatory processes, elements of the models created throughout the projects, including cognitive maps, matrix analyses, role-playing game results and action plans, also provided a range of data for evaluating the co-engineering and participatory water management processes. The co-engineering episode interpretations and impacts presented in the following chapters of the book have been triangulated (Yin, 2003) as far as possible using a variety of these sources, and interpretations cross-verified by a number of the co-engineering groups' members and external evaluators.

6.4 MONTPELLIER PILOT INTERVENTION TRIAL: DESCRIPTION

The Montpellier pilot intervention was proposed to trial a participatory modelling methodology for decision-aiding in the water sector. The abstract case was based loosely on the real situation of the water basin around the city of Montpellier in Southern France, as presented in more detail in Appendix D. The rapidly growing urban area of Montpellier is situated between mountains and the Mediterranean Sea, with the River Lez running through it. Low-lying areas of the city suffer from flooding. Moreover, tourist resorts near the sea receive large tourist influxes in the summer months when water is scarcer.

Agricultural production and ecological protection are also important regional issues.

Preliminary examination of the components of a participatory modelling process for this abstract water management context was carried out with a group of nine university students, composed of four males and five females aged between 18 and 35. The trial involved a series of seven three-hour workshops over a period from mid-October to mid-November 2005. The students had diverse academic backgrounds and were recruited for the 'research' project through advertisements in the universities in Montpellier. They were each paid a small amount of money to cover their attendance costs.

A number of qualitative and quantitative methods were then chosen or designed, in an attempt to maximise knowledge production (Nonaka and Takeuchi, 1995), and used to explore the elements outlined in the decision-aiding process model (Tsoukiàs, 2007) for an abstract problem of water management on three spatial scales: the students' lives; the local neighbourhood; and the region or water basin. These methods included: cognitive mapping of the *problem situation*; a variant of Ackermann and Eden's (2001) 'oval mapping technique' for the objectives and *problem formulation*; UML (Unified Modeling Language™) conceptual modelling for an MSExcel spreadsheet model, which was the basis of a role-playing game for scenario exploration, the *evaluation model*; and periods of debate and individual reflection for the final management decisions, the *final recommendations*. Because of time constraints, mostly individual final recommendations on actions, rather than collective ones, were completed. The process and summary of workshops (WS) are represented in Figure 6.2.

Owing to the exploratory nature of the trial, extensive evaluation was carried out through the process, including 15 questionnaires for the participants, with a range of closed and open questions that examined the 'ENCORE' dimensions: external; normative; cognitive; operational; relational; and equity, as presented in Section 5.5. These questionnaires explored the context, objectives, process and results of the test. Audio and video recordings also aided process evaluation. Further information on the pilot trial is given in Appendix D.

6.5 MONTPELLIER PILOT INTERVENTION TRIAL: LESSONS LEARNT

One of the objectives for the pilot trial was to *'learn by doing'* about method selection, evaluation procedures and other process constraints and opportunities in a relatively risk-free research environment. This objective was mostly achieved, although a multi-criteria evaluation method was not trialled, owing to time constraints in the workshop programming. Many of the other proposed methods were tested with varying degrees of success. The level of success for each method could be partially gauged from the participants' evaluations and from the project team's observations. In particular, the adaptation of the oval mapping technique (Ackermann and Eden, 2001) to the nested scales of the problem situation appeared particularly adapted to encouraging a creative working and learning environment, as well as producing some coherency of problem analysis.

The organisers learnt much from the very intense process, especially as the 'low-risk' environment allowed open testing and debate over methods that would have been difficult to achieve when higher political or resource risks were involved. For example, the advantages and disadvantages of having a non-working simulation model were openly debated with the students in Workshops 6 and 7. Students in Workshop 6 thought that the situation was advantageous for them as it brought the organisers and participants to a more equal level of debate and had potential for exchange and learning. As one of the participants put it: '*The other workshops have provided you [the researchers] with something, but in this session we've gained something.*' Difficulties that arose during the series of workshops related to managing time constraints, emergent co-engineering team conflicts, problem and model complexity and methodological changes en route. These changes were necessary with the abstract subject matter because the real water politics and specificities of Montpellier's physical region were to be avoided and there was an uncharacteristic stakeholder group with all participants being students.

Insights linked to the issue of models and how a 'scientific' view of the organisers could be brought into the process will be briefly outlined in the next sections. With respect to the other challenges of the intervention, several points worthy of further thought and discussion are outlined in Appendix D, including: how time management can be improved; finding optimum levels of procedural and model complexity; and determining an adequate balance between participant world views and information submitted for group analysis by outside 'experts'.

6.5.1 Model or no model?

The original design of the participatory modelling methodology specified that a model, probably a simulation model, would be constructed for running scenarios for the role-playing game or 'role analysis and exploration'. Issues in the co-engineering team, including time constraints, meant that the model could not be designed or constructed as planned. Although the simulation model was almost operable in the final session, it had not been adequately tested to meet the needs of the role-playing game. This meant that a number of adaptations to the workshop methods were required. For Workshop 6, when the use of the role-playing game and model had been planned, the activity of

Students

18–35 years old – various disciplinary backgrounds and levels of study: economics; psychology; accounting; law; information and communication sciences; arts (languages, history and geography); water engineering; and agricultural engineering

Scientific experts

Roles in: process design; evaluation; modelling; and facilitating

WS1: Introduction and problem situation

Personal introductions; process outline presentation; personal objectives setting; individual brainstorming on water 'stakes/ objectives', 'uses', 'factors (actors/processes)', 'problems' & 'management'; start of small group spatial cognitive mapping – 'situation model' building

WS2: Problem situation and formulation

End of 'situation model' building; presentation of models to other groups; start of 'problem formulation' small group cognitive map building for the scales 'life/house', 'village/ neighbourhood' & 'region/water basin' – definition of objectives & plans/strategies

Preparation of an expert scientific view of the problem situation

WS3: Problem formulation and conceptual model design

End of 'problem formulation' exercise – definition of actions/means (resources), actors & constraints/risks for carrying out the plans/strategies; presentation of 'expert' situation model; group discussion; start of UML modelling (actors/objects/processes) in small groups

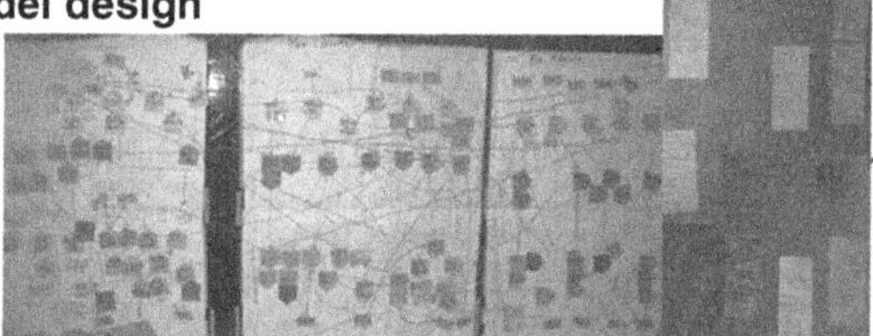

WS4: Conceptual model design and construction

Presentation of extent of existing computer model; choice of most important actors to be represented in the model; interaction game to define the model's dynamic relationships – between actors/objects, objects/objects, actors/actors

WS5: Role-playing game design

Game payoff design and construction – individual role 'satisfaction level matrix' and collective gain definition (choice and specification of equity model); definition of game scenarios and game rules

WS6: Model testing and preliminary role analysis

Group discussion on model hypotheses; individual analysis of the manager's role; development of indicators for 'good management'; model use and group discussion on management plans for one regional scenario

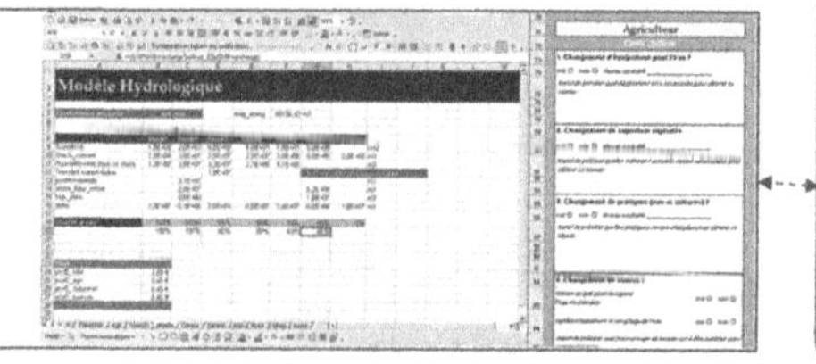

Preparation and testing of the computer model and role-playing game – based on the group's conceptual design

WS7: Role-playing game and final recommendations

One computer model-aided role-playing game round; two further role-playing game rounds without the model support; development of individual action plans and collective management strategy suggestions

Process evaluation, support provision and computer processing of workshop outputs

Figure 6.2 Implemented participatory modelling process for the Montpellier trial.

'model testing', which had not been used to any great extent previously, was reinstated. The activity took the shape of a large group discussion on a number of model hypotheses and how such models can and should be used. Some hypotheses that had been added to the model, such as that 'graduation' at university – one of the group's chosen indicators for student satisfaction – can be linked statistically to a poverty index of the suburbs where the students live, were discussed and accepted. However, others, such as the cost of ecologically friendly products, were not thought to be as reasonable from the students' points of view and so were changed in the model, as mentioned in Appendix D.

The 'role analysis and exploration' also took place in two phases, supported each time by a partially operable model. The first phase was the design of a management scenario through discussion, supported by the hydrological simulation model, in Workshop 6. This was then followed by the role-playing game of Workshop 7, only the first round of which was supported by the model. These activities are further explained in Appendix D.

One of the surprising elements that emerged from this intervention was that the participants did not consider that the non-functioning simulation model had led to process failure. It seemed from the participants' comments and evaluations that increased levels of learning and interest were generated during periods of debate and collective interaction when unaided by the model. Such a surprising insight leads to a large range of new research questions as to why this could be the case, and the implications that it could have on the design of participatory modelling processes for future water management-related interventions. A hypothesis that could be drawn from this might be that the model constrains the types of participant thinking and scope of actions possible, relative to the problem areas under investigation, and that this limits or channels participant learning.

Another hypothesis is that having too much information on the assumptions and uncertainties in models, especially complex ones, reduces the participants' trust in the results, as they realise the consequent weaknesses and uncertainties in the models' outputs. This raises another potential question about how much modelling carried out by participants is valuable or useful to create the agreement and trust required to work together to come to collective decisions. If interventions are carried out in truly complex, uncertain and conflict-ridden situations, is a final working simulation model useful, if its inputs, dynamics and results carry such high uncertainties?

A number of responses seem worthy of further investigation. Firstly, simulation models are often entirely forgotten when urgent decisions are required, which is the general perspective adopted for a number of the problem-structuring methods by theorists and practitioners (Rosenhead and Mingers, 2001b). Secondly, different types of model could be constructed and populated in a participatory manner, such as multi-criteria analysis matrices following the phases of problem situation and formulation, to avoid excessive numbers of assumptions in the design of complex simulation models. In such cases, the first phases will play an important role in building the group relations and trust that could be necessary to legitimate the choice of mathematical supports behind the matrix techniques. Thirdly, the idea of designing a 'role exploration through scenario analysis' after the phases of problem situation and formulation could be conserved without preserving the aim of developing the underlying simulation model. As such, a process and 'game' could remain more abstract, and issues of uncertainty could be more explicitly included and discussed with fewer data and conceptual modelling needs.

System dynamics modellers tend to oppose removing the use of simulation models, as they have found that human beings are not capable of mentally calculating complex feedbacks and impacts of actions (Forrester, 1992b). To create their models successfully, however, also requires much time; moreover, the models are normally very data hungry, which is not always practical or feasible. When causality or consensus cannot be established for the purpose of such modelling, even stronger arguments exist against developing simulation models in participatory processes. Other researchers, like Bots *et al.* (2011), advocate that simple hydrological models, rather than complex simulation models, can still be effectively employed in participatory decision-aiding processes if the participants understand them and trust that they can be useful for a specific purpose or enquiry into dynamics of the water system. These potential alternatives for model use could provide valuable possible directions for research related to participatory modelling exercises for water planning and management.

6.5.2 Providing a 'scientific' vision

A further difference between the designed and implemented methodology in the pilot trial was the place given to 'scientific' or 'outside expert' knowledge. In the planned methodology, there was little intention of specifically introducing any external information on water or its management into the group. Rather, the preference was to elicit and exchange the participants' knowledge within the group. This was to appreciate the participants' visions of water and to discover how the group's capacity to learn and understand water management issues would evolve when only procedural methods were provided. However, during the first couple of sessions, a number of participants asked when and if their visions would be corrected, and they requested more statistics on water use and management. After some reflection by the co-engineers, the information asked for was provided in Workshop 3, as might usually be provided by water experts in the 'real' world. The participants then discussed this provided expert knowledge and expressed their opinions of its validity. In particular, one of the students who came from a technical background thought the information provided by the co-engineers was more 'realistic'. Others did not take the same position, appearing to realise that some of the information was simply another alternative and potentially complementary vision to their own. Their perceptions of the external information provided by the co-engineers are briefly outlined in Appendix D.

In Workshop 6, during the discussions on the model and on the role of the manager following explanation of the scenario, a number of questions of a more technical or managerial nature were posed by the students and could not be definitively answered by the students within the group. In this case, the senior researcher, who was acting as the principal facilitator, took a more equal role in the discussion by providing whatever information was asked for when he was able to and by posing more questions. The fact that his own knowledge was not being hidden but shared with the students appeared to please the participants and a very lively, open and more focused debate ensued, spanning about two and a half hours.

The effect of and need for 'scientific' or 'expert' knowledge is further discussed in Appendix D.

6.5.3 Use of the decision model

Although the Tsoukiàs (2007) model was used in the collaborative design of the methodology, it was not explicitly considered during the 'en-route' design and adaptation process through the implementation. The final summary of the elements assessed, and how the sets evolved through the process, are outlined in Table D.3.

From this analysis, it was observed that most elements of the model were included or elicited in some form throughout the participatory modelling process. However, owing to the 'abstract' nature of some of the exercises and the creation of the role-playing game to examine the roles of other actors, it was difficult to define whether elements, such as the preference criteria, were made explicit. Such factors depend on whether it is the students' real preferences or the designed preferences of the regional actors in the game that are under investigation. As previously mentioned, because of a lack of time for the final multi-criteria analysis, the participants' personal preferences related to action choices were not specifically elicited.

The students found the process of defining indicators to measure progress towards their stated objectives quite challenging. This may have been because of the predominance of non-technical backgrounds or the students' lack of experience in modelling. Skill in modelling usually hones a capacity to define indicators. In the workshops, the modellers in the co-engineering team had to keep probing the participants to further specify their indicators so that they could be quantified. In the end, many had to be created behind the scenes for the simulation model and populated with data. The improvement of such elicitation practices in future exercises is likely to involve similar problems unless longer educational exercises on indicator definition and development are carried out with participants from a 'non-modelling' background. This raises the question of which elements of the decision-model are the most important that the participants should define during participatory modelling, to develop final recommendations that are widely accepted. Another question is how the links between the elements from the decision process model manifested themselves, such as those relations and evolutions shown in the model of Ostanello and Tsoukiàs (1993). A final question that arose was whether relationships and their meanings were specifically defined throughout the participatory modelling process that could usefully 'operate' on the available information and transform it into final recommendations or decisions.

6.5.4 Innovations from the pilot trial

From a perspective that is external to the group involved in the intervention, a certain amount of knowledge constructed through and after the process was made explicit during the trial of a participatory modelling methodology with a diverse group of university students. The participatory process involved a previously untried combination of adapted problem-structuring methods, conceptual modelling for computer simulation models, role-playing game design and payoff construction for a multi-scale and a multi-role analysis activity of water management scenarios on the individual and collective level.

Examining the process from an operational perspective, it will be seen that a number of aspects of knowledge creation from the participants' and co-engineers' perspectives have already been investigated in previous sections, demonstrating that the form of collective action that took place during the participatory modelling process did encourage some knowledge creation on an individual cognitive level and a group relational level. More specifically, a 'collective multi-disciplinary student–researcher learning group' was created through this process. In this form of collective action, each individual was able to construct his or her own knowledge reflectively, as a result of method-supported interactions; both with the 'tools' or 'artefacts' used and created through the process, and inter-personally within the interaction space.

It could be imagined that such a form of collective action could be recreated as an advantageous method of providing university-level education linked to water management and participatory processes. Running a participatory modelling process as a type of 'non-traditional' education programme could be envisaged, especially for research-oriented universities or international Masters degree programmes. In such a programme, it could be considered that there are no real 'teachers' but rather only 'mutual learners', with different existing bodies of knowledge to be investigated, exchanged and constructed. Many improvements and adaptations for this purpose could be imagined, including that the students could design part of their own payoff (marks for the programme) and play a larger role in evaluating the process, perhaps through a journal of observations and final process analysis. Depending on the range of students and their background disciplinary skills, it could be envisaged that a few students in the group take the role of building a computer model if it is required.

As a result of analysing conflicts from this process within the co-engineering team, it could be concluded, as outlined in Section D.5, that team-building exercises between the members of the project team who have never worked together before would benefit from more time, to avoid conflict management situations. For extremely time-limited projects, co-engineering a model with design team participants who have never previously worked or modelled together before is unlikely to succeed. Designer-facilitated processes are suggested when insufficient time is allocated for transferring understanding and skills to the facilitator, so that he or she is able to effectively facilitate the required modelling activities.

6.5.5 Issues requiring further research and reflection

Despite not being a 'real' intervention into inter-organisational decision-aiding in a complex, conflict-ridden and uncertain water planning and management situation, the trial was still useful for investigating water management issues with a group of people often marginalised from such debates in France. The trial helped to build a number of important insights into the act of running participatory modelling processes and their impacts. A number of the observations and results from the process evaluations further helped to raise important questions for consideration in future real or abstract applications of participatory modelling processes for collective decision-aiding.

Issues requiring further investigation include:

- To what extent do co-engineering participatory modelling processes involving project team members from different institutions exhibit different forms of collective action from those where the team members come from a single institution? To what extent are similar forms exhibited?
- What levels of model and process complexity are required to aid collective decision making adequately, from both the co-engineering team's and the participants' points of view?
- Is the use of multiple methods an advantage or detriment to participant and project team member technical capacity, interest, learning and cognitive load levels?
- To what extent can simulation models constrain or aid participation, learning and creativity?
- To what extent can external information be included in such processes? When and how?
- To what extent can technological supports and analysis, such as DecisionExplorer® versions of cognitive maps, be more effectively integrated into such resource-stretched processes without losing the relational and kinetic properties of the group work with paper supports?
- How can more effective studies or controlled processes be constructed if hypotheses related to participatory modelling processes are to be systematically tested? Would more abstract or carefully controlled experimental approaches be more effective?
- How can the time required for participatory modelling processes be decreased without constraining creativity or overloading the project team members and the participants?
- How can the evaluation process be improved to maintain its relevance to studying targeted areas, help to enhance a maximum number of reliable insights on non-targeted areas and cut time costs?

6.6 GENERAL CONCLUSIONS

This chapter has laid out the framework for the case interventions to be studied as part of this book. Following the definition of the principal research questions and propositions, information on the case selection and a brief background to the cases, including data sources and interpretation schemes, has been provided. This is part of the preparatory work required to complete the fifth objective of this book: *to outline the lessons learnt through the individual and comparative intervention case analysis to determine to what extent co-engineering can critically impact on participatory modelling processes and their outcomes*. An overview of the lessons learnt from the pilot intervention case carried out with university students in Montpellier, France, that were used to inform the next two interventions in Australia and Bulgaria, has also been provided related to this objective. These lessons are summarised here.

Through the Montpellier pilot trial intervention, a number of positive points and surprising insights arose, especially in terms of rapid '*learning by doing*' for both the participants and facilitators, and *insights into the co-engineering process* surrounding the participatory modelling process implementation, that have been highlighted in this chapter. One of the surprises was that the unfinished simulation model did not cause the failure of the process or disappointment amongst participants that the organisers expected. Rather, it led to the insight that the model may not be one of the most vital parts of the process and that *questions need to be asked about the advantages of including simulation models in complex and time-constrained processes for aiding collective water management decisions*.

Furthermore, although the methodology was not originally created as an educational programme, it is believed that there could be merit in adapting the participatory modelling process to suit water education or more general operational management courses on participatory methods and group work. As the behaviour and work of a group of students was studied in this test, discussions based on the educational value of such a programme are considered more immediately relevant than those relating to real multi-stakeholder water management interventions.

As a result of this participatory modelling process trial and the presentations and summary report of preliminary insights created from it, I appeared to have gained sufficient legitimacy from certain managers' points of view in France as an 'expert' in running participatory modelling processes, and was considered to be competent to intervene in future real-world complex, conflict-ridden and uncertain water planning and management situations. In Australia, this experience was never a requirement. To work in real-world cases, the only approval needed was that of the Australian University's Ethics Committee to conduct research involving human beings, which was obtained for this research project.

The first 'real' research intervention case on aiding the creation of the Lower Hawkesbury Estuary Management Plan in Australia will be outlined and discussed in the next chapter.

7 Creation of the Lower Hawkesbury Estuary Management Plan, Australia

This chapter presents an intervention based on the adaptation of a participatory modelling methodology to a 'participatory values-based risk management approach' (Daniell *et al.*, 2008b). The purpose of the intervention was to aid collective decision making for the creation of the Draft Lower Hawkesbury Estuary Management Plan in New South Wales, Australia. This process, driven by local government, included three interactive stakeholder workshops. These were based on stages of a generalised 'participatory modelling process to aid decision making' and on the Australian and New Zealand Standard for Risk Management (AS/NZS 4360:2004), as well as on an external scientific and legislative review. A range of stakeholders, including representatives from state and local governments, the water and sanitation authority, local industries and community associations, and residents took part in the process stages of: 'initial context establishment', including the definition of estuarine values, issues and current management practices; 'risk assessment' based on stakeholder-defined values (assets) and issues (risks); and 'strategy formulation' to treat the prioritised risks as input to the estuary management action (or 'risk response') plan. The plan was then opened to broad stakeholder consultation and has now been accepted. Many of the plan's actions have since been implemented. Preliminary evaluation results appear to demonstrate that the process was efficient in its use of time and budget and has a number of other potential benefits, which will be identified in this chapter, along with lessons learnt and questions arising in need of research.

7.1 LOCAL CONTEXT AND OBJECTIVES: ESTUARY MANAGEMENT IN THE LOWER HAWKESBURY RIVER

7.1.1 Local system and governance context

The Lower Hawkesbury River and its Estuary (Figure 7.1) are located on the northern fringe of the Sydney Metropolitan Area, dividing Sydney from the Central Coast Region of New South Wales in Australia.

The estuarine region has a warm temperate climate (Miller and van Senden, 2003) and contains a large percentage of bushland (native forest), much of which lies in National Parks adjacent to the waters and is currently protected from land development. The region has many areas of intense scenic beauty and is of important ecological, economic, cultural and social value. The estuary supports a few small foreshore settlements, which provide the oyster, prawn trawling, fishing and tourism industries with necessary infrastructure and access for their activities. Most of the urban, industrial and agricultural land uses are located further up the estuary's tributary creeks.

The region is currently attempting to cope with a number of important pressures, including: high population growth, with an estimated 15% increase over the 10 years to 2006, and holiday season population influxes; pollution from a variety of sources, including runoff from urban and agricultural areas, discharges from sewerage treatment plants (STPs), boat discharges, toxic substances found in boat anti-fouling paints and slipway scrapings, construction and dredging activities; pest, disease and aquatic weed infestations and outbreaks; unnatural flow patterns, for example due to STP inflows and water extraction; controlled burning and bushfires; and intensive recreational use (Forrest and Howard, 2004; HNCMA, 2005; HSC, 2006b; BMT WBM, 2008).

Recent major issues for estuarine management have included the 2004 outbreak of QX disease in the Sydney Rock Oyster population, causing high mortality rates and substantial economic losses (DPI, 2006); outbreaks and growth of the aquatic weed, *Caulerpa taxifolia*, since 2000 (Kimmerikong, 2005); toxic algal blooms, which pose threats to a number of aquatic organisms, the oyster industry and recreational water users; and contentious issues related to estuarine inflow qualities and quantities from STPs, on-site sewerage treatment systems and stormwater runoff. Future management is likely to be impacted by similar issues and the effects exacerbated by climate change mechanisms (CSIRO, 2007).

Estuarine management practice in Lower Hawkesbury is subject to a large variety of policies and statutory controls. In Australia, the majority of responsibility for management of

Figure 7.1 Location of the Lower Hawkesbury Estuary near Sydney, Australia.

natural resources is vested with the state and territory governments by the Australian Constitution (Bellamy, 2007). The planning and management of water and land resources are carried out by multiple state agencies under state legislation. Local governments, which are the principal means for communities to express their identity, enhance their well-being and care for their local environments, rely on both local community charges and the largesse of state governments to carry out these functions.

In Australia's multi-level governance systems, policies created at the international level or at the Australian Government level are typically translated into state-level policy and legislation after the CoAG agreement on principles, as outlined in Section 2.1.2. One of the main exceptions is the new *Water Act 2007*, which may be enforced at the federal level. However, its relevance to the estuary in the short term is likely to be limited to complying with the new provisions related to the Australian Bureau of Meteorology's access to water information. Therefore, the majority of estuarine management considerations for the Lower Hawkesbury Estuary fall under at least 12 relevant pieces of state government legislation and a range of policies, including state environmental planning policies (SEPPs), as well as falling under local government development control plans (DCPs) and local environment plans (LEPs). Other policies and plans that are relevant to this estuary's management include the Hawkesbury-Nepean Draft Catchment Action Plan 2006–2015, which sets the direction for investment priorities. Further information on relevant legislation and policies applicable to the Lower Hawkesbury Estuary is available in BMT WBM (2008).

Related to the multiplicity of regulations, laws and policies that have a bearing on estuary management, a large number of actors are responsible for ensuring compliance with these instruments, including several local governments, state government departments and the Hawkesbury-Nepean Catchment Management Authority. Furthermore, other regional stakeholders and estuarine users, such as industry groups, and recreational associations and users, also play a significant role in estuary management through their own actions or by their work in local Estuary Management Committees, which develop sub-regional estuary management plans in areas under local government control. Existing estuary management plans of this type in the Lower Hawkesbury Estuary include the Berowra Creek Estuary Management Plan (HSC, 2002) and the Brooklyn Estuary Management Plan (HSC/WBM, 2006). However, the efficacy of overall management in the Lower Hawkesbury Estuary was considered to be limited, owing to policy fragmentation and a lack of coordination of management actions (Kimmerikong, 2005). The region was therefore in need of an integrated multi-institutional, multi-stakeholder agreed and adopted plan for action, in order to ensure a more sustainable future for the socio-ecological estuarine system under current and new challenges.

7.1.2 Proposal and objectives for a regional estuary management plan

The issues outlined in the previous section led to a proposal by the Hornsby Shire Council, (HSC) to create a Lower Hawkesbury Estuary Management Plan (LHEMP). It was intended to be one of the first broader scale estuary management plans (EMPs) of its type to be implemented in Australia. This initiative also followed recommendations from the *Hawkesbury-Nepean River Estuary Scoping Study Report* (Kimmerikong, 2005) that, to improve effectiveness, estuaries should be managed relative to catchment boundaries or by a 'whole-of-estuary' approach rather than by an approach based on administrative local council area boundaries. It was considered that developing such an approach would '*be more strategic, would facilitate an understanding of the links between issues, allow priorities to be identified, and enable more effective and efficient management of issues by improving exchange of information and coordination of activities*' (Kimmerikong, 2005). Prior to the proposal for a regional plan on the Lower Hawkesbury River, only around fifty percent of the estuary and tributary creeks were covered by EMPs based on the NSW Estuary Management Program Guidelines (NSW Government, 1992).

To include the other parts of the estuary in the Lower Hawkesbury River not encompassed by a plan of management, the Hornsby Shire Council (HSC) decided to fund the enlargement process.

It was considered that this planning process, if carried out in a participatory manner, would help to meet the needs of:

- Capitalising on previous work, such as the existing Hornsby Shire Council's and Gosford City Council's estuary planning, monitoring programmes and numerous regional studies;
- Allowing the collective analysis and sharing of knowledge of the estuary and its surrounding communities from a range of different perspectives (stakeholder communities', government representatives' and scientists'), to aid future visions of sustainable development of the estuary and how these could be achieved through good quality planning and management strategies;
- Showcasing the region's proactive approach to supporting research and 'best practice' participatory processes (including their continuous evaluation) as an example for other regions to follow to improve their own estuary planning and management processes.

Compared with the existing small-scale estuary plans developed for parts of the study area, it was thought that creating the LHEMP would help to ensure:

- Improved management and greater estuary protection;
- Better use of local and regional knowledge;
- Improved strategic goals and objectives, based on a system-wide understanding of the estuary;
- The consideration of all values and issues related to the Lower Hawkesbury River rather than issues confined to local areas;
- More efficient and effective use of government resources;
- Greater potential to access and integrate funding and research opportunities; and
- Creation of opportunities for projects and community groups to address similar problems in different parts of the estuary.

7.1.3 Research questions and objectives

These objectives posed at least one major question. How could the regional estuary management plan be created to achieve these objectives? As previously stated, this plan was to be one of the first of its kind attempted in the world, let alone Australia, and therefore had no example to follow. Innovation based on sound research was required.

To undertake intervention research in the creation of the LHEMP, a number of research questions arose based on the previous theory and pilot study discussed in Chapters 3 to 6.

- Could a participatory modelling approach be used to aid collective decision making, improve coordination between actors and manage conflicts to aid in the creation and acceptance of an agreed regional estuary management plan?
- To what extent would the approach need to be adapted in its implementation from its original designed form to meet the project partners' and stakeholders' objectives?
- To what extent could a participatory modelling process for planning prove to be more efficient than traditional planning approaches?
- What are the advantages and limitations of using a participatory modelling process in such a planning context?

To respond to these questions, the following objectives were proposed for the LHEMP intervention:

1. Create a regional estuary management plan using a participatory modelling approach, with actions collectively agreed upon by major stakeholders;
2. Assess the capacity of a participatory modelling process to produce the plan in the local context related to its efficiency, effectiveness, efficacy and other effects, such as learning, innovation and conflict management; and
3. Gain a greater understanding of real-world institutional and water issues in the Australian context, including the positions taken by, and constraints placed on, researchers, community stakeholders, private businesses and governments in planning decisions.

7.2 PROJECT CO-INITIATION AND PRELIMINARY CO-DESIGN

The process of co-developing the above objectives and project scoping, a key part of the co-engineering process, will be outlined in this section. This narrative describes how relationships were formed and tasks undertaken to initiate and negotiate the form of the project collectively. It will be followed by another narrative detailing the following preliminary co-design phase of the LHEMP creation project.

7.2.1 Project initiation and scoping

I first met the estuary manager from the Hornsby Shire Council (HSC) in December 2005 at a scientific conference in Melbourne, Australia. It emerged that we had a number of mutual interests. I explained how I was still examining potential case study subjects for my work, while the estuary manager wondered whether there might be any projects in his council area that could be appropriate. After a number of email exchanges, we met again in April 2006, when the estuary manager suggested the possibility of renewing the existing Berowra Creek EMP (HSC, 2002) as a possible project for collaboration. I then wrote a participatory process proposition report on this project in May, which was submitted to the HSC. Elements of this first proposition are presented in Section E.1. After receiving the report, the estuary manager discussed the possibility of using the designed process to create a regional estuary management plan – a project that he had been working on for some time – to set up within and outside his own HSC area. He felt that the process might be better adapted to aid the implementation process of planning mechanisms in a larger portion of the estuary. This change of process scale was readily agreed to, as it suited my research interests to a greater extent than the smaller estuary plan. The estuary manager then wrote a public tender for the project to find a project manager to coordinate the participatory process and scientific and legislative review aspects for the creation of the Lower Hawkesbury Estuary Management Plan, which was released in early August 2006 after revision by me and by his institutional managers in the HSC.

The research project on participatory processes and the need for the chosen consultants to work with me to define an appropriate methodology for the workshops was presented as part of the tender requirements. The process for the plan creation outlined in the tender (HSC, 2006a) was largely based on the methodology outlined in Daniell *et al.* (2006a) and Daniell and Ferrand (2006), and was to include a series of two to four stakeholder workshops and an external document review. The process stages and elements outlined in Figure E.1 were retained, with only some small editing modifications, and appeared as an appendix in the tender.

One of the consultants who applied for the project rang me before submitting his tender proposal to clarify and negotiate the scope and dates of the workshops and to obtain a preliminary agreement for three workshops. The need to adhere to the University ethics guidelines through this process, including how to obtain participants' consent to be involved in this semi-research project, was also discussed. This consultant was part of a consortium of private environmental engineering consultants, BMT WBM, and planning consultants, SJB Planning, which was finally selected in September through the public tender process to run the project in collaboration with the Hornsby Shire Council and researchers from the Australian National University.

A number of people from each of these organisations and the manager of estuaries from the Gosford City Council met in early October. This initiation meeting was organised principally by the HSC estuary manager and included a range of formal presentations and introductions of the project partners, and an outline of the project schedule and objectives, which was delivered by the engineering consultant. Interactions between the participants were cordial and relatively detached with an a priori agreement on what was presented to them. At this stage, the programme was planned to include three participatory workshops in which a broad range of stakeholders would be represented. The workshops were to focus on:

1) Identifying the current estuary management situation and key issues and values;
2) Presentation of synthesis report outcomes, developing estuarine objectives and prioritising them;
3) Development and prioritisation of strategies and actions to work towards these objectives.

An external scientific and management review, the 'synthesis report', was to be completed by the consultants before the second workshop.

7.2.2 Use of the Australian Risk Management Standard

During the initiation meeting, the estuary manager mentioned that he would be interested in using the Australian Risk Management Standard, although this had not been specifically outlined in the tender. To the project team's knowledge, this was the first time that the Australian and New Zealand Standard for Risk Management (Standards Australia, 2004; 2006) had been proposed for use in such a broad-scale, inter-organisational and participatory process and it was felt that this was a suggestion worth pursuing. Previous uses, especially in the water sector, had been run by a single institution with the more specific objectives of operational management risks, and determining water quality risks or health risks (Billington, 2005; Everingham, 2005; Wild River and Healy, 2006).

The Australian and New Zealand Risk Management Standard (AS/NZS 4360:2004), its associated handbook (HB 436:2004) and the *Environmental Risk Management Principles and Process Handbook* (HB 203:2006) define *risk* as '*the chance of something happening that will have an impact on objectives*' (Standards Australia, 2004; 2006). Along with this definition, it is also noted in the Standard that:

- *A risk is often specified in terms of an event or circumstance and the consequences that may flow from it;*
- *Risk is measured in terms of a combination of the consequences of an event and their likelihood; and*
- *Risk may have a positive or negative impact.*

The Standard has been created to help to guide the management of any types of risk and, as such, takes a broad view on the processes required. It considers that: '*Risk management is the culture, processes and structures that are directed towards realising potential opportunities whilst managing adverse effects*' (Standards Australia, 2004; 2006). Furthermore, the *Environmental Risk Management Handbook* outlines a number of specificities or factors that may be considered when taking a risk management approach, which are particularly relevant when dealing with environmental complexity, as was the case for the LHEMP project. These include (Standards Australia, 2006):

- *A lack of data, or limited data sets, and the need to make assumptions*;
- *Natural variability*;

- *Application of immature sciences, with large differences of opinion at a scientific level on the most suitable actions to take or outcome to be achieved;*
- *Long time spans, in which ecological change may emerge slowly, due to delays and a lack of clear or direct links between causes and effects;*
- *Potential effects on the environment and economic welfare locally, and on regional, national, international and global scales, and the potential for irreversible outcomes; and*
- *The complex and extensive web of stakeholders, with the possibility that those with little control over their exposure may be adversely affected.*

Taking into account the considerations underpinning the Australian Risk Management Standard and the needs and objectives of the LHEMP project and associated ongoing management processes, it could be seen that the risk management process provided in the Standard was an appropriate choice of project methodology within which the participatory modelling workshops and other review and synthesis activities could be planned.

The principal elements of the proposed LHEMP risk management process are outlined in Figure 7.2.

The project team's objectives in using this process, rather than the NSW Estuary Management Planning Guidelines (NSW Government, 1992), included:

- To capitalise on existing stakeholder and documented knowledge, including previous estuary studies and management plans;
- To encourage increased understanding, knowledge sharing and learning between stakeholders to enhance future collaborations and the capacity to manage the estuary effectively into the future; and
- To keep the expansion of the estuary management plan as efficient and effective as possible, considering the resource constraints.

The methods used in each workshop were then selected to help meet these contextual objectives and a number of other project team goals, including managing some already identified key conflicts in the region.

7.2.3 Evaluation protocol adaptation and administration

The adaptation of the evaluation protocol was, in this case, another co-engineering exercise, owing to the LHEMP project's participation in the international project, ADD-COMMOD, financed by the French *Agence Nationale de Recherche*. The estuary manager and I had negotiated this participation, and it had been outlined in the tender, as is further described in Appendix E. In this international project, I first worked with other researchers from the CSIRO, Cemagref and CIRAD (French research institutes), and the Australian National University, to help design the evaluation protocol that would be applied in the LHEMP and in a range of other participatory modelling processes for Natural Resources Management around the world. In this design stage, I presented the elements in the protocol outlined in the 'participatory modelling process' section of Table 5.2, in particular the ENCORE model dimensions. The finalised protocol, which was largely developed by the external evaluator for the LHEMP, was predominantly compatible with my original protocol.

To evaluate the LHEMP project, I collaborated with the external evaluator, a colleague from my university, to design the protocol administration. This included co-designing questionnaires for each workshop that would meet both of our research needs and would be considered reasonable by the other project team members and stakeholders. Only minor negotiations were occasionally necessary, to find solutions that would best meet both our objectives. These predominantly related to the workshop time constraints and how the extensive protocol of the evaluator could best be adapted to fit into these constraints without losing vital information needed for international case-study comparison. Questionnaires for participants after each of the three workshops were created, comprising both open and closed questions, and the external evaluation protocol questions for the design team, administered by both the evaluator and me, where applicable. For example, I conducted a group interview with the project team at the beginning of the process, which was based on the questions in the collectively created protocol and recorded for the evaluator. I also conducted a final interview with the estuary manager, with questions posed for both our purposes. At other times, the evaluator was able to carry out the interviews or observations herself, although in the end not as many interviews had been completed as either of us would have liked, largely because of time constraints. Owing to the lack of recording of some informal interviews undertaken by the evaluator, some observations were difficult to validate through triangulation with the spoken content or body language analyses of other evaluators. It was also evident from the use of the same questionnaire evaluation results but different observational perspectives, that some of the conclusions drawn by each observer about the study contained biases towards their own research objectives and interests. Such biases appear inevitable; however, if further interpretations from other process participants and observers could be gathered, more integrated and widely shared interpretations could at least be constructed. An additional external evaluator, my research director, also assisted by video recording the two first workshops, which helped to provide a third perspective. My interpretations in the following sections were all submitted to the other project team members and evaluators for comment, reinterpretation and validation. Likewise, my

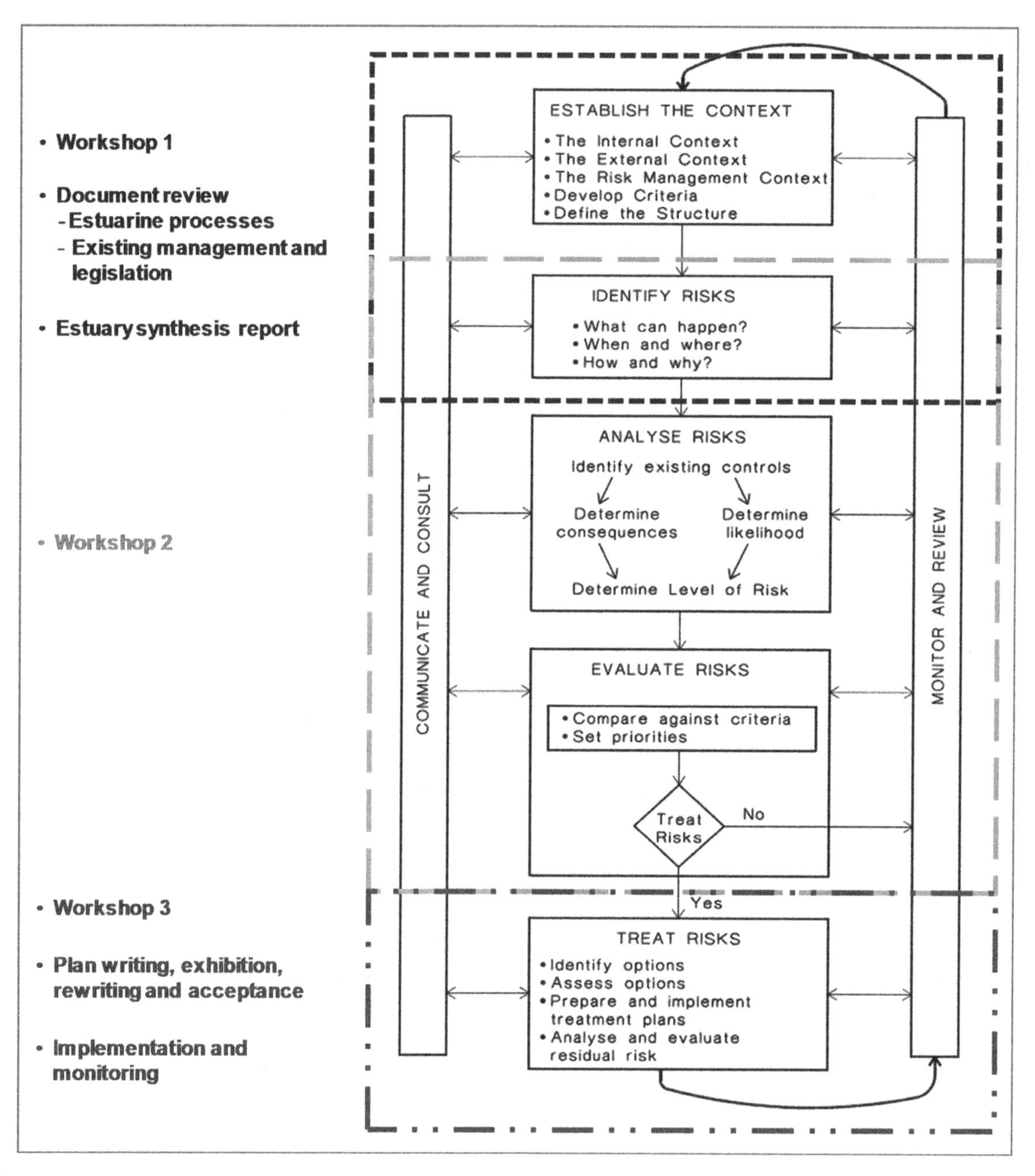

Figure 7.2 LHEMP Participatory Risk Management Process based on AS/NZS 4360:2004.

interpretations of the participants' evaluation questionnaires were all made available for participant comment as part of the Draft LHEMP during its public exhibition period.

7.3 DETAILED CO-DESIGN AND CO-IMPLEMENTATION

This section endeavours to provide a summary of some of the significant negotiation that took place through the co-engineering process. It also includes an in-depth analysis of one key co-design negotiation, originally presented in Daniell *et al.* (2010b), including the impacts this negotiation had on the co-engineering processes and participatory modelling process. The chosen negotiation had a 'normative' and 'relational' basis, where the question of participant exclusion or inclusion from the final two participatory workshops for the LHEMP creation was treated. It provides an illustrative example of the types of difficulty that can arise in managing the co-engineering process and describes negotiation modes that are less or more beneficial for developing collectively supported and positive decisions. It is also noteworthy, because it highlights the importance of conflict resolution in a water-planning project of this type, and demonstrates the need for communication between the parties and continuous reassessment of the co-engineering process and organisational decisions emanating from it. A continuation of the narrative from of all the principal elements of the co-design and co-implementation phases, including both relational and operational aspects, is provided in Section E.2.

7.3.1 Summary of significant co-engineering process events

Compiled descriptions of a number of principal co-engineering process, events and their impacts are presented in Table 7.1. Data sources used to inform the other elements were given in Section 6.3.2. The interpretation schemes used to determine the negotiation modes were presented in Section 4.2.3. These events are summarised on the process timeline for easier visualisation in Figure 7.3.

7.3.2 Negotiation episode analysis: participant inclusion or exclusion

BACKGROUND TO THE NEGOTIATION

The first workshop (WS1 in Figure 7.4) involved a broad range of agency and community stakeholders, and had the purpose of eliciting participants' values and issues of the estuarine region, as will be detailed in Section 7.4. The participatory methods chosen had also been relatively successful in reducing confrontation between community members and a water and wastewater agency over the discharge of tertiary treated sewerage water into the estuary (see Section 7.5.2). Participants had been informed that they would be invited to participate in two further workshops to prioritise their objectives for estuary management and develop and assess strategies to achieve them. A project meeting, following this workshop and a month prior to the second workshop, was held at HSC to determine the details of the combined synthesis report and the contents of the next workshops. It was attended by the estuary manager, the consultant project manager and his environmental scientist associate and me, the researcher of Table 7.1.

DISTRIBUTIVE MODE

However, there then was an unexpected turn of events. In the project meeting following the first workshop, the project manager proposed that the following workshops were not required. He said that an enormous amount of information had been gathered from the participants already, and that it would be more efficient to carry out the rest of the risk-management process and plan development in-house. Surprised at this suggested change of plan, I reacted strongly against the suggestion: such a change would be inefficient and contradicted the project manager's previous assertion that using a participatory process would allow the participants to do the work and make the decisions that consultants would usually have to justify by using multi-criteria analysis (Jones, 2007a). The project manager acknowledged this and instead suggested keeping at least the third planned workshop on strategy building, but proposed that the second on risk assessment could be removed, as it would be too difficult to perform in a large participatory group. I suggested that it would be possible to design an efficient and effective participatory method to perform the risk assessment in a big group, arguing that having larger numbers of participants would increase the knowledge and competency required to complete the risk assessment effectively. The project manager then suggested that a small group of experts, including some from HSC, could be used to create a greater knowledge base. I pointed out that such an approach would have problems of legitimacy among the stakeholders, as risk assessment is a very subjective process. I suggested that the people required to legitimise the plan should be involved in the assessment, especially for the vague risks in the estuarine context. It can be seen here that the negotiation was functioning in a distributive mode, as a 'positional bargaining' approach (Fisher and Ury, 1981) had been adopted. Both the project manager and I were trying to defend our own positions: for the project manager, that there should be minimal participation; and for me, that the workshops should involve all stakeholders. This occurred without our underlying personal interests and needs being articulated.

COMPROMISING/COLLABORATIVE MODE

Throughout this debate, the estuary manager had remained silent, carefully listening to both sides of the argument. Following the last discussion, he reflected to the group his need to have state-agency support for the plan, given that state agencies would eventually fund actions arising from the plan. He suggested that the agencies would appreciate being involved in the risk assessment, and that maybe an 'agency-only' workshop could be planned. I agreed that it could be an option, but worried that community backlash might prove to be a problem when the community representatives found out that they would be excluded, potentially leading to their rejection of the risk assessment. The estuary manager replied that although such backlash was a possibility, he thought that most community members would understand the decision, even if they were not pleased about it, as they recognised the necessity of agency support for the plan. He also added that an 'agency-only' workshop would probably provide a 'safer' place for agency representatives to share their real concerns, as meetings with the community members could be confrontational. At this point, the project manager suggested directly giving the estuary manager the decision of selecting one of four options: (1) no workshop, (2) an 'expert-group' workshop,

Table 7.1 *Description and analysis of significant LHEMP co-engineering process events (the researcher is, of course, the author)*

Negotiation event – people involved	Potential effect on personal objectives or interests	Negotiation mode, outcome and relationship characteristics
Co-initiation phase Change of project scope for participatory process from renewal of sub-estuary catchment management plan to creation of a regional estuary management plan – suggested by estuary manager to researcher	*Estuary manager* Largely positive, as he could use the researcher's skills and the 'latest knowledge and techniques' to help develop an innovative project where there was no precedent of a process to follow, which might increase the potential for external funding and help to improve significantly the coordination of estuary management and the environmental quality of the estuary. He also wanted to introduce an 'independent body of thought' to gain greater accountability for the process. *Researcher* Largely positive as scale proposal and novel nature of the process were much more suited to her research interests.	Collaborative and integrative type of negotiation based on cognitive and operational issues of interest. New scope readily adopted. Increased trust built between researcher and estuary manager.
Co-design phase Method selection for first issues and values workshop – suggested by researcher to consultants	*Researcher* Slightly positive, as the majority of suggestions to use methods to give everyone a voice and to encourage interaction were accepted on condition that the consultants facilitate their use at the workshop. *Estuary manager* Largely positive, as the methods suggested had the aim of decreasing confrontation. This was important for him, owing to existing conflicts between stakeholder groups in the estuarine area. *Consultants* Slightly positive, as despite not having used the methods proposed by the researcher, they would maintain control over the discussions, as they would facilitate the workshop activities.	Integrative, merging on compromising negotiation based on cognitive, operational, relational and external issues. Suggestions for methods quite quickly accepted in exchange for facilitation rights. Some trust and joint working capacity built in the group.
Co-design phase Deletion of final two workshops suggested by consultants to estuary manager and researcher	*Consultants* Slightly positive, as time resources dedicated to performing the risk assessment could be organised on their own terms. However, they would have more work to do, owing to the need to learn about the risk assessment process, as they had little experience of it, and also to back up their analyses to encourage their acceptance by stakeholders and external funding bodies. *Estuary manager* Slightly negative, as he would have to explain the process change to all of the stakeholders, and they would not have the chance to familiarise themselves so easily with the estuary's risk management process. *Researcher* Largely negative, owing to the loss of opportunity to complete the process developed for her research project.	Distributed negotiation based on cognitive, normative, operational and external issues. Suggestion strongly debated and rejected by the researcher. A priori support of this rejection provided by estuary manager (refer to next negotiation). Reduction of researcher's trust in the consultants.

Co-design phase Suggestion of one state-agency-only workshop in between two full-spectrum stakeholder workshops made by the estuary manager to the consultants and researcher	*Estuary manager* Positive, as he would have a greater chance to gain state and regional agency support and funding, and joint ownership for the regional plan. He considered that in the absence of community stakeholders, the agency staff could speak more freely and openly, which would bring a new 'realism' to the plan, whereby agencies could honestly say what they think should or should not occur, thus managing community expectations of what was going to be delivered through implementation of the plan. *Consultants* Mitigated, as they would be under high time constraints that would limit the understanding of the risk standard and assessment process that they could gain before the workshop, although it would reduce their responsibility for justifying risk assessment claims. *Researcher* Neutral, as she saw the advantages of this approach for the agencies and the estuary manager, disadvantages for the community stakeholders and the relatively neutral effects on her own research.	Compromising but integrative negotiation based on cognitive, normative, operational, external and equity issues. Decision offered to the estuary manager by the consultants and the suggestion was upheld. Slight renewal of researcher's trust in the consultants, increase in mutual understanding and learning between all parties.
Co-implementation phase Change in programme suggested by stakeholders to the project team during the risk assessment	*Stakeholders* Positive, as they could collectively discuss the risks that they were most interested in and could gain some ownership over the design and implementation of the LHEMP creation process. *Project team* Mitigated, as they were uncertain how this would affect the workshop timing and the subsequent external scientific validity of the outcomes.	Accommodating and collaborative negotiation based on operational, cognitive, relational and external issues. Suggestion adopted by project team. Built trust between stakeholders and the project team, and amongst the stakeholder group. End result of change maintained tensions between the consultants and researcher.

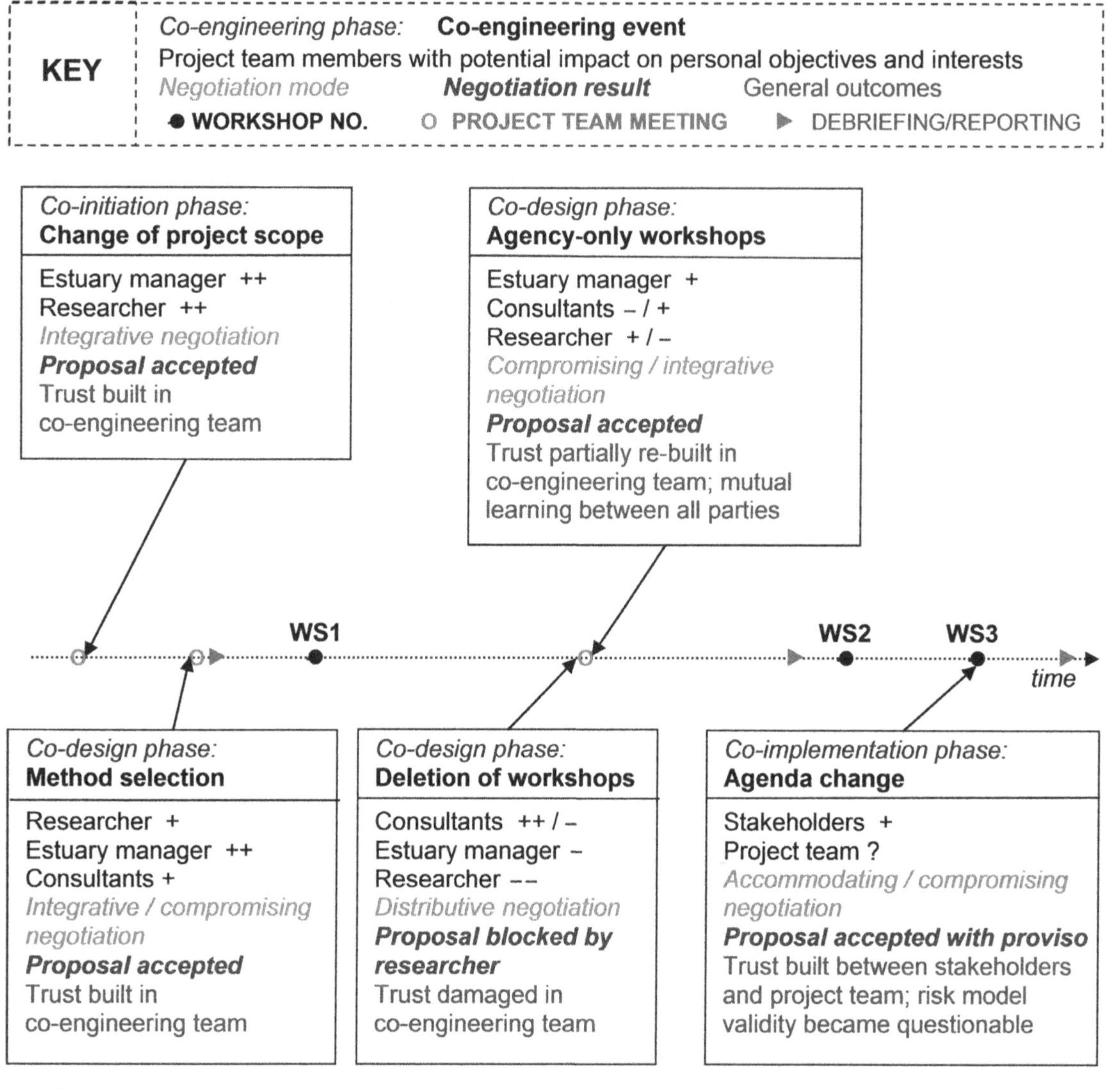

Figure 7.3 Timeline and summary of principal LHEMP co-engineering events. The scale of potential impacts on each co-engineer's personal objectives and interests: + + strongly positive; + positive; +/– mitigated (slightly positive); –/+ mitigated (slightly negative); – negative; – – strongly negative; and ? unknown.

(3) an 'agency-only' workshop, or (4) a workshop with everyone. I agreed with this suggestion. The estuary manager chose the 'agency-only' option, and the discussion moved on to who would be invited. In the end, residents and representatives of local user associations were to be excluded. Representatives of all state and local governments, the Water Authority, Catchment Management Authority, and industry and commercial representatives would still be invited, as they were seen to have the potential financial resources necessary to implement the plan.

FURTHER INTERPRETATION AND *EX-POST* ANALYSIS OF THE NEGOTIATION EPISODE

This last phase of a more collaborative and less confrontational form of negotiation appears to have followed from the estuary manager making explicit his need for the plan to be legitimated by state-agency staff to ensure funding, as well as his recognition of the needs of the agencies and community members. By expanding analysis of this negotiation, using data on the negotiation from my experimental report, an *ex-post* analysis of project documents and emails, and semi-directive and informal interviews that the estuary manager and I had conducted, it could be seen that both the project manager and I highly respected the external evaluator and his needs. This appears to have encouraged us both to become more open to compromise or finding a solution for the workshops that would meet the estuary manager's needs as well as our own. With the need for agency funding support clearly articulated, I was able to look at the overall needs for the process in a different light and began to rethink my own fundamental needs and objectives. As my research project required 'a participatory process' but not necessarily one with everyone participating at all meetings, I saw that

Policy Makers and Managers

State and local government representatives, managers of commercial operations, representatives from the Catchment Management Authority and private water supply corporation

Intermediary Stakeholders

Local government councillors, community associations (i.e. boating, environmental), local residents (including local scientific experts)

Scientific Experts

Environmental engineering and planning consultants, university researchers

WS1: Value (Asset) and Issue (Risk) Identification

Eliciting values, issues (including causes and consequences) with their associated stakeholders and resources and visions and goals for the estuary

- Individual oral presentations, brainstorming
- Small group card classification, mapping, issues/values impact matrices, issue value questionnaires
- Large group discussions

Synthesis Report Production

- Scientific and legislative literature reviews of estuarine processes and current management regimes
- Collate and document the perspectives obtained from WS 1
- Synthesise all information and develop estuarine 'risk tables' (based on AS/NZS 4360)

WS2: Risk Assessment

Assess estuarine risks (related to defined issues) for their consequences on the assets and the associated likelihood of these impacts using the estuarine 'risk tables'

- Determine risk level
- Classify the uncertainty of this prediction
- Evaluate and prioritise risks

Risk Response Plan Production

Develop a coherent table of actions to treat the priority risks

- Condense, sort and analyse stakeholder input from WS 2 & 3 (including risk assessment sensitivity analyses)
- Consolidate workshop production with literature review findings and current or proposed actions of other plans in the estuarine area
- Evaluate each action's potential to reduce estuarine risk levels: identify what actions address which risks and by how much (consequence, likelihood and residual risk)

WS3: Risk Treatment

Define strategies and their associated actions to treat priority risks, as well as stakeholders and resources to carry them out and indicators, monitoring needs and information dissemination strategies to evaluate and improve management

- Strategy mapping and preference distribution on preferred actions

Process evaluation and computer processing of workshop outputs

Exhibition of LHEMP for public comment, plan revision, acceptance and implementation

Figure 7.4 Implemented LHEMP participatory risk management process.

I would be able to change my own objectives satisfactorily if either the 'agency-only' workshop or the 'expert-group' workshop was chosen. At the end of the negotiation, both the project manager and I trusted the estuary manager to make the final decision, as we both knew that the estuary manager recognised our own needs and interests for the participatory process. We also considered that he would make a decision in the best interest of the estuary and the local government, after having assessed the likely impacts of different courses of action outlined in the discussion.

IMPACTS OF THE NEGOTIATION EPISODE

This negotiation episode had a number of effects, which are apparent from evaluation data, both at the level of the co-engineering process and at the level of the participatory planning process (see Figure 1.1). At the level of the co-engineering process, some social learning occurred among the participating co-engineers. The adjustment of my own objectives and perspectives on the participatory process that occurred in the second stage of the negotiation and following it represents one piece of evidence for this social learning. However, the participatory process evaluators also saw that the first confrontational stage of the negotiation had damaged relations and trust between the participants. For example, the principal external evaluator noted tension between affected participants in the second workshop, with the use of body language that appeared partially 'blocking' or negative, even when all remarks and work together remained cordial. I also admitted in an interview with the external evaluator that my trust in the project manager had been reduced, as I did not understand his needs or objectives, but that I was making conscious attempts to try to improve the working relationship.

By reviewing email correspondence during the participatory process organisation and interviews with the project manager and researcher, I realised after the second workshop that the project manager had been under extreme time pressure, owing to a number of other consulting projects that he was concurrently managing. Thus, it appeared that the reasons for wanting to delete the workshops were because of time constraints, and the project manager not being able to dedicate sufficient work efforts to the workshops and to understanding the Risk Management Standard, rather than any fundamental rejection of the participatory process. This furthered understanding, and demonstrates the indirect issues that can be encountered in co-engineering processes.

At the level of the participatory planning process, some of the hypothesised positive and negative impacts of the final decision to hold an 'agency-only' workshop were observed. Firstly, as suggested by the estuary manager, external monitoring showed the choice of inviting only agency representatives created a 'safe space' that was conducive to open, directed discussion of management issues. Participant responses related to this issue to the open-evaluation question, '*Overall, what did you like about the workshop?*' included: '*Good, honest discussion*', '*Open-agency discussion*', '*Different points of view*', and '*[The risk assessment] matrix forced you to work out or question each risk in detail*'. These responses support the view that agency representatives perceived that the workshop provided them with opportunities for frank exchange on management issues, a view further supported by the majority of responses to one of the quantitative questions (to which 13 out of 20 participants responded on a five-point Likert scale – strongly agree, agree, neither agree nor disagree, disagree, strongly disagree), where 11 respondents 'agreed' that the activities in the workshop helped them to share their views and opinions with others, and the other two 'neither agreed nor disagreed.' This level of agreement compared favourably with the other perceived outcomes of this workshop, as outlined in Section 7.5, where further information on these evaluation results is presented.

Nevertheless, as predicted, negative 'backlash' and potential disempowerment of community representatives from the workshop were also observed. One community member phoned the project manager before the second workshop to voice her disappointment about her exclusion and asked if she could still attend. Her request was declined. When the community member came to participate in the third workshop, she described in her evaluation questionnaire how being excluded from the second workshop had been a '*Very disempowering experience*.' She added in the final suggestions section that the government-agency workshop could still have gone ahead, but that the process needed to be explained differently from how it was in the invitation letter, so that the members of both existing local estuary-management committees were not disempowered but retained ownership of the process. Having been contacted by the estuary manager about this participant's disappointment before the third workshop, the evaluator and I were able to focus some of the monitoring activities for the evaluation programme on the previously excluded participants, to mitigate and improve feelings over negative impacts. For example, the evaluator spoke to the community member during one of the workshop breaks to allow the participant to make her feelings about the process known. During facilitation activities, I made extra efforts to encourage her participation, and that of the other previously excluded participants, in the workshop, demonstrating that their contributions were greatly valued. This attention seems to have contributed to the community member's final thought on her questionnaire that '*I am sure the process was well intentioned and I would like to see adoption of the plans by government agencies and the community*.' Additional issues related to co-engineering will be discussed briefly in Section 7.6.3 and further discussion will follow in Chapter 9.

7.4 PARTICIPATORY MODELLING PROCESS IMPLEMENTATION

Following on from the co-engineering process analysis, which focused on roles, relationships and operations that had occurred, in order to realise the LHEMP plan creation process, this section will focus on the implemented participatory modelling process. The descriptions here will briefly highlight the methods used and some of the content results produced through the process. A section on future analyses and research tools that have been

Table 7.2 *LHEMP asset list*

Lower Hawkesbury Estuary assets (stakeholder values)		
Scenic amenity and national significance	Sustainable economic industries	Improving water quality that supports multiple uses
Functional and sustainable ecosystems (including biodiversity)	Community value	Recreational opportunities
Largely undeveloped natural catchments and surrounding lands	Culture and heritage	Effective governance*

* This asset was identified later.

produced in response to issues and needs that emerged in the participatory modelling process is also presented.

7.4.1 Process and results description

The participatory modelling process finally implemented to aid the LHEMP creation can be classified as a '*participatory value-based risk management process*'. This approach, based originally on the inter-organisational decision-aiding process model outlined in Section 5.1.3 and used within the framework of the Australian and New Zealand Standard for Risk Management (Figure 7.2), had the intention of first eliciting stakeholder values and common goals for estuary management, which could be used as a base for reflection for the later definition of improved alternative actions for estuarine management. The elicited values and goals could also be used as the evaluation criteria for the risk assessment part of the process. Therefore, during the first workshop the stakeholders' values and stakes were made explicit in order to use them as a base for finding, evaluating and recommending more desirable management options in the later stages of the participatory process. The stakeholders' 'values' referred to here can take one of two following definitions: firstly, the types of value that are 'held', for example principles, morals, beliefs or other ideas that serve as guides to individual and collective action; and secondly, the types of value that are 'assigned' in line with the qualities and characteristics seen in objects or people, especially positive characteristics (actual and potential) or those that are considered worthwhile or desirable (Mason, 2002). The 'stakes' referred to include the stakeholders' interests or those issues or problems with which they are concerned. This process can therefore be considered as 'value-based' and similar to the improved strategic decision processes proposed by Keeney (1992), which are believed to be superior to traditional 'alternative-based' approaches to decision making (Keeney, 1992). It is noteworthy that the AS/NZS 4360:2004 framework has also been designed and explained with an implied 'value-based' approach to decision making, so it creates a good fit with the decision-aiding process model (Section 5.1.3), where numerous elements of the problem situation and problem formulation are to be theoretically elicited prior to developing alternatives: values may be elicited as part of the set of 'objects' or potentially in the 'resources' set. The process and methods, as implemented in each of the workshops (WS), are presented in Figure 7.4 and briefly described here. Further in-depth information is available in Appendix E, Daniell (2007a; 2007b), Coad *et al.* (2007) and BMT WBM (2008).

The first stakeholder workshop was attended by 30 participants from a wide range of stakeholder groups and state and local government departments. It was used to 'establish the context' of the estuary by eliciting participants' values (assets), goals and issues (risks) related to the estuary (see Figure 7.4), as well as to define the estuarine stakeholders and which resources they possessed or needed, to have an impact on management of the estuary. A variety of individual and group activities were used to elicit and synthesise this information, including: individual oral presentations; individual brain storming; group card categorisation; spatial mapping; issues–values cross-impact matrices; group issue and value questionnaires; and a large group discussion to assemble a list of overall stakeholder values and general visions or goals for estuarine management. In particular, the individual and small group activities had been designed to give all participants the opportunity to participate and to limit undue focus on just a few issues, known regional conflicts and domination of the discussion by vocal participants.

Prior to the second workshop, I analysed the outcomes of the first stakeholder workshop and drew up a report, and the engineering and planning consultants carried out and produced a document review of the current knowledge of estuarine processes, risks and management and planning legislation impacting the new estuarine management plan area. The final list of nine estuarine values (labelled as assets for use in WS2) is shown in Table 7.2. Eight stem directly from WS1 and one more (marked with * in Table 7.2) emerged from the document review process.

These were then used to produce 'risk consequence tables', as outlined in Appendix E (Table E.6), to be utilised in the next workshop. Tables for 'likelihoods', 'risk levels' (based on a combination of consequences and likelihoods (Wild River and Healy, 2006)), 'knowledge uncertainties' and 'management effectiveness' were also produced, as described in Appendix E (Tables E.7 to E.10). This collection of 'risk tables', the document review and the outcomes of the first workshop were then distributed to stakeholders as the synthesis report (BMT WBM, 2007), for their consideration shortly before the second workshop.

The second workshop, attended by 19 participants, was used to: obtain 'agency' (government department, industry and

Table 7.3 *LHEMP risk list*

Lower Hawkesbury Estuary risks (adapted issues elicited from the stakeholders and external review)		
Water quality and sediment quality not meeting relevant environmental and human health standards	Climate change	Residents and users lacking passion, awareness and appreciation of the estuary
Inappropriate land management practices	Excessive sedimentation	Regulated freshwater inflows
Inappropriate or unsustainable development	Over-exploiting the estuary's assets	Inappropriate or excessive foreshore access and activities
Inappropriate or excessive waterway access and activities	Introduced pests, weeds and disease	Inadequate monitoring to measure effectiveness of the EMP
Inadequate facilities to support foreshore and waterway access and activities	Insufficient research	Not meeting EMP objectives within designated timeframes
Inadequate or dysfunctional management mechanisms*		

* This risk was identified later.

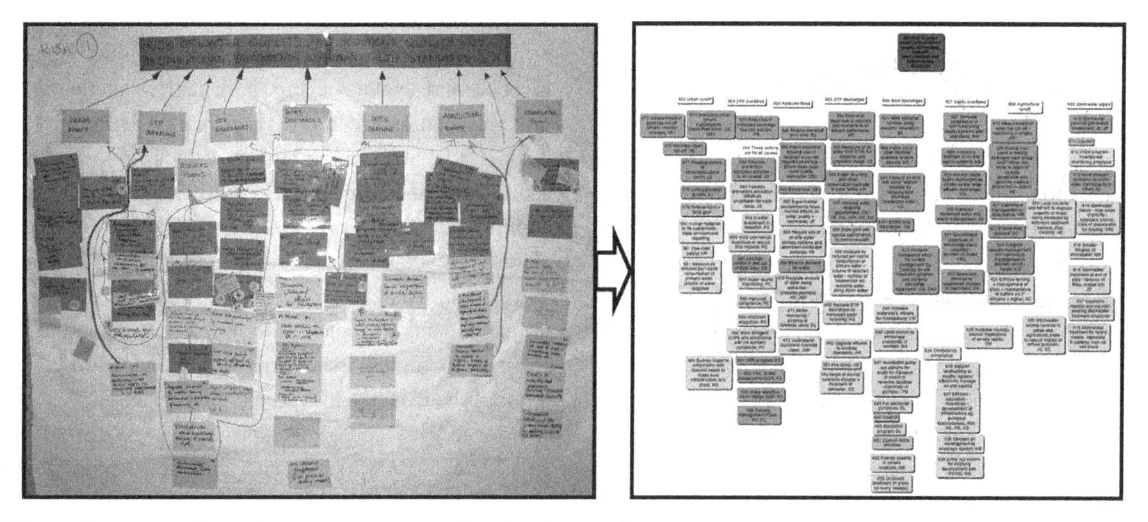

Figure 7.5 Example paper-to-electronic strategy map conversion for plan creation.

commercial representatives) support for the stakeholder-defined values (assets); further identify the risks elicited in the first workshop and an external document review; and then perform a 'risk assessment', to prioritise the estuarine risks for subsequent treatment, as shown in Figure 7.5. Fifteen risks, as outlined in Table 7.3, were discussed and assessed through facilitated large and small group sessions, using the specifically developed risk tables. One further risk (marked with * in Table 7.3) was also discussed and added to the list.

The risk tables were used to help the participants to identify the consequences and likelihoods of risk impacts on nine previously defined estuarine assets, as well as an associated risk level, the knowledge uncertainties related to these classifications, and the level of current management effectiveness of the risk related to each asset. From this information, the priority of the risks – acceptable, tolerable or intolerable – was computed and presented, and the participants given some time to discuss the results. From this assessment, all risks were classified as requiring treatment (tolerable or intolerable). These risk priorities were also reviewed at a later date through a stakeholder email survey, sensitivity analysis and alternative calculations, as outlined in Section E.4, Coad *et al.* (2007) and BMT WBM (2008).

The third workshop was attended by 17 participants representing both agency and other stakeholder interests. This workshop was used to develop strategies and actions for the treatment of all 16 risks, as well as to identify monitoring needs, stakeholder responsibilities and stakeholder preferences related to the proposed strategies and actions. Individual brainstorming on cards of strategies and actions preceded the collective visual 'strategy mapping' exercise for each risk (similar to Ackermann and Eden's (2001) oval mapping technique) and preference distribution. Throughout this workshop, over 900 elements were built into the 16 strategy maps.

As part of the participatory process, participant evaluation questionnaires consisting of approximately 15 open and closed questions were completed at the end of each workshop (50–70% response rate), and related to a variety of areas, including: whether objectives were met; learning outcomes; what was useful; and what could be improved for future workshops or similar processes. External evaluations to further examine the context, objectives, process and results of the project were also carried out, as outlined in Section 7.2.3.

7.4.2 External analyses, final results and possible process futures

EXTERNAL ANALYSES AND FINAL RESULTS

After the final workshop, the strategy map information was computerised using DecisionExplorer® software, as shown in Figure 7.5, and exported to MSExcel to produce a preliminary stakeholder-based risk response (action) table, as outlined in Daniell (2007b).

This preliminary table was then considered and compared with existing management plans and regional strategies by the consultants, and a final table of 'risk-response' actions created. The final planned actions underwent a secondary risk assessment based on the same stakeholder value (asset) list to determine their potential efficacy for treating the estuarine risks (Coad *et al.*, 2007; BMT WBM, 2008). The draft LHEMP was then put on public display and contained 149 strategies for treating the 16 risks. Of these strategies, 32 were outlined as short-listed strategies, which were suggested as having high implementation priority in terms of risk reduction potential (BMT WBM, 2008). These 32 strategies are given in Appendix E (Table E.14). After a few minor adjustments to the plan following comments received, the final plan was accepted by both the Hornsby Shire Council and the Gosford City Council in March 2009.

POSSIBLE FUTURES

As part of this project, a number of potential futures were considered for extending the research part of the work and developing support tools to aid further regional negotiations and management decision making. One possible future is to upgrade and expand the Berowra CLAM, a Bayesian belief network simulation model for 'coastal lake assessment and management' (Ticehurst *et al.*, 2005; Coad *et al.*, 2006) to be compatible with the new information and risk scenarios outlined in the LHEMP process and to expand it to cover the whole Lower Hawkesbury Estuary. Such a tool could be used to investigate management scenarios using the best current expert and stakeholder knowledge available and take advantage of the information received on current levels of knowledge uncertainty surrounding certain estuarine processes and risk consequences and likelihoods. If such a coupling of the risk management process, and in particular the risk assessment results, could be linked to the CLAM, this could open the door for many other regions to develop similar risk assessment processes to improve their own estuarine management models and the other existing CLAMs. The estuary manager has been discussing these possibilities with the CLAM designers and licensed developers over the past few years and hopes to start the expansion when funds become available.

Another decision support tool, which I developed just after drawing up the stakeholder-based risk response table, which could have both practical and research-orientated futures, is an 'actor–action–resources matrix', as shown in Figure 7.6.

This matrix was designed to help analyse the distribution of actors' preferences for the strategies and actions for risk treatment created during Workshop 3, and to show to what extent these actors actually had authority to control these preferred actions or the resources to realise them. The eventual aim of the tool was to discover where there were mismatches between priorities and capacities, so that actors could realise their preferred actions and to determine whether opportunities existed for bi-lateral or multi-lateral negotiations to discuss how any mismatches might be overcome and resources distributed efficiently between actors to meet the most priorities possible. Ideally, actors would be given the opportunity to fill out their own levels of *authority* ('direct authority', 'indirect authority', 'shared indirect or potential authority' or 'potential influence on authority') and their *available resource level* ('available resources', 'obtainable resources', 'scarce or difficult to obtain resources', 'resources unlikely') when stating their preferences on the actions. This would enable actors to obtain adequate information that could then be graphically represented and analysed to aid future negotiations and, potentially, responsibility distributions for the final plan. However, as I only created this tool during my analyses of the workshop outputs after I had left Australia following the end of the workshops, the tool was not tested for the LHEMP. When I next returned to Australia, I discussed the tool and its use with the estuary manager and it was thought that it might have some potential, even if it was only used for internal analyses and estimates of the actors' authority and available resources levels, to help gain some personal insights. I conclude that this simple visual tool could have a future use in aiding multi-stakeholder or inter-organisational planning, but it will require more research and testing, especially on the pertinence of the categorisations.

Finally, another potential future for the procedures used in developing the LHEMP process could be to further analyse the process and determine to what extent it could be improved and

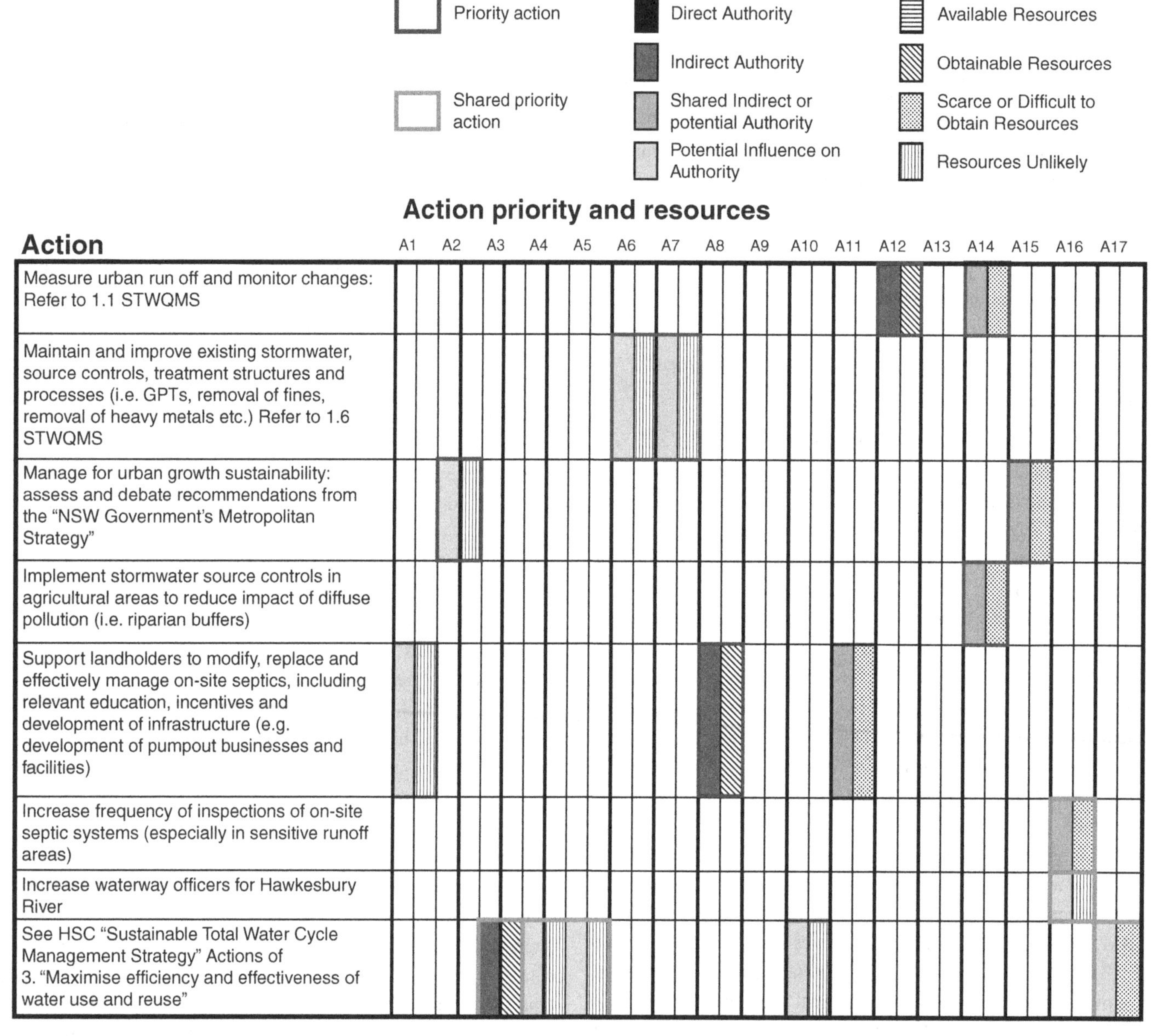

Figure 7.6 Screenshot of the actor–action–resources matrix tool trialled for the LHEMP.

transferred to other estuaries, water management issues or entirely different management issues, such as carbon and energy management, not only in New South Wales and other Australian state contexts, but also in international management contexts.

In looking at possibilities for improvement, perhaps the most pressing issue is to improve the mathematics underlying the risk model, to make the outputs more defensible and meaningful from a decision-aiding perspective. In the LHEMP process, the time and resources to create anything much better than an unweighted average model were unavailable. Methods to elicit preferences and weightings for use in scientifically sound multi-criteria models for use within the participatory values-based risk assessment approach, or 'multi-asset models', would therefore be a useful potential research topic.

7.5 EVALUATION RESULTS AND DISCUSSION

This section will briefly outline the some of the evaluation results obtained related to the LHEMP process and provide a discussion, firstly, of how the decision-aiding model was used and the adequacy of the process, secondly, of a number of selected participant and external observation evaluation results, some related to aspects of the ENCORE model dimensions, and finally, of the overall intervention outcomes in terms of effectiveness, efficacy, efficiency and innovation.

7.5.1 Use of the decision-aiding model

The final summary of the elements elicited from the inter-organisational decision-aiding model outlined in Section 5.1.3 and how the sets evolved through the process are outlined in Table E.4. The model was utilised to a much greater extent in this process to direct my design considerations than in the Montpellier methodology trial. First of all, it formed an explicit basis for the four stage participatory modelling process that was originally proposed to the estuary manager and formed the basis of discussions until the Risk Management Standard AS/NZS 4360:2004 was introduced into the co-design process. After this point, the decision-aiding model was used as a personal check to aid the specific design of the workshop supports and processes, in order to obtain as many elements as possible of the decision-aiding model and to be able to mobilise them in a useful manner. For example, in the first workshop support resources of the 'issue' and 'value' sheets, questions related to resources – in particular management authority and information or data – were added to obtain a set of resources related to the actors and objects mentioned. After each workshop, an inventory of the specific elements that had been elicited or developed from the workshop was drawn up, along with notes on whether they had been adapted or further confirmed and were adequate to aid the collective decision process. For example, one set of elements planned in the second workshop was the participants' weightings on the assets – the 'set of criteria' in the decision-aiding process model. Owing to time limits in the workshop, these elements could not be obtained but the participants did not seem concerned when the subject was raised. Similarly, the operators for the risk prioritisation model, which I had developed, were very briefly discussed, but the participants seemed happy for the 'experts' to create the mathematics and for them not to worry about it – especially as they all may have considered that mathematically they could understand averages! It seemed at this point that I was the only one who was concerned by the potential lack of scientific validity and meaning of the resultant numbers because I had created the model and understood its weaknesses more clearly. However, after the results of the risk prioritisation were presented and discussed with the participants, the project manager showed his dissatisfaction, in part because of one of the counter-intuitive results. Later, after the third workshop, WS3, the project manager took a much greater interest in the mathematics behind the model and adapted it for use in the calculation of the risk reduction potential of the strategies. Surprisingly, the adaptations to the mathematics did not change the counter-intuitive result, and the same critiques about scientific validity and the meaning of the resultant numbers still hold.

Apart from the criteria and operators, most of the other elements from the decision-aiding process model were either developed by, or their contents elicited from, the participants. For example, despite the fact that the dimensions and scales had been developed externally, based on the examples in the risk management standard, the participants populated them with data values for each asset, so they were partly responsible for the results of the model's prioritisation.

A further observation on the use of the decision-aiding process model is that the first 'evaluation model' created and utilised in the process was actually used to define the problem statements, not to examine the effects of alternative actions. Examination of these effects was carried out using a similar evaluation model after the end of the participatory workshops. In other words, the matrix of potential actions and sets of alternative actions for final evaluation were developed after almost all of the elements of the other sets in the decision-aiding model had been elicited. This highlights the iterative nature in which the decision-aiding process model may be employed, so that each set does not need to be elicited specifically one after the other. The fact that the values elicited as part of the 'objects' or 'resources' set then became the 'points of view' to be used for the evaluation model confirms the idea that this decision process was indeed 'value-based' and not 'alternatives-based'.

7.5.2 Selected participant and external observation evaluation results

This section will outline the evaluation results based on participant and external observations on a few key issues, including: whether the approach used was able to achieve one of its goals of obtaining a common set of values on which the rest of the process could be based; to what extent the methods used were able to manage known conflicts; to what extent the process was able to 'capitalise on existing stakeholder and documented knowledge, including previous estuary studies and management plans'; and to what extent the process was able to 'encourage increased understanding, knowledge sharing and learning between stakeholders to enhance future collaborations and the capacity to manage the estuary effectively into the future'. Further results from the participant evaluations are used to inform the following section on the 'overall intervention outcomes' and more are presented in Appendix E. It is noted that the majority of stakeholders who participated in this process had some previous experience of working in participatory settings and that no one was paid by the project team to participate or cover their costs of attendance.

Firstly, responses from the participant questionnaires were examined to gauge the extent to which the approach used allowed participants to express their values and combine them into a common set, upon which the next stages of the process could be based. The majority of participants in the first workshop, WS1, asserted that this objective had been achieved, although responses from a couple of the local government representatives were more reserved, including: '*Yes – however there are lots of differing opinions on what is important*,' '*There is a complacency within the community that seems keen to portray the ecosystem as healthy despite evidence to the contrary*,' and '*Some issues and values were difficult to confine going from an individual to group situation*.' These responses demonstrate that there did appear to be some normative value-based or potentially cognitive beliefs and world-view conflicts between the participants. Furthermore, it introduces the challenge of defining 'collective' rather than 'individual' values. However, by the end of the first workshop, a set of values had been collectively constructed that received little sustained criticism and did form the base of the risk assessment process for both prioritisation of risks and the individual assessment of the strategies derived in majority from the stakeholders' work.

Secondly, in terms of managing the known key conflicts in the region, one of the management agency representatives in the first workshop, WS1, who had been identified by the estuary manager as likely to come up against the most hostilities, especially from the community stakeholders, stated that it had been '*not too confrontational*'. One of the external evaluators commented on improvements in body language between participants between the first workshop, and the second and third ones, where participants, including those who were originally very wary of the conflicts that could flare up during the process, appeared '*more relaxed, less defensive and more open to contribute to the process*' (White, 2007; personal communication). In a final interview, when asked about how the approach and methods used had helped to manage conflict, the estuary manager also backed up this external observation, saying that he had seen amelioration of the severity of the conflicts that he was aware of at the start of the process and that the '*Workshop structure was very well designed so that we didn't have those conversations in an open manner or else we wouldn't have got anywhere near getting this outcome of the LHEMP*.' He also mentioned: '*I think this process, for one of the first times, pulled [the agency in the conflict] out of 'we do everything internally' to actually do something that was external to their organisation ... this was the first time I think the community had the opportunity actually to work with [the agency] on some issues, so I think it was good for both parties*' and stated that he was very pleased to see these involved parties working together by the end of the process.

Thirdly, when examining the objective of whether the process was able to 'capitalise on existing stakeholder and documented knowledge, including previous estuary studies and management plans', a number of preliminary conclusions can be drawn from the participant evaluations, including: that the workshop process '*Established a range of expertise and views of other government stakeholders*,' (WS2, state government representative); that '*It [the workshop] attracted a range of people with different interests and skills*,' (WS3, community representative); and that through the workshop process '*Good supplementary information was generated that could add value to a comprehensive strategy review*,' (WS3, management agency representative). On the other hand, the same management agency representative from WS3 also noted: '*It is extremely difficult to tap local 'expert' knowledge in a way that is useful and where the data collected can be retrieved*.' However, upon a further *ex-post* evaluation interview of the effectiveness of the participatory workshop process, it was highlighted that the stakeholder community coverage of issues had been better than expected, to the extent that very few actions or important documents covered in the subsequent consultant management literature review had been left out of participant comments (Coad *et al*., 2007; personal communication).

Finally, to determine the extent to which the process was able to 'encourage increased understanding, knowledge sharing and learning between stakeholders to enhance future collaborations and the capacity to manage the estuary effectively into the future', a number of participant responses to both closed and open questions at the end of each workshop can be examined. The collective participant responses to the closed questions looking at the comparative effects of all three workshops on participants are outlined in Figure 7.7.

From Figure 7.7, apart from 'aiding creativity and the creation of new thoughts and ideas' in Workshop 2, participants tended to agree that the workshops helped them to get to know others, share their views and opinions with others and, to a slightly lesser extent, aided creativity and the creation of new thoughts and ideas. These quantitative results were further supported by participant comments, including: '*I was able to listen to and consider other opinions and also had the opportunity to build on other people's basic ideas*,' (WS3: local government representative); '*It [the approach] gives everyone a feeling of 'being heard' and ownership*,' (WS3: community representative); and that the workshop process '*Provided a good ground for cross pollination of ideas and perspectives*,' (WS3: community representative).

The participants rated their depth of learning over the three workshops, and this is shown in Figure 7.8. It appears that the more heavily structured risk assessment process in the second workshop did not seem quite as conducive to learning about any of the three areas: management of the estuary and its surrounding environment; other participants in the group; or participants themselves (or their opinions and practices). However, a number of participants noted in their evaluation forms that they had learnt the most in that workshop about the actual '*risk assessment process*' (WS2: environmental agency representative) and through using it that '*There are many, many, interrelated issues impacting on estuary, regulated (or not regulated) in many ways*' (WS2: state government representative). In looking again at

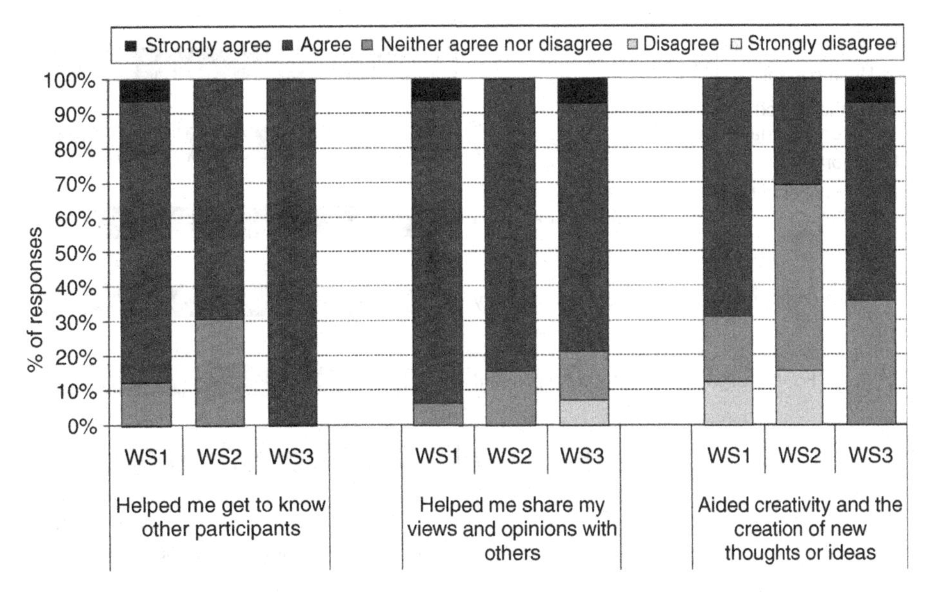

Figure 7.7 Participants' perceived effects of the three workshops.

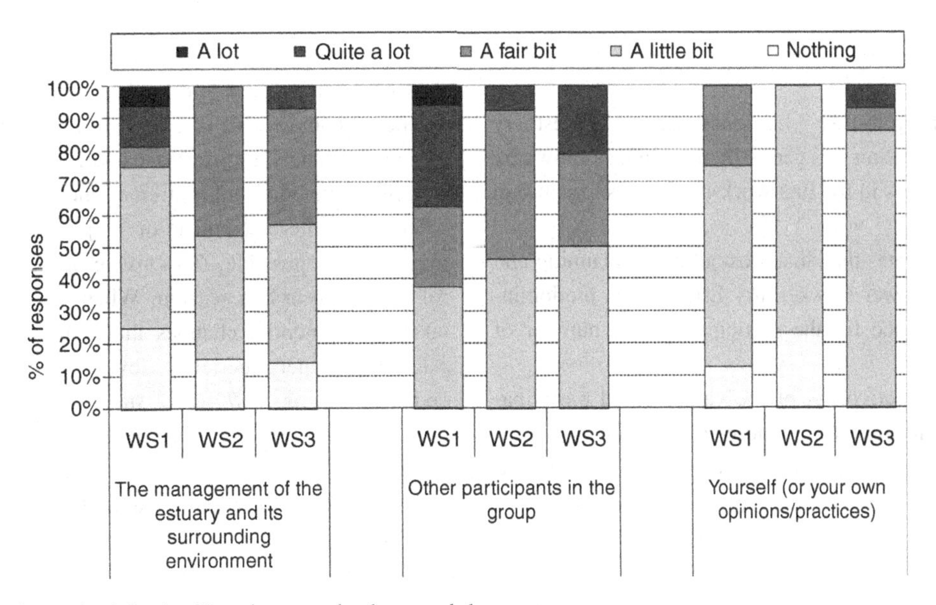

Figure 7.8 Participants' perceived depth of learning over the three workshops.

Figure 7.8, the first workshop appeared to produce the largest learning outcomes related to the other participants in the group and the third workshop's activities seemed conducive to the participants' greater learning about themselves and their own opinions or practices. At the end of Workshop 3, one local government representative stated having learnt that: '*There is no one right way to address identified risks. Collaboration is essential.*' In his final interview, the estuary manager also noted that this process was the first time that many of the agency staff had worked together and with the community and that he was pleased to see them working well with one another. This situation appears to have been translated into 'relational' learning, as Figure 7.8 shows that the participants thought that they had learnt more about other participants in the group, rather than having experienced (on average) more cognitive, normative or operational types of learning. The majority of these quantitative and qualitative results appear to support the hypothesis that the designed participatory value-based risk management process has helped to 'encourage increased understanding, knowledge sharing and learning between stakeholders'. However, whether this will prove sufficient 'to enhance future collaboration and the capacity to manage the estuary effectively into the future' must be assessed at a later date.

7.5.3 Overall intervention outcomes

A number of more general results concerning the effectiveness, efficacy and efficiency of the participatory modelling process implementation will now be examined, followed by a number of innovations that were made throughout the process.

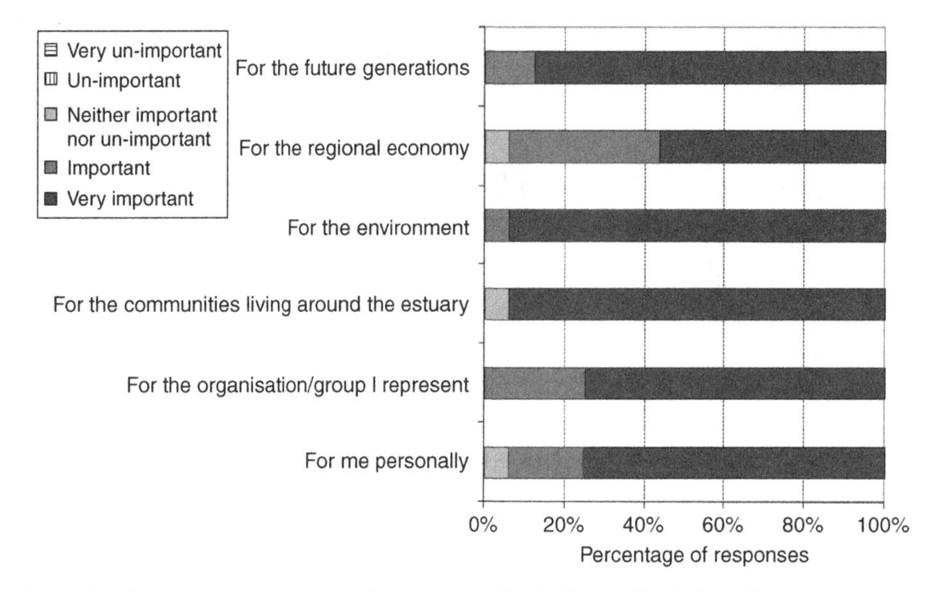

Figure 7.9 Participants' perceived importance of planning and management for the Lower Hawkesbury Estuary.

EFFECTIVENESS

In terms of the decision to take a 'regional' approach to estuary management, legitimisation was generally received, as shown by the participant responses in the first workshop's questionnaire in Figure 7.9.

Figure 7.9 demonstrates that, in almost all cases, planning and management of the Lower Hawkesbury Estuary was predominantly of high importance for the participants on a number of fronts.

Considering the effectiveness of the values-based participatory risk management approach, a number of comments were also gathered from the participant evaluations, including: that the approach '*Supported the understanding that the community is made up of people with different values and perspectives – each needs to be considered and valued.*' (WS1: community representative) and '*Focused participants to common criteria/ objectives*' (WS1: state government representative) – which was a major aim of the 'participatory values-based' part of the risk management approach, as outlined in the previous section; that risk assessment '*can be subjective and outcomes would be very different given stakeholder [community representatives'] participation*'; and that '*Prioritisation of objectives [for estuary management] will be achieved – whether or not this is a true indication of priorities is another matter.*' These comments highlight some of the difficult choices and trade-offs that need to be made within the constraints (i.e. time, budget, existing knowledge and available methods) of the LHEMP planning process. Each spatial and risk scale chosen has its advantages and disadvantages, as do the methods used. For example, as highlighted by one of the participants, participatory risk analysis is often a fundamentally subjective process, thus who participates, and how, can have an important impact on the outcomes. This can be viewed positively or negatively, as the risk analysis process can be time- and cost-effective, especially in cases of extreme uncertainty and complexity, where other more scientific or 'objective' methods of analysis may not be possible. This issue is expanded upon in the following discussion session. When asked about the approach taken and potential changes that could have been made, the estuary manager queried a few potential elements but still ended up stating: '*I think we still used the best tools we could have,*' considering the objectives of, and constraints on, the process.

It was also noted by participants '*that you have a process that you are working on to reach a conclusion amicable to all participants and the community*' (WS1: commercial representative) and that the LHEMP process is a '*very ambitious project but clearly [there are] many stakeholders on board, improving likelihood of success*' (WS2: state government representative); comments that suggest at least a reasonable level of current process effectiveness. However, one community representative in Workshop 3 also noted that: '*Broad input was achieved but truly effective solutions are elusive because underlying pressures can't be addressed,*' referring potentially to the difficulty of working and planning at a regional scale for treating such issues as population growth and climate change causes and effects, where there is a need to work with higher levels of government and international policy makers. Only time will tell if the approach was the 'right' thing to be doing and if the participants and the stakeholders to the process adopt the plan and act to work towards the sustainable development of the estuarine region.

EFFICACY

Efficacy here will be analysed based on the extent to which the research objectives outlined at the beginning of this chapter have been achieved. Firstly, the objective to '*create a regional estuary management plan using a participatory modelling approach with actions collectively agreed upon by major stakeholders*' is still in the process of being achieved. The participatory modelling approach could be implemented in the form negotiated between the research estuary manager and the project manager, with the major deviation from the original methodology being the decision to hold the risk assessment activity as an 'agency-only' workshop. The 'agency-only' workshop did still ensure that the whole of the modelling exercise was carried out in a 'participatory' manner, albeit without all the community stakeholders, and currently appears to have helped a number of 'major stakeholders' to support the process. This issue has been outlined in Section 7.3 and is further discussed in Appendix E. From a recent follow-up, it appears that key stakeholders, including community members, have continued to be supportive of the plan, with many of them now participating in the new Lower Hawkesbury Estuary Management Committee, which was created as one of the first steps of the plan's implementation. Many of the plan's priority actions have already been implemented within the first couple of years since it was drawn up. This successful implementation has been supported by over half a million dollars of funding, mainly from state government sources (HSC, 2009; 2010). Further priority actions are set to be implemented in the coming year, as more funding has recently been allocated.

Next, the objective to '*assess the capacity of a participatory modelling process to produce the plan in the local context related to its efficiency, effectiveness, efficacy and other effects, such as learning, innovation and conflict management*' has mostly been achieved. Evaluations were undertaken through a variety of methods: participant questionnaires; external and internal process and participant observation aided by audio and video recordings; document analyses including emails, meeting notes and workshop outputs; semi-directive interviews with the project team members; and informal interviews with participants, which have allowed the analysis and triangulation of many aspects outlined in the objective. The evaluation of levels of learning and conflict management through the process has already been outlined in Section 7.5.2, and the questions of efficiency, effectiveness, efficacy and innovation are being treated in this section. Further evaluation results are outlined in Appendix E. Effectiveness of the overall plan in improving estuary management on the ground remains more challenging to evaluate, owing to the long timeframes required and the complexity involved in monitoring the effects of implemented actions (Coad, 2011; personal communication). Although some monitoring programmes have already been implemented as actions of the plan to gather data on some aspects of the estuarine system, the ultimate pertinence of the planned actions and of the risk approach will have to be gauged by future evaluation research.

Finally, the objective to '*gain a greater understanding of real-world institutional and water issues in the Australian context, including the positions taken by, and constraints placed on, researchers, community stakeholders, private businesses and governments in planning decisions*' was fulfilled to a large extent. Being able to participate in 'intervention research' meant that I had access and in-depth confrontation with real-world institutional and water issues in an Australian context. Moreover, the negotiations required in a 'multi-institutional' project management team were illuminating and highlighted the different constraints of private consultants, researchers and local government staff, as well as their interactions with the community. Many of these insights have been outlined already in this chapter and will be discussed further in the later parts of this book but here are a couple of brief examples. A common belief still remains in many of the professional sectors that 'objective' scientific knowledge can be obtained to support management decisions, which is a belief not shared by all community groups or researchers. The use of a 'subjective' risk approach, which was used by the managers to add 'objectivity' to the decision-making process, aided in highlighting these issues and conflicts of opinion. In addition, the difficulties in including scientific knowledge in a participatory approach, where there is limited time to discuss any issue in detail, were highlighted. In this case, carrying out the external scientific and management review proved exceedingly valuable for helping to overcome this potential deficit of carrying out time-limited participatory processes for broad-scale, multi-institutional level water planning and management. It was also interesting to understand how Australian laws and policies, such as the NSW Occupational Health and Safety Act 2000, can actually limit community participation in local water management, as the costs incurred by local councils to allow community volunteers to participate in management activities on public lands or waters, such as monitoring or clean-up activities, can often be prohibitive.

EFFICIENCY

Regarding evaluation of the extent to which the LHEMP process was efficient in its implementation and production of results, the answers rely strongly on the metrics used. The LHEMP process took about 18 months to arrive at a draft plan proposal for public comment, which is short compared with the Brooklyn Estuary Management Planning process (a sub-section of the LHEMP area) based on the NSW Estuary Management process (NSW Government, 1992), which took over five years from the start of the Estuary Processes Study to the time that the Draft Estuary Management Plan was made available to public comment (Coad *et al.*, 2007; personal communication). Although the scales and intricacy of assessment are different in these two processes, the outcome of a draft action plan is similar, making this an interesting efficiency comparison. Similarly, total project costs (from a local government point of view) appear favourable compared with other similar scale planning processes (Coad *et al.*, 2007; White, 2007; personal communication). The time dedicated by some of the agency staff to the process was potentially longer than the time they would typically spend on estuary management plan creation activities (1–3 full days), although from their evaluation questionnaire responses, the value of the process mostly appeared to outweigh the time costs. Overall, these evaluations of efficiency on resources of time and money appear to indicate that highly participatory processes can prove to be more efficient than other less participatory and expert-driven planning processes, which goes against the results of many previous evaluations and common perceptions of participatory processes, including specifically 'participatory modelling processes' that are available in the literature. One of the external evaluators also continued to mention how well the risk management process and the methods chosen for it had helped to focus the participants' activity, thereby improving efficiency (White, 2007; personal communication). Despite this apparent efficiency and efficacy, a few agency staff still mentioned to the estuary manager their general intrigue at the lengths the project team went to create such an innovative and participatory process, wondering if it was a little too much for the objective of '*just creating a regional estuary management plan*' (Coad *et al.*, 2007; personal communication). These lengths could potentially now be seen as a valuable use of resources, considering the success the plan has had in attracting significant funds to the estuary and in further building community engagement around estuarine management (HSC, 2009; 2010).

INNOVATION

Through this process, a number of new forms of collective action and knowledge emerged. First of all, a 'multi-institutional level stakeholder working network' emerged through the regional LHEMP planning process, whose members collectively acted to construct the LHEMP, aided by the input of process instructions and artefacts introduced into the interaction space by the project team. In other words, this was highly supported collective action, where the project team largely organised how interactions would take place, by setting and adapting the rules that governed the participants' permissible actions and the objects entering the participants' interaction space. On a few occasions, this organisation was jointly shared by the participants and the project team, such as in the second workshop where the participants negotiated a process change with the project team. By the end of the three workshops, this collective action had resulted in a comprehensive and integrated picture of visions, values and risks associated with the regional estuary management, as well as the potential causes and effects of the manifestation of these risks and many strategies and actions to overcome them – new forms of synthesised knowledge for the regional estuary scale.

Another interesting form of collective action that occurred could be considered as a sub-section of the first form of collective action, when only the agencies (local government, state government, industry, commercial and regional association representatives) were working together on the risk assessment process and showed signs of appropriating the method and using it to drive their own coordination mechanisms and discussions on estuarine management issues. Whether the risk assessment process in itself was useful for structuring this collective action or whether it was just a good excuse for them to benefit from all being in the same room in a 'safe space' away from potentially critical community members and 'overbearing' agency managers, where they felt confident to use their own jargon and debate more 'technical' and 'managerial' issues, such as those related to estuarine water quality, is difficult to estimate. In either case, the workshop succeeded in ensuring a form of collective action that is rarely seen: an inter-organisational and multi-level governance stakeholder attempt at using the Australian Risk Management Standard for broad-scale estuarine risk assessment based on assets defined by regional stakeholder communities. It also led to the definition of the whole process being considered as an example of 'participatory value-based risk management', which will be discussed further in Section 7.6.1.

The second form of collective action and associated knowledge that emerged through the process, related to the first one, was the collective action of the 'inter-institutional project management team' who aided the organisation and collective action of the 'multi-institutional level stakeholder working network'. In this team, relations had been created between local government managers and environmental scientists, private engineering consultants specialising in estuarine processes and management, private planning consultants specialising in statutory and urban planning and researchers specialising in participatory methods and evaluation for natural resources management. Understanding the dynamics of how each member's knowledge, values and previous experience were mobilised through negotiations and translated into individual and collective practice, including synthesised knowledge products, was most illuminating. It was this collective action, along with the collective action of the stakeholder network it was co-organising, that created the new regional draft Lower Hawkesbury Estuary Management Plan in just over one year. A number of insights from this form of collective action will be discussed in Section 7.6.3, and issues specifically related to complexity and its effects on synthesis and integration discussed in Section 7.6.2.

7.6 DISCUSSION AND FURTHER INTERVENTION INSIGHTS

A number of key themes that merit further discussion have arisen from the results of the participant evaluations and other observations of the LHEMP process and its preliminary outcomes. Just three of them will be outlined in this chapter: the advantages and disadvantages of the risk assessment approach; complexity and its impacts on synthesis and integration in projects such as the LHEMP; and general comments on participatory process implementation. Other issues related to the LHEMP are discussed in Chapter 9 and Appendix E.

7.6.1 Advantages and disadvantages of a participatory values-based risk management approach

As outlined in Section 7.4.1, the LHEMP process was specially crafted to meet the needs of the local context and the stakeholders' objectives: in particular, a direct linkage was created between the stakeholders' list of values in the first workshop and the assessment process of Workshop 2, where the values became the 'assets' upon which the risks were evaluated. The approach developed for this process can thus be thought of as 'value-based participatory risk management'. This section will discuss a number of advantages, disadvantages and lessons that have been learnt, which may help to improve the repeat of such a project in a different context.

Firstly, participatory risk assessment is an inherently subjective process (Stirling, 1999), especially in this broad estuary management context, even if it attempts to categorise knowledge uncertainties related to the risk assessments. One of the principal advantages in using such an approach is to aid stakeholders to better understand the nature of risks through developing a common (value-based) assessment of them and then to use this method as a basis for determining priorities for risk treatment. An approach of a participatory nature is also suggested by Klinke and Renn (2002) for use when there is a potential normative conflict related to values amongst the stakeholders owing to ambiguity in the definition of the problem situation and a need for trade-off analysis of the risks and deliberation. As this type of risk assessment can be particularly subjective, it is thought that in many contexts, such as the LHEMP, all stakeholders have just as much potential to contribute to it, especially as some of the assets for which the risks were to be assessed, such as 'scenic amenity', while fundamentally important, are not particularly technical. In particular, it was thought that many of the 'community' representatives in the LHEMP process appeared to possess more in-depth knowledge or scientific expertise on the estuarine system, industries and community values than some of the agency staff external to the estuary. For this LHEMP process, the exclusion of some community members from the risk assessment exercise of the second workshop therefore had a number of ramifications, both negative and positive. The negative results included that some competent local experts could not input their knowledge into the risk assessment process, potentially reducing its effectiveness in terms of scientific basis for risks where consequences and likelihoods could be based on technical knowledge, and in terms of overall community legitimacy. The positive results were the improved relations between management staff in the region, more open and frank debates about management effectiveness, which sometimes do not occur in the presence of critical community members, and improved support of the LHEMP process from some key management stakeholders who had previously been absent from regional management discussions. Ideally, a fourth or fifth workshop to work further through the risk assessments and treatment options with the community and agency stakeholders would have been desirable, but insufficient resources were available for the project.

One of the lessons learnt from this experience is that risk assessment exercises of this nature will always be biased by who participates and the extent of their knowledge – this includes all types of knowledge such as local, technical, legal, managerial or political – so it is important to include the most capable and knowledgeable people, as well as those required to support and legitimise the outcomes of the assessment. What is ideally sought through such a process is an 'inter-subjective agreement' for action. Great care and attention should therefore be taken when organising such a process so that the most relevant participants are able to take part to ensure the success of the assessment results, both in terms of stakeholder legitimisation and scientific validation. However, independently of which group of stakeholders (or even external experts) carry out the 'risk assessment' part of the risk management process, it is thought that the first steps used in the LHEMP process, for establishing the initial context and the definition of values or 'assets', could provide a number of advantages for quality stakeholder participation where the participants have the opportunity to influence the future direction and focus of the planning process. The influence is easy to trace, as the risk analysis subsequent to the initial context establishment is based entirely on impacts to 'stakeholder community' agreed values. This means that the risk impacts examined will be analysed against what values are the most important for the stakeholders.

As this approach, and its application to the complex and ambiguous water sector problems, is still in its infancy, it is believed that further investigation into the theoretical and practical benefits and constraints of the approach is warranted. Potential questions for future research could include:

- How does participation and negotiation during the workshop process shape the final set of community-defined values and to what extent do the participants really share them (i.e. co-construction of 'utilities', as used in some forms of economic and decision theory)?

- To what extent do community and shared risk assessment and acceptance differ from individual assessment and acceptance? For example, there may be only a partial agreement on the values (criteria) and assessment if there are varying views on outcomes (likelihoods, consequences, etc.).
- What effects could changing the order of decision steps in the approach have on outcomes (i.e. defining risks first and then the value criteria for their assessment afterwards)?
- What are the dependencies within the decision process stages and to what extent can causal links between factors be mapped within them or in the real-life contextual situation (i.e. to aid the construction of integrated water decision-aiding models or decision support systems)?

7.6.2 Complexity and its effects on synthesis and integration

Estuary management, like many other management situations in the water sector, is a process characterised by interconnecting and complex problems, which exhibit high levels of conflict and uncertainty, as outlined in Chapter 2. Processes such as that of producing the LHEMP attempt to embrace the complexity of estuarine management, and to work to structure and understand the effects of management regimes on them. To achieve this goal, there is a need to gather and facilitate the integration or synthesis of many types of knowledge: scientific or technical knowledge and expertise; local community and stakeholder knowledge; and managerial, political or legal knowledge. Many different methods may be employed to facilitate the gathering and integration of these knowledge bodies. However, each method will possess its own advantages, disadvantages and introduce a variety of trade-offs, especially related to oversimplification or challenges resulting from too much complexity. In the former case, oversimplification may lead to a loss of legitimacy from many stakeholders' points of view, if their visions are not seen to be taken into account. In the latter case, embracing the 'full' complexity of the estuarine system and its management regimes presents major challenges for integration and synthesis of understanding and information, and could ultimately be an excuse for inaction.

In the LHEMP process, a number of challenges related to embracing the 'full' complexity of the estuarine system were encountered. Within the process, two principal knowledge collection and integration or synthesis methods were used: the participatory stakeholder workshops; and the external scientific and legislative literature review carried out by the consultants. In the case of the participatory stakeholder workshops, an extraordinarily large amount of information was collected and knowledge exchanged in the short time allocated. However, the time constraints, and potentially the methodological constraints, meant that it was often difficult to tap the full expertise and knowledge bodies of the participants. To reduce this problem, it was common for the participants to refer to scientific reports or existing studies that should be considered by the consultant team. Nevertheless, the capacity for the consultant team to carry out an in-depth study of all of the cited documents and to synthesise the perspectives and information in a 'complete' fashion, especially from a time and budgetary perspective, remained somewhat limited within the timeframe of the participatory part of the process. As well as physical resources, finding individuals or small teams with sufficiently in-depth and broad knowledge of these issues and the cognitive capacity to carry out the synthesis is also likely to be a challenge. Nevertheless, in the LHEMP process, the project manager and planning consultants already had so much experience of most of the regional issues that this was not a significant issue for this project. In any case, developing improved methods of quickly tapping existing bodies of tacit and already documented knowledge to aid information synthesis appears to be a topic worthy of research.

7.6.3 General comments on participatory process co-engineering

A small number of more general suggestions about the use of participatory processes could also be made to help improve general understanding and future management and planning projects. Firstly, honesty about the potential positive and negative outcomes of participatory processes is required. This is especially important for the project implementers to explain to the managing institutions and participants. All participatory processes, and the choice of the methods used within them, will require many choices and potential trade-offs, which will have a variety of impacts on the management or process situations, including the possibility of: changed power structures between participants (and non-participants); relationship changes and conflicts; and trade-offs between stakeholder process legitimacy and 'scientific' or 'methodological' validity, as seen from an external point of view. As participatory processes are real-world processes, they will also be carried out under real-world constraints, often time and budgetary constraints. This means that decisions underpinning their design and implementation cannot always be made in collaboration with everyone who would like to be involved or in conformity with an 'ideal' methodological standard, owing to a lack of time and other resources. Last-minute changes or unforeseen contextual constraints are also more than likely to impact the process at some stage of its implementation, but negative impacts may be minimised by flexible and experienced process managers or facilitators, who are used to developing contingency plans and innovating on the spot. It is also acknowledged that many questions remain about the best methods for treating complexity and managing uncertainty and conflicts. This highlights the need for more research and innovative practical trials, like this LHEMP development process, to be able to push sustainable management processes forward and drive continual improvement and learning related to these processes.

From the process of developing the LHEMP, it was also extremely evident that the project team and its individual members often played a critical role in changing or influencing the direction of the process and the design of methods, as demonstrated through the key co-engineering events and their effects presented in Table 7.1 and Figure 7.4. It was evident that personal and common objectives of the project needs changed at a number of points through the process, and divergences in the objectives and representations of the process, its interests and its possible futures under different scenarios were apparent. These divergences had to be negotiated through, despite the fact that a tender had been written and legal contract signed for the project's realisation in a particular form. As in other projects, such as in engineering construction, 'variations' to the contract could be introduced at any time by any party after the signing of the contract and negotiated for until a decision to alter the contract is made. In other words, it is always likely that the co-design phase will continue through the co-implementation phase, to take into account any of the project team members' own or changing objectives. Both task-orientated negotiation and relationship building or restructuring will also occur and have to be managed through the co-engineering process, so flexibility appears to be essential. Further research on both these relational and operational aspects of the co-engineering of participatory modelling processes for water planning and management would appear to be worthy of attention, especially to determine whether the insights obtained from this intervention are generalisable to other co-engineering processes in different contexts or cultures. In addition, the way in which different approaches to co-engineering or a different make-up of actors create different forms of collective action is worthy of investigation. Some of these issues will be discussed further in Chapter 9.

7.7 CONCLUSIONS AND RECOMMENDATIONS

This chapter has outlined the second intervention case of this research programme, the 'creation of the Lower Hawkesbury Estuary Management Plan', in which I worked together with a number of other actors to co-engineer a participatory modelling process as part of the regional planning process in New South Wales, Australia. This work follows on from the work of Chapter 6 to continue the fulfilment of the fifth objective of this book: *to outline the lessons learnt through individual and comparative intervention case analysis to determine to what extent co-engineering can critically impact on participatory modelling processes and their outcomes.* Some of the key lessons learnt and the contributions to knowledge that this chapter provides are briefly summarised here.

The final implemented process of the LHEMP creation is considered to be the first application of the Australian Risk Management Standard to integrated regional scale estuary management. This chapter has outlined how the co-engineering process drove and adapted the participatory modelling process' direction and also why the final co-engineered adaptation of a participatory modelling process based on the decision-aiding model in Section 5.1.3 can be considered a '*participatory value-based risk management approach*'. Results from the use of a co-engineered evaluation protocol for the intervention were presented and discussed, with some of the most important insights being that a number of *new forms of collective action emerged* through the project, including the collective action of an '*inter-institutional project management team*', who in turn aided the organisation and collective action of a '*multi-institutional level stakeholder working network*'.

Many other insights were gained through the intervention, including those about the real-world constraints and priorities of different types of actor (private consultants, government workers and researchers) in the co-engineering processes of participatory processes for natural resource management, as well as the constraints, roles and priorities of a diverse range of actors in estuary management in New South Wales. Related to these differences, *an 'actor–action–resources matrix tool' was created* through the process, in response to a perceived need to examine actors' action priorities and resources relative to one another. However, further research is still required, both on its form and on its usefulness to aid multi-stakeholder negotiations and management coordination.

From the evaluations, one of the most important results of this participatory process, which goes against much discussion in the literature, was that *this process appeared to be significantly more efficient time-wise than other more standard consultant-driven processes* and that it was also considered to be cost-effective. Equally, there were insights and discussion about how the formal division of a working network of stakeholders (through the co-engineering process) into an 'agency' sub-section for the risk-assessment activity of the process had a range of advantages and disadvantages. This discussion and a number of others have opened up many new areas that could benefit from research, including:

- Determining the potential to pre-plan processes with sub-sections of a targeted stakeholder population working both in their own groups and then in larger groupings, and determining in which sections of the decision-aiding process such an approach may be advantageous; for example, creating governance level or smaller spatial groupings of community members (over a territory) who work individually on issues before all coming to work together.

- More in-depth investigation of the role of leadership and team self-management in co-engineering processes and what forms this may take, as well as further investigation into relational and operational needs for effective co-engineering to occur in multi-institutional project management teams.
- How risk management is perceived outside Australia, where there are no norms for its implementation, and what advantages and disadvantages a less standardised approach to risk management presents.

A number of these issues and others will be investigated through the Bulgarian intervention outlined in the next chapter and in the extended comparative discussion in Chapter 9.

8 Flood and drought risk management in the Upper Iskar Basin, Bulgaria

This chapter will outline an intervention based on the adaptation of a participatory modelling methodology to the 'Living with floods and droughts' programme in the Upper Iskar Basin in Bulgaria, to be used for building collective capacity in flood and drought risk management. This year-long process, driven by a number of researchers and regional stakeholders, included two phases of interviews and 15 workshops organised into series for six groups of stakeholders from national-level policy makers and experts to municipal-level government representatives and citizens from around the region. The process was co-engineered to include qualitative participatory modelling activities on: stating expectations, modelling systems and actors, eliciting visions and values using cognitive mapping and causal modelling techniques; developing management options and strategies, framing and assessing strategies using option cards and multi-criteria analysis; and scenario testing of scenarios, risk response project planning and process evaluation. The co-engineering of this participatory modelling process will be presented and discussed in this chapter, along with a range of participatory modelling process evaluation results, some lessons learnt and areas for further research.

8.1 LOCAL CONTEXT AND OBJECTIVES: FLOOD AND DROUGHT RISK MANAGEMENT IN THE UPPER ISKAR BASIN

To understand the current water and risk management situation in Bulgaria and the Upper Iskar Basin, a brief introduction outlining the context of the political and social transitions that have affected the nation's citizens and management systems is given before summarising specificities of the Upper Iskar Basin and current water management governance mechanisms, the 'Living with floods and droughts' research project proposal and the research questions to be treated in this chapter.

8.1.1 Transition in the Bulgarian context

Like many of the other populations in Eastern Europe, Bulgarians suffered from a range of very serious challenges in the early years following the fall of the Soviet Union and the removal of the Communist Party in their own country. In terms of economic challenges with wide-ranging social impacts, early Bulgarian transition conditions included: major declines in GDP and real average wages; high or hyper-inflation of over 200% in the early 1990s; high unemployment levels, reaching almost 30% in 1999 (unemployment was previously unknown under the Communist system); and increasing income disparity between citizens (Svejnar, 2002).

Political upheavals included: the introduction of more democratic election and governance practices, including a new constitution created through private and undemocratic talks between ex-Communist Party leaders and potential opponents (Tanev, 2001); some changes to state administrative structures and responsibilities (Ivanova, 2007), including a decentralisation of responsibility towards local governments (Ellison, 2007) but with inadequate resource transferral (Krastev *et al.*, 2005); multiple changes of government, due in part to large public protests, in the first years largely dominated by ex-Communist Party offshoots (Tanev, 2001); reconfiguration or privatisation of previously state-owned institutions (Peev, 1995); increased power of private interest groups in national and local politics, such as organised crime syndicates, trade unions or NGOs, (Tanev, 2001); and changes to, or the creation of, many laws, including property rights (Peev, 1995).

Further social changes or challenges related to these political and economic upheavals have included: increased differentiation between urban and rural opportunities, social, economic and infrastructural support (Staddon, 1999); food shortages (Tanev, 2001); drastic decreases in birth rates (Vassilev, 2006); increased crime rates (Tanev, 2001); widespread poverty (Vassilev, 2006); large population displacements to urban centres or other countries (Yoveva *et al.*, 2000; Vassilev, 2006); improved individual freedoms (Krastev *et al.*, 2005); and high levels of political engagement in civic society (Letki, 2004).

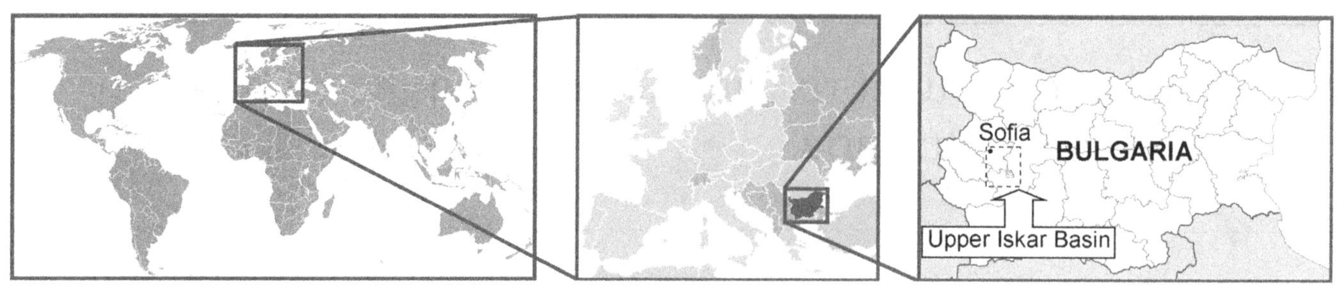

Figure 8.1 Location of the Upper Iskar Basin around Sofia, Bulgaria.

Along with these upheavals, many environmental challenges also emerged, including the extensive air, water and soil pollution problems resulting from years of industrialisation, intensive agriculture and neglect of environmental protection (Carpenter *et al.*, 1996). Extreme climatic events, including severe droughts and floods, have also added to the current list of challenges with which Bulgarians have recently had to cope (Hristov *et al.*, 2004).

Bulgarian society has weathered many changes and challenges through the first transition periods. Unlike many other transition countries, however, the state administration or 'civil service' has not gone through as many fundamental changes. Owing to a reasonably peaceful change from the Communist Party to the democratically elected Socialist Party, representing one arm of the previous Communist Party, in 1990, the majority of the existing Bulgarian administrative services were left in their previous configurations (Meyer-Sahling, 2004). It is only in the recent new transition phase, during the ascension and acceptance into the European Union (EU), that larger administrative reforms have been designed and partially implemented (Dimitrova, 2002; Ellison, 2007). It is, therefore, the challenges of this second phase of rapid transition that the participatory process in the Upper Iskar Basin had to appreciate and accommodate.

8.1.2 Local system and governance context

The Upper Iskar River Basin is located around and to the south of Bulgaria's capital, Sofia, in Europe. The location of the Upper Iskar Basin is shown in Figure 8.1.

Extreme climatic conditions, such as large floods and extended drought periods, have occurred increasingly over recent years in Bulgaria and the Upper Iskar Basin in the region of Sofia. In 1994 there was water rationing, and in 2005 severe floods. There has been debate on whether these 'new' conditions are a consequence of global climate change or merely normal climate variability (Knight *et al.*, 2004; Kundzewicz and Schellnhuber, 2004). Water management in such a context has presented many challenges, not just because of these extreme events or seemingly natural hazards, but also because of the transitory nature of the country's social and political spheres following the fall of the Communist Regime in 1989 and the need to deal with its legacy of heavy industry, widespread pollution and infrastructural system issues (Hare, 2006). State governance structures have remained largely technocratic and hierarchical. With its recent accession into the European Union (EU), Bulgaria must now improve management of its water resources and resolve associated use conflicts between industrial, urban, agricultural, ecological and other human needs in line with EU legislation, such as the Water Framework Directive (WFD). Responsibility for water management in Bulgaria lies at the national and river basin levels, as outlined in the *Bulgarian Water Act 1999*, which is predominantly in line with the WFD (Dikov *et al.*, 2003).

8.1.3 Proposal and objectives of the 'Living with floods and droughts' project in the Upper Iskar Basin

In the Bulgarian water management framework, to improve management of water in the Upper Iskar Basin, which is part of the Danube River Basin, a number of initiatives were proposed as part of the European Integrated Project, 'AquaStress' (www.aquastress.net). These included a participatory risk management process to attempt to support regional co-management of floods and droughts (Ribarova *et al.*, 2006). The general needs for such a project had been identified by the Local Public Stakeholder Forum (LPSF), a diverse group of stakeholders from the Upper Iskar Basin brought together as part of the AquaStress project. These needs were then discussed with the project's 'joint work team', a group of European researchers interested in working in the Iskar region. A proposal to help manage flood and drought risks was put forward by one of the French research partners to test the 'Participatory modelling for water management and planning' water stress mitigation option (Daniell and Ferrand, 2006) which had been previously defined as part of the project. This option was based on parts of the literature review in Chapter 3 and the process of the pilot study is outlined in Appendix D.

At this stage of the project definition, which had been carried out by two French researchers and a private research consultant from a German company, the objectives of using a participatory modelling approach were outlined, as shown in Figure 8.2.

Iskar test site process goals

- Integrate and improve the overall communication between the different actors at different scale levels
- Develop an integrated view of the management system and how it can be sustainably managed over long periods of time
- Develop an integrated view of decision making under conditions of long-term uncertainties; thus formulating answers to the following questions:
  - How does one spend money wisely when deciding between flood and drought management?
  - How do management decisions for flood mitigation affect or constrain drought management and vice versa? How does crisis management affect or constrain long-term management decisions?
  - Are there win–win management strategies that can benefit both flood and drought management over long periods of time?
- Maintain the knowledge of good management across the different flood and drought periods
- Develop a common vision among the stakeholders about living with floods and droughts
- Evaluate management strategies in terms of different indicators, with respect to varying uncertainties and scenarios, rather than provide single definitive answers
- Look at the side-effects of crisis management on the long-term effects
- Assess effects of crisis management on short- and long-term financing of management
- Establish new social contracts and commitments in relation to flood and drought
- Bring stakeholders to consider what could happen in the worst case should there be in the future:
  - No management recognition between drought and flood management
  - No vertical communication and coordination between stakeholder scale levels
  - No long-term consideration of short-term management strategies

Figure 8.2 Objectives of using a participatory modelling process for flood and drought risk management in the Upper Iskar Basin (Hare, 2006). Source: Daniell *et al.* (2011).

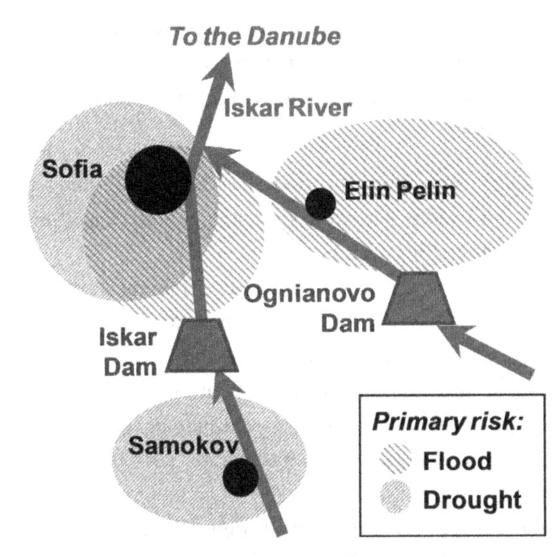

Figure 8.3 Areas of the Upper Iskar Basin considered for flood and drought risk management. Adapted from: Rougier (2007). Source: Daniell *et al.* (2011).

A stylised representation of the Upper Iskar Basin with the areas susceptible to floods and droughts is shown in Figure 8.3. The project was to attempt to take into account and deal conjointly with the management of these risks with stakeholders from all over the region.

8.1.4 Research questions and objectives

Considering these operational objectives for the proposed participatory modelling process, many research questions could also be asked, owing to the innovative nature of the proposed Bulgarian multi-level flood and drought risk management context. As with the Australian process, this proposal was to be one of the first of its kind in Bulgaria and the world, and therefore there was no existing process to base it on. The innovation potential of the proposal meant that challenging research questions were posed, even partway through the process.

In the Iskar context, a number of research questions arose, based on the previous theory and practical work outlined in this book:

- To what extent can participatory modelling improve the capacity of multi-level stakeholder groups to cope with flood and drought risks?
- To what extent can the participatory modelling approach be modified by an intervention researcher entering late in the process, who has different interests and objectives from the other project team members and stakeholders?
- What are the advantages and limitations of using a participatory modelling process in such a research-based exercise?

To address these questions, the following objectives were proposed for the Iskar intervention:

1. Test a participatory modelling process in a 'research' situation to improve the capacity of multi-level stakeholder groups to cope with flood and drought risks.
2. Determine to what extent an intervention researcher entering the co-engineering process mid-way could negotiate, introduce and realise different objectives and protection of different interests.
3. Gain a greater understanding of the links between research and real-world institutional and water issues in the Bulgarian context, including the positions taken by, and constraints placed on, researchers, community stakeholders, private businesses and governments in risk management considerations.

8.2 PROJECT CO-INITIATION AND PRELIMINARY CO-DESIGN

This section describes how the above objectives and project scoping were developed collectively. It is followed by an overview of the co-design process for the Iskar project.

8.2.1 Project initiation and scoping

As part of the AquaStress project, a group of European researchers from a number of institutions both within and outside Bulgaria supported the formation of a multi-level stakeholder group in 2005 to help define major water issues of the region. This group, called the 'LPSF', included participants from national-level policy makers to representatives of citizens' groups, as outlined in Ribarova *et al.* (2008a). From this research-supported participatory analysis, two of the seven largest issues for water management in the region appeared to be the incapacity of both the institutional coordination and the community to cope with flood and drought crises.

In an AquaStress project meeting in Portugal in February 2006, a French research director proposed that a process of participatory modelling could be developed and tested as an intervention research exercise to aid the regional community and to examine these and other issues. The proposition was further discussed and accepted by the LPSF in a meeting in Bulgaria in early June and was tested in an adapted form with a group of Bulgarian students during a four-day 'Borowetz Summer School' (Rougier, 2006). A formal methodological design proposal of the 'Living with floods and droughts' participatory modelling project for the Iskar Basin was then collaboratively created during a meeting in Paris in July 2006 by three European researchers: the French research director; a French Masters student on an internship; and a private research consultant from a German firm (Ferrand *et al.*, 2006; Hare, 2006; Rougier, 2006).

The Bulgarian regional partner coordinating the AquaStress project work in the Iskar Basin – who had a technical engineering background and no previous experience of participatory processes – did not really understand the intent of the participatory modelling process that had been designed by the researchers from the project and had no particular expectations of what results the process would achieve. She trusted the researchers and was willing to find people in Bulgaria to participate in and organise the process and carry out her role in the project, but was not interested in getting involved in it herself (Ribarova, 2008b, personal communication). As requested by the researchers, she managed to find an experienced Bulgarian facilitator with good English skills who would help to run the project.

8.2.2 Use of participatory modelling process methodologies

The methodology for the participatory modelling process was based largely on Daniell and Ferrand (2006), using the 'SAS (system–actors–solutions) integrated model' (Ferrand *et al.*, 2007) as the internal modelling method. The methodology was also designed to be used with a wide range of regional stakeholders, including: high-level national policy makers; private company representatives; non-governmental organisation (NGO) representatives; municipal mayors and council workers; national experts; and citizens from the region, who would take part at different times in a year-long series of interviews and workshops. The process included three main phases:

1. Stating expectations, modelling systems and actors, eliciting visions and values – using cognitive mapping and causal modelling techniques;
2. Developing options and strategies, framing scenarios and assessing strategies – using option cards linked to a system model and multi-criteria analysis; and
3. Testing strategies, process evaluation and planning the future – using the models, a role-playing game and questionnaires (Ferrand *et al.*, 2006; Hare, 2006).

The proposed process is outlined in Figure 8.4.

As can be seen from Figure 8.4, the process included two phases of workshops (WS) and interviews. However, the process had not been designed to be applied to the same group of stakeholders. Rather, the first interview stage, which included approximately 120 interviews, aided the definition of stakeholder groups and allowed interested and competent stakeholders, including unaligned citizens, to be found. These stakeholders were then formed into six groups, stratified according to institutional levels of management, as the mapping in Figure 8.5

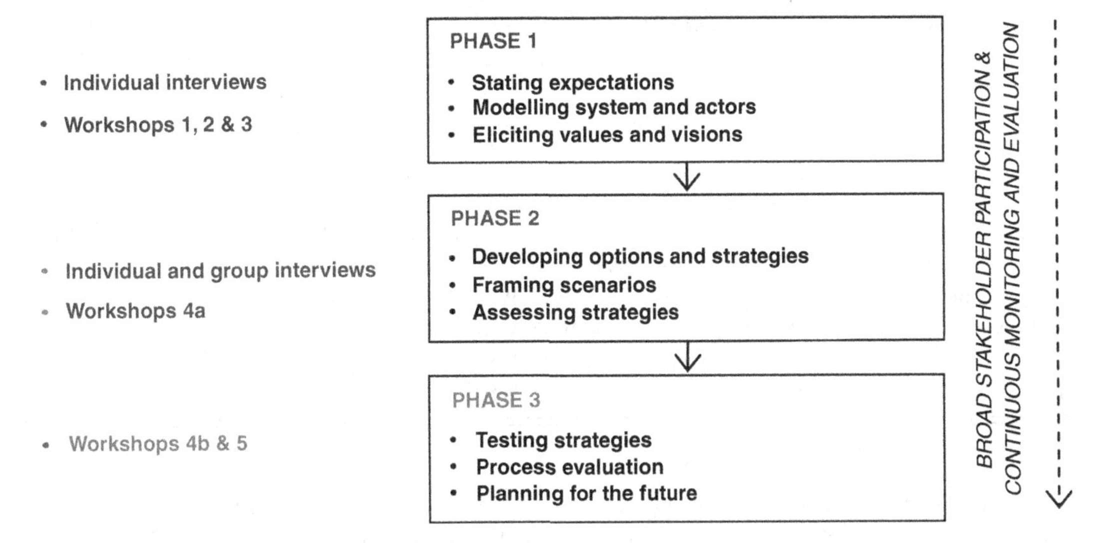

Figure 8.4 Proposed Iskar Participatory Risk Management planning process. Adapted from: Ferrand *et al.* (2006) and Hare (2006).

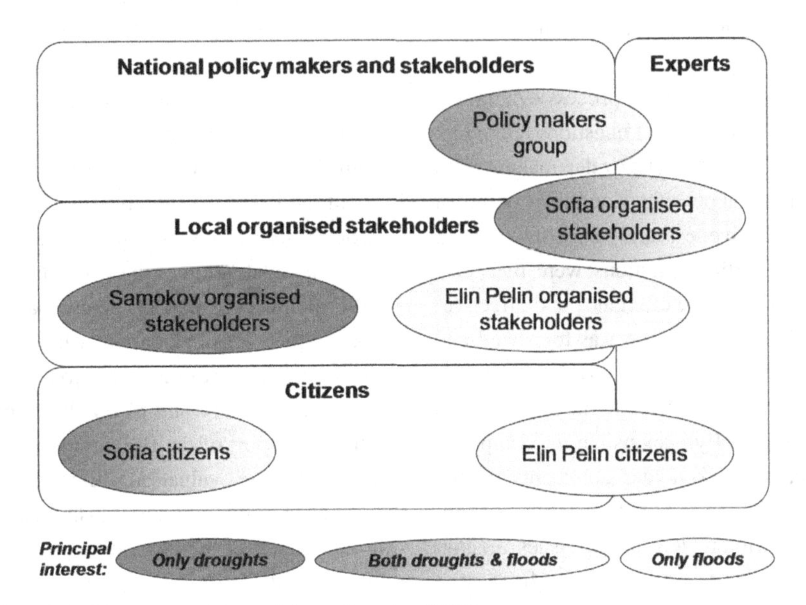

Figure 8.5 Interests of the six stakeholder groups. Adapted from: Rougier (2007).

shows. The participation of these groups in the various participatory modelling phases is outlined in Table 8.1.

The principal differences in participation were that the policy makers had their preliminary activities split into two workshops (WS1 and WS3) and were provided with updates on the other groups' activities in WS3. The second workshop, WS2, was, in fact, a series of four individual workshops for the five other groups, although the Samokov stakeholders and Sofia stakeholders were combined to form one group. Similarly, WS4a was also a series of six workshops for each of the groups. WS4b and WS5 were the only workshops where participants from all of the groups were brought to work together.

8.2.3 Evaluation protocol adaptation and administration

I was not initially involved in the adaptation of the evaluation protocol for the Bulgarian process, as I only entered the process after Workshop 3. However, as I had been involved in the creation of the original protocol for the AquaStress project and had tested the use of the ENCORE model in the pilot Montpellier study with the French research director, the final adapted protocol for the Bulgarian participatory modelling process turned out to be not far from the protocol laid out in Section 5.5.

The Iskar process had a number of types of evaluation procedures. Firstly, *ex-ante* evaluation of the stakeholders'

Table 8.1 *Stakeholder groups taking part in the participatory modelling process stages*

Group name	Phase participation	No. of participants (max.)
Policy makers	Preliminary interviews, WS1, WS3, secondary individual interviews, WS4a, WS4b, WS5	12
Sofia organised stakeholders (LPSF)	Preliminary interviews, WS2, secondary individual interviews, WS4a, WS4b, WS5	10
Samokov organised stakeholders	Preliminary interviews, WS2, WS4a, WS4b, WS5	6
Elin Pelin organised stakeholders	Preliminary interviews, WS2, WS4a, WS4b, WS5	10
Sofia citizens	Preliminary interviews, WS2, WS4a, WS4b, WS5	12
Elin Pelin citizens	Preliminary interviews, WS2, WS4a, WS4b, WS5	6

representation of the water situation and views on water management was carried out through initial interviews, which were administered by the Bulgarian facilitators, the regional partner and a small group of university students.

Secondly, the French research director and the French Masters student had created a closed-response questionnaire on a four-point Likert scale for workshop and ongoing process evaluation. This questionnaire had a number of standard questions related to: knowledge and learning about the Iskar and water management systems; knowledge and learning about others and their points of view; process efficiency and effectiveness; and issues of workshop moderation and organisation. Further questions were then also adapted in each workshop to address the efficiency and effectiveness of the methods used. This questionnaire was translated and administered at the end of each workshop by the Bulgarian evaluator, who had also taken the 'external evaluator' role for the whole Iskar case study. She was additionally responsible for computerising the results and translating stakeholder comments on the questionnaires into English for dissemination to the project team. This all occurred efficiently, without any major issues arising. Some audio and video recordings, as well as photos from each workshop, also added to the data available for process evaluation.

Thirdly, to analyse the co-engineering process, attempts were made to assess project team perspectives and progress after each workshop. At the conclusion of the first workshop, a questionnaire was administered by the French research director and French contract researcher to the Bulgarian facilitator, regional partner, external evaluator and the private research consultant from a German company to gauge their reactions to the process and obtain their suggestions for improvements. After WS2, WS3, WS4a and WS4b, facilitator notes and comments on the workshops were written, as well as some observations by the French research support team. The two large combined group workshops, WS4b and WS5, also had formal, recorded project team debriefing sessions and minutes, which were written by the external evaluator for WS4a.

Finally, *ex-post* evaluations were carried out with WS5 participants in the form of 10 open-response questions, defined by the French research director and me, and translated into Bulgarian by the Bulgarian evaluator. The questions related to perceptions of learning, working with other participants, water practices, process effectiveness and potential improvements, as well as intentions for future involvement in participatory water management projects. I also conducted process follow-up interviews for *ex-post* analysis of the co-engineering process with a number of the project team members: the Bulgarian regional partner; the remaining Bulgarian facilitator (who facilitated in all but the first workshop but had been present to observe it); the French research director and the French Masters student.

Further evaluation of the whole process was undertaken by the LPSF members as part of the evaluation of the Iskar AquaStress case studies. This evaluation, which included questions on context, process and results, was derived from the project evaluation guides (e.g. Jeffery and Muro, 2005) by the Bulgarian evaluator with the support of the Bulgarian regional partner. Further information on these evaluation activities are given in Vasileva (2008).

Because of the required translations from Bulgarian to English through many of the evaluation procedures, there is a significant potential for bias and interpretation difficulties, which are difficult to overcome. I noticed that there were some differences in the Bulgarian translations of the process questions that were given to the evaluator in English (which had been translated from French first and then to English), which really meant that the English equivalent of the questions asked in Bulgarian should have been re-established, to at least reduce the potential for misinterpretation of the Bulgarian responses. In the future, it would prove valuable to have closer collaboration on questionnaire translations, in order to explain and adjust questions simultaneously in both languages to obtain the best match possible, especially as some concepts have no direct translations between languages.

In an attempt to remove significant biases and misinterpretations, my interpretations in the following sections were all submitted to a number of the other project team members for comment, reinterpretation and validation.

8.3 DETAILED CO-DESIGN AND CO-IMPLEMENTATION

This section endeavours to provide a summary of some of the significant negotiation events that took place through the co-engineering process. It also includes an in-depth analysis of one key co-design negotiation, originally presented in Daniell *et al.* (2010b), including the impacts this negotiation had on the co-engineering processes and participatory modelling process. The chosen negotiation had a 'substantive' basis, where the question of changing the problem scope for the Iskar participatory modelling process was treated. A continuation of the narrative form of all the principal elements of the co-design and co-implementation phases, including both relational and operational aspects, is provided in Appendix F. Further information on the process can be found in Hare (2007) and Rougier (2007). Because I only took part in the co-engineering process only after WS3, all information on the first three workshops and two interview phases have been reconstructed from project documents, notes, meeting minutes, photos, audio recorded debriefing sessions and *ex-post* interviews with project team members. These sources are then supplemented by my own observations and notes from the co-design of WS4a onwards.

8.3.1 Summary of significant co-engineering process events

Compiled descriptions of a number of principal co-engineering process events and their impacts are presented in Table 8.2. A number of other co-engineering issues arose through the process, such as the cancellation or avoidance of running public education courses on specific water management issues or using external expertise for developing management options and assessing final strategies, as well as the omission of certain results in the process, such as the citizens' interview results, which were not explicitly reused (Ferrand, 2008; personal communication). As I have less information and fewer sources of data on these aspects, they have not been detailed here. Data sources used to inform the other elements were given in Section 6.3.2. The interpretation schemes used to determine the negotiation modes were presented in Section 4.2.3. These events are summarised on the process timeline for easier visualisation in Figure 8.6.

The next section will provide in-depth analysis of the negotiation episode that comprised the event to 'change final workshop content' shown in Figure 8.6, which took place prior to the final workshop of the Iskar participatory process.

8.3.2 Negotiation episode analysis: proposed problem scope change

BACKGROUND TO THE NEGOTIATION

The negotiation episode of interest in this section took place prior to the final workshop of the participatory modelling process, almost a year after the start of the process. The workshop had originally been designed to be an assessment of flood and drought management strategies using a multi-criteria software program developed in another part of the AquaStress project. However, I was concerned that this multi-criteria assessment might not be the best design, firstly, because of a lack of different clearly defined strategies that had been developed through the other stages of the project, and, secondly, my belief that the process should produce something useful for participants, so that they would see the benefits of participatory processes and not be disappointed by the academic outcomes (see Barreteau *et al.* (2010) for discussion on this point).

I first raised these concerns over the direction of the process at the debriefing meeting of the second-to-last workshop with the regional partner, facilitator, contracted researcher and a number of the regional partner's colleagues. At this meeting, it was clear that everyone supported the idea of making the last workshop interesting, so that the participants would appreciate it, although they were not sure how. These issues of creating a positive end to the process were discussed further with the research team and the French institution's AquaStress project manager, who arrived the next day and was taken on a field trip to see some of the stakeholders and hear stories about their recent flood events. Having met these citizens and seen their everyday difficulties, he supported the idea of attempting to improve the participants' on-the-ground situation and capacity to cope with floods. At this stage, the Bulgarian regional partner mentioned the possibility of there being Bulgarian European structural funds available to help with regional development, and wondered whether the last workshop could be organised to be the first stage of designing a proposal to obtain some of these funds for flood mitigation and adaptation projects.

A week later, the regional partner had travelled to Brussels for a meeting with the European Commission. She informed the researchers that there would be structural funds for risk management and, in particular, flood management. She asked whether the final workshop could create an action plan for the flooded region, which was also on the list of priority regions for structural funding. It is the negotiation and decisions following this specific proposal that will be analysed here.

DISTRIBUTIVE AND COMPROMISING MODES

In France, I was attracted to the action-planning proposal. The research director was less enthusiastic about it when we met to discuss it. He explained that some of the original research objectives and the contractual commitments for the AquaStress project would not be able to be achieved if the programme were to be changed. In particular:

1. The multi-criteria analysis tool created in the AquaStress project would not be tested;

Table 8.2 *Description and analyses of significant Iskar co-engineering process events (the author is here described as 'the Australian researcher')*

Negotiation event – people involved	Potential effect on personal objectives/interests	Negotiation mode, outcome and relationship characteristics
Co-initiation/co-design phase Decision not to take a random sample of stakeholders for the citizens' groups as was highlighted in the stakeholder analysis – practical reasoning outlined by Bulgarian regional partners to external European researchers	*Bulgarian regional partners* Largely positive, as they could find more interested, committed and trustworthy stakeholders through their own personal networks and partially random 'snowball' interviewing system (approach a random stranger in the field for the first interview, then ask them if they know other people who might be interested in being interviewed and involved in the project), thus assuring that they would participate well throughout the process and ensuring operational success in terms of timing and budget of research project. This process would also save the regional partners money and time. *External European researchers* Negative, as research-driven stakeholder analysis and recommendations were not fully considered; for example, societal groups, such as ethnic minorities, local NGOs, village business owners, factory workers and farmers, as well as some important sectors, such as tourism, fisheries and recreation, were largely underrepresented in the process, which would bias and prevent integrated management of the system. Additionally, the geographical scope of the project was being changed to include another tributary to the Iskar River.	Pragmatic choice of Bulgarians with small elements of compromising but integrative type of negotiation, based on cognitive, relational, normative, operational and equity issues of interest. The external researchers had little capacity to regulate the initial choices but continued the negotiation through the process. Some efforts were made by Bulgarian regional partners to include stakeholders whom they would not have personally chosen, so some of the stakeholder groups (policy makers and LPSF in particular) were quite representative of major interests. Bulgarian partners held most of the power over whom to include, owing to language issues, time and presence in Bulgaria so their decisions were upheld. Operational aspects of the process were largely successful. Representation of stakeholders, especially in the citizens' groups, remained a point of sustained tension but with mutual understanding. Increased learning and appreciation of cultural differences on the external European researchers' side. Some strong work opinion differences. Conflict remained work-related and personal trust remained largely intact between most team members.
Co-design phase/co-implementation phase Decisions on how to deal with project team member changes, new roles and responsibilities half-way through the project, when the lead Bulgarian facilitator disappeared from the project and one French process designer whose research contract expired was to be replaced by the recently involved Australian researcher	*Bulgarian regional partner* Largely negative as, with the lead facilitator gone, the project manager would have to take on a facilitator's role or find someone to train, on top of all her other duties. She also worked well with the French contracted researcher and would have preferred that he stayed and offered his good process preparation skills and project responsibilities until the end of the process. *French research director* Mitigated, as although upset that they had lost a facilitator	Distributive negotiation based on relational and operational issues of conflict. The Bulgarian research partners divided the required facilitation, translation and design duties amongst themselves and formed a good, cohesive team with little conflict. The first new project team (minus the Bulgarian facilitator) had some initial issues and successes in the new design and facilitation roles, which resulted in one of the stakeholder groups asking to have another workshop to finish their strategy building, as they found it interesting (but time

	and some of the initial project results, he would have fewer conflicts over process design decisions and better working relationships with the remaining project team members. Similarly, he preferred not to attempt to re-renew the French contracted researcher's position as the Australian researcher replacing him would be easier to work with and obtain results. *French contract researcher* Largely negative, as he was unable to resolve relational conflicts with the French research director. This placed him in an awkward position for maintaining close personal and good working relationships with the other project team members. *Australian researcher* Slightly negative as she would have to take on extra unplanned responsibilities that would have a negative impact on her research deadlines and would lose one of the members of the project team with continuous knowledge of the process and with whom she had developed good relations and trust. On the positive side, she would have more opportunity to adapt the end of the process to help meet her own research needs and become more integrated into the project team.	consuming) and wanted to improve their strategies. Working on the final workshop design (without both the Bulgarian facilitator and the French contract researcher) saw a levelling and changing of power relations and roles, with the Bulgarian regional partners taking a much more active negotiation role in the design on an equal footing with the Australian researcher and French research director. This change in project teams led to their greater collective ownership and self-investment in ensuring the success of all parts of the process (design, content, operations and outcomes). Relationships between the French contracted researcher and his Bulgarian colleagues were only renewed a year after he left the project.
Co-design phase/co-implementation phase Decision to change the objective of the final workshop from performing a multi-criteria analysis on regional strategies for flood and drought co-management to developing an integrated flood management plan for a sub-region of the basin – Bulgarian regional partners' suggestion to French research team and stakeholders	*Bulgarian regional partners* Largely positive as they could help a group of the most enthusiastic stakeholders in the process to produce some concrete outcomes and provide aid for receiving structural funds, rather than just evaluating general regional strategies and not having any strategy for putting actions in place. *Australian researcher* Positive as, ethically, she wanted to be able to help the stakeholders achieve their goal of being part of the project, rather than letting them just be involved for the researchers to 'test' their process and achieve their own scientific goals. It seemed to her that the targeted group of stakeholders had so much hope that good things would come from the process; yet without changing the direction of the project she was afraid that little would	Compromising/collaborative negotiation based on normative, operational, relational and equity issues of conflict. Consensus on the process change was reached and the action planning workshop was considered a success by those involved. Some other stakeholders were just absent. An example of empowerment, apparently stemming from this project, appeared, as one of the citizens of an area subject to flooding, who participated in the process, decided to run for mayor in her municipality and was elected. Relations and trust were further built between the project team and participants. A meeting with many more local stakeholders was held a number of months later but there appears to have been little advancement on the ground since.

Table 8.2 (*cont.*)

Negotiation event – people involved	Potential effect on personal objectives/interests	Negotiation mode, outcome and relationship characteristics
	come from the evaluation, as it could be too technical and uninteresting for some of the stakeholders and they could have been disappointed with the final process and disillusioned with participatory processes. *French research director* Mitigated, as the change would mean difficulties in reaching research objectives of the evaluation of conjoint regional flood–drought management strategies. However, owing to a number of other process changes and lack of the creation of a role-playing game as planned, this proposition was less likely to achieve useful research outcomes and so he was not strongly opposed to change in principle to create an action plan, but still opposed the lack of consideration of drought management and the exclusion of some stakeholders from the final workshop. *Stakeholders* Slightly positive overall, as concrete outcomes for some stakeholders could be envisaged, despite some others not gaining any extra advantages from the end of the process, as their stakes were not as high, relative to questions of flooding in the sub-region.	Good working and personal relations were cemented between the project team members. Only slight tensions remained between the excluded stakeholders and the Bulgarian regional partners, which did not appear detrimental to continued support of the overall process.

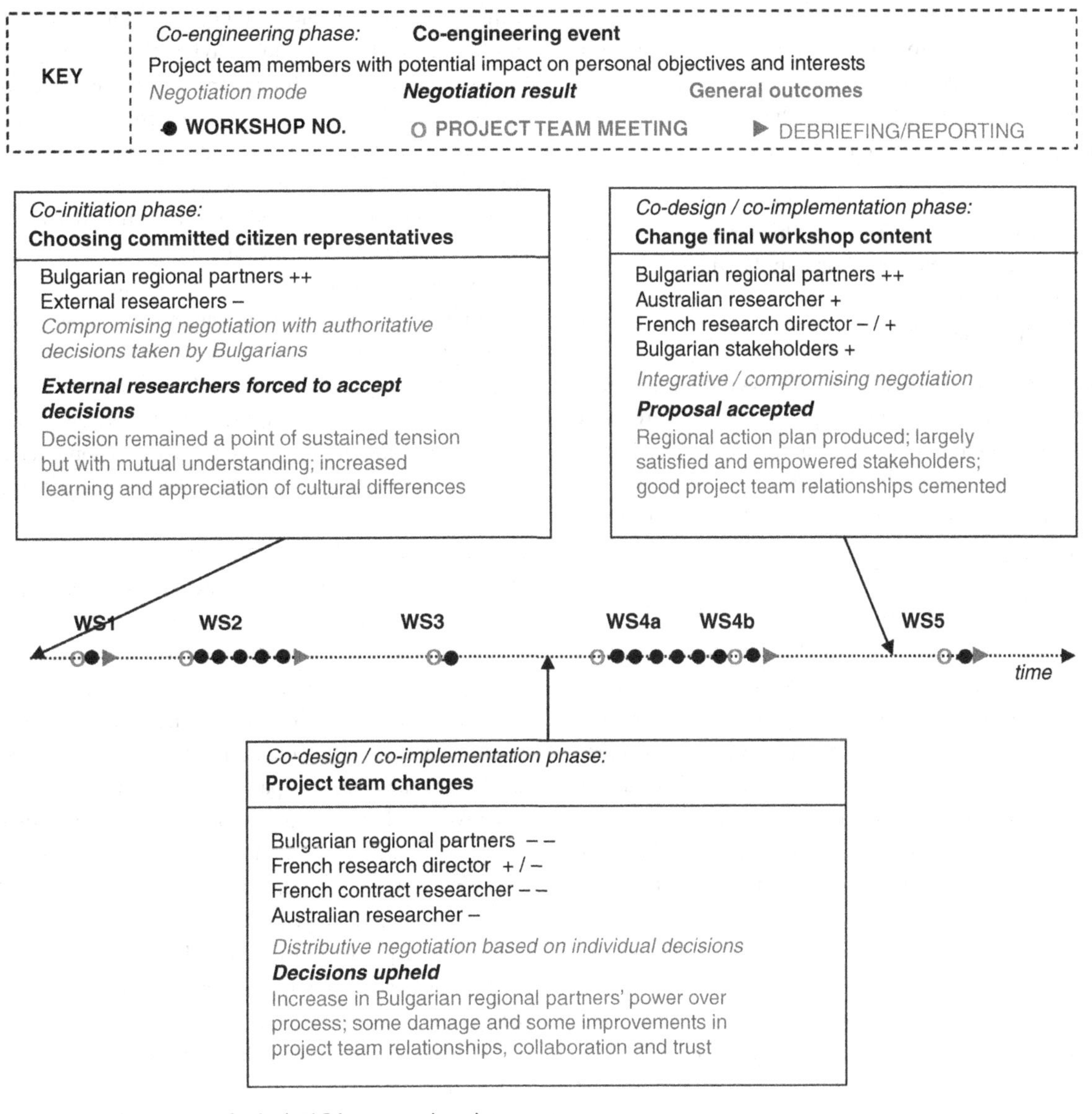

Figure 8.6 Timeline and summary of principal Iskar co-engineering events.

2. The planned area to be covered by the action plan (Elin Pelin) would not be of as much interest for all of the stakeholders involved in the project; and
3. Drought management would not be treated to the same extent as had been planned.

Therefore, this change would constitute a major alteration in the problem scope of the process, from co-management of floods and droughts, to just floods. Considering these issues and the other more positive aspects of the proposal, the research director and I discussed the proposition with the AquaStress project manager. The project manager viewed this proposed change positively, as he understood the regional issues at stake, having recently been to see the problems first hand, and hoped that it would lead to greater Bulgarian support and appropriation of the participatory process. He also stated that he trusted our judgement and would, therefore, support the proposal if the research director wanted to adopt it. The research director finally decided to accommodate the views of the rest of the project team and adopt the proposal, with certain provisos, despite his concerns. This was a significant compromise on his scientific research objectives.

COLLABORATIVE MODE

Following this decision, my work with the research director became more collaborative, as we collectively sought possible solutions for participatory workshop design and method choice to meet as many of our objectives as possible. After our discussions, I attempted to design a process for the workshop that would achieve a variety of remaining objectives, both scientific and operational. The research director made minor additions to the workshop design and sent it to the Bulgarian regional partner.

DISTRIBUTIVE MODE

The regional partner's reaction to the design of the workshop was initially lukewarm, as she did not appreciate the plan to divide the participants into subgroups to deal with certain sectors of the flood planning. The subgroups proposed were:

1. Construction and infrastructure;
2. Planning, management, decision infrastructure and monitoring;
3. Education, empowerment and capacity building;
4. Crisis management and action plan; and
5. Remediation and insurance.

The regional partner also thought that other activities related to the AquaStress objectives and to completing the work required for the structural funding plan were unnecessary. The French research director and I carefully responded, explaining that her arguments would support neither her own goals nor the needs of the AquaStress project.

ACCOMMODATING MODE

Giving in to the researchers, at least for the time being, the regional partner wrote back, '*We will do the workshop as you wish. I shared with you my 'feelings' about it, but as I repeated many times before – this is your study, you are the experts. I provide only local help.*' She also mentioned that it would be best to organise the rest of the workshop programme when the researchers arrived in Bulgaria, as she thought that they were more efficient and understood each other better when discussing matters face to face.

COLLABORATIVE MODE

A few days before the final workshop, I arrived in Bulgaria and was able to work with the Bulgarian regional partner and facilitator to better explain the reasoning behind the design proposal and find mutually acceptable solutions. The majority of the proposal was kept, but the largest change was to shorten the workshop, to keep the policy makers at the whole meeting. After his arrival a day before the workshop, the French research director was able to make minor additions to the activity supports, and to check and add questions to the end of the process evaluation questionnaire that I had developed to meet some of our research objectives. The team agreed that all participant groups were to be invited to the flood action-planning workshop by the regional partner, even though the flood component was not their main interest. In this setting, all parties learned more about the perspectives and needs of the others and worked together to find more mutually beneficial solutions to the issues discussed.

FURTHER INTERPRETATION AND *EX-POST* ANALYSIS OF THE NEGOTIATION EPISODE

An *ex-post* analysis of the Bulgarians' role in the co-design decisions indicates that this was one of the times where the Bulgarians took a much more forthright role in debating the design of the workshop, having learnt through the rest of the process. Both the Bulgarians and the external researchers believed that they could work as equals, which aided the collaboration. This is in stark contrast to the position that the Bulgarians had taken at the beginning of the participatory modelling process when, as outlined in Section 8.2.1, the regional partner had stated that she did not understand the proposed process and had no particular expectations for the results. Through the process, the regional partner and second facilitator developed their own objectives for the process, and became stakeholders in its co-design and co-implementation. In their *ex-post* interviews, the external researchers commented that the Bulgarians held the ultimate power of decision over implementation, as the participatory process was carried out in Bulgarian.

IMPACTS OF THE NEGOTIATION EPISODE

This negotiation episode, and the decision to implement the action-planning activity in the final workshop, had a number of effects both at the co-engineering level and at the participatory planning level (see Figure 1.1). At the co-engineering level, the negotiation episode enhanced social learning and trust among the co-engineers. The Bulgarians' learning focused on the attention required in the design of participatory methods to meet specific objectives, including their own objectives. The external researchers' learning was centred on the needs of the local stakeholders and constraints for providing them, such as the overarching need for further funding and capacity building if any concrete actions were to take place. Trust was built through mutual respect in attempting to achieve both local and scientific objectives and in seeing the achievement of predicted outcomes of the co-engineering choices on the planning process.

The impacts of the negotiation episode on the participatory planning process included the development of an action plan for a suite of projects that could be used for structural funding proposals, as will be detailed in Section 8.4. The objective that the participants would view the process as having been positive was also achieved, as all participants gave positive evaluations of the process in the final questionnaire. From one of the quantitative questions (to which all 28 participants present at the end of this workshop responded on a four-point Likert scale – entirely agree, agree, disagree, or entirely disagree), 15 respondents 'entirely agreed' and 13 'agreed' that 'the meeting was important and deserved to be held', which was a similar level of positive evaluation to the previous workshops, as can be seen in Section 8.5.2. On the qualitative side, positive responses to the question of whether the process had helped manage water in the Iskar Basin included: '*Without any doubt, this process is helping the improvement of the whole area. It is a golden chance to discuss and identify the problems and, based on this analysis, the most appropriate and suitable actions and activities can be undertaken.*' On my side, I succeeded in meeting all of the objectives required for my own research project. However, the negotiated change of problem scope meant that several objectives of the AquaStress project were not achieved, as had been predicted by the research director.

8.4 PARTICIPATORY MODELLING PROCESS IMPLEMENTATION

Following the co-engineering process and negotiation event analysis of the Iskar process, this section will concentrate on the implemented participatory modelling process. The descriptions here will briefly highlight the methods used and some of the content results produced through the process. Further information is available in Appendix F.

8.4.1 Process description

The participatory modelling process was carried out from October 2006 to October 2007 to address the issue of 'Living with floods and droughts' in the Upper Iskar River Basin of Bulgaria. Over 120 paid participants were involved in either the interview processes or workshops, including: national ministers; policy makers; private company representatives; NGO representatives; municipal mayors and council workers; national experts; and citizens from Sofia, Samokov and Elin Pelin. The process participants and general content are presented in Figure 8.7.

The participatory modelling process used for risk management, shown in Figure 8.7, was more elaborate in design than the Australian LHEMP process, with approximately 60 stakeholders divided into six groups taking part in a series of 15 workshops, individual interviews and evaluation exercises over a one-year period. The other 60 participants were only involved in the initial interviews. The process included: cognitive mapping of the current management context and physical system, incorporating flood and risk drivers and impacts (see Hare (2007) and Ribarova *et al.* (2008b) for further details); values, visions and game-based eliciting of preferences for actions; strategy development, evaluation and robustness analyses; as well as the originally unplanned production of a prioritised list of projects for the region of Elin Pelin. Participant voting on projects also took place, with the aim of developing proposals to obtain Bulgarian structural funds in order to implement priority projects. All of the participatory process activities with participants were carried out in Bulgarian. The preliminary workshops were carried out in the six separate groups consisting of: policy makers; national experts and organised stakeholders of Sofia; Sofia citizens; Elin Pelin mayors and organised stakeholders; Elin Pelin citizens; and Samokov organised stakeholders and citizens, as presented in Figure 8.5. The last two workshops combined all six groups and involved approximately 35 participants each. The final development of projects for the action plan was created under five areas by 'task-force' groups in the final workshop, to ensure sufficient and concrete specification of required projects: three for preparedness planning involving construction and infrastructure; education and capacity building; planning, management, decision infrastructure and monitoring; one for times of crisis (crisis management and action plan); and one for reconstruction after disasters (covering remediation and insurance). In total, 24 distinct projects were proposed, along with who should be responsible for carrying them out and where and over what period they should take place. Throughout the process, computer processing was used to convert the paper-based interview and workshop results and to perform translations from Bulgarian to English. The software used included CmapTools (Novak and Cañas, 2006), to transfer and analyse the cognitive mapping outputs; Protégé (Gennari *et al.*, 2002), for managing ontologies; Microsoft Excel, for the assessment matrices, action plan projects and evaluation results; and Google Maps, for spatial mapping of the proposed projects.

Extensive evaluation, including written questionnaires with 65–100% return rates, facilitator and observer reports, and a number of interviews were carried out to assess the impacts and efficacy of the 'participatory modelling' process. A few of these results are presented later in this section and will be further analysed in the discussion. More information on the evaluation results can be found in Vasileva (2007; 2008).

8.4.2 Example content results and discussion

SITUATION ANALYSIS

The initial phase of the process involved a number of cognitive mapping exercises, as outlined in Hare (2007), the results of which can be analysed to gain an overall perspective of the interests and knowledge of the participants at the beginning of the process. The cognitive mapping exercises were undertaken with the objectives, amongst others, of representing individual and group views on, and the relations between:

1. Drivers of floods and droughts;
2. Impacts of floods and droughts; and
3. Actors responsible for change in the system.

There were six kinds of stakeholders involved in this first set of exercises, as defined in Table 8.3.

Not all of the groups outlined in Table 8.3 participated in the cognitive mapping process in the same way. The policy makers (divided into two groups, A and B) and mayors first took part in individual cognitive mapping interviews followed by a phase of group model building (Vennix, 1996) to produce a joint cognitive map. The experts and council workers also developed joint

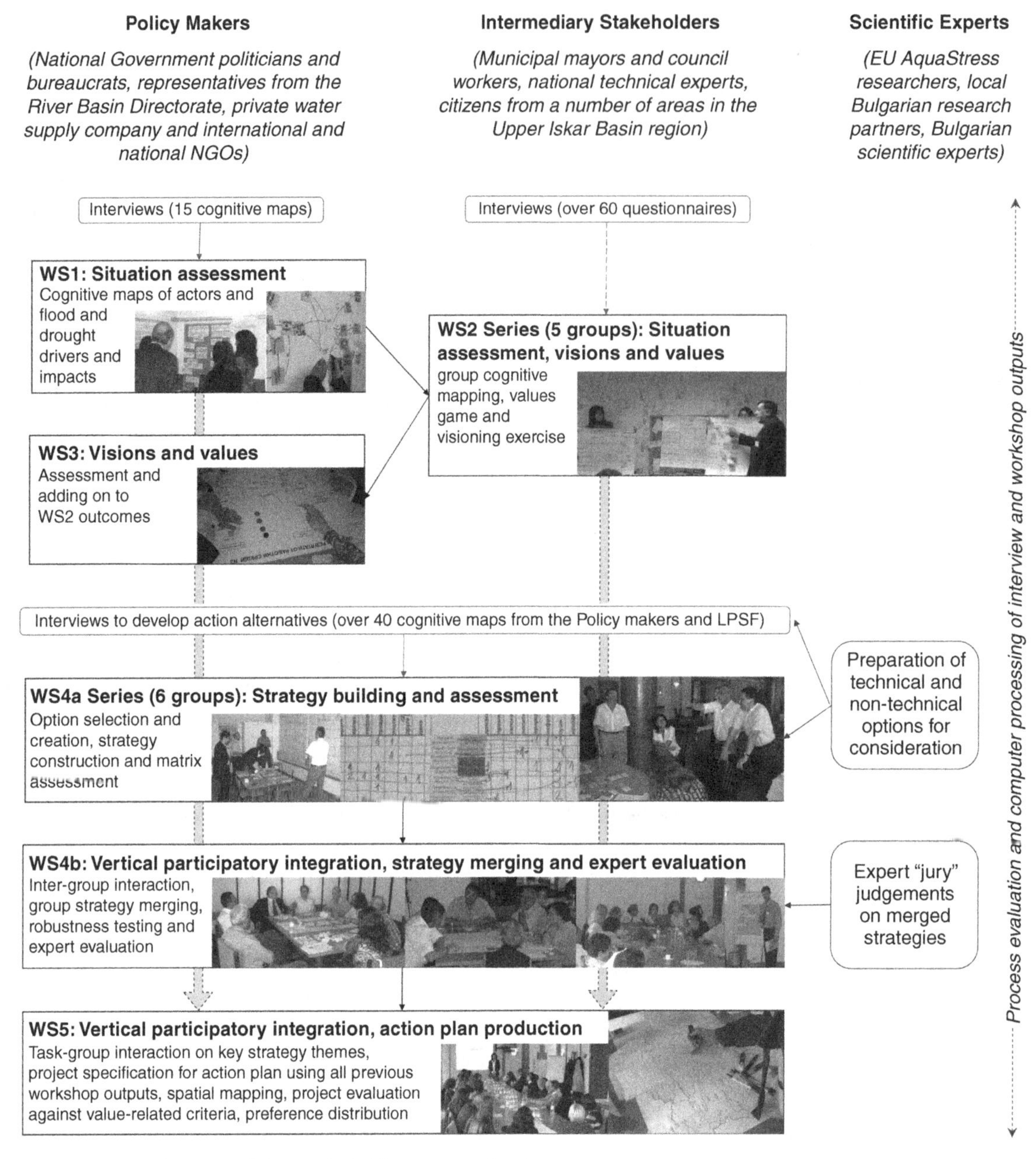

Figure 8.7 Implementation of Iskar Participatory Risk Management planning process.

cognitive maps and the industry representative created an individual cognitive map. As outlined in Section F.1.1, the citizens did not directly develop their own cognitive maps; rather, individual semi-directive interviews were undertaken and the results were then computerised into a cognitive maps-style format. One example of a policy makers' (from group B) cognitive map in its final computerised form is shown in Figure 8.8.

The process used to create these maps is outlined in Hare (2007). A number of photos showing the policy makers' workshop for the joint cognitive mapping production are shown in Figure 8.9.

Seven joint cognitive maps on floods and droughts of the type shown in Figure 8.9 were created by the different stakeholder groups or reconstituted from the citizens' interviews. These maps (policy makers A, policy makers B, mayors, council workers, experts, industry) and the citizens' interview responses were analysed further to study the perception of the participants on flood and drought drivers and impacts. The Bulgarian regional partner and I analysed most of these results together in Bulgaria, to cross-check translations and counting (with the exception of the citizens' interviews). The most common drivers, as identified by the different stakeholder groups, are presented in Figure 8.10. The drivers were divided into more technical issues of the type

Table 8.3 *Groups of stakeholders taking part in the preliminary interviewing and cognitive mapping process*

Group name	Description of group members	Total number in group
Policy makers	One parliamentary representative (from the Commission of Environment and Waters); the Vice Minister of the Ministry of Disasters and Accidents; the Director of the River Basin Directorate (Danube); representative heads of departments from the Ministry of Regional Development and Public Works, the Ministry of Health, the Ministry of Education and Science, the Ministry of Economy and Energy, and the Ministry of Agriculture and Forestry; as well as NGO representatives from Care and the Bulgarian Red Cross	10
Mayors	Mayors from villages with the worst flooding problems: Lesnovo; Ognianovo; Ravno Pole; and Golema Rakovitza	4
Council workers	Vice Mayor of Elin Pelin municipality; the Lead Engineer of Elin Pelin municipality; and the municipality urban planning expert	3
Experts	Scientists in water-related fields from the Bulgarian Academy of Science and the University of Architecture, Civil Engineering and Geodezy in Sofia	4
Industry	Head of the Water and Energy Department in the biggest industrial enterprise in the region – the metallurgical plant, 'Kremikovtzi'	1
Citizens	Representatives from the local villages and the town of Elin Pelin	100

often taken into account in hydrological models (light shading in Figure 8.10) and less technical, socio-economic related drivers (dark shading in Figure 8.10).

From Figure 8.10 it can be noted that all of the groups discussed the technical factors of 'natural climate variability' and 'hydrotechnical infrastructure management'. Another observation of note is that the experts and industry focused predominantly on the technical aspects, with the experts including all of the factors often used in technical models, with a couple of small exceptions. Most of the socio-economic factors were only discussed by the policy makers, council workers and citizens. The groups of policy makers were also the only groups to discuss financing and legislation enforcement. The council workers had the largest coverage of concepts, while the mayors concentrated more on the particular issues of their villages. The citizens did not identify public awareness as an issue.

Next, Figure 8.11 shows the stakeholders' perceptions of flood and drought impacts, again with the more socio-economic issues in dark shading and the technical-environmental issues in light shading.

From Figure 8.11, it can be observed that the impact classifications were more homogeneous than for the drivers. Of particular note is that all of the groups consider well-being reduction as an impact of floods and droughts, but that neither the experts nor industry noted the potential health impacts that result from floods and extended drought periods. Land-use impacts were especially mentioned as an effect of droughts, although not at the local municipality level; yet only one group of policy makers made the link to the issue of population displacement related to this issue. More groups took note of the damage to private rather than public infrastructure from floods, with only the local authorities and citizens raising the public infrastructure issue. The experts did not mention either private or public infrastructure that was not linked to the water systems, or ecosystem impacts, and did not focus on the governance challenges raised by emergency situations. This may be because most of the experts were highly technically trained water engineers or hydrologists specialising in hydrotechnical management and modelling, and preferred not to address the issues outside their competency. Alternatively, they may simply have had a narrow view of the issues involved.

IDENTIFICATION OF VISIONS AND VALUES

During the second series of workshops and WS3 for the policy makers, participant values and visions for the future were elicited in a number of ways. The workshop began with a 'preferences elicitation' game, where each group member, and then small groups, were to distribute a certain amount of money over their preferred economic sectors (agriculture, households, industry and nature) and the different geographical regions of the Upper Iskar Basin (Samokov, Sofia and Elin Pelin). The instructions to participants were: '*If the European Union decided to invest in three little projects of 10 million euros and one big one of 30 million euros in water management, choose where you would want these projects to be implemented,*' (Rougier, 2007). The averaged and accumulated results of the seven groups' results are represented in Figure 8.12.

These results demonstrate that the stakeholders overall have a clear preference for the protection and enhancement of the

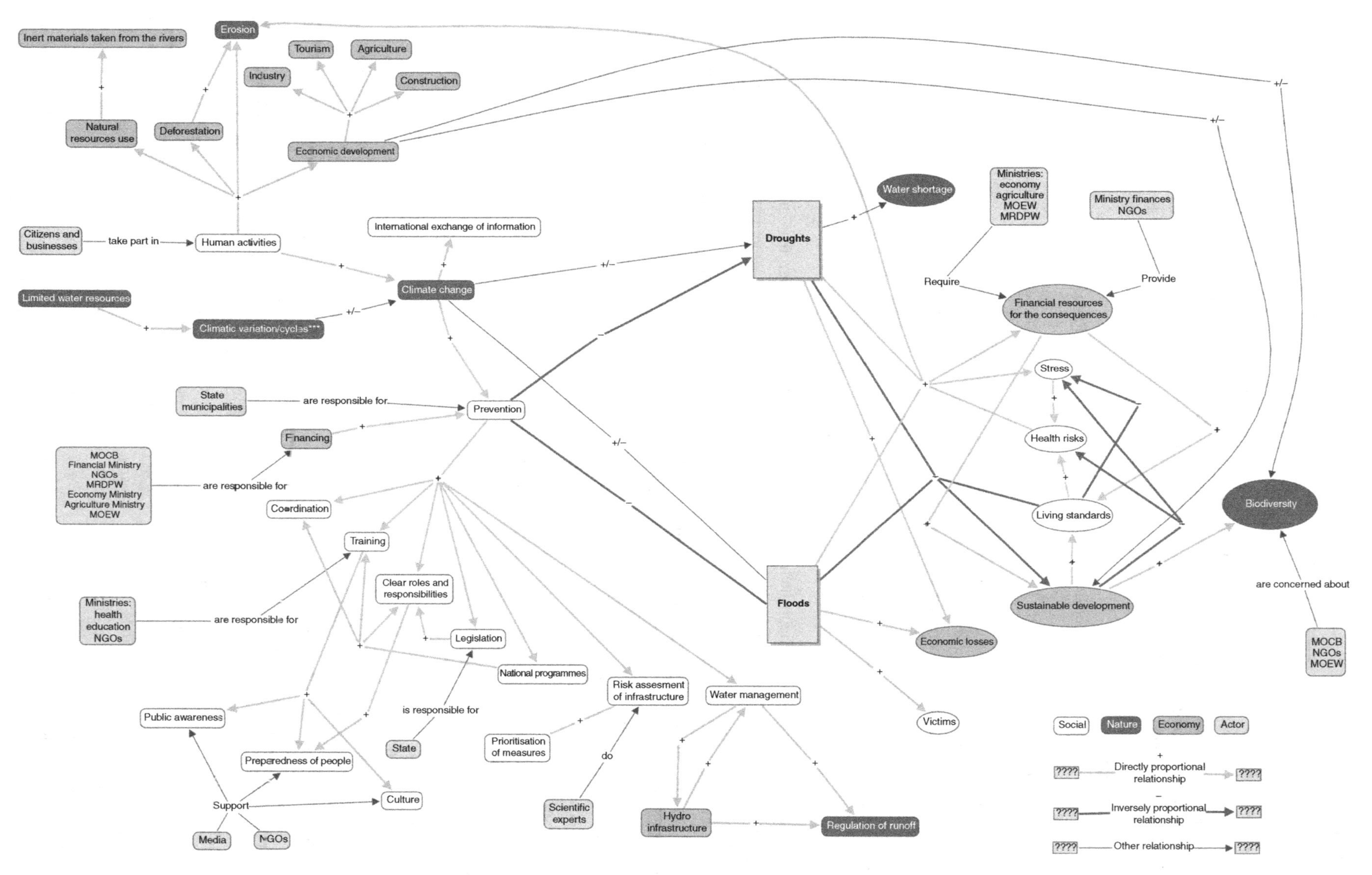

Figure 8.8 The cognitive map developed by the policy makers' group B: the drivers are on the left side of the boxes labelled 'floods' and 'droughts', and the impacts are on the right.

Figure 8.9 Creating joint cognitive maps in the policy makers' workshop (Ribarova *et al.*, 2008b).

Drivers of flood and drought risks	Policy makers A	Policy makers B	Mayors	Council workers	Experts	Industry	Citizens	Total
Natural climate variability	■	■	■	■	■	■	■	7
Hydrotechnical infrastructure management (reservoirs, dykes, irrigation channels)	■	■	■	■	■	■	■	7
Topography					■	■		2
Vegetation cover					■			1
Land use type and management				■	■	■	■	4
Building of infrastructure	■						■	2
Water management		■			■		■	3
Industry		■		■		■		3
Global warming/climate change		■		■			■	3
Polluted/congested riverbeds			■	■		■	■	4
Deforestation		■	■	■			■	4
Legislation	■	■		■				3
Financing	■	■					■	3
Legislation enforcement, monitoring and risk assessment	■	■					■	3
Crisis management system	■							1
Human activities and behaviour							■	1
Public awareness	■	■		■	■			4
Clarity of role and responsibilities	■	■	■	■			■	5

Figure 8.10 Flood and drought drivers, as identified from the seven groups' cognitive maps. Light shading indicates technical drivers, and dark shading indicates socio-economic drivers. Source: Daniell *et al.* (2011).

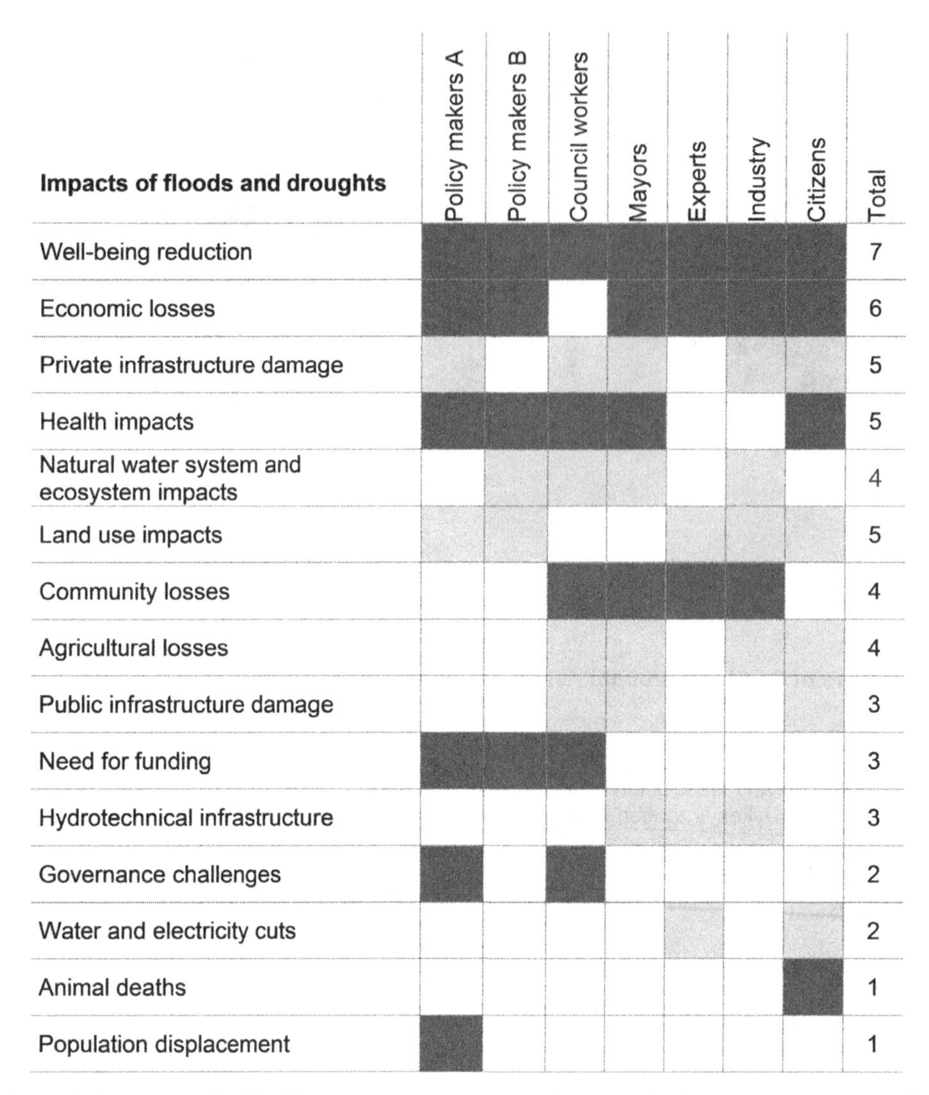

Figure 8.11 Flood and drought impacts, as identified from the seven groups' cognitive maps. Light shading indicates technical drivers, and dark shading indicates socio-economic drivers. Source: Daniell *et al.* (2011).

natural environment, in particular in the upstream areas. There also appears to be a strong preference for the reinstallation and financing of agriculture in the Elin Pelin area, and investment in industry in the Sofia region. Of all the geographical regions, Elin Pelin appeared to draw the overall preferences for funding, although these results may be biased by the participation of two groups from this region, in particular the group of organised stakeholders, who were the only group to distribute all of the money within their own area.

For the elicitation of visions, stakeholder groups were asked to think about positive and negative futures for 10 years' time, both individually and as a group. This exercise produced a list of visions from the six groups, which the Bulgarian facilitator and I later classified into eight categories of values that should be preserved or enhanced. The list of values is presented in Table 8.4.

The visions from each group were collected under each of these values and were presented back to, and used by, the participants in the final mixed workshops, for evaluating proposed projects. Further comparative discussion on these values with the Australian intervention will be outlined in Section 9.1.3.

FINAL RECOMMENDATIONS

Final recommendations built upon many of the previous participants' work and the computer-processed results outlined previously. The flood and drought risk-management options were constructed through cognitive mapping interviews and group work, the creation and multi-criteria assessment of group-created strategies using these management options were merged and their robustness was tested in the first combined group meeting (WS4b). Analysed results from these strategies have not been presented here, as much of the groups' work was not readily comparable, owing to changes in facilitation and techniques for strategy building (WS4a). The most important results from

Table 8.4 *Value categories elicited from the stakeholder groups' visions*

Value categories for visions (in no particular order)		
'To feel secure and healthy' (Enhanced well-being)	Sustainable agriculture	Treated potable water and treated wastewater
'To share our lives' (Enhanced community capacity)	Preserved ecosystems	Effective water supply
Effective management	Sustainable economy	

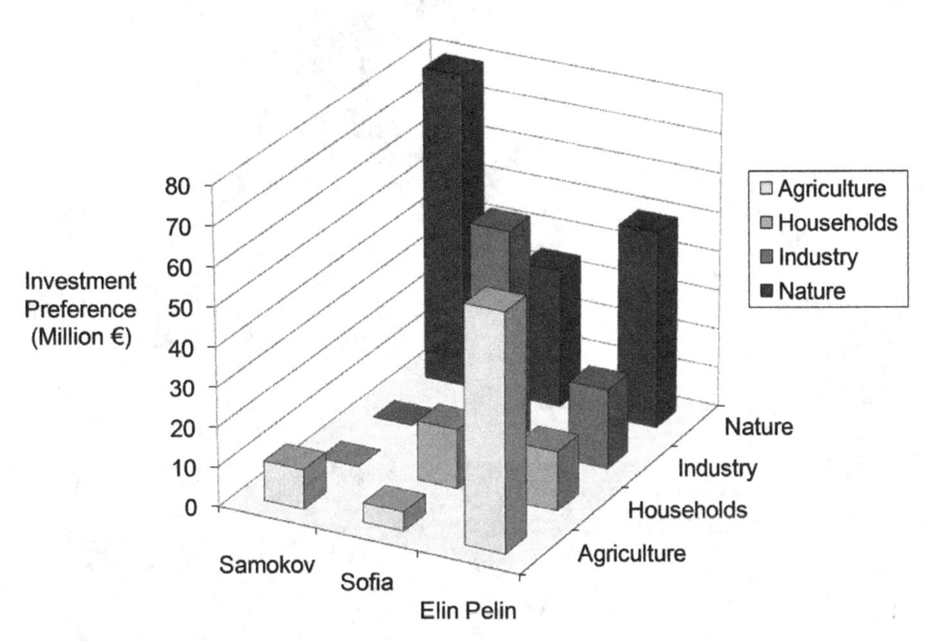

Figure 8.12 Accumulated results of the preference elicitation game of the stakeholder groups. Adapted from: Rougier (2007). Source: Daniell *et al.* (2011).

WS4b are considered to be the relational aspects of the vertical group integration that took place, rather than the content of the final four strategies. Moreover, the subtlety of many of the results of these workshops remained in the conversations that took place in Bulgarian and were thus difficult for a non-Bulgarian speaker to appreciate. Further discussion and evaluation of these workshops is presented in Section 8.5. The Bulgarian facilitator, Bulgarian regional partner and I worked together to present as many of the previous results from each workshop as were relevant to the final action planning session programmed for the last workshop.

The final flood risk mitigation projects defined and evaluated in the last workshop as part of the risk management plan are shown and discussed in Section F.3.2. Each of the final proposed projects was evaluated on its potential to support the list of eight values, presented in Table 8.4, as well as on criteria of implementation problems that the project would be likely to encounter (i.e. costs and infrastructure, social and institutional or uncertainties in the execution). From these evaluations, it was shown that the vision of 'to feel secure and healthy', which would enhance well-being, would benefit people the most if all the projects were implemented, followed by the visions of 'effective management' and 'to share our lives (enhanced community capacity)'. The most likely costs to be encountered were to be 'costs and infrastructure', followed by 'social and institutional'. It remains to be seen, especially if the structural funds cannot be obtained, whether these prioritised projects will be able to be realised either completely or partially.

8.4.3 External analyses, final results and possible process futures

During the final workshop, the Bulgarian evaluator computerised the results of the action plans, to aid final discussion of the projects. Just after the workshop, the French research director transferred the spatial distribution of the projects stemming from the spatial mapping exercise and the computerised plan into Google Earth. Images of the spatial mapping, the final plan of prioritised projects and Google Earth representation are shown in Figure 8.13.

The final spatial layout suggests that one possible future for the project would be to investigate further the coherence of the projects, if they were all implemented. Coherencies could be assessed in terms of time, space and over other factors, such as

Figure 8.13 Transfer of the final prioritised projects in their spatial layout to Google Earth.

feasibility of ongoing maintenance costs. The purpose of the action plan creation was partly to help develop projects that could be submitted to receive Bulgarian structural funds. As these projects are written, this representation in Google Earth could help in determining whether the proposed projects are feasible over the selected territories and could also lead to assessment of their impacts on other projects. For example, 'development of a flood model' for a particular zone may be a prerequisite for the flood retention basins to allow 'storage of high waters during floods', if the design of such a project is to be well informed.

Another possible future for the project is further to support the citizens and organised stakeholders of the Elin Pelin area to continue to work collectively to cope with flood risks. This possible future has already progressed, as a meeting was held to present some of the results of the AquaStress project in the Iskar, at the request of some of the Elin Pelin stakeholders. During this meeting in May 2008, the French research director and I also presented some aspects of flood risk management that are carried out in France and Australia at the local level, a number of photos from which are shown in Figure 8.14.

The determination of the stakeholders from Elin Pelin to continue collaborations with the organising team of the Iskar process to help them to work together was evident, since the 20 participants of the meeting were not paid to attend. Out of this number, half were stakeholders and the group included more local mayors, who had not originally taken part in the Iskar participatory process. After the presentations, discussion was lively, and some of the stakeholders were particularly interested in how voluntary emergency service groups, such as those in Australia, were set up, and how something similar might work and be funded in Bulgaria. The next stages of any possible future for this collaboration are yet to be defined.

Figure 8.14 Elin Pelin follow-up stakeholder meeting.

At the national policy level, another potential project on participatory method use for climate change integration policy was mooted as a potential future proposal to the AquaStress project. This project ended up being supported by the European PEER network, although the results are yet to appear.

8.5 EVALUATION RESULTS AND DISCUSSION

This section will briefly outline some of the evaluation results obtained related to the Iskar process and provide a discussion of them: firstly, on how the decision-aiding model was used and the adequacy of the process; secondly, on a number of selected participant and external observation evaluation results, some related to aspects of the ENCORE model dimensions; and finally, on the overall intervention outcomes in terms of effectiveness, efficacy, efficiency and innovation.

8.5.1 Use of the decision-aiding model

A final summary of the elements elicited from the inter-organisational decision-aiding model outlined in Section 5.1.3, showing how the sets evolved through the process, is outlined in Table F.1.

In the Iskar process, the decision-aiding process model was only explicitly brought into use when I became involved in the process, despite the fact it had also underlain the participatory modelling process proposal on which the Iskar process had been partially co-designed, as explained in Section 8.2. With the model in mind, a number of questions were asked of the French researchers relating to the contents of the Iskar process, including elicitation of stakeholders' objectives, whether values had been elicited and used for criteria, and what type of final recommendations, if any, were to be sought at the end of the process. Before the fourth series of workshops, many of the elements related to the problem situation and problem formulation meta-objects had been elicited and discussed by the stakeholder groups. In particular, a set of actors and objects, predominantly drivers and impacts of floods and droughts, was defined in a participatory manner through the first interviews and first two workshops. Some preference criteria from the evaluation model meta-object had also been elicited formally through the game described in Section 8.4, with the results shown in Figure 8.12. Follow-up workshops and interviews with stakeholders for the definition of actions and multi-criteria evaluation were also planned. Much of the decision-aiding model's elements were designed to be elicited and structured in a participatory manner.

The cited project objectives did not include 'aiding decision making' to help reach a specific decision, such as elements in a plan. Rather, the objective was to, '*Evaluate management strategies in terms of different indicators, with respect to varying uncertainties and scenarios, rather than provide single definitive answers*,' so the form of 'final recommendations' was ill defined. Similarly, the issue of what the participants could take away at the end of the process was not clear. Nor was it clear whether the process was being carried out to help the Bulgarian participants meet their expectations, or whether this was purely a research exercise. If it were the latter, significant ethical questions would have arisen.

Because of the apparent lack of strongly shared objectives in the project team, apart from implementing the planned process according to the schedule, it appeared that encouraging the continuation of a decision-aiding process based on the model elements would not deviate too much from what was planned but would help to provide further structure. It would also allow the Bulgarian and Australian processes to be more closely compared.

Before the fourth series of workshops, one of the particular sets of elements and relations of the decision-aiding model that I had discussed with my French colleagues was how, if at all, the participants' visions and preference distributions were to be used in the following stages of the process. Whether and how the formulation of participant value-based criteria could be elicited or re-established from this previous or future work or could be used for evaluating the actions – the 'set of dimensions' in the evaluation model meta-object – was of particular interest. Another element of the decision model that was likely to be lacking was the uncertainty structure. Although the concept of uncertainty is implicit in the definition of 'risk', the nature of these uncertainties did not appear to have been treated explicitly in any detail, in particular, not to the same level of formalisation as in the Australian LHEMP case.

In the set of workshops labelled WS4a, some of the dimensions against which the actions were measured were the same as those used in the value-elicitation game; yet they were again externally imposed rather than discussed with the participants as being the criteria of most importance to them. The meaningfulness and importance of these evaluations for the participants who were to use them was therefore questionable. There had been so much variation in the set of scales used to measure the impacts on the set of dimensions, that the meaning of the results for the project team, especially the external French researchers and me, remained quite cryptic and not as useful as intended. The design of WS4b was therefore adjusted to address this problem, and there was an opportunity to introduce discussion and reflection on management strategy uncertainty. This uncertainty structure was based on the robustness of flood and drought risk management strategies when faced by different extreme events.

Prior to the final workshop, an analysis of the decision-aiding model elements that had been elicited and whether these had been created in a participatory manner was undertaken by the French research director at my request. Elements or connections between them, which I thought might be lacking, were confirmed as such. The design of the final workshop thus tried to address some of these weaknesses by reintroducing the participants' own visions in the form of eight value categories to be used for final project evaluation. Many of the previous workshops' element sets, such as cognitive maps, management options, visions and strategies, were organised so that the participants could at least use their own sets of operators to consider the information in their creation of final recommendations of projects for flood risk management in the Elin Pelin region. This retreat to intuitive methods had not originally been planned, as the cognitive maps had been envisaged originally as providing possibilities for 'calculation', but this turned out not to be the case.

The change of project direction, as outlined in Section 8.3 and Section F.1.5, although effectively annulling the provision of final recommendations for one of the problem statements, led to a prioritised set of final recommendations in the form of 24 flood risk management projects. By the end of the process, the decision-aiding model had been used and had affected, in part, some process design choices.

8.5.2 Selected participant and external observation evaluation results

This section will outline a number of the evaluation results based on participant and external observations on a few key issues related to the ENCORE model dimensions outlined in Section 5.6. In particular, levels of participants' perceived cognitive, normative and relational learning will be outlined. A few of the participant's perceptions of their own practice and preference changes, as well as a few potentially worrying lessons that some participants took away from the process, will also be highlighted.

Firstly, the participants' perceived depth of their learning relative to a number of areas over the six workshops is shown in Figure 8.15.

From Figure 8.15, it appears that the majority of participants learnt slightly more over the full workshop process about other stakeholders' points of view and relations than about floods and droughts or the impacts of certain flood and drought management options. In particular, it appears that the WS3 for the policy makers' group was significantly polarised towards learning about others' points of view and relations, and especially not towards learning about floods and droughts. This result is particularly important for the project team, as WS3 had been designed with the principal objective of sharing, discussing and building upon all of the other stakeholder groups' representations and visions of flood and drought risk management (Rougier, 2007; personal communication). Similarly, WS4b had been designed with the specific objective of helping the stakeholders get to know each other better, and this was the only other perceived learning result, where 100% of participants agreed that they had learnt more about other stakeholders' points of view and relations. These results demonstrated the potential capacity for co-engineering workshops effectively to allow specific pre-set shared objectives to be met.

From the qualitative responses to the final questionnaire, it was also evident that much was learnt about the importance of prevention and preparing for floods and droughts as part of risk management strategies, as almost a quarter of stakeholders mentioned this aspect specifically when asked about the most important things they had learnt through the process. Examples of such responses included: '*The prevention activities in all directions are a very important issue here; and also the good maintenance of rivers, river beds, dikes, etc.,*' '*I understood that we can reduce the damage effects from the floods if we are prepared for it,*' and '*How to prevent floods and droughts.*' Whether the translation of this final statement does it justice is difficult for a non-Bulgarian speaker to ascertain, but in its current form this comment lacks a little subtlety; as one of the objectives of the workshops was not to teach or encourage the understanding that floods and droughts can be prevented, but rather to encourage the

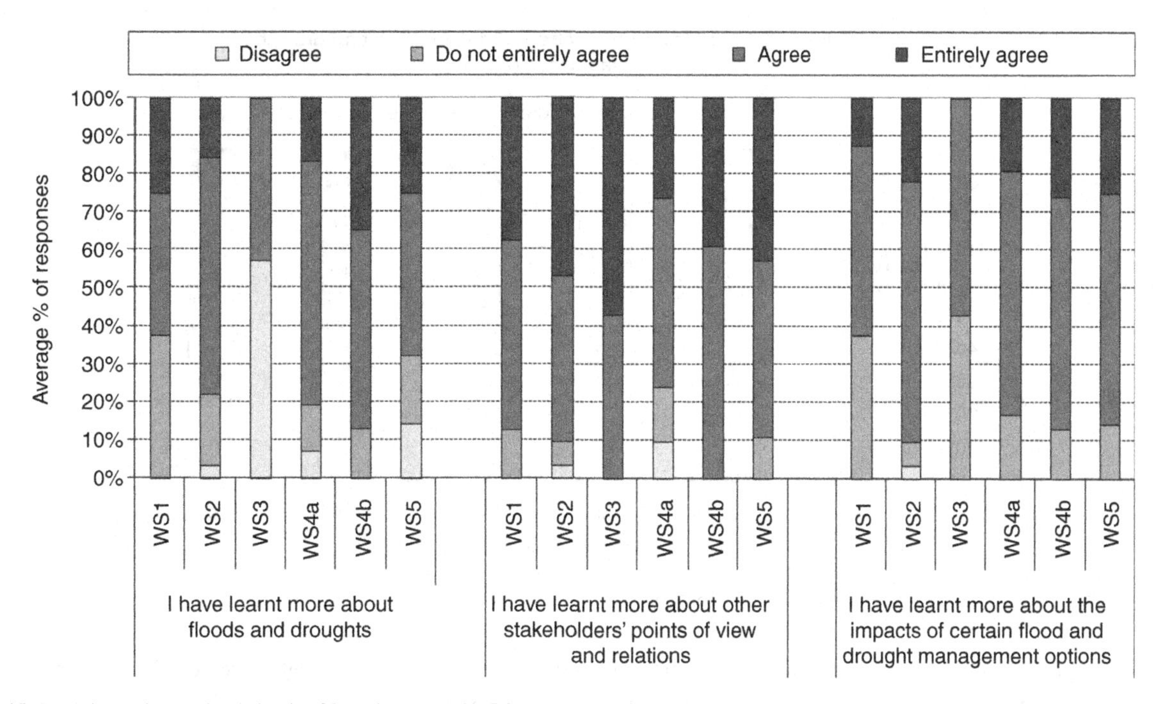

Figure 8.15 Participants' perceived depth of learning over the Iskar process. Source: Daniell *et al.* (2011).

understanding that risks can sometimes be more effectively managed by use of adequate preventative measures and planning that will act on flood and drought drivers, instead of just treating the impacts of the crisis events.

Other subjects of learning mentioned by stakeholders included work methods and experiences of the group work, e.g. '*The new method of working,*' and '*The shared experience of the participants in the process.*' Learning about problem identification was also mentioned a number of times, including that participants had learnt about '*Identification and better understanding of the problems,*' and '*The different factors that influence floods and droughts; team work which provides better solutions.*' Another person commented, '*I met different people during the F & D project with different points of view, opinions and ideas. These contacts and joint activities enriched my thorough vision and knowledge about the discussed problems.*' These comments also provide support for the quantitative results in Figure 8.15, demonstrating in a different manner that cognitive, relational and perhaps even normative ('better solutions') types of learning occurred through the process.

On the issue of 'best' potential flood and drought management options, Figure 8.16 shows the participants' perspectives on whether their personal opinions on the best management options have changed throughout the process

From Figure 8.16, it can be seen that, except for WS3, approximately half the stakeholders in each workshop thought that they changed their opinion on the best flood and drought management options for the Upper Iskar region. This seems to indicate that there was continuous normative change for some participants throughout the series of workshops. As the responses to the questionnaires in Bulgaria were anonymous, it cannot be ascertained whether the same participants kept changing their opinions or whether all of the stakeholders had changed their perspectives at some stage. However, as WS1 and WS3 were only for the policy makers, we can see that at least half of them changed their opinions at some stage through the process (in WS1).

In terms of external impacts of the process, it appears that there was a transfer of knowledge about work methods to at least one of the participating stakeholders (a member of the project steering stakeholder group – the LPSF) who plans to reuse them in other work outside the process: '*Personally, I was glad that I had the opportunity to be part of the Bulgarian team of the project. For me, this participation brought about very useful professional experience, enriched my knowledge and expanded my professional contacts. I think I can use some of the approaches seen during the project work in current work.*' In fact, this summary presents well a whole range of perceived personal changes in the cognitive, operational and relational spheres, as well as being the point of view of at least one participant.

The only dimension of the ENCORE model (see Section 5.5) that was not specifically evaluated by the participants was the equity dimension. However, from external observations, it could be considered that this process was able to provide different groups of stakeholders with a relatively equal opportunity to participate and exchange their ideas and perspectives with others, and that the exchanges and time given to different stakeholder groups to participate was somewhat equitable, with the exceptions that the policy makers' group were given one extra

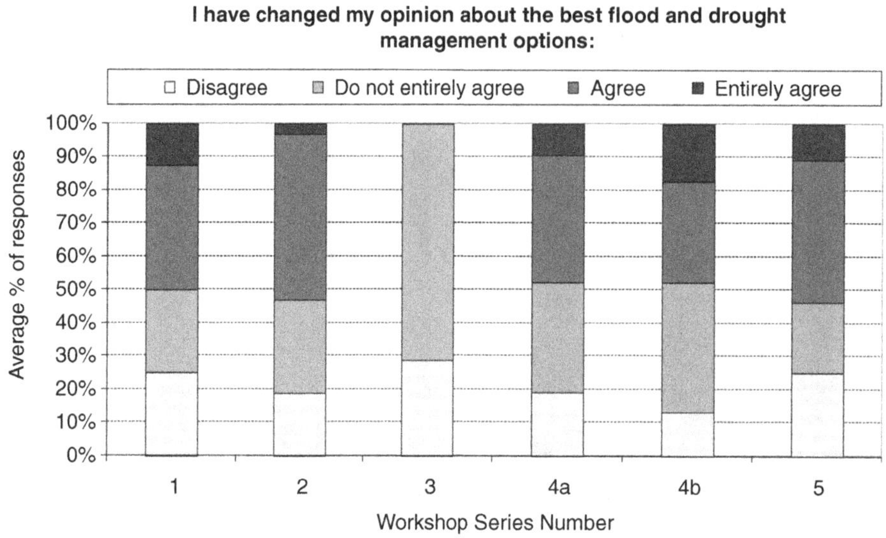

Figure 8.16 Changes in participants' opinions over the workshops.

workshop, they were paid more and their meetings were often in more comfortable settings. The equity dimension was particularly evident in WS4b and WS5, where different stakeholder types, from citizens to policy makers, were able to discuss issues of floods and droughts freely with each other, and from the authority of their own knowledge. For example, in one of the groups in WS4b, it was possible to see two citizens whose homes had been flooded in the Elin Pelin region taking a fairly large amount of the group's time to discuss openly their concerns and thoughts on management needs with high-level policy makers and municipality-level stakeholders. The opinions given by all of these stakeholder types appeared to be mutually respected by the others in the combined workshops. In other words, there were equitable exchanges in these interaction spaces. Considering typical hierarchical and technocratic power structures in the Bulgarian political and water management decision making sphere this in itself is a very interesting result of the process: that some greater forms of equity between stakeholder types participating in the creation of flood and drought risk management strategies were able to be fostered by the process.

8.5.3 Overall intervention outcomes

A number of more general results concerning the effectiveness, efficacy and efficiency of the participatory modelling process implementation with its associated co-engineering process will now be examined, followed by a number of innovations that occurred through the process.

EFFECTIVENESS

In asking whether creating the Iskar process was an appropriate initiative, we find that legitimisation from the stakeholders involved was overwhelmingly received. The stakeholders' opinions on whether the meetings should have been held are highlighted in Figure 8.17, which demonstrates that the stakeholders all perceived that the workshops were important and deserved to be held. Whether this would have been different if they had not been paid to attend remains unknown.

These results are supported by the qualitative participant responses to the final questionnaire, with many statements that consider the importance of the issues and the effectiveness of the process management that allowed fruitful and useful working time. For example, the stakeholders noted that: '*The meeting today was very useful; the project as a whole is up-to-date!*'; '*Excellent organisation; useful information and fruitful joint work,*' '*The process was very well organised and managed. I hope that the overall results from this investigation will lead to positive practical effects and results,*' and '*It's my first time participating in CS3 [the Iskar process] meetings today. I'm very much impressed by the organisation of the meeting and the way it was conducted and the important issues.*'

Considering the choice of using the process as an investigative decision-aiding exercise and not a definitive decision making or planning process with enforceable decisions at the end, it appears that this may have aided the lack of serious conflicts and allowed easier interactions and learning to occur. For example, one of the members of the LPSF (project team member and stakeholder participant) stated that:

> *Real conflicts were not encountered, since the project was not aimed at financing of specific activity but rather the developing of models and schemes for management or support. The project did not raise any obligations for the parties involved; only recommendations were given, which could not lead to serious conflicts. However, if real*

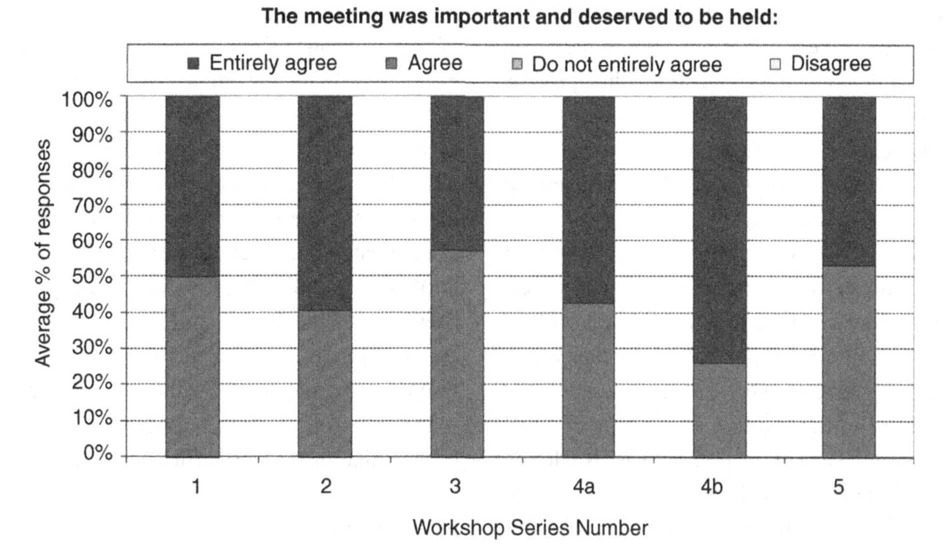

Figure 8.17 Participants' perceived importance of the participatory Iskar process. Source: Daniell *et al.* (2011).

financial resources had to be allocated (especially as a grant), I'm convinced that there would have been real conflicts.

Therefore, for the purposes of encouraging viewpoint exchange, management option investigation and a multi-level participatory process experience, the choice of a non-obligatory framing of the process appears to have been warranted.

EFFICACY

Efficacy here will be analysed based on the extent to which the research objectives outlined at the beginning of this chapter have been achieved. Firstly, the objective to *test a participatory modelling process in a 'research' situation to improve the capacity of multi-level stakeholder groups to cope with flood and drought risks* appears to have been met to a reasonable extent, as the participatory modelling approach could be implemented and evaluated in the form negotiated between the researchers of the AquaStress project, the Bulgarian regional partners and the LPSF. Some results of this evaluation have already been shown and discussed and further general evaluation results will be outlined in this section. However, three main issues could shroud the complete realisation of the objective as stated. Firstly, the classification of the 'research' situation became less evident towards the end of the process, when real participating stakeholders' needs and objectives were considered to be equal to, if not more important than, some of the remaining research needs or objectives. In other words, it is very difficult to 'test' a participatory modelling process with real stakeholders on real problems without the process having to be considered en route for its effects on the involved stakeholders. Significant ethical concerns may arise if research and stakeholder objectives are not compatible. This issue will be discussed further in Section 8.6.3 and Section 9.2.2. Secondly, and related to this aspect, the capacity and willingness of the project team to implement the process and develop the required results for use in later stages, as originally planned, was a real challenge. The participatory modelling process was itself changed in a few small ways and unexpected types of result emerged compared with its planned form through the implementation. These changes, for example syntax differences between the models and a lack of computer processing time, meant that certain activities foreseen by the French research director, in particular having 'calculable' qualitative models for analysing resources and the coherency of actions (Ferrand, 2008; personal communication), were not really achieved. This meant that the 'participatory modelling process' could only be considered to have been partly tested. Thirdly, measuring coping capacity is difficult theoretically, as it may take the next crisis to find out whether this participatory modelling process (amongst potentially other initiatives) has in fact had a physical impact on stakeholder coping capacity. Despite these issues, it appears that at least the first steps towards developing coping capacity have been aided by the participatory modelling process, as participant comments such as this one on whether the process has helped manage water in the Iskar Basin demonstrate: '*Without any doubt this process is helping the improvement of the whole area. It is a golden chance to discuss and identify the problems, and based on this analysis the most appropriate and suitable actions and activities can be undertaken.*' Further qualitative participant responses on this topic are provided in Section F.4.

Next, the objective to determine to what extent an intervention researcher entering the co-engineering process mid-way could negotiate, introduce and realise different objectives and the protection of different interests was largely achieved. As discussed in Section 8.5.1 and Appendix F, I was able to negotiate the introduction and consideration of the decision-aiding process model into the process half-way through to help meet my own research objectives, which led to a few changes in the final implemented participatory modelling process. My involvement was likely to have led in part to the format of the strategy merging and robustness analysis in WS4b, the reconstitution of participant value-based criteria for project assessment in WS5, and the overall task-force group-based design of the final workshop to aid the efficient production of the Elin Pelin risk response plan, which I had proposed based on previous project categories used in the Australian LHEMP case. The final change in the type of 'final recommendations' given at the end of the process was also driven in part by my objectives and ethical considerations, as I supported the idea of providing the stakeholders with their desired, more concrete outcomes and the clear potential to continue their work together in the future. The alternative planned outcome was to achieve some other research outcomes for the AquaStress project, potentially to the detriment of the positive perspectives on participatory processes that had been built up by the involved stakeholders in the previous parts of the Iskar project, as is explained in Appendix F. Therefore, I could, to a large extent, protect and meet most of my own objectives and interests and influence the final realisation of the participatory process and its outcomes, even with my late entry into the co-engineering process.

Finally, the objective of gaining a greater understanding of the links between research and real-world institutional and water issues in the Bulgarian context, including the positions taken by, and constraints placed on, researchers, community stakeholders, private businesses and governments in risk management considerations, was fulfilled to a reasonable extent. Being able to participate in 'intervention research' from partway through the project process meant that the external researchers had access to and in-depth confrontation with real-world institutional and water issues in the Bulgarian context in a strongly research-oriented process. The opportunity to work and develop close personal relationships with Bulgarian colleagues was particularly valuable from this perspective for a more in-depth understanding, at least from a few individual Bulgarian perspectives, of both general social and political issues in Bulgaria, the constraints and opportunities for researchers in Bulgaria, as well as the specific water and risk management issues. It was also valuable to be present to witness stakeholders, from policy makers and business leaders to citizens, working together and this provided many insights into the potential openness and capacity of Bulgarians to treat these issues. Although the legacy of previous political regimes and the very difficult transition period has left many challenges, in particular personal and public economic issues, pollution, changed social structures and degraded infrastructure, it appeared from the participatory process that this second phase of transition into the European Union holds much hope; especially considering the enthusiasm and personal capacities that the stakeholders portrayed through the participatory process to adapt to and learn new ways of interacting and working together through the investigation of the flood and drought risk management issues. The only potential issues related to not gaining an even greater understanding of the Bulgarian context could be attributed to the lack of a thorough capacity to appreciate the subtleties of the Bulgarian-language content of the workshops and the relatively short periods that were spent in Bulgaria through the process (rather than living there full-time).

Overall, from a research objective perspective, it can therefore be seen that the process efficacy was quite high.

EFFICIENCY

In the view of the large majority of stakeholders, it appears that all of the methods through the participatory modelling process were both efficient and effective, as depicted in Figure 8.18.

From the stakeholders' perspectives shown in Figure 8.18, it appears that the group model building in the first workshop was considered to be one of the most efficient and effective methods used throughout the process, based on a percentage of responses ($n = 8$). The efficiency and effectiveness of this particular activity was also echoed by the private research consultant who designed and aided the Bulgarian regional partners with the implementation and who thought that the quality of the models was up to the best that he had seen, despite them only having been built in just over one hour (Hare, 2007). The 'strategies merging' and 'external jury strategy evaluation' from WS4b were also particularly well considered, based on larger numbers of respondents ($n = 23$). Those aspects with the most mitigated support for efficiency and effectiveness were the 'strategy robustness analysis against extreme events' (17% not really convinced, 39% entirely convinced) and the methods of the second workshop series (9% not really convinced, 22% entirely convinced).

Despite a small number of stakeholders not being entirely convinced of the efficiency and effectiveness of certain methods, the efficiency and effectiveness of the whole participatory modelling process appears overwhelmingly to have been considered reasonable. As stated by an LPSF member in the final written evaluations of the Iskar case study for the AquaStress project (which included the participatory modelling process and other activities): '*The methods and the methodology as a whole were efficient enough. Having in consideration the large number of people involved in the activities, it was hardly possible to find a more efficient way of achievement of the tasks,*' (Vasileva, 2008).

Looking more closely at the co-engineering process, it is difficult to evaluate the efficiency of the co-initiation, co-design

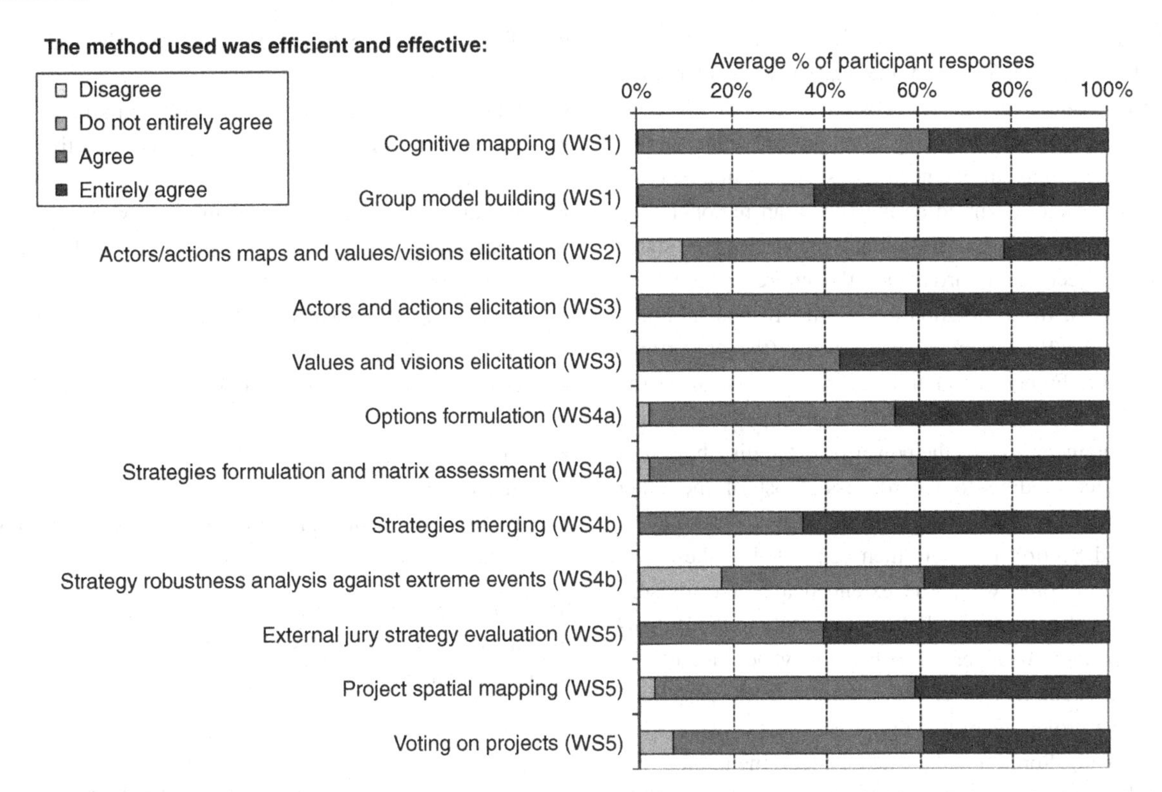

Figure 8.18 Participants' perceived efficiency and effectiveness of the participatory Iskar process. Source: Daniell *et al.* (2011).

and co-implementation process on a comparative basis, as there are few, if any, processes with which it could be compared. It could be argued that having fewer people in the project team at all times could have further improved the co-engineering process efficiency in terms of time and money spent on travel and coming to collective decisions; yet whether fewer people would have been able to provide the same level of up-to-date design knowledge, synthesis of results, logistical support and translation capacity is also largely unknown.

Similarly, the participatory modelling process efficiency in terms of budgetary outlay could have been improved if participants had not been paid, but this might have been to the detriment of the process outcomes if more stakeholders decided not to attend. Overall, the process, which lasted only one year, appeared to be efficient in the general outlay of resources, including time and money for the results obtained; in particular, the high levels of participant learning and the capacity of the different level stakeholders to work together. This view is, of course, highly subjective but as the process is innovative in its own right, especially in the Bulgarian water management context (Vasileva, 2008), more objective comparative evaluations are difficult to make.

INNOVATION

Through this process, a number of new forms of collective action and knowledge emerged. As in the Australian process, a 'multi-institutional level stakeholder working network' was developed, but this time it was in Bulgaria and extended from the national ministry institutional level to unaligned citizens. The diversity of different stakeholder types who came to work effectively together could be considered as larger than in the Australian process, owing to the larger number of unaligned citizens and the presence of parliamentarians and ministry chiefs. This collective action was largely brought about by a network centred on the Bulgarian regional partners and the LPSF members. The collective action was also based on the specific issues of conjoint flood and drought risk management and helped to build shared knowledge collectively on many aspects of the risks and their management, including the current management situation, drivers and impacts of floods and droughts, stakeholder preferences for management investments, visions for the future, technical and non-technical options and integrated strategies for dealing with the risks through prevention and crisis planning activities, in-crisis management and post-crisis remediation phases, and group work methods that can be used for stakeholders to work effectively together. Furthermore, a subset or extended network of collective action also emerged around the creation of the Elin Pelin prioritised projects for risk response, with further local stakeholders becoming interested in working with each other and being further supported by the project team. Qualitative evaluation responses in support of the innovative nature of the project and collective action, such as '*The project is unique, ingenious and very interesting*,' are outlined in Appendix F.

The other innovative but connected form of collective action that emerged through the process was the 'multi-institutional and multi-cultural co-engineering network', which co-initiated, co-designed and co-implemented the participatory modelling process. The collective action network appeared to have developed a continuously evolving core project team for the workshop process, plus an extended advisory and implementation aid network, again centred on the Bulgarian regional partners and, in particular, the operational manager. The stable core of the project team throughout the entire participatory modelling process, from initiation to the final results summary meeting, were the Bulgarian regional partner (the operational manager) and the French research director, who were also both heavily involved in other aspects of the AquaStress project and who both maintained most of the power over the distribution of the EU's project funding to the process. Most of the other members through the life of the process already described in Section 8.2 and Section F.1 were then connected to these two core members in some way. The extended aid and advisory network, on the other hand, was linked principally to the Bulgarian regional partner. It appeared to be largely because of her leadership and management of the networks and operational tasks within the project team that the project was considered to be so successful. Some of the stakeholders mentioned the strength of the organising team in leading the project to success, for example:

> *The trust between us, a public council [LPSF], and the organisers predetermined to a great extent the good work of the whole Bulgarian team. From the beginning to the end, it was one of the most important factors for the success as a whole. As far as the public authorities are concerned, the fact that they agreed to take part in some of the project's activities inspired confidence in me that things will change and they will hear the opinion of other stakeholders in the water stress. I think that such projects, having wide public participation, could change the attitude of the state institutions towards civil society. As far as the foreign participants are concerned – from the very beginning we accepted them as professionals, who had the motivation to work with us honestly and openly. This was maintained during the whole period.*

This comment also demonstrates that trust seemed to be an important aspect for the stakeholder and that it was maintained between the participants and organisers throughout the process. The complete transcription of all quantitative and qualitative evaluation results for this intervention case is provided in Vasileva (2008) and associated MSExcel spreadsheets.

Overall, from these evaluation results and analysis, especially based on participant perceptions, it appears that the following factors led to the creation of team spirit and effective work towards managing the flood and drought risks in the Iskar Basin:

- Trust (positive personal emotion towards, and confidence in, others);
- Opportunities (need for work, goals, importance of issues);
- Professionalism (task focus plus positive working relationships);
- Effective communication (stems from work methods);
- Leadership (effective organisation and relationship management);
- Conflict management (which leads to obtaining goals and social learning);
- Diversity of skills and knowledge (to cover issues and tasks);
- Interaction time (adequate for building trust and meeting goals); and
- Payment of participants (incentive and adequate resources for attendance).

Some of these factors and further issues related to the project team, its complexity and its success will be discussed in Section 8.6.2 and in the next chapter.

8.6 DISCUSSION AND FURTHER INTERVENTION INSIGHTS

Drawing from the explanation, results and preliminary discussions in this chapter on the co-engineering of the participatory modelling process for the Upper Iskar Basin, this section will extend the discussion briefly in three areas: the advantages and disadvantages of the qualitative modelling approach; co-engineering team complexity and efficacy; and balancing research, operational and participant objectives and interests. All of these aspects will also be expanded upon in a comparative discussion of both the Australian and Bulgarian cases in the next chapter.

8.6.1 Advantages and disadvantages of the qualitative modelling approach

The participatory modelling process in the Upper Iskar Basin used a range of modelling methods, as outlined in Figure 8.7. Most of these methods were of a qualitative nature, including cognitive mapping, causal group model building, multi-criteria analysis, robustness analysis and spatial mapping, as well as many group discussions and other collective activities, such as the preference distribution game and the expert jury. Using such a process of qualitative modelling and associated activities had a number of advantages and disadvantages worth considering for future processes and research, which will be outlined here.

Firstly, for the advantages, the range of methods used in the initial interview series appeared to work effectively for aiding individual stakeholder reflection for forming their own ideas and building modelling skills before meeting other stakeholders. In the following group activities, collective 'buy-in' to the next qualitative modelling methods appeared strong, seemingly as stakeholders had already had some training in the use of the types of method (i.e. cognitive mapping), were adequately aided by the facilitators, and did not require a high level of numeracy. The highly visual qualitative methods chosen therefore seemed easy for the stakeholders and facilitators to appropriate and could be used to represent and link many types of knowledge (expert, local, political, judicial, etc.). Representing such a range of knowledge types might have proved more problematic if quantitative or calculable modelling methods had to be used with the stakeholders, potentially leading to more 'black boxes', hidden calculations or data manipulation required by the project team, which have been shown in other participatory processes to negatively impact stakeholders' trust in the models (Bots *et al.*, 2011). Nevertheless, considering the level of investigations for flood and drought risk management that took place in the Iskar process, the qualitative modelling techniques that allowed stakeholders to outline their perspectives and did not present numerical answers seemed adequate; in particular, as political-level decision making is often based on good arguments, the majority view or other negotiated interests, all of which the qualitative process was able to support. However, for the final project-planning workshop, quantitative cost–benefit estimates of potential decision options would have been helpful, although even these could have been carried out up to a certain point without complex numerical models.

As for the potential disadvantages of the qualitative participatory modelling techniques, one of the issues encountered was linked to the process of stakeholder and facilitator appropriation of the modelling methods. It appeared that, as the methods were appropriated, the designed syntax of the models was often slightly adapted or modified (Rougier, 2007; personal communication). This led to a range of challenges in the process, which included the incompatibility and reuse of models, as had been foreseen, and the fact that qualitative or tendency 'calculations' using the models could not be performed so that results processing and synthesis activities were more problematic. In particular, the original 'situation models', the joint cognitive maps of flood and drought risk perceptions and actors–actions models of the management situation (WS1 and WS2), had been appropriated and adapted in different ways, so that there was not a model of the physical water and flood and drought risk management systems – i.e. the hydrological and other physical systems (e.g. economy, infrastructure, social, land use) and current actors' management actions impacts on them. Rather, there was a mix of actor networks, current and potential management actions and risk drivers and impacts, which were difficult to reconcile into one model that was not entirely recreated by the researchers in charge of the results processing. This meant that this work was a challenge to use, as intended later in the process, to analyse management options' impacts on the Iskar system. In the end, a researcher-recreated model was given for use in the final workshop. Another potential disadvantage of the use of qualitative participatory modelling methods, already alluded to in Section 3.3.4, is that it can be exceedingly difficult for the human brain to intuitively consider complex feedback mechanisms, which are likely to be present in a case as complex as the Iskar. This means that many of the multi-criteria analyses and assumptions intuitively made by stakeholders who neglected complex feedbacks could lead to predicted outcomes that might be unlikely to occur. Moreover, with respect to the stakeholder evaluation estimations, attempts were not made to estimate equally the knowledge uncertainty linked with the likelihoods of consequences from potential particular flood and drought events related to the management options, which could have been completed within the qualitative method used as in the Australian LHEMP (see Section 7.4).

It could be useful to consider and analyse the issues of modelling methods, in particular the issue of model syntax appropriation and adaptation in other settings, in order to determine how future use of participatory modelling results could be improved and complex feedback mechanisms be taken into account, without losing the 'collective buy-in' to the overall participatory process. Likewise, examining the circumstances or problem situations in which qualitative or quantitative modelling methods are more suitable, and to what extent the order of deployment of certain methods affects the process outcomes, warrants further research. This particular issue related to the differences between the Australian LHEMP process and the Iskar process will be discussed further in Section 9.1.2.

8.6.2 Co-engineering team complexity and efficacy

From the Iskar project, it became evident that co-engineering teams and their extended networks that participate in a limited number of the co-engineering activities can be particularly complex. The Iskar project presented the interesting feature of having part of its co-engineering team, the LPSF – the regional stakeholder group – initiating the participatory modelling process and some of the overall process design choices, participating in the co-initiation phase, and then being involved as participants through the participatory modelling process and continuing in an arbitration and evaluation role over the process. The interactions of co-engineering participatory modelling processes can therefore be multi-level and complex. The complexity of the Iskar co-engineering team and extended working network also meant that it was difficult for some members to ascertain the extent of the network and who was actually playing a significant role in the co-engineering decision-aiding or decision making processes. However, the majority of the co-engineering team and extended network members became closely centred on the Bulgarian regional partner over time, so she had the opportunity to lead and manage almost the whole network through the process with the aid of close individual personal relationships with each member, what are called 'dyads' or one-to-one relationships (Yammarino, 1994). Lack of knowledge of the other members' influence on the co-engineering process, therefore, did not seem detrimental to the project, as at least someone was coordinating and leading almost the whole network. The only exception to this leadership appeared to be over parts of the other associated research institutions, which were managed by the research directors. This in itself led to significant co-engineering, as it was most of these research groups who held the majority of the process design knowledge.

Moving further on, to the efficacy of the co-engineering process, there were a few particular elements in the Iskar process that made for interesting challenges. Firstly, there was a range of languages (predominantly Bulgarian, English and French) commonly used by the members at different stages of the process, which sometimes affected the capacity to effectively communicate with each other. Misinterpretations or difficulties in understanding subtle differences in meaning were reasonably common, not only because of the language issues, but also because of cultural differences, including social and political norms and different uses of body language. However, the Bulgarian regional partner argued that it was not so much the cultural differences, but rather the personal ones that were more challenging to accommodate and that the co-engineering team members had needed time to get to know and understand each other better (Ribarova, 2008b; personal communication). Despite this assertion, it was clear that during and after the process, the translation issues between three working languages led to significant filtering of results and increased the uncertainties in their interpretations.

Next, the Bulgarian co-engineering process members showed a strong capacity to take on new process roles and perform them well, despite having very little previous experience in them. For example, through the process there was much training, transfer and adoption of implementation and design skills, such as facilitation techniques and method design for participatory modelling from the external European researchers to the Bulgarian regional partners and their associated networks. This was a significant advantage for the efficacy of the process, as the Bulgarians appeared very competent in these new roles, which meant that there was a near total appropriation of the process from external interveners. Yet, this also proved challenging for some of the external researchers who lost control of much of the process and could not verify or attempt to modify the workshop directions if the content was not appropriate, or could be considered erroneous, based on external expert knowledge.

Related to this last issue, it seems pertinent to ask why this appropriation of the co-engineering process and the final implementation of the participatory modelling process was held to be so successful in a country that has had so little previous experience and skills in these approaches. In particular, in a country that still has a fairly strong state structure, with strong traditions in technocratic management and a belief in expert knowledge and state social security and protection, it would seem that such a process would be likely to encounter very serious challenges in its implementation. However, on digging deeper, the surprising success could be attributed to a number of factors (on top of the high level of the Bulgarian partners' leadership), just two hypotheses for which will be mentioned here.

Firstly, considering the technocratic traditions of the country, could having the participatory process run by a group of engineers have added to its legitimacy? Of the final core co-engineering team, including the designers, facilitators, evaluators and translators, there was only one easily distinguishable member who could not be considered as an engineer based on her initial university training (the young Bulgarian facilitator) although even she mentioned in an interview that she typically thought logically (unlike the original Bulgarian facilitator who seemed to work more on relations and feelings) and that to facilitate she just needed someone to tell her the goal of the exercise and she would find a method of achieving it (Popova, 2008; personal communications). Considering this, it should be asked whether this stability of having traditionally 'technically' or 'operationally' minded individuals running the process added to the stakeholders' confidence in the process; and similarly, whether it increased the other project team members' belief in the capacity of one another to quickly develop the required skills for facilitation, which are usually partially developed in engineering training, where strong bases in design and specification development, group work, project planning and management are common. It is just the extra interest in the personal and emotional processes of others, effective communication, political awareness, viewpoint sharing and perhaps some grounding in research ethics that may be required. However, some also argue that many female engineers are more likely than their male colleagues to have developed some of these characteristics naturally (Pease and Pease, 2000; Knight *et al.*, 2002), so may have already developed many of the skill sets required for effective participatory process co-engineering.

Secondly, it appears from the literature that despite Bulgaria's strong state structure, it is one of the Eastern European countries that has had (in 1993) the highest levels of citizen political engagement (higher than countries such as the UK and the USA, population percentage-wise) and previous Communist party membership (prior to 1989), which both appear to have positive effects on the potential democratisation of society and future citizen political involvement (Letki, 2004). In other words, compared with some other countries, particularly in Eastern Europe, Bulgaria appears to have a naturally higher potential to foster successful participatory methods, which may also explain why there have been recent participation stories in Bulgaria, other than this one, in the domains of urban planning, energy and nature conservation (i.e. Watson, 2000; Staddon and Cellarius, 2002; Brinkerhoff and Goldsmith, 2006; Nakova, 2007). However, through historical analyses of previous types of water use or irrigation associations in Bulgaria, it is also debated that citizen self-help, and the establishment of bottom-up collective action, has rarely been seen in this sector and that there still seem to be impediments to re-establishing such user groups (Theesfeld and Boevsky, 2005). From the analyses of the Bulgarian project, this potential difficulty was also apparent, especially when working with some of the citizen groups who did not seem naturally inclined to being able to help and coordinate themselves and instead asked for continued external support to help them to manage their problems. Much further local-level capacity building still seems to need support and encouragement. Investigations into what kinds of education programme might be needed for this capacity building and how a growth in volunteerism in the country could be supported could prove fruitful in this endeavour.

8.6.3 Balancing research, operational and participant objectives and interests

The Iskar process was primarily started as a research project where participants were paid under a 'research participation contract' with the AquaStress project so that they would attend and where the participatory modelling process could be tested in a real complex situation but with low real stakes, as no definitive decisions were to be sought at the end of the process. Most of the project team members were also paid by the AquaStress project, although some had more in-depth knowledge and allegiance to its objectives than others. In particular, the Bulgarian regional partners and the French research director knew the project and its research and operational testing needs better than some of the project team members, who had other or multiple allegiances. At the beginning of the process, the French researchers had the most well formulated research objectives related to the participatory modelling process and the Bulgarian project team members had the clearest operational objectives – to have the process well implemented in time and without trouble or conflict. For the Bulgarian regional partners, having confidence in people's capacity to participate well was paramount to wanting to invite them into the process (Ribarova, 2008; personal communication). Some of the Bulgarian stakeholders in the project also had clear objectives and expectations, many of which were not originally planned as project objectives by the project team, such as the Elin Pelin citizens' hope to continue to work towards physical results that would be implemented to improve the flood risk management situation. These objectives were also brought in late in the co-engineering process after WS4b by some of the project team members, including the Bulgarian regional partner, facilitator, French contract researcher and me, who wanted to give something back to the participants who had clearly hoped for something extra out of the process. As described in detail in Section 8.3.2, this led to issues of negotiating and trading-off between aims of the AquaStress project, for example, to use the multi-criteria tool and to continue to concentrate on the evaluation of conjoint flood and drought management strategies or to try to fulfil the new objectives of some of the project team members, based on the idea of sustaining participants' hopes and making their participation experience as positive as possible in the remaining process time. Other scientific and operational aims of the AquaStress project that were not realised because of a range of co-engineering process factors, included reusing the citizens' interview results as synthesised elements in the process, the running of public education courses on aspects of water management, which were not performed before the Workshop Series 4a was planned, and the lack of external expertise from the AquaStress project in developing the management options and judging or assessing strategies. Another aspect of this issue was that the original research aims of the participatory modelling project, provided in Figure 8.2, did not appear to be particularly homogeneous or even potentially coherent, as the different researchers involved in the process were all interested in studying the project and its outcomes in different ways, often based on different underlying research paradigms and questions of interest.

In light of these issues, it is likely that more time needs to be taken within the co-engineering team to reconcile or discuss potentially divergent or conflicting operational, research and participant objectives at each stage of the process, to try to better reconcile, juggle or balance them. For example, sometimes research objectives may need to be downgraded or entirely different questions asked that would fit within the operational time allocated, the human capacity available and the participants' interests. It could also be postulated that the flexibility and adaptation required to run successful participatory processes, especially from the overall stakeholder and project team perspectives, lend themselves better to some types of research, which require rather less control or reliance on results or model validation that they may be unable to obtain, except at the expense of the participants. Clarifying and discussing research objectives in an ongoing manner with participants and operationally inclined managers could help allowances to be made in operations and by participants when the proposed activities or issues, such as validation or keeping to the suggested syntax rules, are not highly important for them but they understand better the reasoning behind the proposed activities. This may also help to reinforce the idea that they are performing research together or with each other, rather than that the researchers are performing research 'on' the participants. In other words, the ethics of research and the appropriateness of different types of research questions and approaches require further examination, an issue that will be partially discussed in the next chapter.

8.7 CONCLUSIONS AND RECOMMENDATIONS

This chapter has outlined the second intervention case of this research programme, the co-engineering of the 'Living with floods and droughts' project in the Upper Iskar Basin in Bulgaria, where I worked to co-engineer the participatory modelling process with many other co-engineers and stakeholders from partway through its implementation. This work follows on from the work of Chapters 6 and 7 to continue the fulfilment of the fifth objective of this book: *to outline the lessons learnt through individual and comparative intervention case analysis to determine to what extent co-engineering can critically impact on participatory modelling processes and their outcomes.* Some of the key lessons learnt and the contributions to knowledge that this chapter provides are briefly summarised here.

The final implemented process is thought to be one of the first such vertically integrated participatory modelling processes for flood and drought risk management, in particular in a country with very little previous experience with such participatory processes. This chapter has outlined how the co-engineering process drove and adapted the participatory modelling process' direction and how the Bulgarian regional partners' leadership played such a large role in the success of the process. Results from the use of a co-engineered evaluation protocol for the intervention and other observations were presented and discussed, with some of the most interesting insights being that a number of *new forms of collective action emerged* through the project and could be investigated, including the collective action of the '*multi-institutional and multi-cultural co-engineering network*'. This network had a more diverse membership and initial skill base than the Australian equivalent outlined in Chapter 7.

Many other insights were gained through the intervention, including those about the real-world constraints and priorities of different actor types (researchers of different nationalities and, to a lesser extent, national-level policy makers, municipal government representatives, private company representatives and citizens) in the co-engineering processes of participatory processes for water management, as well as the constraints, roles and priorities of a diverse range of actors in Bulgarian flood and drought risk management.

From the evaluations, one of the results of this participatory process which goes against much discussion in the literature was that *this process appeared overwhelmingly successful from both stakeholder and project team perspectives despite it being implemented in a country with few established skills and experience in such multi-level approaches.* Various reasons and hypotheses for this outcome were postulated and other insights into the strengths and weaknesses of the qualitative participatory modelling approach taken were outlined, along with discussion on the complexity and efficacy of co-engineering the participatory modelling process, as well as on balancing research, operational and participant objectives and interests. These discussions and a number of others have opened up many areas that could benefit from research, which include:

- Examining the ethical issues associated with the co-engineering of participatory modelling processes, including: the types of research question and approach that could be appropriate and inappropriate; the cultural and contextual differences in ethical norms and their potential effects on the co-engineering of participatory processes; and questions of legitimacy for intervention and procedures for monitoring and legal recourse in the event of extreme, negative effects of co-engineering or unethical behaviour in such processes;
- Further in-depth investigation and comparison of the role of leadership and team self-management in co-engineering processes and the forms this may take in different contexts, as well as further investigation into relational and operational needs for effective co-engineering to occur in multi-institutional project management teams and their extended networks, where there are a range of divergent objectives; and
- Investigating the particularities of co-engineering participatory modelling processes for water planning and management, as opposed to other co-engineering processes in different sectors, such as the health or energy sectors, or with different ends, such as product design.

The first two of these issues and others will be investigated in the next chapter, which will treat the extended comparative discussion. Important areas in need of further examination will be outlined in the research agenda presented as part of the conclusions in Chapter 10.

9 Intervention case analysis, extension and discussion

The two intervention cases in Australia and Bulgaria, presented in Chapters 7 and 8, have only been analysed in their own contexts. The aim of this chapter is to compare and contrast the findings deduced from these interventions and the evaluation protocol, to uncover further insights and to test the hypotheses on which this book is based (Section 1.4). Discussion of the results from the two case studies will focus on: context effects; participatory modelling methodologies; the co-engineering team processes and effect of divergent objectives and leadership; participatory process ethics; and participant evaluation results. Validation of the models and protocols used through the intervention research and subsequent analyses, including an inter-organisational decision-aiding model, a participatory structure model, the evaluation protocol, and legitimisation of the intervention research findings, are then also discussed. These are followed by analyses of, and insights from, extension case studies from Germany, Algeria and a Pacific Islands nation, the Republic of Kiribati. Finally, suggestions for co-engineering best practice are proposed.

9.1 COMPARATIVE INTERVENTION RESULTS

From the descriptions in Chapters 6 to 8, including Table 6.1, the principal differences between the LHEMP and Iskar processes were:

- The Australian project was a *management-driven process* driven by water managers working for a democratically elected body with legislated management responsibility. The process had the goal of a specific output, a plan designed to improve estuarine management, the LHEMP, which was open to participation by stakeholders and included research to ensure the best possible outcome. In contrast, the Bulgarian project was a *research-driven process*, with research objectives rather than operational output goals, and had the potential of aiding local stakeholders to improve their capacity to manage their water systems under extreme conditions.
- The Bulgarian process was more operationally complex and longer than the Australian process, although the spatial areas of interest were similar.
- Both organised participatory water management processes had a similar number of institutional levels participating. In Bulgaria, the interviewing processes and the large number of preliminary workshops were carried out with stratified Bulgarian stakeholder groups. In the Australian process, mixed stakeholder workshops were held from the beginning, rather than just at the end of the process, as in Bulgaria.
- The Bulgarian participants were all paid by the research project team to attend the workshops and a different rate was applied to different stakeholder groups. In Australia, no one was paid by the project team to attend the workshops.
- In Bulgaria, the project team considered that there was a need to build the local resident and stakeholder capacity separately from the policy makers and regional stakeholder groups, to enable them to work together later on a more equal knowledge and skills base. All groups of different backgrounds and affiliations were led through the same process separately, before being merged. In Australia, with its long history of stakeholder participation, including in the targeted estuarine region, all stakeholders demonstrated a similar capacity to participate and work together from the beginning of the process.
- The membership of the Australian project team remained fairly constant in each phase of co-engineering process, whereas the Bulgarian project team underwent a number of significant changes in membership through the co-design and co-implementation processes.
- In Australia, only one language was spoken. The Bulgarian intervention was conducted in Bulgarian, with partial translations into French and English for the project team.

9.1.1 Co-engineering role comparison results

In both the Australian and Bulgarian processes, a number of people played different and often multiple roles in the process. In total, approximately 15 people were involved in the Australian

Table 9.1 *Comparative role distribution in the co-engineering processes*

Co-engineering process role	Number of people	
	Australia	Bulgaria
Principal process and method designers	3	5
Facilitators and mediators. *In brackets: number of these with recognised training or certification in this role*	5 *(0)*	9 *(1)*
People playing a significant organisational or logistical role. *In brackets: number of these specifically working in a translator role*	9 *(0)*	12 *(6)*
Project team members playing a significant technical analysis or modelling role. *In brackets: number of stakeholder process participants also taking on such a role*	6 *(38)*	7 *(~60)*
People playing a significant role in the evaluation design, implementation and analysis	3	6
Project team members playing all the above roles at some stage. *In brackets: total involved in all above roles except evaluation*	1 *(2)*	1 *(3)*
Total number of people playing at least one role in the co-engineering process, including project initiation.	15	30

process' co-engineering, and 30 in the Bulgarian process. The approximate role distribution is outlined in Table 9.1.

There were many co-engineering events, requiring a variety of decisions for the initiation, design and implementation of the projects. Many of the aspects that had to be considered by the project team in just the (co-)design phase are highlighted in von Korff *et al.* (2010). A number of other key co-engineering events are summarised for the Australian process in Figure 7.3 and for the Bulgarian process in Figure 8.6.

In both processes some of the 'scientific experts' acted as facilitators for the workshops, with their role being to aid the other stakeholders to work together and create or elicit the desired information required for the next steps of the processes. At the beginning of the process, some of these experts had little or no experience in facilitation but most adapted quickly to the role. However, depending on the professional and disciplinary backgrounds of the facilitators, neutrality over the content was variable. In a few cases, external observations of some facilitators who possessed high levels of knowledge about water management revealed that they occasionally presented their own views on the content or acted as 'gate-keepers' on which views would be given space in the collective visions. The ratio of facilitators to participants was typically no larger than 1:8, except for plenary and large group discussion sessions.

In the Australian process, the members of the project team were unchanged during the design and implementation of the participation, and there was a single focus on the required output. During the Bulgarian process, the project team varied throughout the year and for specific workshops. This led to a number of last-minute deviations from the original designed methodology, including changing its underlying objectives. For example, the inter-organisational decision-aiding model (Section 5.1.3), which I introduced just prior to the Workshop 4a Series, had a number of subtle ramifications for the subsequent process design, including the use of previous elicited values (from WS2 and WS3) as evaluation criteria for the project proposals' prioritisation. Although the Australian project team's membership and the overall goal remained constant, last-minute process changes occurred, which included excluding the non-agency stakeholders from WS2, as well as in-workshop programme changes suggested by participants. It would seem that no matter how well structured and clearly defined the objectives of a process, flexibility and developing contingency plans are essential in both co-engineering and participation.

9.1.2 Comparison of the proposed participatory modelling methodologies

One of the major differences in the two proposed processes was that the Australian project was a process with a specified output goal, the draft LHEMP, but which, because of its novelty, sought to incorporate research into the process. In contrast, the Bulgarian project was a research-driven process, with research objectives rather than specified operational, planning or management output goals, but which had the potential to increase the capacity of local stakeholders to manage their water systems under extreme events. This important procedural difference led to many more negotiations over the research agenda and process methodology required in the Australian case. The estuary manager and consultants employed as project managers had legal responsibilities for the project outcomes and risks. These risks included failure to produce a widely endorsed plan through the proposed participatory process or, worse still, increase in conflict through using the participatory process. This legal responsibility was one of the reasons behind using the Australian Risk Management Standard as a basis for the process, as it was considered that the outcomes would be more defendable to local government councillors and senior managers, state government agencies and the public.

Table 9.2 *Comparative stated collective values underlying water management (in no particular order)*

List of estuarine and surrounding community values – Australian process	List of river basin and surrounding community values – Bulgarian process
Scenic amenity and national significance	'To feel secure and healthy' (enhanced well-being)
Sustainable economic industries	Sustainable economy
Improving water quality that supports multiple uses	Treated potable water and treated wastewater
Functional and sustainable ecosystems (including biodiversity)	Preserved ecosystems
Culture and heritage	Effective water supply
Community value	'To share our lives' (enhanced community capacity)
Largely undeveloped natural catchments and surrounding lands	Sustainable agriculture
Effective governance	Effective management
Recreational opportunities	

In the Bulgarian case, although there were many discussions on how to design the process, most of the design choices were left to the European researchers, in part owing to their funding of both the process and the participants, who were paid to cover their attendance costs at workshops. The specific choices of method implementation were left to the Bulgarian local partners because of language difficulties. Unlike the Australian case, the adoption of a standard calculation approach to risk assessment, such as that proposed in the Australian Risk Management Standard, was not advanced for treating 'risks'. Instead, it was proposed to draw upon a range of methods, including scenario analyses, role-playing games and multi-criteria assessment methods.

In both projects, specific methods and tools used in the participatory workshops and for analysis by the research teams were to be negotiated and selected within the project management teams as the processes progressed. Unlike the LHEMP project, the inter-organisational decision-aiding model was not specifically considered in the preliminary design phase of the Bulgarian process, but was rather introduced and considered in the later implementation stages of the process when I entered the process, as will be further discussed.

In both interventions, the methodologies were re-designed through the co-implementation process. One change that occurred in both was that multi-criteria analysis approaches were suggested. These were to be mathematically-based matrix assessments based on decision-aiding theory such as the ELECTRE, PROMETHEE and AHP methods (Saaty, 1980; Brans and Vincke, 1985; Roy, 1985). However, in the end these were adapted in the participatory context to more rudimentary forms of matrix analysis. This occurred during the value preference elicitation games and the options and project assessments in the Bulgarian case. There were a number of reasons for this, including: to aid stakeholder comprehension, because the majority of the project team members had insufficient proficiency in the methods to make the underlying mathematical assumptions understandable to their colleagues and participants; and, owing to a lack of time, to gain and sort weighting or rank preferences. In the Australian case, a simple weighted-average approach was used, which Bouyssou *et al.* (2000) claim may compromise the real 'meaning' of the final numbers. Strangely enough, when this aspect was discussed with project team members and process participants, it incited little or no reaction or interest. It appeared in this case that, as long as the project team members were seen to have a legitimacy to manage the process and the underlying mathematics, the final results would be accepted, as long as obvious discrepancies between instinctive and calculated ranks could be logically argued. This insight is drawn in particular from the discussion and later acceptance of the low prioritised ranking of the 'water quality' risk in the LHEMP, which was instinctively labelled as of high or medium priority by all participants (refer to Appendix E for further details). Lack of multi-criteria analysis application in the Bulgarian case resulted from time constraints and doubts by some project team members of the usefulness and interest of such an approach for the final workshop of the process.

9.1.3 Further comparative notes and selected evaluation results

In both of these processes, the 'evaluation model' artefacts of the inter-organisational decision-aiding model (based on Tsoukiàs, 2007) were not constructed as a separate phase but, rather, were co-constructed along with the problem situation and problem formulation elements. For example, the criteria for the 'multi-asset' risk assessment in the Australian case were developed directly from values elicited in the 'problem situation' construction. The lists of common values elicited from participants and used in different ways in the two processes as a part of the 'evaluation models' are given in Table 9.2 and will be discussed briefly in Section 9.2.1.

One of the other large procedural differences was that the participants of the Bulgarian case built a range of individual and collective causal 'situation' models and other linked-factor cognitive maps of actors and their current actions as part of the 'problem situation' and 'problem formulation' phases of the decision-aiding process. In the Australian case, although a range of participative methods, such as spatial mapping of issues and issue–value cross-impact matrices, was developed, such causal linkages were only elicited informally in speech or in written group questionnaires, although they were also defined in the synthesis report (BMT WBM, 2007) by the engineering consultants. Rather than using causal models for analysing the influences of different factors on the estuarine system's behaviour, subjective collaborative decisions based on available and shared knowledge in the agency group (WS2) were elicited using the 'risk tables' as part of the 'problem situation' and 'problem formulation' phases of the Australian decision-aiding process.

Another potentially important difference in the processes was that most of the Australian participants were seasoned 'participators' and appeared to have a marginally more sceptical opinion about the possible positive outcomes of participation than did the Bulgarians, who for a large part had never before participated to a similar extent in 'multi-level' participatory water management analysis and decision-aiding processes. For the Australians, this participation experience meant that they were more aware of the underlying constraints of the participation process design used. However, as can be seen from a selection of questionnaire responses to similar questions posed to participants at the end of both processes, given in Table 9.3, many more positive outcomes, including participant learning and increased understanding, were still achieved, even though the participants exhibited some cynicism due to their previous participation experiences. The comparative evaluation results have been manually matched, where similar responses were available, to demonstrate common participant perspectives in different contexts.

The implications of these context differences and the similarities between elements of the qualitative evaluation responses will be discussed further in the following section.

9.2 PARTICIPATORY MODELLING PROCESSES IN CONTEXT: DISCUSSION —

This section will concentrate on formulating insights related to just two key areas:

1. The importance of context – the value and constraints of designing and implementing participatory risk management approaches in different regulatory and political environments.
2. Unintended ethical issues that can arise when working in 'real-world' management situations.

9.2.1 The importance of context: the value and constraints of participatory risk management approaches in different regulatory and political environments

Australia has a long history of participation and participatory approaches in water and natural resources management. The use and acceptance of risk management approaches to decision making is also common, as evidenced by the existence of the Australian and New Zealand Standard for Risk Management, and its accompanying handbooks, including one specifically designed for 'Environmental Risk Management' (Standards Australia, 2004; 2006). Although some participatory water management processes in Australia may not be specifically designed on coherent spatial or administrative scales with carefully developed decision-aiding methodological knowledge, there appears to be a common acceptance by and a general capacity for Australians from many walks of life to participate in them voluntarily when required, often by local, state or national regulations or legislation, whether or not they agree with the underlying purposes. Such familiarity exhibited by many participants and managers with regard to participatory approaches, including those focusing on 'risk' or 'asset' management, could mostly be considered as a positive element of the Australian regulatory and political context. However, it also presents process designers and implementers of participatory processes with a range of challenges to attract and keep the participants' interest and to achieve useful, timely and concrete outcomes. Creating efficient and effective processes and publicising them appropriately is increasingly becoming a necessity in Australia, if participation of the required individuals and organisations for achieving change is to be assured. In the LHEMP process, it was considered that the introduction of the use of the Risk Management Standard in a non-traditional domain, such as regional-scale estuarine management, and the participatory process with a workshop dedicated to just working with policy and managers, provided the necessary 'draw card' to help obtain agency, community and funding support of the LHEMP process (Coad *et al.*, 2007; personal communication), which has since been translated into funding and community support for the plan. Although presenting positive outcomes, these choices also had, and could have had, more negative ramifications, such as alienating some members of the community or encouraging a return to 'technocratic' and non-participatory management, an issue which is discussed further in Appendix E.

Unlike current management systems in Australia, the Bulgarian water sector has long been characterised by technocratic management systems. Since the fall of the country's Communist regime, the former rural community structures, based on agricultural work and equipment sharing in villages, have also been dismantled, leaving rural populations with fewer services and

Table 9.3 *Comparative evaluation – selected qualitative questionnaire responses (Daniell* et al., *2008a)*

Estuarine risk management process, Australia*	Flood and drought risk management process, Bulgaria
What are the most important things you have learnt throughout the (workshop*) process?	
'*The multi-faceted nature of environmental issues*' (WS3)	'*The basic and most important issues and problems, which are connected to floods and droughts.*'
'*There's lots to do – where will the $$ and political/management come from?*' (WS3)	'*I learnt more about the role of the different institutions in the field of water management. Actually, I understood that the region of Iskar basin is not ready yet to cope with these problems.*'
'*There is no one right way to address identified risks. Collaboration is essential.*' (WS3)	'*The most important thing I've learnt is that there are always two different points of view and they are equally important.*'
'*Many different views (understandably). Has helped me to formulate and form up my own opinions.*' (WS3)	'*I met different people during the 'flood and drought' project with different points of view, opinions and ideas. These contacts and joint activities enriched my thorough vision and knowledge about the discussed problems.*'
'*A range of challenges to the estuary exist and are ever evolving*' *(WS3)*	'*The floods cannot be predicted but the risk and the bad impacts can certainly be prevented and the appropriate measures for their reduction can be undertaken in time.*'
How did the (day's activities*) workshop process help you to work with and (relate to*) communicate with the other participants?	
'*Each workshop has increased my awareness of these processes and issues associated with presenting, managing such a process. Got to know and hear more from other participants*' (WS3)	'*In a very positive way. Every participant has the opportunity to enrich his knowledge about the problems of being a member of a large group with different people. The motivation to work in the best possible way is quite bigger when you are a member of a team.*'
'*Helps develop a team mentality*' (WS3)	'*By creating friendly and comradely relations in the team.*'
'*Gained a better understanding of individual agency responsibilities and knowledge with regards to the estuary.*' (WS2)	'*It helped me to understand better how the institutions with affiliation with water and water problems are functioning.*'
'*Good open and honest discussion, effective facilitation,*' (WS2) '*not too confrontational*' (WS1)	'*The joint work had very positive influence upon all the participants. The discussions were open and straightforward, without confrontations or conflicts.*'
How do you think this process is helping to better manage (the estuary*) water in the Iskar basin? (If it is not, please also state why.)	
'*The process provides a focus for the estuary, brings all these parties together to at least discuss and endeavour to try and plan/improve the estuary*' *(WS3)*	'*Without any doubt this process is helping the improvement of the whole area. It is a golden chance to discuss and identify the problems, and based on this analysis the most appropriate and suitable actions and activities can be undertaken.*'
'*Will only help if it doesn't end in a report that isn't widely communicated and adopted*' (WS3)	'*I really cannot understand how the results from our work – strategies, plans, information data base, will be used later at higher level – institutions and legislation.*'
'*Getting different groups (government + community) talking together and operating under agreed framework*' (WS3)	'*The project provides an excellent opportunity to put all stakeholders in the region around the table – managers, common people and experts.*'

collective capacities. Until recently, there has also been little concern for the environmental or social impacts of management decisions and infrastructural projects. Although there is some evidence that Bulgarians are active participators in some sectors of social community life (Letki, 2004), there are few, if any, prior examples of participatory multi-level inter-organisational water or risk management processes that have been carried out in the country. Early assessments in the AquaStress project by European researchers also highlighted that the Bulgarians encountered had little knowledge about participatory processes and their potential to aid the Upper Iskar Basin's water management (Hare, 2006).

Another interesting difference between the two countries' contexts was the familiarity with the concept of 'risk'. Early in the Bulgarian process, most attention focused on issues of better dealing with 'crises' of floods and drought, with relatively little consideration of the need for pre-emptive local community planning to reduce community vulnerability through capacity building. It was rather considered that it was the government's job to 'protect' them from flood and drought events and to reduce their susceptibility to such hazards. However, later in the process, sufficient learning appears to have taken place so that participants began to understand the concept of 'risk' and the need to develop a more holistic response. This was made evident by the 13 pre-emptive projects put forward in the prioritised proposals in the final workshop. Despite the previous lack of experience in managing or involvement in participatory water management processes, the Bulgarians exhibited great proficiency in adapting and working effectively in them. Unlike in the Australian process, there was rather less cynicism surrounding the use of such a process and there was apparent sustained interest, perhaps because participants were being paid to attend. Considering the high levels of acceptance and proficiency in participating in this process, it could be suggested that further participation initiatives in the Bulgarian context may have a good chance of succeeding, if the initiators have sufficient skills and legitimacy to coordinate such a process. Financial resources may also help.

It was also interesting to note the similarities and differences in values elicited in the two processes, as shown in Table 9.2. In both countries, common values, such as economic sustainability, ecosystem health, the importance of community and effective governance or management, were made evident. However, a number of differences were also observed. The 'effective water supply', 'to feel secure and healthy' and 'sustainable agriculture' categories of the Bulgarian case stood out. Deficiencies of safe food and water in the recent past have caused Bulgarians enormous stress and suffering. Therefore, the important values elicited from the Bulgarian participants, linked to these basic requirements of life, are on quite a different level of needs (Maslow, 1943; Beck and Cowan, 1996) from the 'scenic amenity' and 'recreational opportunities' values outlined by the Australian participants.

Despite the obvious contextual and procedural differences of the two projects, there were still a number of close similarities in outcomes noted by the selected participant responses to the evaluation questionnaires, as outlined in Table 9.3. This leads to partial support of the general assertion that carefully designed and implemented participatory risk management approaches, regardless of context, are likely to aid learning, appreciation of common and divergent views on complex problems and support inter-organisational and multi-stakeholder coordination, which could help to aid future water management outcomes.

9.2.2 Potential unintended ethical issues arising in 'real-world' management situations

Although the idea that there are many ethical quandaries that might be encountered when embarking on participatory research programmes is in no way new (Cahill *et al.*, 2007; Sultana, 2007), it appears that in certain research and cultural contexts there is still minimal attention paid to them. In both the Australian and Bulgarian projects, a number of different ethical issues requiring consideration arose, two of which will be further discussed here.

It is possible that research agendas will not be fully able to comprehend or accommodate the local needs of involved communities, stakeholders and project managers before a participatory process is embarked upon. Such participatory processes have the possibility of instilling hope in the minds of participants that the process has the potential to 'make a difference' in the lives of the local inhabitants, even though such expectations may not have been intended by researchers running a 'research project'. This was the case in the Bulgarian intervention. Researchers are then put in a position where ethically they must reassess whether their own plans for obtaining certain preplanned research outcomes, if such proposals were made, are more important than the hopes of the participants. Such considerations were required in the Bulgarian case before the last workshop. After much discussion between the project team members and their institutional superiors, the process was changed to help the participants develop a list of prioritised project proposals to manage flood risks, which could be used for Bulgarian structural funds for a sub-section of the Iskar Basin; rather than completing the planned research programme that had been agreed upon by the European Union research project commission.

Grand plans for inclusive participation programmes, initially agreed upon in principle by project managers and researchers and communicated to participants, may strike obstacles when underlying managerial process objectives surface partway through the process and require a reduction in planned numbers or types of participants. Such a situation could disempower participants by omitting them, resulting in a boycott of further participation initiatives, or cause much larger impacts, such as negative publicity, distrust of the organising institutions, and mistrust of any further participatory initiatives in the region (see Barreteau *et al.*, 2010, for further discussion on this point). There are, therefore, ethical considerations when changing the structure of participation in the course of a process. Such a change occurred in the Australian process, but with apparently limited consequences, thanks to the way in which the process changes were managed.

A range of other ethical issues needs to be considered, depending on cultural norms. Issues such as the preservation of participant anonymity, the use of photos, audio and video recordings, and storage, distribution and publication of this information need to be critically examined within the cultural context. Cultural differences in ethics principles were quite noticeable between the Australian and Bulgarian processes. For example, when the Bulgarian researchers were asked if the debriefing session could be audio recorded, as required by Australian university ethics procedures (Daniell *et al.*, 2009), they did not

Table 9.4 *Two sets of questions to investigate for participatory water management*

Stakeholder participatory planning process for managing water systems	Co-engineering process for organising and managing the participatory process
Who ought to be considered as a stakeholder of the water system under investigation?	Who ought to be responsible for organising and managing the participatory process?
Why ought a water plan be created for the system?	How ought the scope and purposes of the water management plan be decided?
What ought to be the goals of the water plan?	How ought the decision be made on who ought to participate and when?
What ought to be the actions to achieve these goals?	Which participatory methods and resources ought to be used and why?
How ought progress towards these goals be measured?	Who ought to design, implement or facilitate the use of these methods with the participants?
Who ought to be responsible for funding, resourcing, implementing and monitoring these actions and when?	Who ought to analyse and synthesise the results stemming from the participatory process?
How ought a plan be adjusted based on these evaluations?	How ought the evaluation of the process take place and who ought to be allowed access to the raw data and final results?

understand why anyone would ask such a question. It seemed normal to the Bulgarians that the session could be taped, if this was required by someone they trusted. In contrast, in the Australian process, one of the participants wrote to the researcher to check that ethics clearance had been obtained from the University Ethics Committee (which it had been). In Bulgaria, trust appeared to be the driving force of local ethical standards. In Australia, adhering to correct administrative procedure, including commonly agreed ethical standards, was the key. The cultural differences surrounding ethics in participatory projects require careful attention when working in cross-cultural or even interdisciplinary research teams, to ensure that all those involved in the process develop a mutual understanding and adhere to adequate and suitable ethical standards. Common questions requiring investigation for managing participatory projects where conflict or ethical dilemmas may surface are highlighted in Table 9.4. These have been divided into two sets, with one focused on the stakeholder participatory planning process and the second on the co-engineering process.

The questions in Table 9.4 take the 'ought' form highlighted in Ulrich's boundary questions, provided previously in Table 4.3, to focus on the ethical and critical reflection dimensions required. All of these questions were posed and negotiated through to a varying extent in both the Australian and Bulgarian cases presented in this book. Extending discussion of ethics and frameworks for supporting critical reflection on these questions is beyond the scope of this section, but can be found in Daniell *et al.* (2009). Further information, case studies and tools for working through water conflicts and consensus building around these kinds of questions are provided in Delli Priscoli (2003) and Delli Priscoli and Wolf (2009).

Certain aspects related to the co-engineering questions and issues in the project teams will be outlined next.

9.3 REFLECTIONS ON CO-ENGINEERING

Elements from the relevant literature and process outlines of the Australian and Bulgarian interventions, given in Chapters 7 and 8 and Appendices E and F, will be broken down into a number of sub-sections: dealing with divergent objectives in multi-institutional organising teams; operational differences in the co-engineering processes; relational differences in Australian and Bulgarian project teams; and the importance of the co-engineering processes and project team constitution on participatory water management process outcomes.

9.3.1 Dealing with divergent objectives in multi-institutional organising teams

Firstly, it was found that working with inter-institutional co-engineering groups involved a range of issues not often consciously considered by observers or participants of participatory modelling processes for inter-organisational decision-aiding in water planning and management. Different project team members and participants held a variety of objectives for the process design, which were not necessarily shared or coherent, and were not anticipated in the co-initiation phase of the participatory water management process project. The introduction of the Australian and New Zealand Standard for Risk Management (Standards Australia, 2004; 2006) after the tender process for the LHEMP in Australia was one such example, as neither the chosen project managers nor researchers had used it before. Likewise, the interactions of the variety of different skills, resources, values and preferences that each project team member possessed commonly influenced the final process co-design and co-implementation. This was particularly evident for major process adaptation decisions, such as deciding to make LHEMP Workshop 2 'agency' only. In this case, project team members held different viewpoints and preferences on how the process should have been carried out, and in the end the funder (the estuary manager) was required to end the negotiation by making the final decision (see Table 7.1 and Section 7.3.2). Both co-engineering processes required continuous negotiation or other forms of decision making, such as consensus building. On some occasions, vetoing behaviour by certain project team members was also observed; this behaviour was typically exhibited by the employer, funding institution or legally responsible project managers, who used it to help carry out their dedicated responsibilities or contractual commitments to external institutions, for example to the EU in the Bulgarian project, or to satisfy superiors to whom they were accountable, such as the councillors and council managers in Australia. The exertion of this power was also seen to have negative effects on the cohesion of the project teams if the decisions were not supported by other key project team members, as occurred on a couple of occasions in Bulgaria, where research objectives under the project contract were not entirely supported or even understood by other team members. At other times, the project team members 'on the ground' simply took executive decisions without first consulting all other project team members. This most commonly occurred with the Bulgarians in Bulgaria, where language and culture were barriers for the external researchers.

From these interventions, insights were also gained on the significance of the 'co-engineering' process, rather than the 'engineering' of the participatory water management process, and this was accentuated when the team members had divergent objectives or hard-to-reconcile issues. Such divergent objectives were either manifested, often in the form of incompatible or competing arguments in negotiations (Figure 7.4 and Figure 8.6), or remained latent. In the latter case, it was observed that one party would drop an objective or attempt to ignore his or her own issue if he or she was not interested in entering into a negotiation or conflict situation in order to attempt to achieve or resolve it. Some design choices related to the introduced methodologies in the Australian intervention appeared to hide these types of behaviour. For example, the choice of a plenary session at the end of the first workshop raised some concerns from the researcher but did not entail in-depth negotiations, as group cohesion seemed to be a more important issue at that early stage, especially after negotiations earlier in the day on the other participatory methods to be used. Another possibility was that actors could attempt to work silently towards their own objectives in the hope that those actors with divergent objectives would forgo their own. This was evident to a certain extent in the Bulgarian process, related to the change in the final workshop content and the will of a number of project team participants to help the stakeholders of a particular sub-region. The emergence of this bias and objective was traced back to early in the process through the *ex-post* interviews. In the Bulgarian case, the initial methodology suffered drastic modification during the participatory process, owing to new visionary objectives that emerged during the process. These were partly supported by alternative participatory methods proposed and implemented in the final two joint workshops. Manifest conflicts, depending on their importance and impacts on the activities that the project team was to perform, required some form of management to allow work towards central objectives to continue. Many of these conflicts and their impacts were illustrated in Figure 7.4 and Figure 8.6.

9.3.2 Operational differences in the co-engineering processes

The processes varied significantly in the institutional make-up of the project teams. The Australian team had a more institutionally diverse, pragmatic and multi-accountable make-up, with the local government manager, private engineering consultants and university researchers working together. This had a significant effect on power relations and created a need to reconcile a greater diversity of requirements and objectives. In particular, the local government manager had, on the one hand, to keep his hierarchy and the elected councillors satisfied and, on the other hand, to keep the stakeholders content, to ensure future collaborations and funding opportunities. The private consultants had their profit margins and reputation to maintain and enhance, and their client, the HSC, to satisfy. The researchers were less accountable to others, as they were externally funded, had research objectives to attain and theoretically had less attachment to the context, its stakeholders and the outcomes.

In the Bulgarian context, the project team members shared some task-based objectives, at least at the start of the project, as they were all from research institutions being funded by the European Commission to carry out the Iskar project. However, there was a much greater cultural and linguistic diversity, with French, Bulgarian, English, German and Australian researchers working together, thus creating specific issues and team dynamics. In particular, there was a need for the complete transfer of the implementation and en-route re-design during interviews and workshops to the Bulgarian team, as they were all carried out in Bulgarian. In fact, this proved a valuable opportunity to enhance local capacity building, despite some frustration expressed by a few researchers because of a lack of translation facilities and in-workshop control. Such a capacity building process was absent from the Australian case, as external consultants and researchers were largely responsible for these aspects, and local stakeholders were already fully aware of participatory processes. Therefore, *language barriers may in fact present some opportunities for greater skill transfer and local capacity building, rather than simply presenting challenges.*

Attendance fees constitute another operational difference between the two cases. In the LHEMP project, no compensation was provided to participants, creating some inequity between institutional representatives attending on duty and local community stakeholders volunteering their time, which may have contributed to the perceived cynicism. In the Iskar project, different rates of compensation were provided, depending on the status of the participants, creating some bias in the participatory process and perhaps in the participants' opinions of the process but ensuring at the same time a widespread attendance.

Linked to these characterisations of diversity in our two cases, as well as those from the literature (e.g. Page, 2007), I would suggest that diversity and creative tension among project group members internal and external to the water management system provided the necessary opportunities for innovation and collective knowledge creation required to manage complex adaptive systems, even though this increased the need for more concerted efforts in also managing the co-engineering process. In particular, the two co-engineering processes were quite complex in their role distributions, with a large number of people contributing directly to the project teams or their extended network, and key individuals playing multiple roles, as shown in Table 9.1. In the LHEMP project, I took on all of the noted roles at some stage through the process, while in the Iskar project, the regional partner took on all of the roles. What was observed from both of these processes was that more conflicts took place as project team members played the same roles, but whether the co-engineering was more generally constructive in its attitude to conflict (see: Eisenhardt *et al.*, 1997) and added to the quality and results of the process, or was destructive, is difficult to ascertain. Whether this complexity could be effectively reduced, and whether this would be desirable, as some team management literature suggests (e.g. Katzenbach and Smith, 2002; Gratton and Erickson, 2007; Gratton *et al.*, 2007), would be a useful topic for future research. It would be equally interesting to investigate what models of co-engineering groups appear to be the most effective in promoting effective participatory water management outcomes, in other words who 'ought' to participate in which roles in co-engineering processes. It could also be of interest in future processes to determine whether another level of participatory management, that is, a 'co-engineering' of the co-engineering process, and hence the use of specialised participatory methods, could be used to work more effectively with this co-engineering group diversity. This issue will be further discussed in Section 9.5, with an example of a potential method of co-engineering support provided through the Algerian extension case.

9.3.3 Insights on team structure and relational characteristics

In the LHEMP project, one essential factor of cohesion was a compelling performance purpose shared by the principal co-engineers: the creation of a legal and sustainable management plan for the estuarine region. This common purpose appeared to guide the collective and task-based work of the group, despite some of them knowing very little about one another. It also appeared that the estuary manager would at times exhibit his transformational leadership capacity (Bass and Avolio, 1994) to motivate the group to keep focus on their vision. While most decisions were taken after face-to-face meetings, phone conferences or emails, some of them emerged from individual initiatives, creating, in some instances, tensions or power shifting in the group. This issue was probably reinforced by the fact that three of the co-engineers had overlapping expertise and similar higher education levels, both factors being challenges for developing effectively performing teams (Gratton and Erickson, 2007). However, this did not appear to be damaging overall, as team members exhibited strong mutual and individual accountability for their work and managed to avoid open conflicts (Eisenhardt *et al.*, 1997).

In the LHEMP project, despite divergent individual objectives and interests, the overall performance and outcomes of the co-engineering process represented was greater than the sum of individual contributions. In other words, the participatory process that was finally implemented was indeed the result of joint efforts and largely integrative negotiations. It could therefore be considered that the core co-engineering group mostly followed the 'real-team' discipline described by Katzenbach and Smith (2002), despite its short lifespan. This in itself appears to be an interesting finding in that some team members can know little about one another and may not have built very high levels of mutual trust over a short period of time (Covey and Merrill, 2006) but can still manage to work together and obtain outcomes in line with an objective, although not necessarily with all of their personal objectives. It could be considered that, although the estuary manager was not managing the project team, his natural leadership capacities and good social skills encouraged the successful performance of the co-engineering team. The estuary manager typically led by example and readily showed his respect for the work of each of the project team members and stakeholders in a firm, measured and quietly supportive way, which led to strong levels of mutual respect between him and all of his co-workers. These leadership characteristics and the common goal of all team members wanting to ensure that the LHEMP became a legal document appeared to ensure the project's successful outcomes.

In the Iskar project, some quite different group dynamics occurred early on, but these had evolved quite significantly by the end of the process. Early on, there was a lack of a common vision between project group members, but a clearly defined schedule of tasks to complete was available. This lack of a simple and clear vision was probably linked to the two-headed leadership of the project structure (the local regional partner and the French research director) for operationalising research tasks, and led to predominantly transactional leadership style characteristics being exhibited (Bass and Avolio, 1994) at the beginning of the project. Initially, it was the local regional partner's university chief who was considered the champion by participants; yet not far into the process, the regional partner's enthusiasm and leadership emerged. Both the French and Bulgarian leaders had contract workers for the project under their jurisdiction, which made the project's accountability quite complex. A natural inclination of the regional partner to exhibit 'team player' and transformational leadership styles (Bass and Avolio, 1994), in particular to support and inspire the young facilitator, also appeared to exist at the beginning of the process. Likewise, pockets of mutual accountability between some of the Bulgarians and French researchers appeared to exist but these were not widespread and only some lines of individual accountability were maintained. Therefore, originally, the co-engineering group could not be considered as a 'real team' but rather as simply an 'effective group' (Katzenbach and Smith, 2002). However, once the local regional partner had helped to create an almost common vision for the participatory process, it was much easier to mobilise almost the whole co-engineering network around the core project team and to work collectively towards the vision aided by her transformational leadership capacity (Bass and Avolio, 1994). The regional partner also showed her personal appreciation for each of the project team members and stakeholders involved in the process and she organised many social events to help build personal relationships. This effort and the subsequent affective links that were built between her and many of these people helped to develop a strong group of project supporters, workers and promoters to ensure the success of the process.

In summary, both co-engineering processes did reach the form of a real team that was able to work collectively to provide high quality participatory water management process outcomes. However, in both cases, the common compelling performance purpose required to create this real team was secured by one team member capable of mobilising transformational leadership capacities (Bass and Avolio, 1994), even though parts of the visions had originally stemmed from each leader's interactions with stakeholders. Each team was also surrounded by an extended network of co-engineers with more limited or individual roles in the process, which also required coordination and management. Therefore, *effective leadership to support coordination or a strong common purpose for intervention, or both, appears paramount for effective co-engineering.*

9.3.4 Importance of co-engineering and team constitution on participatory outcomes

Despite both interventions being initially based on similar methodological approaches, having been adapted from the integrated methodology proposed in Chapter 3 and the pilot test in Montpellier, group constitutions and dynamics had important effects on the co-engineering processes and the outcomes of their resulting participatory frameworks. In particular, the distribution of roles, relational ambience and divergent individual goals were of prime importance in 'setting the scene' for the stakeholders participating in the water management forums and further driving the processes in the direction of their interests. Thus, it is essential to recognise that not only must the resulting participatory process be scrutinised, but also the co-engineering process itself. This issue becomes even more crucial when some of the co-engineers can exert a strong influence on water management outcomes and implementation.

In the LHEMP project, the estuary manager who set up the project had already reflected on and refined his reasoning over his decision to bring together private consultants and researchers in the project team. They were supposed to guarantee independence from the local government in the eyes of participating stakeholders. Obviously, this independence was relative, as some team members depended financially on the local government, and the estuary manager participated directly in, and often led, the co-engineering team. The estuary manager's leadership over the process included the selection of team members for their perceived skills and knowledge, with the objective of creating an innovative and effective process. Effectiveness was initially thought to be an outcome of collaboration, although it can be seen that many factors, including negotiation capacity, influenced the co-engineering process and the ability of the team members to work collectively towards a compelling performance purpose.

Although there were elements of independence in the Iskar project linked to non-Bulgarian researchers involved in the design and organisation processes, most final decisions were made by the local co-engineers. In particular, they strongly influenced the selection of, and maintenance of, relationships with participants, drawing on their personal and professional networks. However, some independent or less strongly attached stakeholders were also involved. It could be thought that such a bias would have constituted a major problem for ensuring process effectiveness; yet this would not be a fair assessment of the Iskar process. It is most likely that these privileged relationships contributed greatly to the appropriation of the participatory outcomes and the overall satisfaction of the participants. It is even more remarkable, if it is considered that almost all of the local co-engineers and many participating stakeholders had little experience or confidence in participatory approaches prior to this project. It could be suggested that the current level of uncertainty experienced by most Bulgarians through their country's recent transitions, associated with the legitimacy offered by a project financed by the European Union, probably helped many participants to consider these unknown management practices.

It is worthy of note that concrete and largely consensual outcomes were achieved through the modification of a principally academically designed process. It is also importantly that *leadership over the process design and content slowly shifted from the foreign researchers to the local team members, who finally had the majority of control over, and appropriated almost all aspects of, the co-engineering process.*

9.4 MODEL AND PROTOCOL VALIDATION

From the research interventions, a number of remarks on the operational validation of the inter-organisational decision-aiding model based on Tsoukiàs' (2007) and Mazri's (2007) descriptive participatory structure model can be made. Likewise, comments on the intervention research and evaluation protocol will be outlined.

9.4.1 Inter-organisational decision-aiding model operational validation

In both the LHEMP and Iskar interventions, I used the inter-organisational decision-aiding model as a framing tool to shape the needs of the participatory workshop activities and external analysis of final recommendations stemming from the workshops. I also reused it *ex-post* to describe and analyse the elements obtained through the different workshops and activities of each case, as presented in Appendices E and F and discussed in Sections 7.5.1 and 8.5.1. In the LHEMP case, it played a central role through the process, as it underlaid the original participatory modelling methodology proposition, and was reconsidered during each negotiation on methodological changes and after each workshop to determine the required content elements of following activities. In the Iskar case, the model was used in a similar manner but only after I joined the process prior to the fourth series of workshops. The model also became a communication and analysis tool between researchers, as the elements elicited and still to be obtained in the process were discussed and noted by the French research director of the Iskar project and me before the final workshop. This allowed adaptations of content objectives for the design of the final workshop to be made and implemented.

As the model could adequately be used to describe the content of the decision-aiding processes in these two projects and the pilot test, it could be considered to be operationally validated. It is worth noting that it was sometimes difficult to assign elements in the 'set of resources' and 'set of objects' categories, as some elements elicited from the participants, such as the value 'good water quality', could be assigned to either or both categories depending on the analyst's interpretation of the participants' meaning.

The model was also used in a normative manner, as described above, to guide the construction of participatory methods and my understanding of how the decision-aiding process was progressing and what should be carried out. Although the model was not originally designed for this purpose, it appeared to be useful in this role, especially to drive me towards efficiently aiding the participants to reach a set of final recommendations as an outcome of the decision-aiding process. Especially in the Iskar process, the normative position that I took, based on wanting to use and elicit all elements of the model through the process, led in part to the change in the final workshop design. It is noted that the process, as a research project, had not originally been designed as a decision-aiding process but rather as an exploratory capacity-building exercise. Along with the Bulgarian regional partner managing the project, who wanted the participants to obtain something concrete at the end of the process, my position can be seen to have significantly influenced the final direction of the project. Whether, ethically, this was a reasonable use of the model and power over process design is debatable. However, having clear theoretical and normative objectives for the process – to elicit all the models' elements in the most adapted participatory nature possible and use them collectively to construct a set of final recommendations – seems to have had predominantly positive outcomes in terms of the effectiveness, efficiency and efficacy (see Sections 7.5.3 and 8.5.3) of the decision-aiding processes. It could therefore be suggested that normative use of the model has also been partially validated for aiding the design and implementation of effective, efficient and efficacious inter-organisational decision-aiding processes for water planning and management.

9.4.2 Participatory structure model validation

The participatory structure model outlined in Mazri (2007) was published too late to be considered or used in the two interventions outlined in this book. However, it appears to provide a pertinent and interesting alternative descriptive model with which the cases can be examined. In particular, the model comes with clearly outlined conditions of validity, which are based on the validity claims of Habermas (1984) and Webler (1995), linked to Habermas' (1984) theory of communicative action and how a process of rationalisation and legitimisation may occur in an ideal speech situation. It could be enlightening to examine the extent to which the two co-engineered participatory processes actually adhered to, or could be described by, this model and the validity conditions.

Through the critical review of pluralist and contextual approaches to decision-aiding interventions in Section 4.1.4, it has already been mentioned that some authors (e.g. Taket and White, 1993; 1996; White and Taket, 1998) do not support Habermas' normative position on aiming to work towards an 'ideal speech situation' with its associated validity claims, as they believe that other mechanisms or types of pluralist stances would better advance participatory practice. However, as it is not the prime purpose of this book to advance this underlying philosophical debate, it will simply be noted that Mazri's (2007) model was developed with a pragmatic objective in mind and it still seems of interest to determine to what extent his model can be validated or invalidated through *ex-post* analysis of the research interventions in this book. This includes the extent to which communicative or instrumental rationalities were exhibited and legitimised through the participatory processes. The processes in this book could likewise be analysed with respect to whether the principles of discordant, pragmatic or critical pluralism (see Section 4.1.4) were adhered to; however, as what has been reviewed of these positions does not provide easy-to-interpret validity conditions for a model of a co-design process for participatory structure design and is, moreover, outside the book's stated objective of investigating co-engineering, this will be left for future research.

In the following section, only the LHEMP case will be analysed in detail related to Mazri's model (shown in Figure 3.7), as I was not present at the beginning of the Iskar case when the participatory structure was first debated among the co-engineers. Furthermore, the research-driven process did not originally have a strong 'decision-aiding objective', so the concept of 'decision maker' used in the model was more difficult to define until later in the process.

CO-DESIGN PROCESS CONFORMITY TO MODEL

In the LHEMP case, the prime decision maker on the participatory structure could be considered to be the council's estuary manager, as he had found the funding for the planning project and acted as the 'client' with whom I, the 'analyst' in Mazri's model, would work.

When I first met the estuary manager, we discussed the characterisations in *Step 1*, as outlined in Appendix E. It is noted that at this stage the scope of the project was to draw up a smaller estuary management plan. The scope was later renegotiated, as outlined in Table 7.1. The *intrinsic characteristics* of financial, technological and knowledge *resources* that existed and resources required (i.e. human resources to run the project, expert knowledge on participatory process design and estuarine processes, and management and models to aid management decisions) were discussed, as well as the estuary manager's *stakes* (i.e. maintaining estuarine environmental quality and adequate professional recognition). Similarly, for the *extrinsic characteristics*, the *actors* required to be involved in the participatory process were outlined (i.e. the existing estuary management committee members, the water utility and other state and local government representatives), and the *subjects of debate* and *objectives for the participatory structure design* were elicited from the estuary manager and discussed, as outlined in Section E.1 and predominantly presented together in Table E.1.

During this discussion of the subjects of debate, the aspects in *Steps 2* and *3* were discussed simultaneously. The *resources* required and *stakes* for each subject of debate were mostly considered relative to blocks of actors (i.e. community estuary committee members, government and agency managers, scientific and planning experts, the private sector and the general public). The prime factor of 'competence' (the validity principle related to resources) that was discussed was adequate knowledge of the subjects of debate, as it was assumed that all actors would have the other talents (such as linguistic capacity and interactive competencies) required to participate in debates. As for stakes, it was considered that most actors would have a stake in the new planning activities, but less interest in the review and analysis activities, such as project costing. Therefore, a broad and representative range of actors would be *implicated* in most of the new planning activities, and some actors *consulted* for the review and analysis activities. The outcomes from this analysis are provided in Table E.1, with many subjects of debate marked for full implication of actors, a few of which were to be completed by scientific experts (consultants) in consultation with some actors with adequate knowledge resources, and one subject marked for optional inclusion in the process.

At this stage, the estuary manager was satisfied with the vision and so I went on to design and outline a proposition for a participatory process structure, *Step 4*. As the Mazri model formalisms were not considered at this time, the proposition took the form of the staged participatory modelling process description and timeline, where the elicited subjects of debate were bundled into categories based on the inter-organisational decision-aiding model discussed in the previous section. All actors were to have the opportunity of participating in the debates on each subject as part of the participatory modelling exercise. This proposition was then submitted back to the estuary manager and his superiors in the council.

It is noted that the LHEMP process does not conform to the Mazri model entirely, as it was here that the vision of the decision maker appeared to have been enriched and there was *iteration back to Step 1*, rather than before the construction of the proposition. This reflection led to the estuary manager reconsidering his own objectives for the participatory process and the match of the actors' *resources* and *stakes* to the scope of the project (*Step 2*). It was here that he considered that the participatory structure might be an improved use of resources if the scope of the project was expanded to be a regional estuary management plan (*Step 3*). Brief negotiations, outlined in Table 7.1, resulted in a second participatory structure (*Step 4*) agreed on by the estuary manager and me and this was put at the disposal of all actors (*Step 5*) as part of the public tender to find a project manager to manage the process.

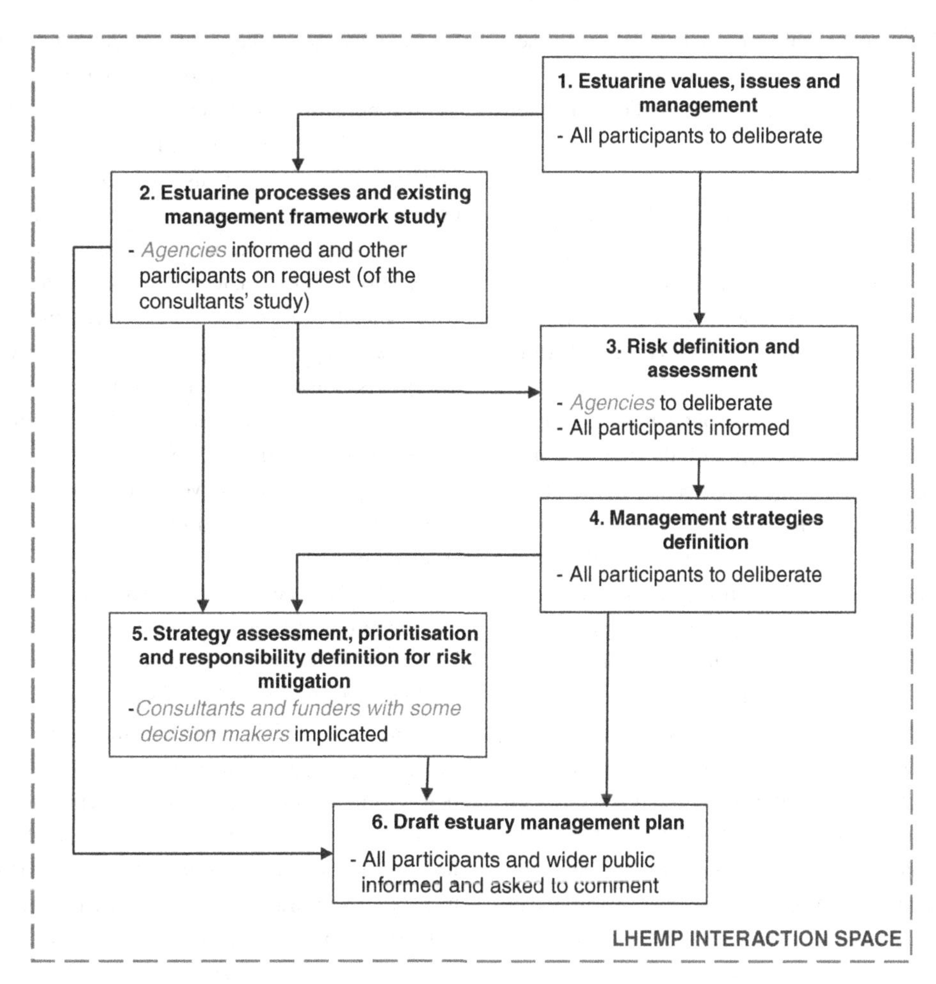

Figure 9.1 Final LHEMP participatory process structure.

After selection of the project manager and associated consultants, an iteration was then made to *Step 1*, as there was theoretically another decision maker who would decide on the participatory structure to be implemented. The *subjects of debate* and associated *participation levels* to be included in the participatory structure were renegotiated, especially due to human, financial and time resource considerations. This meant that not all actors would be implicated in the final 'strategy assessment', shown in the organisation model in Figure 9.1. The model was agreed upon by the LHEMP project team and presented in the first participatory workshop on the first object for debate, but those parts in Figure 9.1 now labelled in italic text were originally to include all participants. This corresponded to another *Step 5*, where the actors whose representations were considered could comment on the structure.

The participants accepted the process structure in its given form in the workshop, but, as outlined in Table 7.1, a negotiation between the project team members after the workshop resulted in the change to have just the agencies participate in the next workshop. After this point, the structure remained stable and the participatory processes around the objects of debate were carried out related to their dependence through time, as represented in the model in Figure 9.1.

Although the structure of the participatory process remained stable, the Mazri model could also be used in a similar descriptive manner to outline the evolution of subjects of debate in the workshops, who the decision makers were and how the participatory structure evolved. As this process has already been analysed in content using the inter-organisational decision-aiding model, as shown in Tables E.4 and F.1 and discussed in Sections 7.5.1 and 8.5.1, further analysis will not be pursued here. It will rather simply be noted that *the actor–action–resources matrix tool* developed in response to the analysis needs in the LHEMP (refer to Section 7.4.2) to perform the tasks related to the fifth object of debate, the definition of responsibilities for actions of risk mitigation, adequately *could be used as an analysis tool by the decision maker(s) and researcher to inform Steps 1 to 3* for those potentially broader-scale participatory structures needed for in-depth negotiations over responsibility distribution between actors. The individual negotiations involving a small number of key actors with adequate resources and stakes between them to deal with a number of actions would then form the individual subjects of debate, which would again have to be organised based on their dependencies (*Step 4*). In this latter case, the plan of action responsibility negotiations submitted to the concerned actors would form *Step 5*.

MODEL VALIDATION AND LEGITIMISATION

Looking next at whether the Mazri model appears to have been legitimised and validated in its role as a descriptive decision-aiding model for participatory structure co-design, the LHEMP case provides a number of positive conclusions to these conditions.

Firstly, from the previous descriptive use of the model, it can be seen that the described process elements of the Mazri model closely fit the co-design process of the participatory model structure that was finally implemented in the LHEMP case. This is with the exception of the feedback loops appearing to occur after Steps 4 and 5, rather than after Step 3. The LHEMP process therefore provides one empirical validation of the Mazri descriptive model, with the exception of the placement of the feedback loop.

Now looking more closely at whether the model validity principles of competence, equity, efficiency and legitimacy were operationally adhered to in the creation and implementation of the model, it can firstly be considered that the efficiency criterion was largely adhered to, as outlined in Section 7.5.3. On the principles of equity, competence and legitimacy through the design and implementation processes (related to the co-engineering team members and subsequent actors involved in the later rounds of model implementation in the LHEMP interaction space), it appears that the methods used to aid debate over the objects in Figure 9.1 allowed the largely competent actors to equitably present a range of arguments with the four different underlying claims to validity. Actors' claims of: comprehensibility; truth; truthfulness (or sincerity); and rightness (or appropriateness), were typically respected by other participants and debated openly. In some cases there were exceptions to these conditions, that, when in action, underlay the ideals of Habermas' (1984) 'communicative rationality'; occasionally examples of 'instrumental rationality' (Habermas, 1984) were present in the interaction space. When considering the discourses present in the interaction space:

> *Communicative rationality insists that, out of this Babel of perspectives, a reasonable course of action will emerge. But instrumental rationality insists that the costs of communicative action in time and money are too high, that people are confused about their own real needs, that impulses and emotions override rationality in public debate, and that good action depends on expert guidance.*
>
> (Killingsworth and Palmer, 1992)

For example, at the end of the second agency workshop on risk assessment, one of the participants remarked that the group of participants had the collective potential to advance, 'If we could just keep the emotion out of it.' Elements of underlying normative support for an instrumental rationality, rather than supporting the ideal of communicative rationality to its end point of emergent action, were also present in participant evaluation comments, which outlined a greater need to rely on the external scientific analyses and perspectives, and not just to consider the workshop outputs. In terms of the competence principle, this was typically adhered to in the interaction space's workshops, with the exception perhaps being the required activity of small group risk assessment in Workshop 2, where knowledge resources were inadequate or too unevenly spread between participants to complete the exercise effectively. In this exercise, some participants were not able to use scientific truth claims but had to rely on other forms of justification for their risk assessments.

Despite these occasional deviations from the model principles and communicative action in the interaction space, it appears that legitimisation of the participatory structure and its results was well constructed throughout its design and implementation process, with many diverse participants (including decision makers) actively participating and collectively supporting even intuitively odd results following open debate (refer to the discussion on the prioritisation of the water quality risk in Appendix E). This legitimisation appears to have also continued after the process to support the draft and final plans, plus their implementation; potentially highlighting that participants could see their undistorted views and interests preserved in the final document.

From this brief analysis of model validity and participatory process legitimisation, it appears that Habermas' norms of communicative action were predominantly upheld through the participatory structure, apart from a few small deviations that could be considered to distort communication (Dayton, 2002), and that collective legitimisation of the structure was largely obtained. This provides further general support for the normative use of Mazri's model.

9.4.3 Evaluation protocol operational validation

As part of this research work, an intervention evaluation protocol for studying and comparing co-engineering processes was proposed in Section 5.5, following a review of participatory process evaluation literature.

In each case, the elements of the protocol were adapted, and in some cases debated with other project team members, before application. In particular, the questions asked of participants were co-engineered with 'external' evaluators who also took process observations, as outlined in Sections 7.2.3 and 8.2.3. This co-engineering and the resulting protocol appears to have lived up to its expectations, since each process could be carefully analysed and compared on a range of specific issues, as demonstrated through the process outlines presented in Chapters 7 and 8, based on the protocol. Even some almost identical questions could be asked of the participants at the end of the processes in the two countries, permitting cross-cultural evidence on the co-engineering of participatory modelling processes leading to some very similar outcomes from the participants' perspectives, as shown in Table 9.3.

One part of the protocol that was not adhered to in either case was the idea of having *ex-ante* and *ex-post* participant evaluations for each participant workshop. From the pilot trial, this ideal strategy was found to be too time-consuming and not a very active and interesting activity for use at the start of participatory workshops. Moreover, debating with other co-engineers on the value of such intense evaluation in real-world decision-aiding processes with a variety of constraints is challenging. Just *ex-post* evaluation at the end of each workshop therefore appeared adequate for the interventions. If *ex-ante* evaluation is to be used in future participatory modelling processes, more interactive and interesting methods, rather than written surveys or interviews, will probably need to be explored.

Access to a few project team members for interviews was also difficult to obtain owing to time and other constraints, so the cross-validation of interpretations of the co-engineering processes remains incomplete. The co-engineers not only included the 'core members' of the design and implementation teams, but there were also a number in networks around the two process leaders, and it was a challenge to define people in these groups. Co-engineers in these networks were not interviewed as part of the protocol either, but some of their views could be obtained through the core members, debriefing notes or project evaluation questionnaires for both cases.

Apart from a few small issues and adaptation of the protocol to fit the constraints and expectations of the other process co-engineers, the protocol could be considered to have been operationally validated. Replication also appears unproblematic.

9.4.4 Intervention research legitimisation

As described in Sections 7.5.3 and 8.5.3, the participatory intervention research process adopted for this research led to a number of innovations in the form of aiding the emergence of new types of collective action and actionable knowledge. Through both interventions, 'multi-institutional-level stakeholder working networks' for their own regional contexts in Australia and Bulgaria were created around water management issues through the participatory processes. Each of these networks also had sub-networks, in which other forms of collective action evolved – the agency network in Australia, involved in the risk assessment, and the Elin Pelin regional network in Bulgaria, involved in creating the prioritised risk response projects and seeking out structural funding and further organisational aid.

New forms of collective and productive action also evolved in the co-engineering teams and extended networks in both countries: an inter-institutional one in Australia spanning researchers, consultants and government managers and an intercultural and inter-institutional one in Bulgaria.

The emergence of this collective action also led to the creation of actionable knowledge, which I could reapply, continuing to learn and co-constructing more actionable knowledge through each intervention, including through the pilot trial. Learning about the advantages of dividing actor groups for various reasons (i.e. allowing more open communication, preliminary capacity building and diminishing power differentials) as a result of several difficult but highly useful and personally transformative negotiations, was a particularly evident change in world-view and personal norms for me. Other co-engineering team members taking part in the intervention research also underwent second-order personal changes in views and personal norms, such as the Bulgarian regional partner who underwent a transformation from fear of participatory processes and doubt at being able to organise such a process to, three years later, believing profoundly in the benefits of participatory processes for transforming water management in her country. She even mentioned in her final interview that she had learnt that she was more of a '*people person than an engineer*'.

As noted in Section 5.2, one method of aiding the validation of intervention research is to write experimental reports that adopt a critically reflective attitude to the interventions. Chapters 7, 8 and their supplementary information stem from such reports, along with the reporting of the pilot trial. Reports of the co-engineering processes and their targeted participatory modelling process have been produced in the form of conference and journal articles, as well as reports for participants in the LHEMP. These articles, which are cited elsewhere in this book, have all had the chance to be cross-validated to enhance further debate and learning in the project teams, and have been submitted to outsiders for comment and discussion.

From an external perspective, the insights created through the intervention research programme have also already proven of interest to a number of outside researchers and practitioners, with the insights from the LHEMP being used to inform South Australian engineering consultants' work in another estuarine region (Helfgott, 2008; personal communication) and for female mud-crab management by the Queensland Government in Australia (Brown, 2010), which provides at least partial validation for their usefulness.

9.5 EXTENSION CASES AND DISCUSSION

The research presented so far in this book has led to the development of a new way of understanding the participatory water management process. Specifically, there are two decision arenas to manage effectively: the co-engineering arena, where organisational decisions are made; and the participatory planning process arena, where collective decisions on water management plans and actions are made by stakeholders. With this understanding, including that derived through the analyses of the two case studies presented in previous chapters, other case studies can now be viewed through this new 'co-engineering' analysis framework. In particular, this section will look at: a case in Germany that examines how niches for developing participatory water management approaches can be found; a case in Algeria on how co-engineering processes can be better and more formally supported; and the issues of politics in developing participatory forms of water governance, drawing on a case in the Pacific Island nation of the Republic of Kiribati.

9.5.1 Finding niches for developing participatory water management approaches: ecological quality restoration in the Dhünn Basin, Germany

Moellenkamp *et al.* (2010), through their work in the Dhünn Basin in Germany, have looked at how niches can be found or created for informal participatory water management processes amongst formal structures, which can then support the objectives of later formal processes. This research describes the work of a 'trialogue team' that set up a participatory process to support the development of water management options to increase the ecological quality of the Dhünn River. Although not part of the official Water Framework Directive (WFD) processes, the process hoped to support work towards the achievement of 'good status' for the Dhünn Basin. Drawing on published research documents and papers (Moellenkamp *et al.*, 2008; Seecon Deutschland GmbH, 2008; Speil *et al.*, 2008; NeWater, 2009; Moellenkamp *et al.*, 2010; Wupperverband, 2011), some of this case will be reinterpreted through a 'co-engineering' frame of analysis, in order to provide some further insights into the practice of co-engineering and what can be learnt to inform future processes.

The Dhünn Basin is located in the middle of the transboundary Rhine Basin in Europe, in the North Rhine-Westphalia region of Germany. The major water-related challenges of the basin include the potential of not meeting the 'good status' required by the EU Water Framework Directive (WFD) (refer to Section 2.1.3), specifically related to ecomorphological concerns and ecological integrity issues, largely owing to the regulation of the river through a major drinking water supply reservoir and weirs. To meet the WFD requirements, there is a need to bring the fish population back from its decline, which has been linked to cold water temperatures in summer caused by releases of deep dam water. Any changes to dam or weir regulation practices or other envisaged measures would affect a range of stakeholders (Moellenkamp *et al.*, 2010).

CO-INITIATION

The process started from a perceived need for a participatory process to aid in solution development and management recommendations to support work towards 'good status'. However, there was a challenge in determining who had the authority, legitimacy and resources to develop such a process, owing to complex legal and administrative arrangements linked to the WFD and German water planning structures.

Moellenkamp *et al.* (2010) presented a clear analysis of the range of legally responsible ministerial structures for implementation of the WFD and how a support structure with resources and legitimacy (in the eyes of stakeholders) but no legal authority was able to negotiate with these structures and researchers to convene an informal participatory 'pre-process' that could be used to inform the official WFD processes. The final stage of the co-initiation was the signing of a treaty document between the water association and the NeWater research project partners working in the Dhünn Basin. This included a set of objectives to guide the design of the participatory process, as follows (Seecon Deutschland GmbH, 2008):

- Development of approaches to the implementation of the WFD in the Lower Dhünn River system that includes affected stakeholders;
- Joint discussion of possible measures and development of compromise solutions to water use conflicts;
- Concrete results in the form of an outcome document, which outlines and explains the process undertaken and methods used, as well as the range of different opinions and stakeholder interests, and potential for agreements or compromises to be made;
- Facilitate early discussion of possible opposition to later implementation of measures;
- Network formation, long-term cooperation;
- Improve public awareness of the issue; and
- Promote learning and develop a methodological basis for subsequent processes.

PRELIMINARY CO-DESIGN

With the agreement of the regional ministerial structure legally responsible for implementation of the WFD to set up this informal 'pre-process', the co-design process commenced. This had three major groups of co-engineers (what Moellenkamp *et al.* (2010) described as the 'trialogue team'), each of which had different objectives for the participatory process and resources to support it:

- The regional water association (the Wupperverband) had an objective of developing a role for itself in the implementation of the WFD and saw its role as convenor of the participatory process development as a means of gaining some power in the water management governance structures.

It also had key resources of importance, such as access to many stakeholders, including its own members;

- German and Dutch university researchers involved in the NeWater and ACER research projects, who had the objective of encouraging the development of adaptive management practice, and evaluating case studies of this practice for their projects. These researchers had access to financial and human resources through the research projects, as well as scientific knowledge on participatory and adaptive management processes; and
- Consultants from a private research consultancy firm, who were paid through the NeWater and ACER projects to provide process design knowledge, logistics support and facilitation services to carry out the participatory process.

An important aspect of the co-design process, outlined by Moellenkamp *et al.* (2010), was that much energy was focused on minimising the risks of negative outcomes by working to reduce the uncertainties of knowledge about the water system, such as dam management practices, and of the possible results of the participatory process on water management. The former was dealt with in the co-design team by making the key problem focus of the process the dam management practices. The latter required negotiation within the team for the consultants and researchers to convince the water association to take a 'leap of faith' regarding their competency in organising such a participatory process and the use of ongoing evaluation to monitor the process carefully as it was to be implemented, with adequate time being left between workshops for the evaluation results to be investigated and the next stages of the process adapted and redesigned if required. The commonly agreed objectives of the process in the co-design team focused on the capacity, by the end of the process, to recommend river basin management mechanisms to increase ecological quality, support learning between stakeholders and agencies and develop practice in participatory processes and their adoption in future management activities. The co-design process also included interviews and questionnaire surveys of key stakeholders, which were used by the whole co-engineering team as part of the stakeholder analysis to inform the selection of participants for the participatory workshops, as well as to gain ideas on workshop design. The participation plan included three participatory workshops to be held over a period of six months. This preliminary co-design phase took approximately one year to complete.

ONGOING CO-DESIGN AND CO-IMPLEMENTATION

The next part of the process involved the implementation of three participatory workshops – in October 2007, December 2007 and April 2008. They were all designed collaboratively by the whole co-engineering team and were: 1) convened by the water association; 2) facilitated by the consultants; and 3) evaluated by the researchers. Over 60 stakeholders participated throughout the process, representing a broad range of interests that included water management, fisheries, environmental conservation, heritage, city and regional planning, agriculture, industry, recreation, hunting and forestry (Seecon Deutschland GmbH, 2008). The first two workshops took the form of plenary information lectures followed by small group-facilitated discussions and activities, such as participatory mapping, followed by presentations of the small groups' results and a plenary discussion (Moellenkamp *et al.*, 2008). From the participants' listed issues and distributions of preference weightings on policy options gained from these workshops, as well as participants' responses to a questionnaire between the workshops, a document of outcomes was created by the co-engineering team that would form the basis of the third workshop. The third workshop then focused on this document and gaining an agreement on its contents, which specified possible measures to achieve 'good ecological status' that could be carried out in the basin. This agreement, as well as an agreement for the water association to run a dam discharge experiment, was achieved by consensus (Moellenkamp *et al.*, 2010). The rest of the workshop was used to plan follow-up activities.

The results of the evaluation questionnaires distributed by the researchers after each workshop helped to monitor the participants' satisfaction with the process and to gain insights into how the workshops could be improved. Some of the results from the first workshops led to the invitation of other stakeholders considered important to the participants to later workshops and to the adjustment of the workshop design to increase time for stakeholder discussion, rather than spending it on plenary presentations. Final follow-up questionnaires, sent by the researchers to participants of all three workshops, helped to gain extra feedback on the effectiveness of the whole process (NeWater, 2009).

The final phase of the co-implementation process involved all members of the co-engineering team completing the process reporting, and there was dissemination of results to ensure their consideration in the follow-up, formal WFD implementation processes (Wupperverband, 2011).

KEY CASE INSIGHTS AND DISCUSSION

This informal process, set up in an identified niche in the formal administrative structures, was later considered by authorities and stakeholders as a naturally sensible part of water-planning regulation, cementing the legitimacy of the water association fulfilling this inter-organisational and stakeholder engagement coordination role (Moellenkamp *et al.*, 2010). The legitimacy for this process was also significantly aided by the agreements that the co-engineers had made before the commencement of the process with the authorities legally responsible for implementing the WFD, including their verbal agreement to consider the outcomes as part of their own official WFD processes. This close and strategic management of the co-engineering process by the key team members appeared beneficial, in particular, for the water association to strengthen its position as a key player in regional water management, as it was able to prove its capacity to lead and aid inter-organisational decision making.

It also appears that the leadership shown by the water association in using participatory processes to shape the WFD implementation processes in the Dhünn Basin, and support by the researchers and consultants, formed an important coalition that was able to bring about this innovation amongst the formal administrative structures in the region. Without key leaders in the water association willing to take a chance and trust the researchers and consultants, it is unlikely that the process could have gained such support from the administration and stakeholders alike. What is of even greater interest is the apparent altering of stakeholder perceptions, so that by the end of the process they considered that the role of the water association as convenor of participatory stakeholders was entirely normal (Moellenkamp *et al.*, 2010). One other factor that appeared beneficial to the effective co-engineering of this participatory water planning process was the clear division of roles in the co-engineering team, as it promoted good working relationships and a mutually supportive and collective capacity to implement workshops.

This case has a number of clear similarities to the Australian estuarine management process, including the co-engineering team being based on a three-pronged structure of a local water management body, consultants and researchers, and the contents of the water management challenge being largely related to ecological quality. Both processes also had the clear objective of coming to a collective agreement or plan, based on a short process of three workshops and some external work carried out by the co-engineering team, which they managed to achieve. This German case therefore provides supplementary evidence of the potential efficiency of well co-engineered participatory processes to drive the development of collective agreements, as well as demonstrating the important role that leadership anchored in local water management bodies plays in having research-supported participatory processes contribute to real-life water management decision making processes and actions.

9.5.2 Supporting the co-engineering process: agricultural water demand and irrigation practice in the Mitidja Plain, Algeria

One of the major questions that arises from the awareness about, and the acknowledgement of, the importance of the co-engineering process to participatory water management is how new co-engineering processes can be better supported or even more effectively managed. Some thought on this question and others important to the design and facilitation of participatory processes has already led to one initiative that provides some useful reflections: a virtual *community of practice* (www.particip.fr), which was established in Montpellier, France, by a group of consultants and researchers with an interest in participatory processes for natural resources management. The aim of the community is to provide an interaction space for organisers of participatory processes to discuss their cases and trial run workshops with other participatory process researchers and practitioners, in order to test simulation tools and participatory methods, as well as their own facilitation skills (Dionnet *et al.*, 2011). In practice, it acts as a support group for one or more of the co-engineers to open up their participatory process and water management case to others to help them more effectively negotiate, design and implement their interventions in the field.

One African co-engineering case study that was supported through this community of practice will be outlined here. Many aspects of this case, which focused on agriculture and irrigation water management in the Mitidja Plain in Algeria and the broader agricultural and cultural context of the region, have been documented in other works (Imache, 2008; Imache *et al.*, 2008; 2009a; 2009b; 2009c; 2010), so here the focus will be on the co-engineering support and how this affected the reflection of the organisers and their co-design and co-implementation of the participatory process.

The Mitidja is a 100-km long littoral plain situated in the centre of northern Algeria. It has a long agricultural history and a Mediterranean climate that shows signs of increasing aridity. Irrigation systems have allowed greater development and intensification of agriculture since the construction of the first irrigation dam in 1937 and subsequent large-scale irrigation projects, which have also served to preserve the groundwater resources of the region for potable water supply (SIRMA, 2011). The area of focus for the case study was the West Mitidja Plain near the capital city, Algiers. The approximately 10 000-hectare agricultural area, with main crops of irrigated fruit trees (55%) and horticulture, as well as rain-fed cereal, provides the bulk of agricultural produce for the four-million strong population of Algiers (Imache, 2008). The area supports approximately 2000 farmers, who originally worked under the socialist farm scheme of the state before its dissolution in the agricultural land reforms in 1987. As the state withdrew from managing public lands, small collective farms (EACs) were created with 3–20 farmers assigned to each 9–50 hectare area, as well as some private farms (Imache *et al.*, 2009c). Under collective management, however, conflicts arose over many issues, including investments, task-sharing and decision making processes linked to water sharing and irrigation network access. This led to a range of situations that could be considered as an informal privatisation of agricultural lands, e.g. informal land division, land rentals and informal land sales, as well as a greater exploitation of groundwater resources through new illicit tube wells and old boreholes (used to source water prior to the establishment of collective irrigation schemes) (Imache *et al.*, 2009b). The informal nature of much agricultural and water use practice, which is typically not formally recognised by the state water and irrigation scheme managers, complicates the determination of agricultural water demand and production, as well as the agreement of beneficial regional solutions to water and land use.

This case of co-engineering a participatory process was initiated through the SIRMA project of water economies in Northern Africa (www.eau-sirma.net; Kuper *et al.*, 2009), as part of a research investigation into individual and collective irrigation water use behaviour, including the effects of institutional arrangements on these behaviours (Imache, 2008). An Algerian researcher, who was carrying out this investigation as part of his PhD at a French Agricultural and Environmental Engineering research institution, had already tried hydrological, land use and economic modelling and analyses with the aid of many field surveys in West Mitidja, in order to characterise water use behaviour and institutional arrangements affecting irrigation practices; yet he saw the potential to extend the analyses and support farmer interactions with the state administration to work towards realistic management options for improving surface water and groundwater management in the region.

CO-INITIATION

It was in this context that the Algerian researcher sought the advice and support of other colleagues in the SIRMA project, and researchers and consultants with expertise in participatory process organisation in Northern Africa, to develop a participatory action–research approach to investigating water demand and use behaviour. After initial discussions with the organisers of the Montpellier-based community of practice (CoP), it was jointly decided that sessions of the CoP could be organised to support the co-engineering of a participatory process that would focus discussion on both land and water management and have the objectives of: 1) improving understanding of the farmers' situations; 2) engaging both farmers and administrators in constructive dialogue and bridging the gap between them and their own visions of water and land management challenges; and 3) on the basis of a common representation of issues, beginning to identify potential options for improving surface water and groundwater management in the region (Imache, 2008). However, despite these agreed objectives for a process, the researchers in the SIRMA project, who came from a variety of disciplinary and cultural backgrounds, had a range of views on how these objectives should be achieved. It was considered by the Algerian researcher that this diversity of views could be discussed by researchers and practitioners of participatory processes in the CoP, and potential designs for the process collectively developed. It is important to note that at this stage the Algerian researcher had no prior experience in organising participatory processes, so was hoping to rely on colleagues with more experience to support his learning and expertise development.

PRELIMINARY CO-DESIGN

The preliminary co-design phase had both normal behind-the-scenes meetings between the key co-engineers (from the SIRMA project and the consulting company) and three formal facilitated co-design sessions in the community of practice to develop and test a proposed participatory process. The proceedings of the community of practice sessions and who was present are outlined in the session summary reports available on the CoP's website, as well as in the researcher's thesis (Imache, 2008) and other papers (Imache *et al.*, 2009b; Imache *et al.*, 2009c).

Of particular interest in the first sessions was the opportunity for the consultant facilitator and researchers from the project to share an understanding of the context of the Mitidja field site with the session participants and for these organisers to reflect deeply on their future convening role and how they might best be able to seek their own objectives, after having heard and discussed the experiences of the other participants in participatory process development and Northern African agricultural development. The first co-design workshop was made highly interactive, with two subgroups being formed to work on individual design proposals for the participatory process, as well as a joint discussion at the end of the session to investigate the reasoning and differences in the proposals. From this process, a number of key decisions were made and issues for the process clarified, including that it would probably be useful to separate the farmers and administrators in the first workshops to allow time to build trust between the participants and process organisers and to see their separate visions of the water management situation, and that it would also be helpful to leave the process open enough for participants to address other non-water related issues of concern to them. It was also decided to organise another session to test the workshop set-up that was to be finalised by the researcher and consultant prior to its application in the field (Imache, 2008).

Of key importance in the discussion and closely connected to the strategic nature of co-engineering was a discussion of how this research-driven process could be supported by the administration and farming champions. The personal networks of the project members were investigated, as well as people whose cooperation would be needed early on in the project, or people to be wary of as they could block the process. Many of the participants of the CoP brought up the challenge of having such processes connect to reality and the troubles associated with attracting sought-after stakeholders to attend workshops. In particular, it was noted that if key community leaders were not engaged in some aspects of the co-engineering (even if only in a supporter or promoter role – see Figure 4.2), then they would probably not use their networks and capacities to encourage other stakeholders to participate in the process. The participants of the CoP thus encouraged the organisers to find some local stakeholder champions to support the proposed process from early on.

The insights from these discussions were worked through before the next two sessions of the community of practice, during which the workshop process was simulated by participants playing the roles of farmers with the Algerian researcher facilitating. The first of these two sessions resulted in important suggestions being put to the organisers, such as: only having farmers from the same EAC attend each workshop; that the different groups of participants could be mixed up in later parts of the process; and concentrating on only the 'participatory scenario' development – one of the two activities planned for a workshop – owing to the tiring nature of the whole proposed exercise. The key question of what the stakeholders were to gain from participating in the workshops was also a key point of discussion – a common co-engineering ethical dilemma – with the focus of the workshop to be designed to maintain their interest and capacity to gain something from it, including being able to bring up their concerns with agricultural and water managers and administrators. The need for a problem focus on land issues, as well as water issues, seemed a key to gaining participants' interests, owing to the challenges of informal land and water use practices, which included land leasing arrangements. This was discussed alongside the need for a coherent research methodology for the completion of the researcher's PhD thesis on agricultural water demand, as well as the role an external evaluator could play in monitoring the process and informing the organisers of any key issues or biases arising through the workshops. The need to re-test an adapted process in another short session of the community of practice was decided on, and it was thought that it could also be valuable to trial the process with students in Algiers, who would be present at the workshops.

Thus, prior to the next workshop simulation (the third co-design session in the CoP) an additional process planning meeting was held in Algiers with SIRMA project members and one of the aforementioned students, in particular, to sort out local logistics for the workshops. The third CoP session, as well as deciding on the final plan of activities for the workshops and further fine tuning the participatory methods (related to the representation of the EACs, constraints and evolution scenarios) to be used in them, was mainly found to be beneficial to the Algerian researcher, who understood better the time constraints of a workshop and how to improve his facilitation skills to deal with certain participant behaviours, and who was then able to distribute time more equitably for individuals to have a say during group discussions. The emphasis on the anonymity of participants in the written results was also made clear, to support the participants' capacity to raise issues that were of an informal or illegal nature.

ONGOING CO-DESIGN AND CO-IMPLEMENTATION

Following the preliminary co-design process, the first phase of the co-implementation process was made up of four workshops with members of four different EACs. All of these workshops were carried out according to the pre-designated plan and facilitated in Arabic by the researcher. There were few major issues, except that one group did not want to discuss a borehole that had been represented on their EAC by the organisers (perhaps because they did not want to admit it was there because of its illegal nature), and a few challenges in supporting an equitable division of time for all participants to speak (Imache, 2008). A number of resulting insights were considered valuable by the organisers, including the major constraints to more effective regional management being considered, such as: constraining land tenure arrangements; lack of access to bank credit; and insufficient volumes of water (Imache, 2008). Many of the participating farmers also found the approach intriguing and were interested in talking to the administration and water managers about their land and water management issues. However, a couple of farmers were not entirely sure what the process would be able to achieve.

These results were fed back into a debriefing and co-design session of the CoP in Montpellier, aimed at designing the next phase of the process. This session was used to simulate a proposed workshop process for the regional-level discussion between administration representatives and farmers and to discuss potential improvements. The discussions provoked a number of changes in the design, not only to simplify the workshop process, but also by arranging more workshops with farmers and administration representatives separately before bringing them together. The proposed arrangement (which was later enacted as planned in the field) was, therefore, to carry out a first participatory scenario exercise with the farmers all together from the four EACs, focusing on challenges and futures at the regional scale (the scale that administration representatives typically deal with), followed by a workshop with the administration representatives, in which they would simulate the situation of farmers in an EAC to help them to understand the issues at the farmers' scale (Imache, 2008) – in other words, this would be a repeat of the role-playing game tested in the CoP, which had the aim of projecting the administrators into the farmers' world. The final workshops could then have mixed participation by farmers and administration representatives, who would have been primed for working through a range of scenarios for the future of water and land management in the region. For this workshop, members of the CoP discussed the importance of finding a neutral meeting place (i.e. not an administrative building or farm), so the organisers decided to find a local community hall.

KEY CASE INSIGHTS AND DISCUSSION

The co-engineered process appeared relatively successful in meeting the initially agreed objectives of the participatory process: to improve understanding of the farmers' situations; engaging both farmers and administrators in constructive dialogue and bridging the gap between them and their own visions of water and land management challenges; and – on the basis of a common representation of issues – to begin to identify potential options for improving surface and groundwater management in the region. In particular, through the workshops, it was possible to identify and prioritise issues in the water and land management systems from both farmers' and administrations' points of view, as well as to outline similarities and differences in these points of view. Although the issue considered of highest importance by most of the farmers was the collective ownership status of the EACs, the administrators considered that the main issue was the need for greater farmer supervision and support by administrative structures. Other issues were seen to be important for both, including the lack of available water resources provided by the collective irrigation system and the lack of easy access to bank credit (Imache, 2008).

To deal with the commonly perceived issue of a lack of water, one proposed solution was the installation of individual water meters, which farmers would support as it would provide incentives to save water and not be billed on the normal network billing package (Imache, 2008). Once trust was established in the group discussing the difficult issue of illegal groundwater use, one possible option to increase control over the current situation was to start regulating this groundwater by granting borehole or drilling permits. The issue of land tenure was also addressed, with a growing recognition by some participants of the important role that informal lessees of land play in promoting innovation in agricultural practice, such as the use of drip irrigation to grow high-value greenhouse crops that make agricultural production in the region profitable, and also a recognition of the negative impact of often exploiting groundwater resources and using pesticides systematically, practices which damage soil fertility in the longer-term (Imache *et al.*, 2009b). Finding a way to formalise the lessees' existence therefore seems an important topic to address in future processes.

This case study also provides a number of insights of procedural interest, including that the community of practice sessions appeared to be a successful mechanism for supporting the organisation, co-design and facilitation of a participatory process by a researcher with almost no previous experience of such approaches. The simulation of facilitation practice in a relatively safe environment, as well as the in-depth discussions of hypotheses and typical organisational dilemmas of participatory processes, appeared to be of particular importance for honing the researcher's skills and intervention design. Continuous evaluation and reporting on both the community of practice support of the co-engineering process and the participatory process with stakeholders in the Mitidja Plain also added to the transparency and reactivity of the participatory process' co-engineering, and gave the co-designers not involved in the field work the opportunity to learn and debate co-engineering practice. Strategic issues, such as who to invite onto the co-engineering team, were also openly debated and documented, making this a valuable case from which to learn about one possible means of supporting co-engineering processes.

Comparing this case with the Bulgarian case, it was possible to see that in both processes previously untrained facilitators with little experience of participatory processes were able to quickly learn about the necessary skills and fill the required facilitation roles, through supportive co-engineering processes. In both cases, the language barrier between the advising participatory process researchers and consultants, and the participant stakeholders in the field, appeared beneficial for this learning and appropriation of participatory practice by the facilitators who spoke the local languages. It could have otherwise been too easy for the external project team members to have taken on the facilitation role. Since these processes, both facilitators have continued to work on promoting and implementing participatory water management practice, with processes continuing in Northern Africa with the Algerian researcher and the Bulgarian facilitator being involved in projects in South Africa, where she has in turn had the opportunity to work with local people to develop their facilitation skills and knowledge of participatory processes through river basin game development and use (Ferrand *et al.*, 2009a; 2009b). Another point of similarity with the Bulgarian case was the decision to separate groups of stakeholders from the beginning of the participatory processes to allow them time to develop their own visions of the situation and become comfortable with the participatory styles of working and methods, before integrating the groups, to continue to build the potential for social learning and exchange. In both processes, this seemed to have functioned according to the desired objectives of designers and appeared to be an appropriate and non-threatening way to introduce participatory practices into social and political contexts where they are rarely used. In particular, creating seemingly safe environments for exchange also allowed challenging topics, such as illegal groundwater use in Algeria or corruption in Bulgaria, to be raised and discussed to some extent. These cases therefore provide confirmation of the potential successes that such staged participatory approaches can have and the lessons that might be proposed for other cases with similar aspirations in the future.

Finally, the method of supporting the co-engineering process presented through this case study in the form of a community of practice (Wenger, 1998) is just one of the potential means that could be envisaged for supporting successful co-engineering. One key question that remains in this area is whether greater transparency of the co-engineering process and agendas of the process organisers is often likely to be beneficial or not. Proponents of increasing this kind of transparency have also developed other tools, such as personality profiling and co-design management guides, that could be used by co-engineering team members to encourage comprehension of divergent viewpoints and provide a basis of understanding them for managing these divergences (e.g. Kolfschoten *et al.*, 2004; D'Aquino, 2007; Dandy *et al.*, 2007; Gratton and Erickson, 2007; Gratton *et al.*, 2007; Brown, 2008; D'Aquino, 2008); yet it could also be imagined that training some key co-engineers in negotiation practice could help them to further their capacity to implement participatory processes in tricky political situations. Power, politics and ethics in co-engineering processes and in the transition processes towards improved forms of water management more generally remain a key area where further reflection is required.

9.5.3 Politics of developing participatory forms of water governance: participatory water management practice and policy development in the Republic of Kiribati

Developing participatory water management practice is a complex and often highly political process. In some contexts, participatory forms of water management, especially those that include NGOs and community members, are seen as a threat to existing forms of water governance, power relationships and legal and regulatory structures designed to maintain clear lines of accountability. This means that establishing participatory water management is often met with resistance, in particular from technical government administrations and large private companies who often do not see that the potential shift in power over water management priorities is likely to meet their interests. Such resistance is also often based on deeper issues of how individuals understand the world to work, or a clash of 'ways of knowing' outlined by Schneider and Ingram (2007), related to the water management challenges at hand.

In this book, co-engineering, or the collective initiation, design, implementation and evaluation of participatory processes, has been the key frame through which cases of developing participatory forms of water governance have been analysed. In the cases presented so far, the co-engineering teams were able to work collectively to achieve jointly developed goals. However, it is important to acknowledge that in many other cases there may be alternative ways of developing more participatory forms of water governance, where 'co-engineering' has not actually been a goal of organisers and advocates of such participatory approaches. In particular, it is common to see some individuals, who are interested in promoting participatory water management practice, pursuing their own agendas i.e. not working together in pre-designated projects or programmes but against each other on alternative visions of the future.

For example, the literature on 'water policy entrepreneurs' (Huitema and Meijerink, 2009; Meijerink and Huitema, 2010) provides many insights on these processes, including that entrepreneurs (either individuals or institutions) can: develop new ideas, such as participatory, integrated or adaptive water management (as seen from the local context); build coalitions with other actors and sell ideas; recognise and exploit windows of opportunity; recognise, exploit, create, or manipulate multiple venues for developing their water policy ideas; and orchestrate and manage networks. In particular, the role that incentives and donor institutions can play in policy change in low- and middle-income countries is seen as important for instigating policy change, but has been less successful in enduring effective implementation of the new ideas. In many cases, it can produce perverse outcomes, e.g. the diversion of human resources from on-ground implementation activities to applying for more funding for the next rounds of policy and plan development. Specifically, institution building has been a challenge, with the typically short funding horizons of donor projects not conducive to long-term support of individual and institutional capacity building.

One specific example of such challenges and the potential ways that some attempts at co-engineering of participatory water management processes might not live up to expectations, in addition to the politics, ethics and complexities in implementing participatory water planning and management in low-income countries, can be seen in the Pacific Island nation of the Republic of Kiribati and its densely populated low coral atoll, Tarawa.

Among all the countries with water management issues, problems facing small island countries, and in particular, remote, densely populated, low coral atolls, are some of the most critical in the world (Carpenter *et al.*, 2002). Fragile groundwater lenses under coral atolls are easily polluted and rendered unsuitable for potable water or sanitation, for example by: climate events, such as droughts, storms and tidal inundation; excess water pumping or poor groundwater supply system construction; or the pressures of human development, including agriculture and waste disposal. Water management difficulties are also accentuated by conflicts between traditional resource rights linked to land ownership and the demands of urbanised societies, risks of climate change related to sea-level rise, and the scarce human resources available on islands to have the technical and leadership capacity to develop and implement effective water management strategies (White *et al.*, 2008; White and Falkland, 2010).

The Pacific Island nation of the Republic of Kiribati includes 32 low atolls and one raised island stretched over three million square kilometres of ocean in the centre of the Pacific Ocean The 17 km^2 urban centre of South Tarawa houses over 43% of the nation's population, with densities as high as 15 000 people per square kilometre (White *et al.*, 2008). Households use multiple sources of water for drinking, which include: treated water, piped from groundwater reserves (67%); local groundwater from household wells (72%); rainwater (43%); and, more recently, bottled water (4%) (Daniell *et al.*, 2009). The use of forms of improved sanitation is a key issue, with only one-third of households having access to the sewerage system and many preferring to follow cultural traditions of open defecation on the beach or elsewhere, even if alternatives are available. Large numbers of free-ranging domestic animals, particularly pigs, and the encroachment of settlements on groundwater sourcing areas, also add to the risks and incidence of water contamination. Partly because of these water system conditions, the nation is plagued by endemic water-borne diseases, with infant death rates due to water-borne diseases and incidences of gastro-enteritis tragically high, especially in urban areas.

Over the past four decades, warnings about the extreme risks of using local wells to source groundwater have been widespread (AGDHC, 1975; Shalev, 1992); yet progress towards developing and providing alternative and culturally appropriate systems of drinking water supply and sanitation remains slow. The lack of speed of this progress cannot necessarily be associated with a lack of effort, considering that the Republic of Kiribati has a long and well documented history of water and sanitation projects and policy initiatives seeking to improve the health and development potential of the I-Kiribati people, funded predominantly by a range of donor agencies, including the Australian and New Zealand Governments, Development Banks, the European Union and its Member States, the Global Environment Facility and United Nations groups, such as the UNESCO International Hydrology Programme. These have typically been supported by regional organisations, such as SOPAC (now the Applied Geo-science and Technology Division of the Secretariat of the Pacific Community), the Republic of Kiribati's National Government and a range of consultants, researchers, technical experts and NGOs. If anything, there have often been too many projects carried out at the same time by different groups with competing visions for the future of water management systems on the islands and this does not lead to collectively beneficial outcomes for the islands' inhabitants. Governance of this large range of actors with an interest in Kiribati's water future often runs up against issues of a lack of coordination of water management-related programmes, lack of cooperative approaches to management and not having clear national priorities and plans to which donors' funds and projects can be aligned (White, 2011; personal communication). In addition, one of the major challenges in improving the water management situation has been a reticence of many involved in the islands' water management to focus on understanding and promoting the types of behaviour change required to support sustainable water practices and to improve human health, as many potential technical water system options to support water and health improvements are ineffective if communities and administrations do not understand them and how they can fit into their ways of life.

In recent years, some projects have attempted to treat this issue head-on, for example the participatory groundwater modelling process carried out with some community members and administration staff as part of one Australian–French jointly funded water and sanitation project (White *et al.*, 2002; Dray *et al.*, 2006; 2007), which was designed to promote understanding and collective learning about the connections between human behaviour, management practices and groundwater resources. Although appearing promising from a number of fronts, the co-engineering of this process failed to successfully manage blocking dynamics of the project by members of a concurrently run water and sanitation project, leading to project delays and unwillingness of some administration representatives in the National Government to want to recognise the recommendations stemming from the participatory process (Dray *et al.*, 2007; Jones, 2007b; Moglia *et al.*, 2008).

Another recent programme, the Kiribati Adaptation Program (KAP) of the Global Environment Facility, also set out its aim of supporting the development of a National Water Resources Policy and a 10-year National Water Resources Implementation Plan (Government of the Republic of Kiribati, 2008a; 2008b) through a participatory approach with key government ministries, NGOs and community organisations, as well as with widespread public consultation. Initial public consultation carried out as part of the programme showed that freshwater was the highest priority of those surveyed. In a next stage, the research consultant contracted by the KAP and responsible to the lead water agency for the development of the policy and plans, wanted to develop a whole-of-government and community approach to freshwater and management (White, 2007). Owing to the known reluctance of the government ministries in the water sector in Kiribati to engage with communities, the research consultant proposed the formation of a more formal 'National Water and Sanitation Coordination Committee' (NWSCC), which could bring together all key players and parties with an interest in water and sanitation.

However, this proposed approach to engagement between government ministries and community members was short-lived. At the inaugural meeting of the NWSCC, the issue of inclusions of NGO and community organisations, such as the Council of Churches and the Chamber of Commerce and Industry, was raised, but was soon quashed, with the secretary of the lead water agency stating, '*No NGOs; water is government business.*' This posed a significant politically tricky and ethical issue for the research consultant, as the lead agency, to whom he was answerable, had effectively removed NGOs and community organisations from the formal participation process, despite the terms of reference for the policy and plan development calling explicitly for the coordinated participation of all key stakeholders, which he also considered to be an important element of the process (Daniell *et al.*, 2009).

As one of the objectives of the policy and plan development process was to obtain consensus for a widely acceptable and more implementable plan, the research consultant reacted to this situation by re-engineering the participatory process to have an informal participatory process linked to the formal government process. In the process that was eventually implemented, the research consultant had to first develop each iteration of the policy and plans with the government-only NWSCC, then negotiate separately with NGOs and community groups to gain their acceptance of the draft proposals before the next iteration. This effectively doubled the original planned length of the process but finally achieved the required consensus of all stakeholders, so that the policy and plan could be forwarded to the Cabinet for endorsement. The finalisation of the process was also significantly aided by the appointment of a new lead Ministry Secretary, who immediately recognised the value of having community representatives on the National Water and Sanitation Coordination Committee. He appointed Church, Chamber of Commerce and NGO representatives to the Committee who were able to galvanise the process and bring it to a swift conclusion where the developed policy and plans were submitted to Cabinet (White, 2011; personal communication).

KEY CASE INSIGHTS AND DISCUSSION

As can be seen from these two examples, although on paper many donors and government administrations may espouse the importance of participatory approaches to development and water governance, in practice many simply do not support the approaches and often work to explicitly block them from being implemented; instead preferring the traditional model of deciding on a project and approach and implementing it with little input from communities, often leading later to unusable or unmaintainable systems. Policy change and development of stakeholder-inclusive water planning processes still often face uphill battles, but even so can be worked towards through imaginative, individual entrepreneurship and co-engineering. In such situations, as in the Republic of Kiribati, it can be considered that one of the most important needs to work more constructively through this widespread impasse of a lack of community engagement in water management is behavioural change and education programmes. These could most effectively be developed over the coming decades by working with children and their teachers (White, 2011; personal communication), hopefully leading to much improved health outcomes and development potential for the nation.

This case presents a number of quite different issues, dynamics and complexities not encountered in the other cases outlined in this book, especially related to the challenges of meeting the Millennium Development Goals in a world of competing donor, government, external and stakeholder priorities and preferences of ways of working. On a more positive note, it does also display how some clear guidance by donors for participatory approaches can be translated into on-the-ground practice, even if it is in a less than ideal form. Clearer and real incentives from donors for governments and stakeholders to take on participatory approaches might further enhance the process, but whether it will lead to longer-term behavioural change is less certain. For example, even with the monetary incentives in the Bulgarian case (Chapter 8) for the ministries, NGOs, industry and community stakeholders to work together to the end of a water-planning process, the lack of further incentives, time and local capacity to self-organise meant that implementation has proven difficult. Of greater importance than external incentives for change is likely to be working on in-country behavioural change and the building of capacity in both levels of government and local communities in water science, technology and participatory forms of exchanging and working, so that they can play 'enabling' roles for the development of more widespread sustainable and participatory water management practices. Enabling or even merely supporting the development of arenas for discussion and collective decision-aiding, whether they be for on-the-ground water management actions, participatory water planning or co-engineering, will be a vital way for the world to improve its water governance into the future.

9.6 BEST PRACTICE GUIDELINES IN CO-ENGINEERING PARTICIPATORY MODELLING PROCESSES FOR WATER PLANNING AND MANAGEMENT

From the intervention and extension cases of this book, their comparative analysis and discussion in this chapter, and the earlier review of theory and practice of operationalising decision-aiding models and processes in Chapter 3, a set of 'best-practice' guidelines will be proposed here for co-engineering participatory modelling processes. Although they stem in part from interventions in the water sector, it is suggested that they may be more widely applicable to other domains. The participatory structure model (Mazri, 2007) or 'participation plan' development guidelines (Creighton, 2005) and inter-organisational decision-aiding model (based on Tsoukiàs, 2007) that are drawn upon are considered to be applicable to a range of domains. The three principal objectives of the co-engineering phases, and their suggested corresponding activities, are shown in Figure 9.2.

Through these co-engineering process stages, it is suggested that a critically reflective attitude relative to required co-initiation, co-design and co-implementation decisions be adopted by analysts and other project team members and stakeholders, along with an open attitude to fellow project team member and stakeholder needs, views, values and aspirations. These two different task- and relationship-based attitudes can then assist these participants' readiness to aid the uncovering and debate about boundary judgements (Midgley, 2000) relative to the problem scoping, choice of participants and choice and design of methods.

As these guidelines are newly constituted from experience and theory, they will require further operational validation in new intervention cases.

Co-initiation process: determine grounds for a mutually beneficial collaboration

i) Initial discussions with clients, potential decision makers and other stakeholders or analysts

ii) Constitute a representation of each participant in the potential project team's objectives for the collaboration and determine whether a coherent and shared vision for collaboration can be found
- This may include using Mazri's (2007) model steps 1–4 or 'decision analysis and process planning' questions from Creighton (2005) as a base to discuss each participant's perceptions on the project's needs and may require some problem structuring methods or planning methods to be chosen or used by the analyst(s)

iii) Determine personal skills and other resources required to design, implement and evaluate the envisaged participatory project proposition and whether the project team requires extra members to provide these
- These may include content, process, people or strategic management capacities and people should ideally be found with complementary skill sets to aid team functioning

iv) Find people exhibiting these characteristics and committed to the project outcomes to join the team and establish all project team contracts (including ethics requirements) as required

↓

Preliminary co-design process: develop a participatory modelling plan proposal

v) Repeat steps (i) and (ii) as necessary with new project team members

vi) Critically discuss and negotiate the choice, mixing or design of appropriate methods and participatory modelling methodologies for the context and objects of debate in the proposal
- This process can be further structured by considering the elements of the inter-organisational decision-aiding model's meta-objects (Section 5.1.3, based on Tsoukiàs (2007))

vii) Critically discuss the outcomes of the last stakeholder analysis and decide upon how to invite participants or leave participation open for each object of debate and planned activities

viii) Finalise plan (Mazri step 5) within project team and invite comment and stakeholders to participate

ix) Carry out 'implementation planning' (Creighton, 2005) based on expected participation expectance
- This will need to include collectively determining the meeting agendas, roles of each of the project team members in the implementation process (according to skills, legitimacy considerations, constraints and team member relationships) and meeting logistics

x) Determine whether skills or resources are missing and find people or resources to take part in the implementation process. Assemble implementation team to build necessary relationships and to discuss compelling performance purposes. Provide training if necessary

↓

On-going co-design and co-implementation process: implement and adapt the participatory modelling process as required

xi) Introduce project plan to participating stakeholders, invite comment and carry out contractual or ethics procedures as required (e.g. on payments, ownership of intellectual property created, use of evaluation results, recordings and storage of sensitive information)

xii) Implementation as planned and en-route adaptation, if required, of methods, participating stakeholder groups, project team roles and agendas, in order to take into account unforeseen constraints, opportunities and participants' requests, critically discussing changes with other project team members and stakeholder participants if possible
- This may again include repeating steps 1 to 4 of Mazri (2007) with adequate participatory modelling support methods, and the collective construction of the four meta-objects in the Tsoukiàs (2007) model adapted to the inter-organisational context

xiii) Carry out all data processing in a timely fashion, including evaluations from meetings, synthesis with other required or requested information. Distribute through process as required to stakeholders to allow review and distribution to organisational associates present in the extended network of the multi-accountable participatory group

xiv) Hold debriefing sessions and project team meetings regularly to discuss updates and reassess work towards the common vision or other objectives, in order to resolve any arising issues as quickly as possible. Finding means of working with divergent objectives or interests may require negotiation

xv) Support public diffusion and discussion of groups' and decision makers' final recommendations and other project outcomes, including evaluation results, as well as revision processes as required. Continue to monitor outcomes and reflect on lessons learnt to inform future work

Figure 9.2 'Best-practice' guidelines for co-engineering participatory modelling processes derived from intervention research practice and theory.

10 Conclusions, perspectives and recommendations

To meet the new challenges of environmental governance in today's increasingly interconnected world, there has been a push for broad-scale multi-level participatory processes to aid the decision making processes required for adaptive management of socio-ecological systems. Such decision-aiding processes and, more specifically, participatory modelling processes in the water sector, are typically organised or 'co-engineered' by several agencies or actors, owing to their size and complexity. Strangely, these co-engineering processes and their impacts on participatory water management processes have previously received scant attention, leaving a large gap in current knowledge.

10.1 CONTRIBUTIONS TO KNOWLEDGE

In an attempt to fill this gap, this book aimed to *investigate the impact of co-engineering on participatory modelling processes used for inter-organisational decision-aiding in water planning and management.* Part I of the book focused on arguing why this aim is pertinent for research in the field of water management by critically reviewing water governance structures and theoretical and practical decision-aiding, participatory modelling and co-engineering literature, followed by the development of theory and protocols required for an intervention research programme on the topic. Part II then focused on outlining the lessons learnt from a French pilot trial, and Australian and Bulgarian intervention cases, their analysis and comparison. Extension cases from Germany, Algeria and the Republic of Kiribati were used to delve deeper into issues of successful and less-than-successful co-engineering practice around the world, as well as to provide potential directions for future work. Many insights were provided through the chapters of this book on both the theoretical and practical intervention sides of co-engineering participatory modelling processes for water planning and management. In these conclusions, just six of the major contributions to knowledge from this research will be outlined, as they link specifically to the six objectives and research hypotheses set out in the introduction (Sections 1.4 and 1.5). Further insights have led to the proposals for key areas of future research and practice outlined in Section 10.3.

10.1.1 Investigating the increase in water management complexity

It is commonly claimed that water management is becoming increasingly complex, owing to a number of factors, such as population growth, climate change and conflict over scarce resources. However, others question this claim, citing past conflicts fuelled by similar factors and failed civilisations that were unable to address water management problems of a similar complexity that are faced today. In the face of this debate, the result of which could significantly steer the approach taken to aiding current water management, the first objective of this book was to critically review past and current water governance systems, their management priorities and strategies, to examine whether water management has become increasingly complex. This review was carried out in Chapter 2. Drawing on past and present Australian, European and international governance systems and management priorities, as well as strategies that have seemingly produced consistently negative water management results, it was demonstrated that today's water management is indeed becoming increasingly complex. This complexity is driven by a number of key factors that were not present to the same extent in past societies. The introduction of the 'sustainability concept' and associated themes, such as 'integrated water resources management', into legislation, policy and the public's mind has created a range of considerations above the technical and economic issues of water management that were predominant in the past, and these include environmental and social factors. These factors, and the fact that they may be perceived or constructed differently by different people, have driven the need for public participation in many water management decision and modelling processes, as technical water managers have been seen to be unable to represent the interests and multiple views of stakeholders. More uncertainties are also uncovered as greater quantities of information become available and the rate of technological change increases. Multiple layers of legislation and decentralisation of water management, and broader and more diverse networks of people, owing to globalisation and technological-aided interaction, have also led to a broadening and dispersion of resources and power over water management issues. This means that water management is no longer the preserve of a single agency but requires collaboration and negotiation with a broad range of players. Such complexity was not present to the same extent in past societies. Today, this new complexity underlies the need for improved inter-organisational and multi-stakeholder decision-aiding processes. This is to support collective work by actors from different administrative levels and spatial scales on the water planning and management problems that affect them. This critical review therefore also confirmed the initial assumption that the '*increasing complexity of water-related problems has contributed to the need for improved inter-organisational decision-aiding for water management and planning*'.

10.1.2 Demonstrating the usefulness of operationalising decision-aiding process theory

Water management has a long history of using operational research theory and modelling techniques to aid decision making processes. Much of this theory has remained strongly in the technical domain, although attempts to use integrated assessment techniques, participatory modelling or 'soft' operational research methods, such as Checkland's (1981) soft-systems methodology or cognitive mapping, have become more common in the water sector. In line with the second objective of this book, *to critically review decision-aiding theory and methods, including participatory modelling, and how they could be used to improve water planning and management*, the second major contribution of this book was to demonstrate that much theory exists in the operational research, policy and engineering domains, which could be useful for further aiding inter-organisational decision-aiding processes for water planning and management. In particular, there are strands of decision-aiding 'process' theory and models that appear to be adapted to current participatory process decision-aiding needs, but have been rarely put to use. When combined with an appropriate range of participatory modelling methods, integrated and theoretically founded methodologies for organising participatory decision-aiding processes for water management can be created to work towards specific objectives, such as process efficiency, knowledge elicitation, conflict management or equitable distribution of input. One example of a methodology based on a critical review of participatory modelling methods was proposed and then later trialled and evaluated in the Montpellier pilot trial intervention. An adapted inter-organisational decision-aiding process model based on Tsoukiàs (2007) was developed and used as the basis of my two interventions in the LHEMP in Australia and the Iskar project in Bulgaria. The model significantly aided the structuring of the decision-aiding processes and provided two cases for its operational validation: this is in itself a contribution to knowledge, as the model had not previously been used in practice. Decision-aiding theory and different participatory modelling methods, which were chosen or specifically designed for the intervention contexts, achieved their aims on a number of occasions, such as conflict management or gathering stakeholder knowledge quickly in the LHEMP. These examples outlined the usefulness of explicitly using decision-aiding process theory and methods to improve participatory process interventions for water management. Further *ex-post* operational validation of the Mazri (2007) decision-aiding model for participatory structure design, with the exception of one feedback loop (see Section 9.4.2), also contributes to knowledge in the domain of operational research.

10.1.3 Revealing the two processes to be organised: the participatory process and its co-engineering process

From the critical review and operational use of decision-aiding theory, models and participatory modelling methods, it was further demonstrated that the system of organisers or 'co-engineers' of these participatory decision-aiding processes, and their roles as analysts, modellers, facilitators, method designers or project sponsors, constituted a large knowledge gap that required further research to fill it. This led to the third objective of this book, *to develop a definition of, and critically review, the concept of co-engineering as it relates to the organisation of participatory modelling processes for water management*. This objective was fulfilled through Chapter 4, where co-engineering was defined as having the three principal stages of: co-initiation, co-design and co-implementation, in which a number of people work collectively to realise a participatory modelling process. A trans-disciplinary review of literature relevant to these three phases and the overall co-engineering process then demonstrated that a broad range of potentially relevant literature was available for studying their different aspects, but that little work had previously focused on the equivalent of the whole co-engineering process for participatory modelling processes. The literature review also found no evidence to refute the assumption that *participatory modelling processes used for inter-organisational decision-aiding in complex water management contexts are co-engineered*, but had a number of practical cases that supported it. Assembling a research programme, including presenting large knowledge gaps in the co-engineering processes from the literature review, was a preliminary contribution to knowledge. In particular, gaps in the combination of operational and relational aspects of co-engineering processes were identified. A significant contribution of the work in this book was to bring to the fore the issue that there are two processes to understand and to organise when carrying out participatory modelling for inter-organisational decision-aiding in water planning and management: the co-engineering process of the project team members and the participatory modelling process of the stakeholders. This insight led to the proposed central hypothesis of the research that *co-engineering can critically impact on the participatory modelling processes and their outcomes.*

10.1.4 Developing a framework – research approach, decision-aiding process model and evaluation protocol – that allowed the investigation of the impacts of co-engineering on participatory modelling processes

The gaps in the operational and relational aspects of co-engineering processes and the overall potential criticality of co-engineering on participatory modelling processes were studied and narrowed through the consequent intervention research programme. The research objects of the co-engineering and participatory modelling processes have never previously been studied in relationship to one another and so there was a clear need *to formulate an intervention research programme and evaluation protocol for investigating co-engineering of participatory modelling for inter-organisational decision-aiding in water planning and management.* This was the fourth objective of the book, which was achieved through Chapter 5. More specifically, the objective was achieved by: defining a generic participatory intervention research process; presenting and expanding the Tsoukiàs' (2007) analyst–client decision-aiding process model to the multi-stakeholder, inter-organisational or 'multi-accountable' group situation, so that it could be considered as a basis of elements on which to build participatory modelling processes for collective decision-aiding in the intervention cases; and outlining and analysing a review of a number of existing bodies of literature on participatory process-related evaluation, which was prepared to aid the development of an evaluation protocol that was relevant to an investigation of the impacts of co-engineering on participatory modelling processes. This evaluation protocol outlined a collection of leading questions and factors, to focus research investigations through the three phases of the co-engineering process, the participatory modelling process and the overall outcomes of the impacts of co-engineering on the participatory modelling process. This protocol was therefore able to aid the comparative analysis of research interventions by driving the elicitation of structured information from the different cases. It could also act as a guide for critical reflection by co-engineers of participatory processes.

10.1.5 Confirming the hypothesis that co-engineering can critically impact on participatory modelling processes and their outcomes

The contribution of this research framework was significant in allowing the investigation of the Montpellier pilot trial and the Australian and Bulgarian intervention cases. It largely permitted the fifth objective of the book: *to outline the lessons learnt through the individual and comparative intervention case analysis, so as to determine to what extent co-engineering can critically impact on participatory modelling processes and their outcomes* to be achieved.

Some of the key lessons outlined through Chapters 6, 7, 8 and 9 on the individual cases and their subsequent comparison were that divergent interests of the project team members and a lack of a clear overall purpose goal made co-engineering particularly challenging. In these cases, conflicts and negotiations could occur, the results of which typically had critical impacts, either positive or negative, or both, on the participatory modelling processes and the next stages of their co-engineering. For example, the decision after the intense negotiation in the LHEMP case on what changes, if any, would be made over which stakeholders to invite for the second workshop had a number of both positive and negative impacts on the participatory process. Most of these had been successfully predicted in the negotiations as the possible outcomes of the different decisions, which supported the central hypothesis of the research that *co-engineering can critically impact on the participatory modelling processes and their outcomes.*

Likewise, in the Iskar case, negotiations over process changes and subsequent method choice and design to achieve pre-planned objectives had critical impacts on the participatory modelling process – both positive and negative, depending on the different stakeholders' or co-engineers' points of view. For example, the choice to pursue the creation of projects for structural funds, rather than carry out a test of a multi-criteria method in the last workshop, achieved its objective of keeping the participants enthused and positive about their first participatory process experience, but compromised the achievement of other research project objectives, as had been predicted. Moreover, the Iskar case highlighted what an extra co-engineer entering the participatory process is able to negotiate to redirect the process to achieve his or her own objectives. It is therefore evident from both cases that purposeful action by co-engineers can often lead to achievement of their objectives and can critically impact the participatory modelling processes. It was coming to these decisions and managing the co-engineering team to make their impacts as positive as possible for all concerned, that was shown to be a key lesson.

From the two interventions highlighted in this book, it was concluded that *diversity and creative tension between project team members internal and external to the problem context system provide the necessary opportunities for innovation and collective knowledge creation required to manage complex adaptive systems, even though this increases the need for more concerted efforts in also managing the co-engineering process.*

Extension cases then provided further insights into other co-engineering processes and how the co-engineering processes may be supported. These included the use of a community of practice in the Algerian case – a group of researchers and practitioners with interest and experience in participatory process organisation – who could act as a sounding board for advice and work together to simulate participatory workshops, in order to inform and improve support to the co-engineering process. A number of other key lessons highlighted through the book will be used as a basis for the definition of priority areas for future research in Section 10.3.

10.1.6 Creating a set of best practice guidelines for co-engineering participatory modelling processes

The range of lessons learnt through the research interventions, in combination with the critical literature reviews in the first part of the book, were necessary for the creation of a set of best practice guidelines for co-engineering participatory processes. The creation of the set of guidelines given in Figure 9.2 separated the suggestions into the three commonly observed phases and objectives in research interventions, as noted in some of the literature: the co-initiation process, where the objective is to determine grounds for a mutually beneficial collaboration between co-engineers; the preliminary co-design process, where the objective is to develop a participatory modelling plan proposal that is accepted by the co-engineers and proposed to the stakeholders; and the ongoing co-design and implementation process, where the participatory modelling process is implemented and adapted as required or driven by the co-engineers. These suggestions given for best practice achieve the first section of the final objective of the book, which was *to propose suggestions for future best practice, new perspectives and priority areas in need of further research in co-engineering participatory modelling processes for inter-organisational decision-aiding in water planning and management.* As they are one of the results of this research programme, the guidelines will require further testing. This could be just one priority area of research. Further new perspectives and priority areas for future research will be outlined in Section 10.3, to achieve all the objectives laid out for this book.

10.2 LIMITS OF THE RESEARCH

Although the research so far presented in the book has, to a large extent, fulfilled the objectives laid out to achieve the aim of *investigating the impact of co-engineering on participatory modelling processes for inter-organisational decision-aiding in water planning and management* and provided evidence to confirm the formulated hypotheses, the study could be extended much further. The research has been limited in its scope through the chosen interpretation and analysis schemes. To investigate co-engineering processes, a number of co-engineering events were analysed as negotiations, and relational aspects were examined, drawing upon leadership and team building theory. These choices, although providing a number of interesting insights, mean that the investigation of co-engineering remains partial. Many other concepts or bodies of theory could also provide other types of insight into the two interventions and extension cases, based on the available data. These alternative analyses could include, but are not limited to, investigating the co-engineering process and the participatory stakeholder processes in terms of:

- The boundary judgements and the boundary critiquing activities or actions that took place (Ulrich, 1983; Midgley, 2000);
- Power–knowledge relations (Foucault, 1977; 1980; 1982) or an adherence to principles of discordant or practical pluralism (see Section 4.1.4);
- Finer analysis of the processes of communicative action that took place, for example using the framework laid out by Forester (1993) for studying planning activity;
- Network analyses, dynamic relationship construction (information and trust networks) (e.g. Lazega, 1998);
- Finer analyses on individual interests, coalitions and how power was wielded throughout the broader participatory process to achieve certain results in the plan (e.g. Sabatier and Jenkins-Smith, 1993);
- Further socio-political and institutional analyses and comparison of administrative and legislative frameworks, opportunities and constraints placed upon both processes by the contexts (e.g. Ostrom, 2005; Poteete *et al.*, 2010); and
- Finer analyses of working and potential advantages, disadvantages and improvements that could be made to each participatory modelling method used, for example on the meanings and use made of them by participants or on the content they were able to treat, structure or had to omit.

The research was also deliberately limited in its analysis of alternatives to participatory modelling. In particular, comparative evaluation of different participatory methods and participation levels in different contexts was not systematically investigated here. For example, the political philosophies behind, and many methods associated with, deliberative forms of democracy (Dryzek, 1990; 2010; Elstub, 2008), such as citizens' juries, parliaments or consensus conferences and deliberative polling, have not been explored in detail here and could form the basis of future projects. Another limit, which was alluded to in the scoping of the book, is that the co-engineering process examined through the interventions and in the literature was said to end at the end of participatory process implementation. In some cases, it could be considered that this process is followed by a co-management phase, which the project team also takes a part in organising. This aspect was not investigated in this research to any great depth, but some evaluative work on it in previous participatory modelling projects has been carried out by the ADD-COMMOD project (Etienne, 2010).

It is also noted that, owing to the intervention research approach taken in this work, the hypotheses and assumptions acted to guide intervention choices, with the eventual aim of either confirming or refuting them with available literary and practical case evidence. Therefore, they were not specifically designed to be systematically tested and refuted in carefully pre-planned and controlled experiments, as in experimental economics. Real-world interventions are rather more haphazard and cannot be easily controlled, especially from an ethical point of view. This means that the general findings and lessons developed from this research, although appropriate to fulfil the objectives of this book, should be critically considered for their generality past the limited number of cases presented. Only further research will help to confirm or provide contrary evidence to the insights and lessons learnt through this research.

10.3 KEY AREAS FOR FUTURE RESEARCH

Through the theoretical analysis and intervention research programme of this book, many insights and ideas emerged, which have led to a range of new avenues for research. Those insights appearing to open up possibilities for the most substantial research programmes will be outlined here, along with a few key questions that could drive future work.

10.3.1 Co-engineering of participatory processes

The insight stemming from the work in this book, that *there are two collective processes to organise for broad-scale inter-organisational water planning and management processes: the participatory stakeholder process; and the co-engineering process of the organising group*, opens up a range of areas for future research investigation. Many of the potential research questions that could drive these investigations were formulated in the evaluation protocol in Table 5.2. Their application could be extended to develop research programmes to investigate current co-engineering of a range of inter-organisational participatory processes with different purposes and in a range of domains. Starting with attempts to further outline current co-engineering practice in the water sector, this could be expanded to look at a range of environmental governance and other sectors, such as energy, health, transport, land management and food production, to determine key similarities and differences in the processes. Of particular interest are the questions of who drives these processes, who should be driving them, to what extent they are successful in achieving their aims, and how they could be improved.

The primary point of investigation, based on the discussion in Section 9.3.4, could be to examine the advantages and disadvantages of having co-engineers who are independent of the local context and participating stakeholders. Where there is heightened conflict, known biases of local organisers and potential mistrust of the local managing authorities, it could be hypothesised that a more independent approach, with researchers or external consultants paid by a more neutral authority, could improve results. This was one of the issues in the LHEMP case. However, if one of the aims of running a participatory process is also to build local capacity and coordination, then there may be disadvantages in maintaining an independent approach. Originally in the Iskar process, the non-Bulgarian researchers were concerned about the lack of independence of their local Bulgarian colleagues from the stakeholders in the process, and in particular the language barrier, which meant that the Bulgarians inexperienced in participatory methods had to take responsibility for recruiting stakeholders, facilitating the process and translating and interpreting the raw data produced. What was perceived in the end was in fact that *language barriers were instrumental in ensuring transfer of participatory process organisational, facilitation and design skills, and in building capacity and appropriation of the process by the local stakeholders.* For development work, especially, this insight could provide the basis for future research on how to co-engineer participatory processes that could be more self-sustaining when independent researchers or consultants leave the project.

Another question of interest for research on a more theoretical level is: in what circumstances could a co-engineering group or team be reclassified as just an 'engineering' team? Although co-engineering in this book was originally defined as engineering that must be performed 'together' or 'with' others, the essence of co-engineering may require a more subtle definition for future work. From the insights gathered in the intervention cases, *it appears that the differences in accountabilities and interests of the project organisation members may be the key to whether participatory processes are co-engineered or just engineered.* Therefore, when all objectives for the project are shared, the individual group members are all accountable to the same person or organisation and their interests are aligned, co-engineering may be able to be reclassified as engineering. When might such a case occur in complex water management situations? One potential case could be when the leadership over the co-engineering team and process is so strong that it manages to align interests and create allegiance to a leader or a clear team performance purpose that breaks down other accountabilities. Glimpses of such a case started to emerge at the end of the Iskar process. Another example would be if whole government or consultancy groups were dedicated to organising participatory processes. Further research is needed to investigate the possibilities of such cases and other issues, including the ethical and legal ones, associated with them.

10.3.2 Model building and use in urgent decision-aiding processes

What types of model are appropriate to build and use for urgent multi-stakeholder or inter-organisational decision-aiding? In both the Montpellier pilot trial and the Iskar intervention, researchers had the aim of creating causally calculable participatory-built models. This was not satisfactorily achieved in either case, partly because of time limitations, leading to a number of important insights that could spawn future research programmes. In the Montpellier trial, it was observed that a fully operable computer simulation model was simply not necessary for the students to react to management challenges through the role-playing game and to debate and agree on management solutions. Moreover, it was found that some participants had an extreme dislike of numbers and had difficulties in interpreting numerical model outputs; yet had little trouble in understanding most system dynamics using diagrammatic and linguistic supports. Potential underlying differences in communication preferences and numerical and linguistic capacities of different participants in multi-stakeholder processes could be a preliminary reason to rethink the objectives of the types of participatory modelling used for decision-aiding processes. The issue of how modelling methods might be adapted to allow the equitable participation of stakeholders – who have different capacities and communication-style preferences – in the participatory processes is in need of further research.

One important insight from noting the difficulties in producing and using calculable models in the Iskar process was that *as participants increase appropriation of qualitative modelling methods, which is typically a positive outcome of the process, they often adapt the syntax to match their own understanding and preferences.* This then reduces: the possibility for reuse of models; using the models for calculation (if the new syntax does not allow it); and comparative understanding of models developed by different groups, especially when time is limited for syntax re-conversion. Urgency and the types of model proposed for construction thus appear to be key factors affecting the capacity to build functioning and useful calculable models in a participatory setting. One area requiring future research is therefore to determine to what extent it is actually desirable to build calculable (quantitative or qualitative) models in urgent decision-aiding contexts; especially when causalities are uncertain and disputed. Other methods, such as the participatory risk assessment used in the LHEMP, could be evaluated for their comparative desirability and appropriateness to the management situations under investigation. Otherwise, completely different approaches may be taken, such as using existing models for responding to clearly specified questions, or not building models at all and remaining at the level of discourse. This area, of required levels of participation in modelling and types of models used, is still in need of a lot of further investigation.

Related to this issue of participation levels is the question: to what extent is it valuable to divide large multi-accountable stakeholder groups into similar knowledge or authority blocks during participatory processes, rather than keeping all stakeholders together? Preliminary findings from the interventions of this book highlighted the fact that grouping stakeholders by institutional level or resource similarity level (e.g. knowledge or authority over decision making), and adapting methods to their needs could provide forums for more open and non-confrontational discourse than when mixing them; this was particularly true for the LHEMP. However, this limited the capacity to build an integrated shared vision through the whole decision-aiding process and collective work between the levels and different stakeholder interest groups. It also appeared that dividing groups could create or reinforce existing inequities in the social contexts and that groups of people could be disempowered by being excluded, if the processes were not adequately managed. For example, the last-minute change of programme in the LHEMP process to run an 'agency-only' workshop disempowered at least one excluded member, created inequities in the inputs of some participants to the process and did not allow equivalent levels of learning and knowledge sharing among the different stakeholder groups. In the Iskar process, divisions of groups ended up reinforcing inequities on a time and monetary level, as the costs of the project and number of workshops were divided unequally to keep the high-level politicians adequately informed and to show adequate respect to them, to ensure their continued participation in the project. On the other hand, the carefully planned division of groups in the Algerian case seemed to work effectively to build a shared understanding of issues at multiple scales of management. From these insights, it can be seen that there are certain advantages and disadvantages in dividing groups, and further research and systematic evaluation of other interventions in different contexts with varying levels of collective participation in the different process stages are required.

10.3.3 Participatory risk management

In Chapter 2, it was asked what type of approaches might be appropriate to avoid repeating the same water management mistakes of the past and to treat the challenges better. Building on what was proposed in the literature, it can now be suggested, based on the experiences in this book, that *participatory risk management approaches, based on collectively defined participant values and including mechanisms for adequate information flow between the stakeholders and external networks of the multi-stakeholder group, and adaptive updating of assessments, appear to be appropriately suited to many of today's messy water governance challenges.* More research is required to validate this proposition further. Related to this complex proposition, a few key points, which are also areas for future research, are:

- Firstly, to examine the extent to which water management situations exhibiting conflict over values, interests and representations of the problem situation can benefit from highly structured and well organised participatory modelling interactions. From this research, it was shown in the LHEMP process that effective co-engineering and structuring of the preliminary workshop activities, where there were pre-existing conflicts between stakeholders, allowed successful management of these conflicts in a non-confrontational manner and enabled the stakeholders to come to a set of shared values that all of them were willing to support as the basis of the risk assessment, even if some personal discrepancies existed. The extent to which such structured processes can be replicated, in particular at larger spatial scales, requires examination.
- Secondly, to examine the extent to which participatory risk management approaches can be used to structure complexity and work under a range of uncertainties, including those related to causal system relations, possibilities of future hazard events, knowledge and political priorities. Both the LHEMP and Iskar projects took a participatory risk management approach, which allowed the groups to work with many uncertainties and structure their understandings of the complexity. Considering the broad differences in these approaches, including that the LHEMP process was more formalised, there is again a need for further research to improve and trial the repeatability of the risk assessment approaches taken, especially the Australian process, as the methodology could be more easily replicated and improved. Initial take-up of the approach in other settings, such as for mud-crab management (Brown, 2010), appears to indicate that there is good potential for adaption of this approach for a range of applications.
- Finally, to determine the extent to which developing value-based participatory risk management processes is a worthwhile way of ensuring meaningful participation of stakeholders in the problem-situation stage of the decision-aiding process that is likely to have an impact on the final outcomes. The example of the LHEMP process showed that developing a commonly acknowledged set of values, which the estuary management was to preserve and enhance, provided a solid structure and vision for the rest of the process, even when the community stakeholders were not involved in the risk assessment. Using these values directly as the criteria in the 'multi-asset' risk assessment process meant that agencies were assessing risks based on collective stakeholder-defined values or 'assets'. Despite visions and values being individually elicited in the Iskar process, the links to communally agreed-upon values, except in the last workshop's evaluation where the values had been analysed and grouped by the analysts, were not evident, and the vision and significance of the process and actions appeared harder for the participants to understand. These intervention observations appear to support Keeney's (1992) views on the advantages of value-focused thinking and his preferred design of decision-aiding processes, with value elicitation being the first stage. Further research and empirical validation of value-based risk management processes relative to other processes, for example situation-model-based processes, are still required.

Issues relating to the multi-accountable nature of groups in participatory risk management will be addressed in the next section.

10.3.4 Decision-aiding theory and models for multi-accountable groups

Through the theoretical reviews and practical case interventions of this research, it was proposed that *when resources, including knowledge, decision making authority, money and time, are typically dispersed over a wide range of organisations, including stakeholder groups, participatory processes can offer the possibility of coordinating and capitalising on the efficient and effective allocation of these resources.* It is further suggested in this book that participatory processes in the context of these 'multi-accountable groups' are not just useful for dealing with conflicts, but are a necessity for efficient and effective resource use and water management outcomes when resources are widely distributed, as in many westernised democracies. Preliminary evaluations of the LHEMP and Iskar processes provided some insights into their efficiency and effectiveness in capturing and integrating knowledge resources; yet much more research on more efficiently and effectively allocating other types of resource, such as finances and decision making authority, is required. To begin this work, the actor–action–resources matrix negotiation support tool (Section 7.4.2) was developed in the LHEMP with the aim of supporting effective and efficient allocation resources and responsibilities on priority actions for water risk mitigation management. It has yet to be tested in its proposed role and so requires further investigation through practical application and evaluation, leading to its eventual improvement.

Another related insight from this research was that the co-engineering of decision-aiding processes for multi-accountable groups requires flexible means of open participation and the creation and dissemination of intermediary feedback documents, as well as adequate and timely general updates for participants, which they can distribute to their own networks or institutional superiors for opinion. For example, multi-accountable group processes are likely to need to leave themselves open to newcomers or replacement representatives, who may be interested institutional superiors or colleagues. In the experiences outlined in this book, including in the German extension case, these types of addition or exchange occurred late in the processes, with short notice. Keeping all of these potential participants or decision makers up-to-date – by providing them with the means of having even partial input or awareness of the meta-objects produced through the decision-aiding process – may help those in the extended network of the multi-accountable group to legitimise the process and support the collective decisions, the problem situation representation, the problem formulation, the evaluation model and the final recommendations.

Multi-stakeholder participatory modelling and participatory processes in general are often considered inefficient in terms of time and other resources for the ends that they achieve (Korfmacher, 2001). From the intervention research examples outlined in this book, it is argued that *using good decision-aiding theory, as a part of effective co-engineering efforts, can support efficient and effective participatory modelling processes.* Using available theory, decision-aiding models and knowledge of practical advantages and disadvantages of different decision-aiding methods was a key in both research interventions to being able to creatively choose and construct adapted methods to work towards efficiency, efficacy and effectiveness of participatory modelling processes. Good theory, whether it is tacit or explicit, is invaluable for effective practice, so it makes sense to use the best available decision-aiding theory to inform participatory modelling practices. This opens up a large research field on reviewing, using and creating decision-aiding theory and models appropriate to aid multi-accountable groups. The expansion and use of the Tsoukiàs (2007) decision-aiding process model was one example in this book of what research in this field could constitute. Operationally testing the Mazri (2007) participatory process design model and the co-engineering 'best-practice guidelines' proposed in Section 9.6 are other immediate research needs.

10.3.5 Rethinking water management and decision-analyst education

Water managers, and in particular environmental engineers, hydrologists and other scientists and professionals who commonly work in water planning and who are likely to find themselves in co-engineering roles, would most probably benefit from additional training in decision-aiding techniques and theory. Many of these managers are taught to use a host of different statistical and numerical modelling techniques; yet they are currently lacking in an understanding of 'soft' problem structuring methods and multi-criteria or multi-objective analysis techniques. As has been demonstrated in this book, messy water management situations are characterised by conflict, as well as complexity and uncertainty. It is especially the conflict over values, representations, interests and objectives of stakeholders, and those in the project teams, that necessitates adequate problem formulation and adapted modelling methods and group-work techniques for aiding current water management decision making. The challenge for potential decision-aiders working in water management contexts is to emphasise arguments for particular theories and work towards collective co-engineering group understanding, learning and legitimisation of the chosen theories and methods. Potential decision analysts thus require not only solid theoretical grounding for their role, but also a range of operational, organisational and relational teamwork skills to be able to perform effectively as part of the co-engineering team, especially by using negotiation skills, creativity, flexibility and emotional awareness. If the decision analyst is incapable of working adequately as part of the co-engineering team, then the models and theories for process improvement are unlikely to be used, so it is worth further researching these aspects of co-engineering roles and how to set up and maintain effective co-engineering teams. In the push to consider sustainability in engineering curricula in some Australian universities, progress has been made towards teaching 'systems approaches' to messy engineering problems and to consider concurrently the economic, social, environmental and infrastructural aspects of engineering problems (Foley *et al.*, 2003; Maier *et al.*, 2007). In some cases, this has even been extended to online role-playing simulations of international water conflict situations, to aid student learning on the diversity of underlying views, values and issues that must be considered in an interactive environment, and to evaluate how they would act in such a situation (Maier, 2007). What is missing, however, is adequate theory and methods to then aid these budding engineers in improving their decision-aiding capacities when they find themselves having to work in, and especially *organise*, planning and management initiatives in such messy real-world situations.

Research is required into what new additions to higher education curricula might better equip water managers and decision analysts to perform well in co-engineering roles, as the need for the organisation of broad-scale participatory process and stakeholder network management increases. This might include: training in the participatory use of the Australian Risk Management Standard; developing a basic knowledge of the concept of 'boundary critique' (Midgley, 2000) to help facilitate and understand the consequences of different choices of system boundaries for management; and experience in using a range of problem-structuring methods (Rosenhead and Mingers, 2001b), from simple card techniques and cognitive mapping to the more complete soft systems methodology or strategic choice approaches that could prove invaluable for their professional practice. Likewise, learning the theory behind (and application of) a small range of multiple-criteria decision analysis techniques (see Bouyssou *et al.*, 2000; Belton and Stewart, 2002 for an overview) in messy problem situations would probably prove a useful investment of education time, as the mathematical principles and the construction of value, utility or preference functions in the urgent 'on-the-job' environment might not be learnt, owing to lack of time, as was discovered during the two interventions in this book.

Carrying out water management courses in the form of the Montpellier pilot trial with some extra theoretical grounding on the methods used could be an effective way of introducing participatory modelling and methods as a programme of 'learning by doing'. Some courses on co-engineering, of the type already run in Australia and the Netherlands with both Masters level students and government officials, which include co-engineering negotiation simulation exercises, could supplement planned participatory process education and training programmes. Adequate education programmes in all of these domains for co-engineering participatory processes for water management would be beneficial for many professionals and researchers and will then hopefully lead to some of them pursuing the other key areas for research, highlighted throughout this section.

10.4 EPILOGUE: THE FUTURE OF INTERNATIONAL WATER GOVERNANCE SYSTEMS AND THE COMMONS

It was noted in the review of international water governance systems at the beginning of this book that some water experts and scholars are now dubious about the utility of the international-level water meetings and forums that have proliferated in recent years for helping to solve today's 'water crisis' but do not appear to be improving the situation on the ground, as a billion people still lack access to safe drinking water and 2.6 billion lack access to adequate sanitation (Gleick *et al.*, 2006; UNDP, 2006).

Like these water meetings, participatory modelling processes are often suggested to be too costly and inefficient for their concrete results (not including the networking opportunities). However, in this book it has been demonstrated that, with good decision-aiding theory and effective co-engineering, participatory modelling processes can be efficient and effective in achieving the principal planned outcomes. Could the same be true for international water governance meetings? Could serious attention be given to using the best possible decision-theory and highly skilled co-engineers with exemplary knowledge and practice in operational, organisational and relational management to co-engineer efficient and effective participatory international water meetings? Might some more truly insightful management ideas and concrete methods of allocating available resources emerge, to really improve the priority areas of world water management? Could the time of 100 world hydrological experts be better spent than in sitting together in a room to check and discuss every paragraph of a report? Maybe alternative ideas for international water meeting structures could lead to an improved use of so much of their collective knowledge and hope, to make a difference through their work.

One of the problems of supplying water to those in need is that international governance mechanisms often have limited applicability, as regional- or national-level administrations are responsible for this task. Multi-level participatory processes, up to the national level in some cases, were implemented through the interventions in this book. Could this be extended to the international level for some critical regional cases or are there other more effective mechanisms that could be collectively developed for supporting these regions?

In a rapidly changing and connected world where climate change, population growth, natural disasters and water and food scarcities are likely to cause increased human suffering, perhaps it is time to put adequate energy and resources into enabling both the highest-level world politicians and water experts, as well as the most vulnerable and marginalised people and many other levels in between, to build their own capacity to communicate and act together to manage increasingly scarce water resources collectively for the benefit of current and future generations. This will be no easy task, but with much more applied research, energy and practical intervention, it may be possible to organise today's struggles more effectively to govern the commons.

Appendix A Understanding water and its management

This appendix supplements Chapter 2 by extending the brief descriptions of water planning and management previously provided. The investigations presented here on water, its place in the world, and associated means of planning and managing it, were carried out prior to the development of Chapter 2 and have significantly informed the views presented in the book.

A.1 STATE OF THE WORLD'S WATER

Water sites are sources of life, and the regeneration of life in all its forms

Lingari Foundation, 2002

A.1.1 The substance and location of water

Life and water and are inextricably linked. All known forms of life require liquid water to survive in the long term, owing to its unique chemical properties (Ball, 2001). Water, the molecular compound H_2O, can act as an effective solvent, owing to its high dielectric constant, its capacity to form hydrogen bonds and its wide temperature range in the liquid state, which corresponds to many current environmental conditions on the Earth's surface (Rothschild and Mancinelli, 2001). It also plays an important role in most metabolic processes as a reactant or product and constitutes a large proportion (60–95%) of all living organisms (Pimentel *et al.*, 1997).

Water possesses many other interesting properties compared with other chemical substances, including that it is densest at approximately four degrees Celsius and expands as it reaches its freezing point. Ice is therefore able to float on liquid water, a chemical property that has aided the capacity for life to evolve and remain protected in large water bodies, such as lakes and oceans. This is because, even if lakes or oceans freeze over, the densest water, at around four degrees Celsius, will remain close to the bed or ocean floor, providing habitable living conditions for a variety of organisms. In its gaseous state, water also plays a key role in supporting many of Earth's life forms. Water vapour in the atmosphere is one of the most important and abundant greenhouse gases which regulate the Earth's temperature (Forster *et al.*, 2007). However, despite its comparative importance to other greenhouse gases, water vapour's influence in future global warming is still surrounded by high uncertainties (Forster *et al.*, 2007) and is the subject of continuing scientific enquiry (Held and Soden, 2006).

Although water is also one of the most common liquids on the planet, it is one of the most scientifically interesting as well, with much about its physical properties, behaviour and origins remaining unknown or under heavy debate. For example, recent scientific exploration has demonstrated that the well-known expression, '*oil and water don't mix*', should soon be qualified by '*under normal conditions*', because if water is degassed, certain oils can be dispersed in it without the use of any other additives (Francis *et al.*, 2006).

Another common belief, that the hydrological cycle is closed and that the water on Earth has been here since the planet's creation, or arrived shortly afterwards in large meteors, has also come under debate in recent years. In 1986, after viewing ultraviolet images collected by NASA's Dynamic Explorer I spacecraft, Professor Louis A. Frank and colleagues hypothesised that water was brought into the atmosphere in an almost constant stream of small icy comets that vaporise upon entering the atmosphere (Gleick, 1998). Since its release, the theory has come under harsh criticism, but recent research continues to support the original hypothesis (Frank and Sigwarth, 2001). With calculations of thousands of such comets entering the atmosphere each day, the entire volume of water on the planet could be accounted for if this rate has been fairly constant since the Earth's creation (Frank *et al.*, 1986).

Of the 1360 million cubic kilometres of water on the planet (Clarke, 1993), approximately 97.5% is found in the oceans and approximately one-third of the remaining 2.5% is fresh liquid water (the rest is found in the form of glaciers, snow or polar ice caps). Of the world's liquid freshwater, it is estimated that 98% occurs as groundwater, with the largest remaining portions being found in either lakes or rivers (Shiklomanov, 1999). Some of the largest stocks of groundwater occur in the most arid regions of

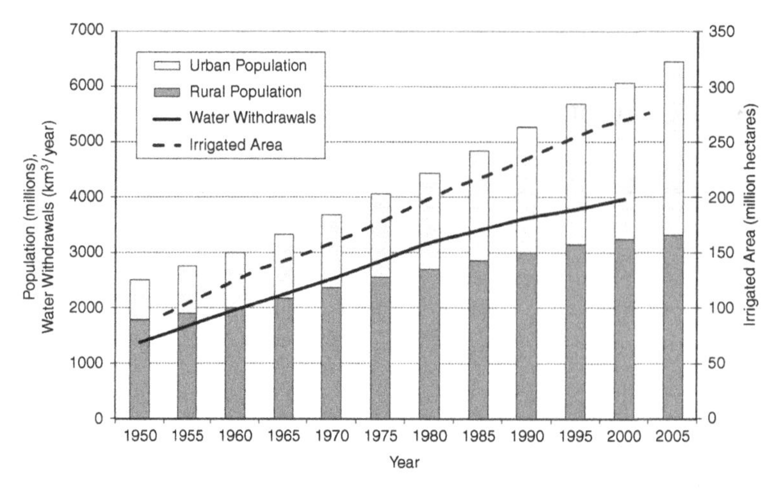

Figure A.1 World trends in water withdrawals, population growth in urban and rural sectors and irrigated area. Developed from: United Nations (2005); Shiklomanov (1999); EPI (2006).

the world, such as Northern Africa (i.e. the Nubian Sandstone and Northwestern Sahara aquifer systems), Australia (the Great Artesian Basin) and the Middle East (UNESCO, 2006). However, the water in such aquifers has residence times of hundreds to thousands of years, with large proportions of the total volumes being largely non-renewable from year to year. It is estimated that the actual quantity of renewable freshwater accessible for human use and consumption is only about 9000–14 000 cubic kilometres per year (Clarke, 1993); supplies that are very unequally distributed spatially throughout the world.

A.1.2 World water use

Approximately 6–8% of all currently available water resources (renewable and non-renewable) are being withdrawn each year, or about 4000 cubic kilometres (Gleick *et al.*, 2006). It is noted that water 'withdrawn' is not necessarily 'consumed' (i.e. it does not leave the freshwater stocks for another part of the hydrological cycle through evapo-transpiration, evaporation or being discharged into the ocean) but that in some cases water may be used and then returned to the freshwater stocks in a similar or slightly modified form (e.g. water from wastewater treatment plants or industrial cooling water can be returned to rivers). Figure A.1 shows the evolution of worldwide water withdrawals since 1950, which have steadily increased along with population growth in both the rural and urban sectors, and the total area of irrigated land.

However, it is interesting to note from Figure A.1 that growth in water use as a percentage of the total population, and thus water use per capita, was on the rise between 1950 and 1980 but has since levelled out and decreased slightly. The same change is observed for irrigated area. Potential drivers for these trend changes will be examined in the following sections of this appendix, but could include: improvements in water use efficiency, i.e. through new irrigation technologies, use of less water-intensive crops and more efficient urban management systems, technologies or use behaviours; or decreasing availability of water, i.e. water 'scarcity' due to a combination of factors to be discussed in the next section.

To better understand how water is used around the world, the aforementioned information on water 'withdrawals' only provides one small piece of the puzzle. This water is principally used for a number of physical or economic production reasons, such as domestic use (i.e. drinking, cooking, sanitation, irrigating parks and gardens); agriculture and food production, in the form of irrigation and processing; and industry, including energy production, manufacturing and commerce. Approximately 70% of total world water withdrawals can be attributed to the agricultural sector, 22% to the industrial sector and the remaining 8% to domestic uses (UNESCO-WWAP, 2003). These statistics vary substantially between country income levels and world regions, with the group of highest-income countries using a much higher proportion of their total water use for industrial (42%) and domestic (16%) purposes (World Bank, 2007). Africa, the Middle East and South Asia, on the other hand, use about 90% of their water withdrawals for agriculture (World Bank, 2007).

On top of these water withdrawals, water has many other uses, purposes and values, which are more rarely considered in typical accounting methods of water use. As well as supporting human needs for life, water also supports all ecosystem life and the services they provide to society. These 'ecosystem services' include:

Supporting services, such as nutrient cycling, soil generation, purification and decomposition processes, seed dispersal and pollination;

Regulating processes, such as hydrological cycle regulation (mitigation of floods and droughts), pest and disease control, climate stabilisation and moderation of weather extremes, maintenance of soil fertility and coastal and river bank stabilisation (a requirement for navigation);

Provisioning services, such as food, water, plant and animal products, genetic material, medicinal products, natural fibres, energy sources, precursors to industrial and synthetic products (i.e. minerals, oils, rubbers, natural chemicals);

Cultural services, such as recreational, intellectual, aesthetic, spiritual and other non-material benefits; and

Sustaining services, such as maintenance of species diversity and key ecological systems, and components required to provide all of the other listed services and other future needs (Daily, 2000; MEA, 2005).

If sufficient water is not provided to maintain healthy ecosystems, then all of these services are at risk of disappearing.

Increasing environmental degradation, some due to overuse or misuse of water resources, and subsequent loss of ecosystem services have recently driven more scholars and practitioners to push for inclusion of ecosystem water use in typical water accounting procedures (Katz, 2006). This has resulted in a number of innovations, including prescribing provisions for 'environmental flows' in a number of pieces of legislation around the world, including Australia (Gardner and Bowmer, 2007). A number of alternative conceptual frameworks for understanding and accounting for water have also been developed, such as the 'green–blue water' paradigm and its associated aim of learning to use the 'green' and 'blue' water flows stemming from precipitation more effectively, to meet both ecosystem and human objectives: *green water* being the water stored in the root zone and all invisible water vapour flows – from either evapo-transpiration, or evaporation processes off blue water; and *blue water* including all remaining liquid waters, i.e. in rivers, lakes and groundwater stocks (Falkenmark, 2003). The approach focuses on bringing land, water and ecosystem integration into management practices and has been championed by leaders, such as the former UN secretary, Kofi Annan, for use as the conceptual base for meeting the Millennium Development Goals (Falkenmark and Rockström, 2006).

Another alternative accounting method is the concept of 'embodied' or 'virtual' water. Although not specifically designed to determine water for ecosystem purposes, the embodied water concept is particularly useful for better understanding the impacts of human consumption and behaviour on the environment. An 'embodied water' footprint is similar in concept to a carbon or ecological footprint (Wackernagel *et al.*, 2002), in that it converts all consumption, production and other activities into one indicator; in this case a quantity of water required to carry out these actions (Dey *et al.*, 2007). Such embodied water footprints can be calculated for almost anything, from goods and services to people and governments, and can look at the first-order water use only or follow the production chain of goods and services back a number of stages (Foran, 2008). For example, Dey *et al.* (2007) calculate that the average water footprint for an Australian is 760 kl per year, only 16% of which is direct indoor water use. The largest percentages of water use are from food consumed (46%) and from the purchase and use of other goods and services (28%). The equivalent concept of 'virtual water' (Allan, 1997) is more commonly used to describe the higher-level trading of water in products between countries, although recent analyses tend to focus more on the agricultural 'green water' than on the whole water budget (Islam *et al.*, 2007; Warner and Johnson, 2007). Although contentious, it is thought by some authors that trade of virtual water could help countries with severe water scarcity, especially to meet their food demand (Dinesh Kumar and Singh, 2005).

A.1.3 Water scarcity

The concept of water 'scarcity' or 'stress' remains highly debated in the literature, owing to its relativistic and ambiguous nature (Wolfe and Brooks, 2003). The scale of analysis, the definition of 'needs' and the perception of social, economic and political aspects may all have an impact on the degree at which water is believed, or calculated, to be 'scarce'. The most commonly used definitions of water stress tend to relate to a physical lack of water per capita, with the UN Food and Agriculture Organization defining water scarcity as less than 1000 cubic metres of renewable accessible water per person per annum (FAO, 2000), a measure that stemmed from the works of Falkenmark and her colleagues, which resulted in the 'water barrier demarcations' and later the 'Falkenmark water stress index' (Falkenmark *et al.*, 1989; Falkenmark and Widstrand, 1992; Falkenmark, 1997; Falkenmark and Lundqvist, 1998). Many other water stress indicators have been developed, including: the 'vulnerability of water systems' indices (Gleick, 1990); the 'SEI water resources vulnerability index' (Raskin *et al.*, 1997); the 'IWMI index of relative scarcity' (Seckler *et al.*, 1998); and the 'water poverty index' (Sullivan, 2001; Sullivan and Meigh, 2007).

The main aim of such indicators is theoretically to allow decision makers at a variety of levels, from international funding bodies to local governments, to make more informed choices for water-related projects and policies. However, the simplicity of the first indicators developed (e.g. the Falkenmark water stress index), which are still the most commonly used, often hide local water scarcity realities and can sometimes undermine informed policy choices because of: the scale of analysis; the focus on only 'blue water' availability and use; a lack of perspective on whether local human needs are fulfilled; and the neglect of water quality issues (Gleick *et al.*, 2002).

Table A.1 *Type classification and description of water scarcities and options for managing them. Adapted from: Clarke (1993); Wolfe and Brooks (2003)*

Type of scarcity	Principal causes	Management style	Typical solutions
Physical or environmental	Physical environment and hydrology, climate (aridity, drought), population	Command and control: supply-side management	Engineering works: dams, wells, distribution systems, desalination, local rain water collection systems
Economic	Prohibitive supply costs, resource depletion	Efficiency and cost acceptance: demand-side management	Economic instruments and technological adjustment: pricing, taxes, fines, water system optimisation
Social or political	Divergent interests and conflict, complexity, ecological degradation, desiccation, scientific uncertainty, unsustainable practices	Democratic discourse and broad policy processes: integrated or adaptive management	New allocation rules and technologies to meet social and ecological objectives: participatory processes, education, water markets, eco-technologies, integration between sectors (i.e. energy and waste management)

Some of the more recently developed indicators have attempted to overcome these issues by including different types of water use data and socio-economic factors in their calculations, with the aim of giving a fuller description of the types of water stress encountered (Rijsberman, 2006). The 'water poverty index', for example, is an agglomeration of a whole range of indicators under the following categories: resources; access; use; capacity (social and institutional); and environment (Sullivan and Meigh, 2007). Other large research projects, such as the European Integrated Project, 'AquaStress', worked to develop and apply these more informed indicators for use at a local scale (Sullivan *et al.*, 2007).

With or without the aid of these indicators, an understanding that water scarcity might be perceived to exist in a variety of forms, and can thus be managed differently, may be of more use to decision makers. As well as 'physical' water scarcity, it can also be considered that different forms of socially constructed scarcity exist; those that could be classified as 'economic' and 'social' or 'political' scarcities are shown in Table A.1. These types of scarcity are also known as first-, second- and third-order scarcities, as described in Wolfe and Brooks (2003).

Any region may suffer from any or all of these three types of scarcity at once, depending on the local drivers for perceived water shortages. By understanding the drivers of the water scarcity, appropriate styles of management can be chosen to mitigate the water scarcity issues, along with a range of potential solutions or policy options. The final 'social or political' scarcity category is increasingly encountered around the world and is the most challenging to manage, owing to its complex, conflict-producing and uncertain nature. As noted in the 2006 Human Development Report (UNDP, 2006):

> *The scarcity at the heart of the global water crisis is rooted in power, poverty and inequality, not in physical availability.*

The purpose of the following section is to highlight how we arrived at this situation of socially constructed and physical water scarcities. This will be achieved by outlining a number of salient cultural and technical evolutions in the water sector, as well as the water planning and management practices that supported them. It is the intention that these brief reviews will demonstrate a number of points, including the fact that water scarcities are enhanced through human choices or management decisions; the observation that evolution in the water sector has not been a linear process – poor knowledge and seemingly sustainable solutions to specific problems have been forgotten at certain stages of history and resurfaced much later when the same problem has reappeared – and the recognition that some ancient cultures implicitly incorporated many of the principles being touted in today's best practice management paradigms into their understanding of the world.

A.2 EVOLUTION OF WATER PLANNING, MANAGEMENT AND DEVELOPMENT

> *The sage's transformation of the world arises from solving the problem of water. If water is united, the human heart will be corrected. If water is pure and clean, the heart of the people will readily be unified and desirous of cleanliness. Even when the citizenry's heart is changed, their conduct will not be depraved. So the sage's government does not*

consist of talking to people and persuading them, family by family. The pivot (of work) is water.

– Lao Tze (Warshall, 1995)

A.2.1 Cultural and technical evolution in the water sector

SUBSISTENCE AND SPIRITUAL WATER

Water means different things to different people. At the beginning of human life, as for all other forms of life, it formed a necessary means of subsistence. Use and perceptions of water have evolved over time and through a number of phases of human development and knowledge acquisition. In the hunter-gatherer and nomad societies that inhabited the Earth for many thousands of years, water played a central role in their organisation, daily rituals and, in some cases, spirituality. It was an element to be respected, and sometimes feared, owing to its capacity to bring life and take it away (i.e. through flood or drought). For most early societies, and some more modern ones, it was also seen to be so important that it took the form of a god, a spirit or other type of deity. To the current day, festivals are held for the coming of the rains in many parts of the world, such as the Indian monsoon parties held for the Hindu water gods (Pearce, 2004) and the Australian Aboriginal songs and dances to celebrate 'country' (Rose, 1996).

In traditional Australian Aboriginal cultures, water is thought to embody a life force or be 'living' (Jackson *et al.*, 2005). The expression, 'living water' (*kunangkul* from the Kimberley region cultures), describes both the cultural significance of water sources, typically groundwater, and their physical properties (Yu, 2000). Australian Aboriginal cultures also do not differentiate water and land as being two separate entities; they are both parts of the living 'country' for which people are responsible and they must be equally cared for and nurtured (Lingiari Foundation, 2002). Interestingly, some other cultures have similar perceptions, with, for example, Bulgarian folklore and the Bulgarian magical perspective of the world treating water from springs as 'alive', and considering it to be the 'pearled blood of the Earth' and having social powers (Fotev, 2004). However, unlike the Australian Aboriginal perspective, the Earth is perceived as a woman, and the rains and Sun's rays the inputs of a man, which together create new life (Sheitanov (1994) as cited in Fotev (2004)).

CONTROLLING WATER: THE BIRTH OF AGRICULTURE

From around 9000 BC in the Neolithic period, societies realised that they could control water to improve life and grow food (Avery, 2005). Such technologies as earthen dams and dug-out canals were developed to trap water and transport it. This agricultural revolution, accompanied by water engineering practices, originated in western Asia or the 'Fertile Crescent', where evidence of irrigation canals 8000 years old can be found in Mesopotamia (WCD, 2000), later moving east through the modern-day countries of Iran and India and spreading slowly through Europe and Northern Africa, then later to Kenya and to the British Isles and Scandinavia (by 3500 BC) (Diamond, 1999). Similar revolutions also appeared independently about the same time (before 7500 BC) in China, followed by South-East Asia, Korea and Japan; and a little later in Meso-America (3500 BC), followed by North America (Diamond, 1999). As well as shifting water to irrigate crops by canals, other types of rain-fed irrigation techniques were also developed by some of these civilisations in about 500 BC (Cech, 2005). These included the mountain agricultural terraces in South-East Asia and Peru, contour-bounding techniques used in Africa and low embankment construction in Northern America, which appeared to have allowed the societies to survive well in seemingly harsh environments (Pearce, 2004). Community organisation in these societies was also often based on the management, construction, maintenance and use of this infrastructure to sustain their agricultural livelihoods (Pearce, 2004).

CREATING CIVILISATIONS: LARGE-SCALE WATER ENGINEERING AND GOVERNANCE

Early forms of writing evolved in the 'Fertile Crescent' societies (Ancient Egypt and Mesopotamia) between 4000 and 3200 BC (Avery, 2005), in China at some stage before 1300 BC (Diamond, 1999), although there is significant scientific debate over how much earlier (Li *et al.*, 2003), and in Meso-America about 600 BC (Diamond, 1999). Along with this advancement, these societies grew into larger and more structured groups, and further developed other bodies of knowledge, e.g. mathematics, geometry and astronomy, as well as the practical governance skills that were required to create larger technologies. The earliest major waterwork is considered to be the 14-metre high Sadd el-Kafara dam in Egypt, dating back to 2950–2690 BC (Mays, 2005), with qanats (see Figure A.2) for the transportation of groundwater being developed by the Persians at a similar time (Burke, 2005).

The design knowledge for the construction of these hand-dug channel systems was slowly exported to many regions of the world, with the system being known by many other names, including: *karez* (Pakistan, Afghanistan, Western China); *foggaras* (Northern Africa); *laoumia* (Cyprus); *falaj* (Oman); *surangam* (India); and *galleria* (Spain, Spanish Central and Southern American colonies). Similar tunnel systems, called 'spring-flow tunnels', but carved out of hard rock, are found in the lands surrounding Jerusalem and are estimated to have been built from about 2000 years ago. Hydrologists examining these

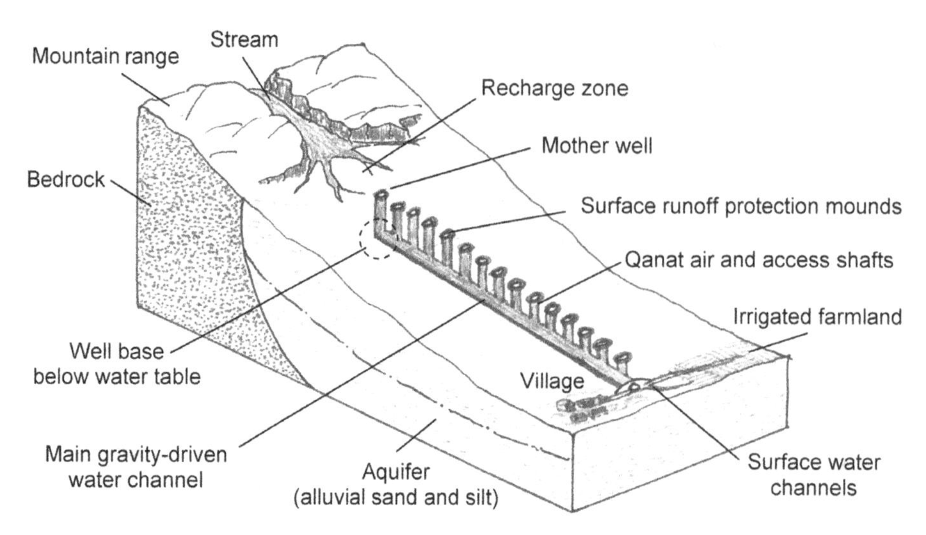

Figure A.2 Example qanat system. Adapted from: Cech (2005).

tunnel systems today are amazed at the hydrological knowledge the tunnel designers must have possessed, as the tunnels exhibit the optimum relationship between height and flow to provide efficient irrigation; the relationship known as 'Darcy's law', which was 'discovered' by the French engineer, Henri Darcy, in 1854. These systems are particularly interesting to analyse, as they are hydrologically sustainable; they only tap the groundwater supply up to the levels of natural replenishment and do not degrade the aquifer (Pearce, 2004).

Following the construction of the first large Egyptian dam, many other dams were soon constructed in Jordan and other parts of the Middle East, followed 1000 years later by China, Meso-America and the Mediterranean region (WCD, 2000). Continuing urbanisation and large-scale rural development for the support of these communities brought many more water supply and evacuation technologies, such as the Roman aqueducts and sewerage systems, '*cloacae*', dating from the sixth century BC (Avery, 2005), as well as the 1800-km long Sui Dynasty 'Grand Canal' in China (China Daily, 2006). To feed growing populations in the cities, agriculture in the rural zones was intensified and irrigation used to augment the natural rainwater supplies and to reduce hydrological uncertainties. One of the largest projects of this type was the Dujiangyan irrigation project in China, which supplied 800 000 hectares of agricultural land and is now 2200 years old (Zhang, 1999). Many civilisations considered the control of water for human needs to be the ultimate symbol of good governance, with the Chinese even having a specific word for this idea, *zhì*, which means both 'to rule' and 'to harness a river'. Wittfogel (1957) considered these societies to be 'hydraulic civilisations', as their leaders were irrigation engineers or water priests rather than military experts. He hypothesised that such civilisations required strong centralised and competent bureaucracies to manage such large systems and that, if capable of managing this challenge, the government would become too powerful, as the population would be totally reliant upon it to maintain their lives. This theory of 'Oriental despotism' has since been widely criticised (Needham, 1959; Burke, 2005) but it still provides some thought-provoking insights into the potential relations between water control and power.

EARLY ENVIRONMENTAL DEGRADATION AND THE INTRODUCTION OF WATER-ALLOCATION LAWS

With the expansion of civilisations and technological developments, new problems surfaced for some communities. Towards the end of the fourth millennium BC, extensive irrigated agriculture in the Tigris and Euphrates river basins had started to cause salinisation of the land. Irrigation water had gradually increased the soil salinity until the mostly wheat crop had to be replaced with barley, a more salt-resistant plant that could cope with the decreased soil fertility. The cultivation of wheat in the region was finally totally abandoned by 1700 BC (Ghassemi *et al.*, 1995). It is easy to acknowledge that such problems are far from just stories of history, as an estimated 3 hectares of salinised land are currently lost for production worldwide each day (CISEAU, 2006).

It is interesting to note that just before the abandonment of wheat in Babylonia, salinisation was probably not the only water-related issue that the civilisation had to consider. Issues of flooding, drought and infrastructure maintenance appeared to be prevalent, as King Hammurabi (1795–1750 BC) established the first 'water allocation law' with references to these issues as part of the Code of Hammurabi, a document encompassing all aspects of Babylonian law (Cech, 2005). Many centuries later (in the sixth century AD) the Romans also developed within their *Corpus Juris Civilius* a framework for water allocations and rights, known as the 'Riparian doctrine'; this was followed soon after by the Spanish and a number of other civilisations (Cech, 2005).

HEALTH AND WATER QUALITY: THE INVISIBLE ISSUES

Apart from a number of ancient myths and suggestions that poor quality water could impact health, the development of methods to improve drinking water and wastewater treatment for health improvement was not undertaken with the same zeal as many other water-related engineering endeavours. The principal water quality concerns in the ancient world related to the taste and visual quality of drinking water. To deal with these issues, a variety of filtering, boiling and simple chemical treatments was used, such as the Egyptians adding alum to aid the removal of suspended solids (Cech, 2005). Roman engineers in the fourth century BC also followed the advice of the Greek physician, Hippocrates, who suggested that 'healthy water' from pristine sources be brought to towns to improve the societies' health (National Driller, 2002). Evacuating waste and stagnant water from human settlements also became a priority for the Romans; this included the draining of swamps and marshlands and the construction of gutters and storm-water and wastewater removal systems for cities. In approximately 200 BC, the Babylonians also passed sanitation laws to prevent wastes contaminating wells and cisterns (Cech, 2005). However, much of the knowledge developed about the importance of clean drinking water and good sanitation practices appears to have been forgotten until recently.

Centuries after the Romans, Edwin Chadwick suggested again that cities had to be cleared of their filth, and provided with clean drinking water and sewerage systems to promote better health for their inhabitants (Chadwick, 1843). One of the first cities to follow the guidance to construct water and wastewater systems was Hamburg, in 1843, which had to reconstruct the city after a major fire (Derry and Williams, 1960). Napoleon III, after confronting health problems in Paris, followed similar steps by commissioning the construction of an elaborate sewerage system in Paris in the 1850s to evacuate the polluted city waters, as well as the construction of new drinking water supply systems, aided by the Seine '*Préfet*', Georges Haussmann (Pinkney, 1955). Although a number of debates had raged in medical circles around this time over the causes of cholera and other diseases being linked to polluted water, it was not until 1854 that Dr John Snow, after studying maps of cholera outbreaks in London relative to the location of the water supply systems, put forward a more persuasive argument that cholera was a waterborne disease (Mara, 2003). In the 1870s, instead of all the untreated waters being returned to the nearby rivers, the sewerage of cities started to be treated in a variety of ways, including pumping it to nearby farms for irrigation purposes and to decompose the organic matter; a treatment practice pursued in Paris, Sydney, London and Berlin (Stevens, 1974). From this time onwards, engineers took over the development of improved sewerage management practices, which ranged from septic tanks to more sophisticated chemical and biological treatments (Beder, 1993).

TECHNICAL EVOLUTION AND THE INDUSTRIAL REVOLUTION

In addition to the static hydraulic structures developed by a number of the ancient civilisations, dynamic mechanical technologies, such as water-lifting wheels and pumps, have played important roles in water management practice. Early examples of these technologies, which have been used since antiquity, especially in the Middle East, the Mediterranean and China, include *norias*, *tambour* and *sakiya* (Magnusson, 2001; Cech, 2005). However, in the Middle Ages, from the late thirteenth to the fifteenth century, such water wheels and pumps became more prevalent and began to be used for more complex urban water systems in Northern German towns and Switzerland (Magnusson, 2001). Steam engines for water pumping then emerged in the 1760s (Avery, 2005). Early industries and manufacturing processes were the next major developments, along with an increased realisation of the economic worth of water. The creation of major cities and the industrial revolution spurred on unprecedented rates of population growth that were coupled with an increased use of water and the need for supply. Much more elaborate supply, evacuation and water treatment facilities became a necessity to cope with these populations and to maintain their health, as well as more advanced farming and irrigation techniques, to supply them with sustenance.

By the beginning of the twentieth century, society had started to create so many dams, canals, levees, dykes and pipelines, that many rivers and water sources had become almost entirely artificial. As society, and especially the engineers and planning authorities behind some of these water management schemes, attempted to control nature and improve human standards of living with this infrastructure, the elevated risks for human and financial loss, and the severity of major natural events, such as floods, became more evident; for example, large loss of life was caused by the breaching of levees and dykes in the Netherlands and the United States of America, as well as by dam failures throughout the world.

AN ENVIRONMENTAL AWAKENING

Throughout the twentieth century, and especially since the 1970s, some of the impacts of these human interventions and population growth, such as serious environmental degradation and declining resource depletion – witnessed for water as supply shortages, quality problems through pollution, and aquatic and land-based ecosystem destruction – spurred new considerations for the environment to be taken into account (Hutton and Connors, 1999). Laws on water were rewritten and water engineers, managers and planners were required to consider and reduce all environmental impacts of interventions (Thomas and Elliott, 2005). Over the past 30 years, the impacts of climatic extremes, such as floods and droughts, as well as increased urbanisation, levels of water pollution, and the use of water for agriculture and industry, have led to more conflicts over rights to water, including rights for the environment, and how water resources should be managed.

THE INFORMATION REVOLUTION

With the advent of computers, communication systems and advanced technologies for water treatment later in the twentieth century, large changes and further development of water supply, demand control and management of water systems became possible. Supply networks could be more easily optimised using computer modelling techniques to provide cost savings and greater efficiency. Dams and other infrastructure could also be designed, modelled and built for larger capacities than ever before, and communication networks allowed the global transfer of information, knowledge and new technologies to occur at an unprecedented rate. However, with the computational gains from computers and the increase in connectivity between people and information databases provided by communication networks (i.e. mobile phones, the Internet), the complexity of decision making has increased (Fischer, 2000). Connectivity has led to a general increase in awareness of the amount of information and diversity of interests that can be quickly accessed, and thus can be taken into account, when making decisions. Modelling and simulation methods, for example, based on advanced theories, such as artificial intelligence, self-organisation and thermodynamics, are also developing and becoming more complex at a similar speed to computational advances.

THE CONTEMPORARY SOCIAL DIALOGUE

In recent years, socially based disputes have arisen over values, power struggles, interests (economically driven or otherwise), and differences in viewpoints. Such subjects as whether water is of a sufficient quality or if it is equitably distributed between people and geographical areas, can spark highly value-charged debates. Protests and social activism over water issues, such as dam construction, have also become more widespread (Hutton and Connors, 1999). Furthermore, the recent creation of the Internet and the realisation that all human activities can have global impacts through economic factors such as trade and societal movement – whether physically, culturally, internally (e.g. through changing beliefs, values, views, relations and practices) or environmentally – has made planning and management for water more complex. This is because not only do the technical, economic and environmental factors have to be considered, but also the social ones. Information transfer now allows local issues to become the international focus of protests; for example the human rights issues of the displacement of the estimated 1.3–1.9 million people for the Three Gorges Dam on China's Yangtze River (Gleick *et al.*, 2006). Contemporary water management is therefore often characterised by a process of deciding how available water should or could be used and shared between a variety of stakeholders and the environment, and conflict resolution amongst these stakeholders; where stakeholders are considered as people, institutions or organisations that have a stake in the outcome of decisions related to water management, as they are directly affected by the decisions made, or have the power to block or influence the decision making process (Nandalal and Simonovic, 2003).

IS SOCIAL CONFLICT OVER WATER A CONTEMPORARY DIALOGUE? EVIDENCE FROM HISTORY

For thousands of years, water has aided peace and war. Rivers, in particular, have often been the centre of rivalries over water use, as water use downstream is affected by upstream users, and water use on one bank can be affected by use on the other bank. It is interesting to note that the meaning given to the Latin word, *'rivalis'*, means *'using the same river as another'* (Wolf *et al.*, 2005). However, whether this implies that rivers incite cooperation or conflict is entirely another question.

It is suggested that since 2500 BC until the current day, over 3600 individual water treaties have been created globally (Wolf, 2002a). Water has been a common source of conflict in transboundary water basins, especially in the Middle East, Asia and Africa, as outlined in the 'water and conflict chronology' (Gleick *et al.*, 2006). In such tensions, water systems and resources can be considered as sources, instruments or targets of conflict (Gleick, 1998). About 720 BC, irrigation systems of the Haldians in Armenia were targeted and destroyed by the Assyrians (Gleick, 1998). A couple of centuries later, a type of 'hydraulic warfare' based on the use of dykes to divert deathly floodwaters onto opponents' villages and fields was used in China (Cech, 2005). In more recent times, conflicts in developed countries have also occurred with a variety of motives. Direct attacks on a water system for survival or political motives included the dynamiting of an aqueduct system being used to divert water away from California's Owens Valley farmers for use in Los Angeles between 1907 and 1913 (Reisner, 1986) and more recent attacks on pipelines and treatment plants in Israel and Palestine (Gleick *et al.*, 2006). In many recent conflicts, such as World War II, the Korean War and the Gulf War, dams, levees, pipes and desalination plants have also been targeted to disable opponents' vital infrastructure (Gleick, 1998). In 2000, an economic motive prompted conflict when French protesters aiming to protect their work rights at a chemical plant dumped 5000 litres of sulphuric acid into a tributary of the Meuse River near the Belgian border. A specialist in social conflicts stated of the event that, 'The environment and public health were made hostage in order to exert pressure, an unheard-of situation until now,' (Cu, 2000). At the same time, on Australia's Sunshine Coast, another potentially economically motivated and previously unthought-of 'cyber-terrorist' attack on a wastewater treatment plant was uncovered. For two months, it had been a mystery to the managers of the Maroochy Shire wastewater system how the plant had been leaking hundreds of thousands of litres of putrid sludge into parks, rivers and nearby hotel grounds, killing marine life and creating an unbearable stench. During his 46th successful intrusion into the plant's computer system from a radio transmitter and stolen computer in his car, the culprit was finally discovered and arrested by the police (Gellman, 2002). These new kinds of water-related conflict show just how interconnected with other interests and different parts of society effective water management is becoming, and hence some of the new challenges that future water managers are likely to face this century.

Although international treaties have been drawn up for thousands of years between two or more parties, recent international laws and treaties have attempted to prohibit the aforementioned uses and abuses of water courses and infrastructure, as well as improve the sharing of water resources between States (Gleick, 1998). These include the:

- 1966 Helsinki Rules governing international waters;
- 1977 Environmental Modification Convention;
- 1977 Bern Geneva Convention;
- 1982 World Charter for Nature; and
- 1997 Convention on the Non-Navigational Uses of International Watercourses.

Indeed, Delli Priscoli and Wolf (2009), in their analysis of water conflicts around the world, find increasing trends towards cooperation in attempting to manage and transform conflicts instead of trends towards heightening of conflict.

CONCLUDING REMARKS FROM THE HISTORICAL ANALYSES

Several important lessons can be drawn from this review, only three of which will be highlighted here. Firstly, although it may be self-evident, we should be aware that human beings, and especially their leaders, make choices that will affect their societies, their futures and the futures of others, such as decisions to centralise governance systems, irrigate or adopt new technologies. Considering the increasing complexity and interconnectedness of modern forms of society and the issues they face, making informed decisions has never been so challenging. This means that there is a need, more than ever before, for improved forms of decision-aiding to help our leaders and societies to make informed choices, in order to aid our current transition to more sustainable forms of development. Secondly, as part of this challenge, there is a need to review our past for potential effective solutions to our current problems or invent entirely new ones if past solutions have consistently performed poorly. It is hoped that this will help us to avoid repeating the same mistakes that were made in the past. Some of these are presented in Chapter 2. Finally, linked to this last point, there is a need to re-examine and work with people from a number of key cultures, whose cumulative and intricate knowledge of the workings of the environment may help us to redefine a more sustainable path of choices for the future. A number of minority indigenous cultures (e.g. Australian Aboriginal) and the ancient eastern (e.g. Chinese, Japanese) philosophies for research may provide the most interesting insights as a first step, owing to some apparent successes in the longevities of the cultures.

A.2.2 Evolution of planning and management theory and practice

Alongside these cultural and technological evolutions in the water sector, planning and management theory and practice have evolved substantially through the phases of human development. *Planning* can be considered as a process of formulating objectives and then actions necessary to achieve them. Such a process may be carried out unconsciously or consciously, although it is the formalised conscious process that is most commonly referred to in the literature. *Management* is closely related to planning, and can be thought of as a process of continuous decision making (consciously or unconsciously) to plan, guide actions, lead, coordinate, control or organise. However, many variants on definitions of planning and management exist and are actively debated in the literature, some aspects of which are well explained in Ansoff (1984) and Mintzberg (1989; 1994). This section will give a brief overview of a number of general evolutions in formalised planning and management theory, followed by a number of specific forms of planning and management that have been used in the water sector.

PLANNING AND MANAGEMENT THEORIES: A FEW HISTORICAL EXAMPLES

The Chinese philosophy of *feng shui*, literally 'wind–water', which focuses on the art of placement related to energies and their influence on a landscape, is thought to be one of the oldest formalised 'planning' type methodologies, dated to about 4000 BC, which dictates how buildings, towns, gardens and other parts of a landscape should be orientated and organised to maximise the beneficial 'feng shui forces', related to both the physical environment and other abstract or mystical influences (Kerr, 2004).

The first forms of organised planning that related to time rather than space are believed to have stemmed from the study of astrological phenomena at a similar time, with the harvest calendars of the Neolithic period representing a good example (Avery, 2005). About the same time, the use of project planning and management principles was evident in the large structures and systems created in a number of civilisations. The pyramids of Ancient Egypt, and large dams and water distribution systems

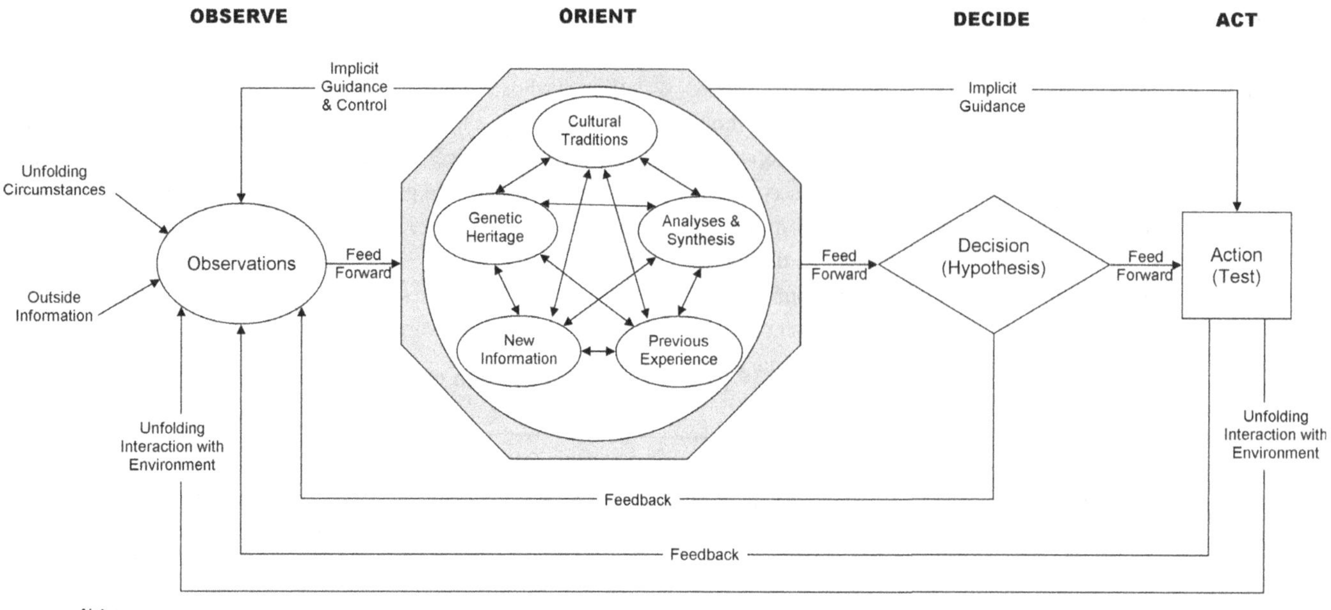

Figure A.3 Boyd's OODA 'loop' from *The Essence of Winning and Losing*, January 1996. Adapted from: Strömgren (2003).

found from those times onwards, were projects that must have required formal planning to have been conceived and successfully constructed.

In the domain of military strategy, one of the first formalised theories of planning and management was outlined in approximately 500 BC in Sun Tzu's *The Art of War*, a text that is still studied today by strategists, planners and managers all over the world (Tordjman, 1996). Different types of planning and management tools have also been invented through the ages, including instruments such as rain gauges, which have been used since 300 BC as a tool for determining government tax collections and aiding planning decisions in India, China and Korea (Cech, 2005).

PLANNING AND MANAGEMENT THEORIES: RECENT EVOLUTIONS

With the Industrial Revolution came the increased need to manage production and workforces, leading to a proliferation of new thinking on viable forms of management and societal structures, such as scientific management (Taylor, 2003), administrations (Fayol, 1918), bureaucracies (see Albrow (1970) for a succinct overview) and organisational models (see Dixon and Dogan (2003) for an analysis of different currents of thought on the subject).

More recently, theories of planning and associated management practices have been developed in many domains, including: management (operational, organisational, and of the society, economy and environment); military strategy; engineering (projects, production and construction); and education. All of these theories tend to delineate, or help to delineate, a certain number of phases in the planning and management process, which are required to designate actions that will help to meet specified objectives. One of the best-known examples of a planning process to help delineate project phases through time is the Gantt chart, designed by the American engineer and management consultant, Henry Laurence Gantt, in the 1910s and later used all over the world in business, production or construction management, including for the management of the Hoover Dam project in 1931 (UNITAR, 2005).

Since that time many other planning and management models have emerged, including systems models, long-range planning, contingency planning, strategic planning and strategic management (Ansoff, 1965; Emery, 1969; Steiner, 1969; Ansoff, 1984; Mintzberg, 1994), many of which are based on Simon's (1960) depiction of the decision process. Although diverse in their content, some of these models take a cyclic rather than linear form, and have been developed for use in a number of domains. For example, Boyd's OODA loop – for 'observation, orientation, decision, action' – was originally developed for military strategy planning but is now commonly used for organisational planning and management (Strömgren, 2003), as shown in Figure A.3.

This kind of planning and management cycle differentiates between a number of phases that could be considered part of a conscious decision making process. It is important to note the possibly incomplete nature of such processes, as feedback loops may occur at a number of stages. Boyd's loop is also typical of many other planning and management cycles, including Deming's (1986) continuous improvement cycle for 'total

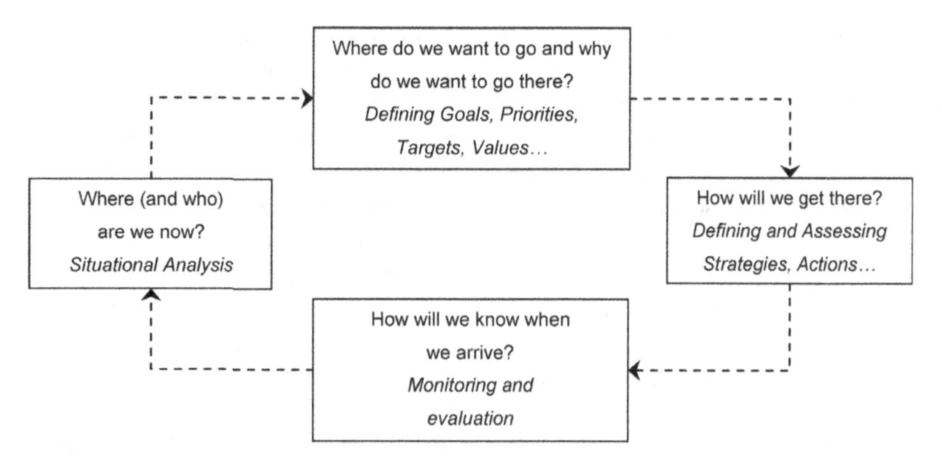

Figure A.4 Questions and activities considered in typical planning and management cycles.

quality management' and Holling's (1978) adaptive management cycle for environmental and ecosystem management.

The new 'best practice' cyclical (or spiral) planning and management processes designed to achieve sustainable development objectives have also specified that ongoing evaluation or monitoring is required as the planned actions are implemented (Walker *et al.*, 2002) if real progress towards such objectives is to be achieved (Bellamy *et al.*, 2001). This consideration, even though not clearly explicated, is still evident in the feedback loop to 'observation' in Boyd's loop. From all of these cycles, irrespective of how many phases they are divided into, a number of key ideas can be summarised by a number of questions, as illustrated in Figure A.4.

A number of other models of planning and management have also emerged in recent years, including participatory planning and communicative action theory, risk management and adaptive complex ecosystem management, and models of 'leadership', as opposed to 'management'. Increased conflict over management and planning values and objectives and a lack of resource control by management authorities, as well as democratic motives in Western countries, have been partially responsible for driving a range of participatory, interactive and collaborative planning processes (Forester, 1989; van Rooy *et al.*, 1998; Forester, 1999). The theory of communicative action depends on the use of language oriented to obtain mutual understanding and decision making through deliberative discourse and interaction (Habermas, 1996). However, increasing awareness about a range of uncertainties as 'risks' and an incapacity to adapt to their impacts have driven the development of risk management processes (Beck, 1992; Slovic, 2000; Jaeger *et al.*, 2001), chaos theories of management (Cohen *et al.*, 1972; Gleick, 1987) and adaptive complex system and transition management processes (Levin, 1998; Cortner and Moote, 1999; Gunderson and Holling, 2002; van der Brugge *et al.*, 2005). This growth in 'leadership' has come from the realisation that management is predominantly not about managing objects, but rather about managing people (Adair, 1983), leading transformational change (Senge, 1990) and learning (De Geus, 1988; Brews and Hunt, 1999). Management and planning in these cases are considered to be 'process-orientated' rather than linear and 'goal-orientated', and place a large emphasis on the creation of knowledge, innovation, creativity and social or organisational learning (Nonaka and Takeuchi, 1995).

PLANNING AND MANAGEMENT APPLICATION IN THE WATER SECTOR

As well as the individual internal reflection processes and individual or collective conscious decision processes on actions required to achieve certain water-related goals in ancient societies, such as digging wells, irrigation channels or situating homes near an already available water supply, there are many ancient examples of water-related and other types of development that inherently required some form of organised planning. Dam building, water distribution and sewerage system design and construction were just some of the projects that could be created through the centralised decision making structures of those times.

Looking specifically at planning approaches used for water management in modern times, Sharifi (2003) identifies three dominant styles of planning that are currently in use: traditional supply-side planning; least-cost planning; and integrated resource planning. Each one of these planning styles targets the three types of scarcities outlined in Table A.1 to a certain degree: physical or environmental scarcities; economic scarcities and social or political scarcities. The third type of planning approach identified by Sharifi (2003) as integrated resource planning is also outlined in detail in Nichols *et al.* (2000), who compare it to the first two 'traditional approaches' to planning.

Traditional *supply-side planning* is a process based on ensuring that a safe and adequate quantity of water can be provided to users. It is often based on the development of extra water storage and usage capacity and does not generally take other goals into account, such as environmental protection or the possibilities of reducing water demands. It concentrates on coping best with physical and environmental (i.e. climatic and hydrologic) uncertainties. This type of planning and its associated management practices are what Wolff and Gleick (2002) refer to as the 'hard' path to water management.

Traditional *least-cost planning* is typically an economically orientated process that focuses on the evaluation of a range of both demand and supply alternatives for water management, which generally includes the pricing of such externalities as environmental benefits and other non-utility goals, such as recreational water uses. These alternatives are then evaluated, and the minimum-cost options selected that are the most robust over a wide range of potential economic futures, in order to cope best with the uncertainty of economic environments.

These two planning styles can also be classified as 'rational' approaches (Bouleau *et al.*, 2005), referring to the Cartesian version of rationality based on Descartes' principles (Descartes, 1637). This type of rationality is also specifically known as 'economic rationality', owing to the ideal economic behaviour, which is assumed in these normative planning approaches (March, 1978). These approaches are typically promoted and used by engineers who have been trained to think analytically and to break large problems down into their simplest elements in order to solve them (Dandy and Warner, 1989; Pahl-Wostl, 2002; Bouleau, 2003a). Considering a number of challenges in the water sector that are implied by the social or political type of water scarcities (refer to the causes in Table A.1), these planning methods have come under intense criticism by many authors (Matondo, 2002; Pahl-Wostl, 2002; Wolfe and Brooks, 2003; Falkenmark *et al.*, 2004), owing to their incapacity to adequately predict human behavioural effects, which do not conform to the economic ideal of behaviour.

One of the responses to such criticism was *integrated resource planning*, the planning systems behind 'integrated water resources management', which typically takes a wider view of water management issues, and can commonly find itself addressing problems of social or political water scarcity. These approaches are underlaid by Simon's (1954; 1977) concept of 'bounded rationality', where human behaviour can be 'rational', but only within the constraints of partial knowledge, cognitive capacity and organisation of memory. In other words, different stakeholders may all have different rationalities and different decision behaviour, based on their own normative positions and constrained environments. Integrated resource planning is therefore based on multiple objectives and constraints, usually derived from the kind of sustainability objectives outlined in Agenda 21 (United Nations, 1992), which focus on the conservation and development of all social, environmental and economic resources for the future, in participation with all related stakeholders. Integrated resource planning is also supposed to help identify and manage all perceived risks and uncertainties and provide coordination between all stakeholders and any regulatory requirements at the scale of management. The objective of integrated resource planning is thus to find sufficing solutions within all system constraints.

Over the last couple of decades, the importance of improving and integrating planning processes along with management procedures has increased, leading to these 'integrated' systems becoming the new 'best-practice' methods to manage water and its related resources. However, this need for such an integrated approach to planning and management has not always been recognised. Despite large bodies of academic and policy writing on the subject, it is only recently, in some countries, such as Australia, France and the Netherlands, that planning has increasingly become the basis for water management (Handmer *et al.*, 1991; Bouleau, 2003a; van der Brugge *et al.*, 2005). Before this time, planning was commonly considered a 'back-office' support role, a process to be completed before the water management process and plan implementation started. Particularly at local levels of water management, this vision of planning still remains a barrier in certain organisations to the achievement of long-term sustainable management of water (Brown *et al.*, 2001). Long-term sustainable management of water and its associated resources requires future thinking, which is closely linked to planning (Torrieri *et al.*, 2002). Determining and shaping what kind of future is desired for water resources should be a shared responsibility, especially in democratic societies. Therefore, rather than leaving this responsibility to a back office, or even one organisation, there is a push to open planning processes up to the participation of all associated stakeholders from the phase of problem formulation and onwards through the full water planning and management cycles. This movement can be considered to represent yet another view on rationality – that of Habermas' 'communicative rationality' (Habermas, 1984). Habermas suggested that a process of 'rationalisation' and co-construction that involves the confrontation of different stakeholders' rationalities through deliberative discourse and interaction could be used to come to commonly legitimised decisions (Habermas, 1996). This underlying change in perspectives for planning is also evident from an examination of engineering textbooks. For example, the need for participation and working with other stakeholders is outlined here, even though 'expert' needs to be interpreted loosely to include local experts as well as other disciplinary experts:

> *In unusual projects, the engineer rarely has sufficient knowledge or training in the relevant specialist fields. It is necessary to gather information rapidly and also to work in association with consultants and experts with specialised knowledge.*
>
> Dandy *et al.* (2007)

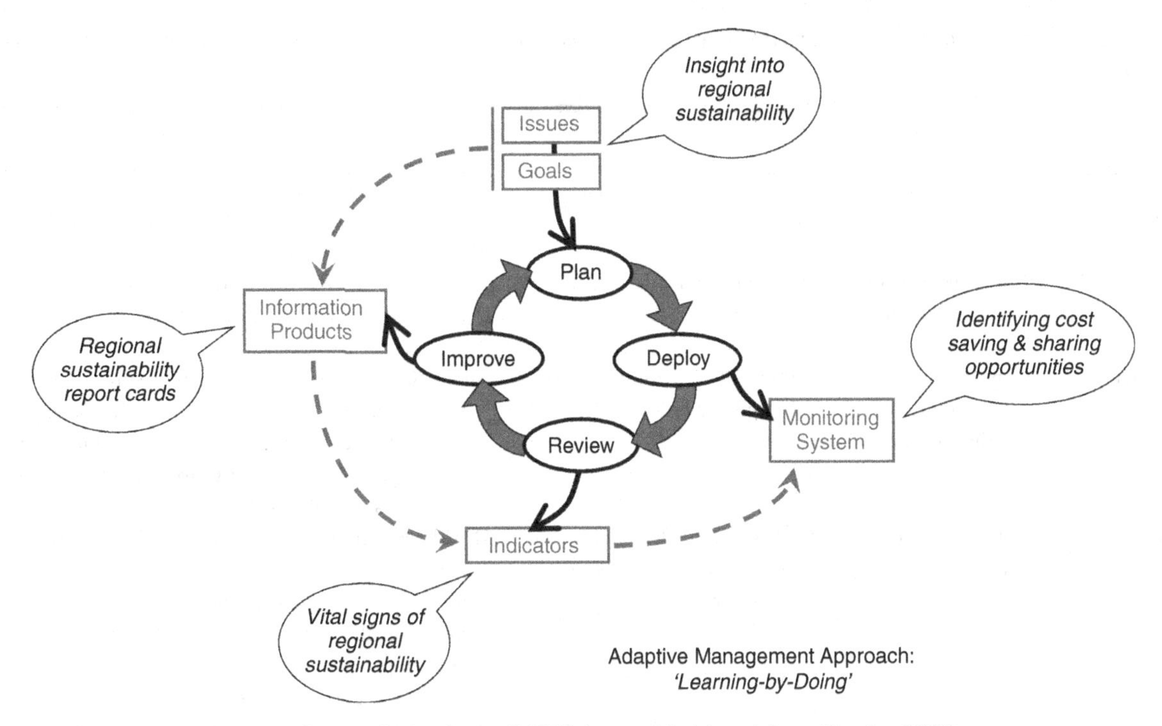

Figure A.5 Monitoring, evaluation, reporting and feedback: the 'MERF Approach'. Adapted from: Fleming (2005).

It is also stressed that stakeholder participation should be explicitly included in planning and management processes from their outset in many international water directives, including the second of the Dublin Statement's (ICWE, 1992) four principles and the Global Water Partnership (2000), where the concept of participation is defined to mean more than simply a process of consultation (Arnstein, 1969).

One recent example of work that uses one of the 'best-practice' planning cycle theories in a participative setting is SKM's 'MERF approach' to plan for the design and operation of information and monitoring systems for sustainable development (shown in Figure A.5), which also involves the water sector. The approach has been applied successfully with participating stakeholders in many regions across Australia, including irrigation communities and industries throughout the Murray-Darling Basin (Fleming, 2005). The planning phase of the approach is based on the 'reverse engineering' of the adaptive management approach or 'learning by doing'.

From Figure A.5, it can be observed that there is a typical management cycle of the Deming or Shewhart type of loop: 'plan–do–check(study)–act(adapt)'. This cycle is to be 'reverse engineered' in the planning phase, to ensure that the tools for monitoring progress have been correctly designed to provide the essential information for adaptively managing the sustainability of the system under study. Such a process is designed to help define exactly what is to be observed after implementation of management options; a situation that typically helps to avoid two problems: on the one hand, that of having no useful information about the effects of management options being obtained, perpetuating a lack of relevant knowledge production; and, on the other hand, that of collecting too much costly data that cannot be adapted into useful indicators and information products for managers (IEA, 1993).

Many other new approaches to planning and management in the water sector are also being theorised and applied, a good summary of which can be found in Pahl-Wostl *et al.* (2008). Apart from these dominant and newer styles of water planning, a number of other more theoretical and less frequently applied types of planning and management have been proposed for use in the water sector, many of which match the recent theories of planning and management outlined earlier in this section.

PLANNING AND MANAGEMENT SPATIAL SCALES

It is increasingly regarded that water planning and subsequent management processes should occur on the geographically cohesive scales for water resources collection areas, i.e. catchments (river reach), watersheds or regions (river) and basins (river system) (Fleming, 2005). The European Framework Water Directive (EU, 2000) also dictates that water resource planning should occur at this largest 'basin-level' scale. However, in some countries, international cases or in local areas where water management authorities and other governance structures and regulatory frameworks have not yet been realigned to these scales, the 'problem shed' scale suggested by Loucks (1998), which implies the scale at which all stakeholders concerned with a particular water-related issue are involved, may prove a more appropriate scale for the planning and management of water-related issues, in order to help stakeholders work through just

Table A.2 *Cultural and managerial evolutions in the water sector matched to predominant cultural value systems of Beck and Cowan (1996)*

Societal structure	Predominant water-use values	Decision making locus	Principal water management style and tools development
Unstructured hunter-gatherer	Subsistence	Individual	Instinctual: uses biological senses and physical skills to find and obtain water for personal life-support functions
Nomadic or tribal	Necessary element for life, spiritual virtues	Individual and tribal elders	Respectful: based on tribal decrees; cumulative knowledge from cultural and spiritual heritage
Local agrarian or feudal societies	Control of water for agriculture, cultural beliefs	Individual and community or empire leaders	Small-scale control: construction of simple water infrastructure – water diversions to canals and pipes; use of simple architectural, topographical and astronomic design principles; planning from cultural design principles and philosophies
Centralised civilisations	Control of water for community enhancement and growth	Civilisation leaders and bureaucratic servants: water engineers	Broad-scale control: construction of large infrastructure and technologies for urban and rural water distribution and treatment: use of more advanced mathematical and mechanical principles and models; objective-based and supply-side planning or engineering problem-solving; water allocation rules
Industrial societies	Water as a resource for development, economic value	Societal leaders and industrial managers: water engineers and economists	System optimisation and water-use strategies by a number of decision makers for efficiency, production and economic gains: use of advanced mathematical and statistical models, economic theory and instruments; strategic and least-cost planning
Modern democracies	Water as a resource for multiple uses, equity	Societal leaders, policy makers, interest groups and individuals	Allocation of water between all uses: political discourse, participatory planning, education, technical advice and scientific knowledge as only one part of the debate
Information societies	Water as an important element of all systems; knowledge is power	Distributed – multiple loci of decision making	Efficient allocation of water for value-adding and maintenance of key resources: integrated resource planning, policy analysis and systems approaches, markets, conflict management and targeted participatory approaches to obtain relevant knowledge for problem solving, risk assessment, integrated simulation modelling and advanced computer and spatial information tools
The future: a global village?	Water as the life-force, synergy and global harmony	Global democracy? Enlightened central authorities?	Global management of water for the common good: synergetic and holistic planning, advanced simulation and organisational concept use (i.e. models of thermodynamics, self-organisation modelling), interconnected global monitoring systems, long-range visioning, global markets and comprehensive water accounting, blending natural and 'smart' eco-technologies to enhance Earth's health and human well-being

some of the possible institutional conflicts related to boundary issues within these geographical regions. Even though the word *problem* is used here, it is noted that there has been a push in certain academic communities to change it to something with less negative connotations, such as *issue* or *dilemma* (Flood, 1998). This choice has been made to provide consistency with principal methodological sources from the operational research community, which are outlined elsewhere in this book.

A.2.3 Piecing the puzzle together

From the previous two sections on the evolution of cultural values and technological development and the accompanying planning and management processes, a general picture of evolutions in the water sector can be developed and expanded upon, as shown in Table A.2.

Table A.2 can be used to help to understand cultural differences resulting from dominant values (Beck and Cowan,

1996), as well as the perceived needs – in terms of water planning, management and technologies – of those value systems. The preferred or usual decision making loci are also outlined. It should be noted that all of these value systems and associated needs for water management are still present in today's world. In each problem situation encountered, whether at a local, national or global scale, all or some of these value systems may be present and may require concerted attention, understanding and adapted management programmes.

Appendix B Understanding decision-aiding

This appendix supplements Chapter 3 by extending the brief introduction to the intervention research approach. Background information on models of the decision process – on which many decision-aiding processes are based – is also presented, along with further information on problem formulation and evaluation models used in these processes, or to aid them. Finally, an overview of currently used methods for decision-aiding in complex, uncertain and conflict-ridden problem situations is outlined.

B.1 INTERVENTION RESEARCH FOR DECISION-AIDING

Adapted forms of research practice are required that can produce useful knowledge to inform the types of collective decision and action that must be taken in the inter-organisational fields of water planning and management in today's interconnected information societies. Of the many available research methods available (with different underlying epistemologies), one is required that allows the investigation of the question presented at the end of Chapter 2: *How and to what extent can decision-aiding processes be carried out in a critically reflective manner so as to learn from the past and adapt proactively in the face of future challenges?* From a review and analysis of possible research positions, it is postulated that adopting an 'intervention research' position may lead to the best-adapted insights. This is largely because of the need to embrace the complexity of messy problem situations and pursue collective action to manage them into the future. Hubert (2002) describes and comments on the utility of three types of research, 'laboratory', 'field' and 'intervention' research, analysing them in terms of their orientation to the construction of knowledge from an objective to constructive stance, and their predisposition to embrace or reduce complexity in, or to study, system behaviour, as shown in Figure B.1. Because of its apparent match to the recently outlined research question, 'intervention research' will be investigated further here.

The *intervention research* approach is commonly used in the management sciences (Berry, 1995) and is aimed at critically co-constructing collective action in the field (David, 2002), through which new knowledge and insights can be created (Avenier *et al.*, 1999). The 'field' in this case can be interpreted as the interaction space outlined in Figure 3.6.

Hatchuel and Molet (1986) consider 'intervention' as:

> *A constitutive mechanism by which a conscious attempt is made to modify organisational phenomena according to some pre-established concepts or models. It is therefore a common means of change and thus the vehicle for the legitimisation of any theoretical tool.*

Related to this definition, Hatchuel and Molet (1986) identify five typical phases in an intervention research process in an organisational setting. These start with identifying 'feelings of discomfort' and 'building rational myths' before the 'interventions and interactions' in the field are put in place. Then follow the elicitations of the participating actors' 'sets of logics' and the final outcomes of 'knowledge vs. implementation'. Each of these phases will be briefly outlined here:

1. *'The feelings of discomfort'*: on primary interactions with the potential clients and organisations, a certain number of current issues and dysfunctions where improvements are needed become apparent to the researcher.
2. *'Building a rational myth'*: the researcher formulates and theorises about the problem situation observed or otherwise 'sensed' in (1), developing a number of theoretical models or tools to understand or describe the organisational structure linked to the problems (and potentially how they could be rectified).
3. *'Intervention and interaction'*: the experimental stage, where stimuli are inserted into the organisational processes, with feedback being obtained from actors in the form of 'reinforcement' (support of or interest in) or 'resistance' to the 'rational myth'; the aim is to create a learning process.
4. *'Portraying a set of logics'*: the inductive phase, where actors give their theories and thoughts on the processes in which they are involved and have an opportunity to enhance or antagonise the 'rational myth', in the process bringing

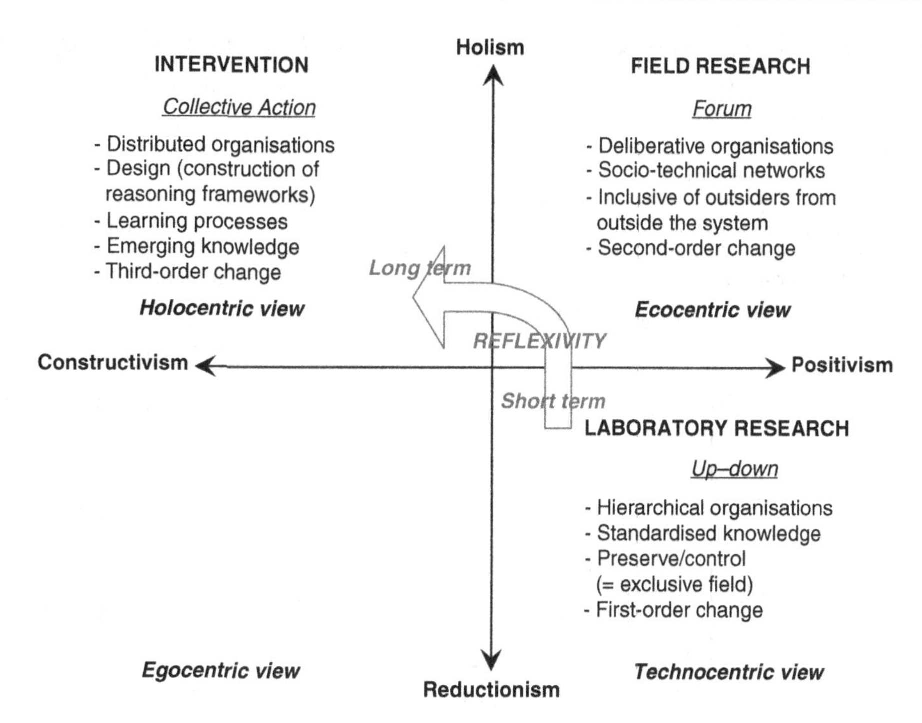

Figure B.1 Research positions based on degree of embracing complexity and epistemology. Adapted from: Hubert (2002), Bawden (1997) and Ison *et al.* (2004).

forward a new vision (a 'set of logics') of the advancing problem situations and organisational structures.

5. '*Knowledge vs. implementation*': the learning process, induced by the introduction of the rational myth into the organisation can bring about a variety of outcomes, including: each actor having developed new theories related to the functioning of the operational systems; and new understandings of the 'feelings of discomfort' about the problem situation first acknowledged at the beginning of the process. The research may end with a 'stabilisation of existing logics' or 'a change of structures and tools' that has been observed relating to the construction of new logics, rather than an actual 'implementation' of the tools of the researcher.

This approach has specifically been designed for organisational settings. How it might work in inter-organisational settings is not specified. Later outlines of the intervention research approach have been less descriptive in the specific steps that need to be undertaken; instead prescribing a set of principles that should be adhered to in creating and undertaking intervention research (Avenier *et al.*, 1999; David, 2000). Such principles could apply to either organisational or inter-organisational settings. Avenier *et al.* (1999) notes, in particular, the need to:

1. Negotiate individual goals of both the practitioners and researchers, as well as a collective project from which they will both benefit: it is in this phase where the researcher must compromise between his or her own research goals (i.e. knowledge creation) and the on-site needs and problems (i.e. action) and decide whether an intervention can prove fruitful for both parties (Arnaud, 1996).
2. Negotiate and decide on the modes of interaction permissible in the organisational structure (for example, co-designing or operating projects) and the structure of the research project (for the researcher and in relation to an outside research team or possible steering group, where ideas and emerging theories can be discussed during the intervention (Berry, 1995)); it is important in this phase to begin to construct a legitimate position, both operational and analytical, between a number of actors and needs that are likely to be diverse and potentially conflicting (Mayer, 1986).
3. Engage in 'ongoing construction' of knowledge, based on a dynamic of oscillation between action and reflection; this should be a collaborative process between the researchers and practitioners, where continuous dialogue and ongoing evaluation are maintained (Couix, 1997), to re-design and re-orientate the loops of action and reflection throughout the intervention. As Moisdon (1984) explains, it will be the reflection on previous stages that forms the basis for new research, analysis, testing and reflection in the following phases. In the intervention process, the researcher should take the initiative in selecting the methods and theories considered to be best adapted to the situations encountered (Berry, 1995).

Although this stance has recently been theoretically specified by these researchers, it has roots in common with a number of

Table B.1 *Formalisation and contextualisation of four types of management research. Adapted from David (2000). The arrows in thick, full lines indicate what is actually carried out during the research; the thin, full lines with question marks indicate what would be a logical continuation of the process, but not carried out during the research*

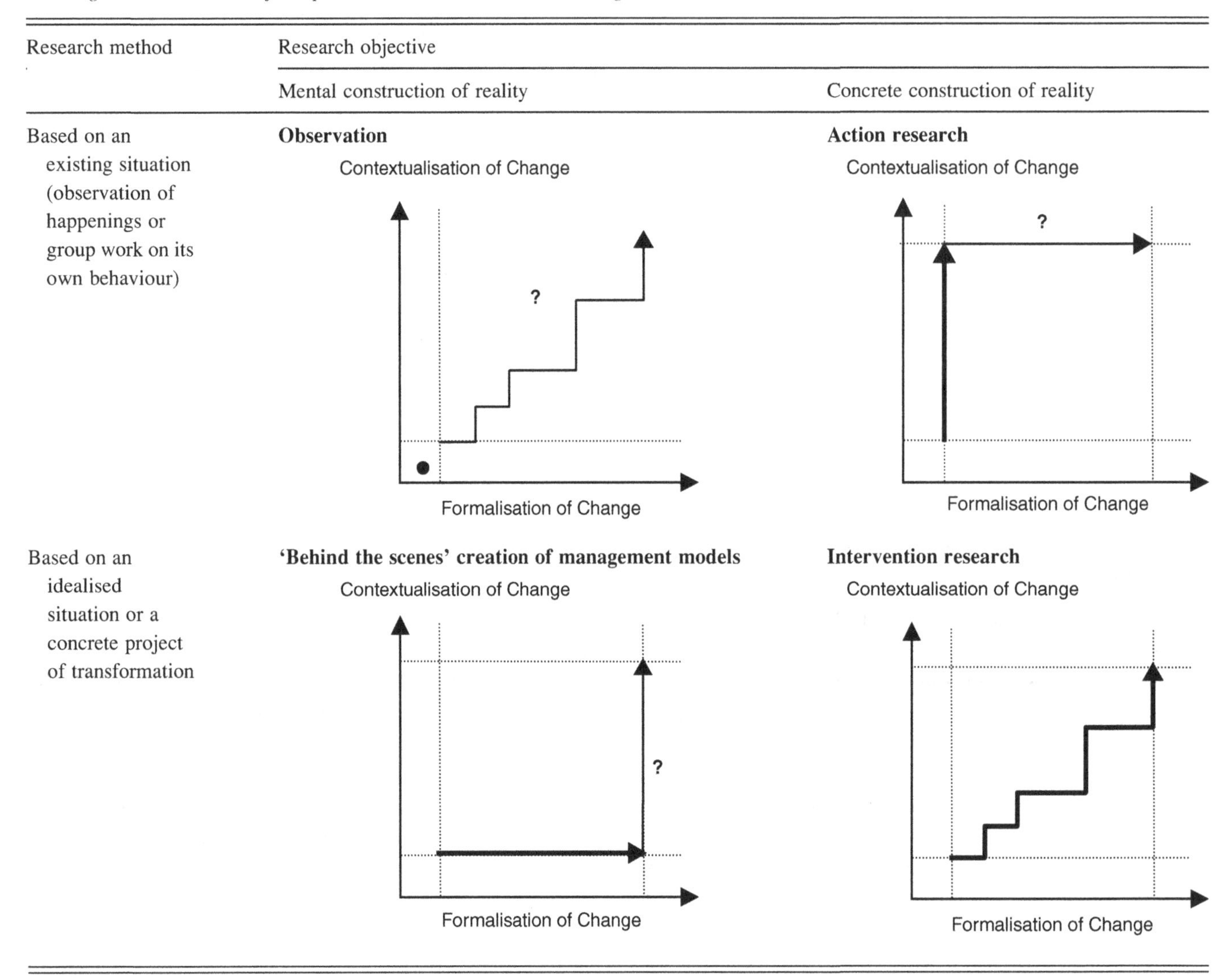

other 'transformative' or engineering research methods of developing knowledge through action in the 'field', which include action research (e.g. Toeman and Thompson, 1950; Lewin, 1951; Susman and Evered, 1978; Checkland and Holwell, 1998), grounded theory (Glaser and Strauss, 1967), enquiry in action (Torbert, 1976), 'direct' research (Mintzberg, 1979b), cooperative enquiry (Heron, 1981), action science (Argyris *et al.*, 1985), clinical research (Schein, 1987), participatory action research (Whyte, 1991), decision-aiding science (Roy, 1993) and the intervention research in Thomas and Rothman (1994).

Taking the traditional views of action research, as described by Lewin (1951) and Argyris *et al.* (1985), as well as other traditional forms of research in the management sciences of observation methods and theoretical model construction, David (2000) believes that the 'intervention research' approach can be considered as a broader general approach that can include and add to all the other research approaches taken in the management sciences. A comparison of the four typical approaches taken in the management sciences is given in Table B.1, and is based on the degree of formal definition of change and innovation and the contextual nature of changes and innovation that occur through and could occur after the research process.

From Table B.1, it can be seen that the intervention research approach consists of continual theoretical reappraisal during the project, which re-orientates the action in the organisational context. This is in contrast with the traditional 'action research' projects that concentrate on obtaining organisational

Table B.2 *Example 'phase'-denoted decision making and problem-solving processes*

Dewey, 1910	Polya, 1957	Simon, 1960; 1977*	Brim *et al.*, 1962	Witte, 1972	Mintzberg *et al.*, 1976	Nutt, 1984
Origin and stimulus (suggestion)		Intelligence	Identify problem	Problem recognition	Recognition	Formulation
Identify problem (i.e. through observation and inspection of facts)	Understand problem		Diagnose problem	Information gathering	Diagnosis	Concept development
Hypothesis formulation (i.e. of possible solutions)	Devise a plan	Design	Generate solutions	Development of alternatives	Search, design	Detailing
Develop reasoning for solutions			Evaluate solutions	Evaluation of alternatives	Screen	Evaluation
Testing of elaborated ideas					Judgement, analysis, bargaining	
		Choice		Choice	Authorise	
Test results guide new observations and experimentations	Carry out the plan and look back (reflect, evaluate)	*Implementation and monitoring	Implement and revise selected solutions			Implementation

action, and then may theorise on how and why the action occurred that way at a later point, and lead to the creation of models for organisational change or decision support that have been theorised (and will hopefully be later applied). Observation, on the other hand, plays a more external role but could induce theoretical and contextual changes at a later date. It is noted that David's comparison is based on stereotypes of management approaches and that in reality there may be a number of hybrid approaches; for example, a number of descriptions of 'action research' processes and the theory behind them are very similar to the 'intervention research' described in the French management sciences e.g. as illustrated in the later sections of Flood (1998).

In any case, it is considered that well-intentioned and theoretically sound intervention research projects, whatever their appellation, can offer many possibilities for the extension of current, theoretical and practical knowledge bodies, and that their application could be beneficial to the water sector.

Within an intervention research project, as reflection is based on constructing collective action, it is advantageous to analyse theoretically how decisions on taking action are typically made and what methods could be used to aid them.

B.2 DECISION MAKING MODELS

Many models of human decision making have been postulated, including those based on theory from philosophy, psychology, mathematics, neuroscience, the management sciences and operational research. The most common similarity is that decision making is considered a process rather than just a moment of choice (Roy, 1993). The large majority of models presume the existence of a number of sequential 'phases' in a decision process between the emergence of an idea and its transformation into some action (Lipshitz and Bar-Ilan, 1996), whereas other researchers have hypothesised that in a number of situations – especially in organisations – decision making is influenced or driven by other factors (Nutt, 2005) or is more disjointed (Braybrooke and Lindblom, 1963). Driving factors could include serendipitous opportunities, where confluence emerges between solutions, need and decision makers' choices (March and Olsen, 1976), or a need for compromise that could lead to bargaining behaviours, political power exertion and truncated searches for information to aid decision making (Cyert and March, 1963; Pfeffer and Salancik, 1974). For the 'phase'-type decision processes, the most basic consider a *divergent phase* of reflection on the decision stage, followed by a *convergent phase* of thought to arrive at a choice (Russo and Schoemaker, 1989; Montibeller *et al.*, 2001). A small selection of such processes from the literature is presented in Table B.2, with the phases running vertically in typical time order.

It must be noted that some of these authors underline the possibility of feedback through and between stages, rather than sequentially. This is especially so for Witte (1972), who analysed all of the phases denoted as emerging continuously through the decision making cases that were studied. Some other authors also question the validity of these processes as to whether they occur in such sequential orders and if they produce 'successful' decision outcomes (Lipshitz and Bar-Ilan, 1996). From these phase-type models, there are typically at least two different phases that occur before the 'choice' and 'implementation' stages, that could be aided. These two phases, the initial awareness and formulation of the problem, as well as the evaluation phase, will next be reviewed briefly.

B.2.1 The need for problem formulation

The formulation of a problem is far more essential than its solution.

Albert Einstein

In its basic form, problem formulation or 'framing' is a process that occurs on a habitual basis in everyday life between reflection and action in a decision process, or between subjects (people) and objects (their outside environment) (Landry, 2000). Many domains have attempted to formalise the concept of problem formulation, especially since the Second World War (Landry, 2000), including operational research, engineering, management, the physical and social sciences, and even literary theory. Exactly what constitutes the phase of problem formulation is not the same in each domain, and can often vary substantially between authors from the same domain. When optimisation theory is examined, problem formulation is generally considered as the selection of variables, objective functions, constraints and models to use in order to perform the optimisation (Dandy and Warner, 1989). This type of problem formulation forms the basis of many closed engineering and water resources management problems, such as irrigation or drinking-water pipe network optimisation and reservoir design. One of the common ways to formulate problems in the field of management is to consider responses to the questions, 'Who? What? Why? When? Where?' and 'How?' This set of questions originally stemmed from a poem of Rudyard Kipling, written in his tale *The Elephant's Child*, and has since been adopted into the management literature as the first step in planning and problem solving (Adair, 1986). Such a process, describing a problem situation and the potential ensuing actions, is not always referred to as 'problem formulation', but comes under a myriad of other appellations, including 'problem structuring', 'problem posing', 'problem specification', 'problem framing/decision framing', 'problem outlining', 'problem definition', 'problem identification and classification', and even 'planning'. The fuzziness of such a concept may thus lead to difficulties for researchers and practitioners in effectively communicating the exact processes they are referring to when such terms are employed. To overcome potential confusion and misunderstanding, greater effort has been recently made in certain organisations or research groups to identify clearly the meaning of 'problem formulation' or any of its variants. For example, Woolley and Pidd (1981) classify four 'streams' of problem structuring – checklist, definition, science research and people – based on their underlying assumptions, to help clarify the implicit processes expected and understood by certain stakeholders or decision analysts.

In a less explicit manner, the RAND Corporation defines problem formulation as the process not only of 'joining the dots' but also 'collecting the dots', referring to the need to determine carefully where, and from whose points of view, a problem definition has stemmed – for example through various forms of social and data analyses, where both tacit and explicit knowledge are examined (Nonaka and Takeuchi, 1995) – rather than simply constructing a seemingly given problem (joining the dots) in terms of components required to formulate potential solutions. The process of 'collecting the dots' is considered to be commonly forgotten or inadequately examined in many problem formulation and solving exercises, which has led to many failures to solve or examine the 'real' underlying problems (Libicki and Pfleeger, 2004).

Following a similar vein, but in the domain of operational research, Tsoukiàs (2007) clarifies that in a decision-aiding process, the first phase of constructing a description of the *problem situation* is an essential activity that can be considered separately from the description of the *problem formulation*, and an *evaluation model*, which is then used to construct a *final recommendation*. In this abstract and more formalised process (in terms of required content in each of the stages), once the situation of the problem has been defined in terms of the actors, participants or stakeholders concerned, as well as their concerns or stakes and the resources or commitment to the process that they are likely to contribute, the second phase of *problem formulation* can then be represented. This is performed in terms of a set of potential actions that could occur within the problem situation, a set of points of view from which these actions could be analysed and a problem statement that explains what operation could be performed by these potential actions. A consensus on the problem formulation must then be achieved with all the actors or stakeholders, before the evaluation model is constructed and a final recommendation decided on. This model has been expanded to the inter-organisational context in Section 5.1.3.

When consensus building is difficult or problems appear to be of the messy type described in Figure 3.1, the case in which independent problems are unidentifiable, and instead a dynamic situation of complex systems of interacting changing problems at a variety of scales exists, it is suggested that 'problem structuring methods' may be more appropriate (Rosenhead and Mingers, 2001b). In these cases, the unstructured problems may be collectively 'structured' and 'managed', rather than 'formulated' and 'solved'. For such problems or situations, including many of those confronted in today's water management and policy making activities, it is now almost unanimously agreed upon that the participation of stakeholders in these situations is required in the problem structuring and later planning and management phases, if adequate management of these situations is to occur (Viessman Jr., 1998). In group situations, where a problem is to be co-formulated, Rousseau and Deffuant (2005) suggest that problem formulation is a continuous cycle, of individual viewpoint construction followed by collective comparison and discussion, which feeds back into the next reiteration of individual and collective problem formulation. This implies that problem formulation is an inherently subjective process, the outcomes of which are likely to be largely related to the person or group of people involved in any such process. This view is backed up, specifically referring to the domain of water resources planning and management, by HarmoniCOP (2005b) and ADVISOR (2004). However, despite these attempts at process classification and clarification, there is still no single agreed-upon definition of exactly what constitutes the process of problem formulation or structuring (Corner *et al.*, 2001) in water management or in other domains.

It is also noted that in certain quickly evolving situations, information used for problem structuring may only remain valid for a limited period of time (Bouyssou *et al.*, 2000) and so reformulation of problems and the applications of subsequent stages in a decision-aiding, management or planning process should also be adjusted and implemented accordingly.

Taking the problem formulation further, it has also been suggested that many messes will most probably require 'reframing' (Wittgenstein, 1953) or reformulation, rather than simply framing of their internal problem areas. Yet this is not linked to validity of information but rather to the need for extensive change to manage certain messy situations and intractable problems. Second-order change, which pulls the issues out of their current frame, rather than first-order change, which treats problem causes or effects in their current frame, is considered necessary for the successful management of such problems (Watzlawick *et al.*, 1974). Such change often requires the creation of new options outside of a pre-formulated set of alternatives (Watzlawick *et al.*, 1974). The example outlined in Box B.1 demonstrates just how such a reframing could occur.

Reframing may be sufficient to dissolve or change the 'problem'. However, if this cannot be achieved or it does not lead to action (Brocklesby and Mingers, 1999), once a problem is structured or restructured there is typically a need to evaluate alternative courses of action, in order to inform choice to manage it. Evaluation will therefore be briefly discussed next.

Box B.1 Reframing problems – driving second-order change

Optimism vs. pessimism – breaking out of the dichotomy

In his first class of the year, a psychology professor was interested in scoping out the current dispositions of his new students as optimists or pessimists. He poured water into a glass in front of him until it was half-way up the glass and then asked them:

'Who considers that this glass is half full and who considers that it is half empty?'

Approximately two-thirds of the class raised their hands agreeing that it was half full and the other third preferred to think of it as half empty.

The professor saw that one student in the back row was looking puzzled and had not raised her hand. He asked her: *'I noticed that you have not answered my question – what do you consider the glass to be?'*

A little taken aback, the student apologetically responded: 'Sorry, sir. I am actually an engineering student who just came along to your class with a friend – I didn't think that it was right for me to participate.'

'Very well,' replied the professor, 'but since you are now in my class I am interested in having a response to my question.'

Much more embarrassed now, she finally responded: 'When you asked the question, I didn't first think that the glass was either half full or half empty – my first reaction was to think that the glass was twice as big as it needed to be...'

B.2.2 Evaluation

Evaluation is another term that has multiple definitions (Dart *et al.*, 1998) and has evolved in a number of forms in different civilisations and disciplines (i.e. public health, organisational sociology, industrial psychology, education (Freeman, 1977)), many of which are dependent on different value systems and corresponding epistemologies (Krane, 2001). Edwards and Newman (1982) identify four principal reasons for carrying out evaluations: for curiosity, i.e. to find something out; monitoring, i.e. to check progress towards goals; fine tuning, i.e. to inform adjustments relative to planned actions; and to inform choice, i.e. decide on actions. In this section on decision-aiding models, it is the 'inform choice' reason for evaluation that is of particular interest, although the other elements are typically required to complement this element. In the 'inform choice' situation, evaluation can be considered as the phase of the decision process that is undertaken once a problem has been formulated. However, some recent literature on evaluation processes considers the problem formulation phase as an important part of the evaluation. In fact, such 'evaluation' appears to encompass an entire decision process. Related to these forms of evaluation, Guba and Lincoln (1989) identify four large 'generations' of definitions and bases for evaluation – measurement, description, judgement and responsiveness – the outlines of which are noted in Table B.3 and include descriptive elements from other sources, including

Table B.3 *Principal types of evaluation considered by Guba and Lincoln (1989)*

Evaluation type	Philosophical or epistemological basis	Objectives	Typical evaluation tools, techniques or models
Measurement	Positivist, objective external evaluation	Quantitatively measure desired variables to obtain 'hard' data; test what is 'true'; capacity to exert authority on the basis of collected 'facts'; underpin causal inferences	Observed values, scoring, quantitative indices, physical measurement (e.g. length, volumetric, chemical testing), closed question surveys, mathematical manipulation and statistical tests Evaluator performs a 'technical role', as in most experimental research designs
Description	Positivist, objective external evaluation	Quantitatively or qualitatively describe progress towards objectives in a formative or ongoing manner; discover the strengths and weaknesses of current strategies relative to stated objectives, provide 'rich' contextual information	Narratives, longitudinal studies of variables (quantitative measured or described qualitatively), interviews, surveys Evaluator performs principally as a 'describer' but retains some of the 'technical role', as in some empirical field or case study research designs
Judgement	Positivist or pragmatist, uses subjective value-laden information in a seemingly 'objective' manner	To review goals, procedures, results and inform choices for future courses of action, to aid decision making to be made in a timely manner with available qualitative and quantitative information	Voting, grading, ranking, standards, commissions, juries, scale constructions, reliance on interpretation of quantitative and qualitative information Evaluator acts as a 'judge'; while also maintaining the describer and technical roles to inform the judgements, as in externally developed cost–benefit analysis (CBA) and multi-criteria decision analyses (MCDA)
Responsiveness	Constructivist, construction of knowledge inter-subjectively, oriented towards collective action	To co-construct and carry out in a reflective and ongoing manner the evaluations desired or required by stakeholders in contextualised 'problem situations'; use stakeholder issues, concerns and claims as a basis for building and negotiating the ongoing evaluation process; focus on reflexivity to provide continuous feedback for informing stakeholder decisions	All measures, descriptions and judgements must be considered relative to their evaluations and the contexts in which they were made; concentration on process – on responsiveness and doing; as in action or intervention research designs; stakeholder dialogue and negotiation, Delphi approaches, MCDA/MAUT type models and integrated assessment models that are directly co-constructed with stakeholders

Edwards and Newman (1982), Estrella and Gaventa (1998), Bouyssou *et al.* (2000), Linstone and Turoff (2002), Croke *et al.* (2007) and Tharenou *et al.* (2007).

Although these classifications are by no means definitive categories into which evaluation approaches and models can be mapped (see Dart *et al.* (1998) for a number of others), they do help to demonstrate the diversity of approaches to evaluation. Other types of evaluation models not explicitly included in the table include a range of other 'formal' methods, which use mathematical models of preferences or objectives of stakeholders and decision makers, such as optimisation algorithms for multiple objectives (Dandy *et al.*, 2007) and automated evaluation methods that are built with explicit decision rules, which may have been previously co-constructed or judged by decision makers or experts, and which may be based on automatically sensed changes to variables, for example from light and heat sensors (Bouyssou *et al.*, 2000). A number of these evaluation methods will be discussed further in the next section, along with methods used for other phases of the decision-aiding process.

The next section will aim to outline and classify these 'problem situations' in more detail, so as to better define the need for future research in the domain of water planning and management. Many methods have been developed over the years to aid different stages of the decision making process for these problem situations. Most have been in response to the challenges affecting the problem situation, underlying societal values and consequent decision process, including aspects of uncertainty, complexity and conflict. Relative to these three principal facets of messy situations, a number of different approaches to managing these aspects will be outlined, along with some of the methods that have been developed for these specific conditions. It is noted that there is much overlap between the facets, and the classification of methods is by no means strict, as methods may be used for a variety of reasons in different contexts. The purpose is, rather, to present some of the wide variety of methods commonly used in decision-aiding practice. Where possible, references to the use of these methods in the water sector will be highlighted.

B.3 A RANGE OF PROBLEM SITUATIONS AND DECISION-AIDING METHODS

A *problem situation* can be described as the context in which decisions need to be made. In the decision making domains of water planning and management, such a context typically has many constituents, including those from the physical or environmental, economic and social or political spheres. Depending on the possibility of nesting elements within these spheres, other contextual factors that should be considered in a problem situation have been outlined by a number of authors:

- Taking a systems approach to analysing the sustainability of water resources systems, Foley *et al.* (2003) outline that both the 'natural' resources (environmental and social) and the 'man-made' resources (infrastructural, technological and economic) should be taken into account;
- Looking at natural resources management, it is suggested that the technical, ecological, economic, social, political, institutional and legal dimensions should be considered (Cumming, 2000; Bellamy *et al.*, 2001; Allison and Hobbs, 2006);
- For participatory socio-environmental processes, Le Bars and Ferrand (2004) suggest that the external, normative, cognitive, operational, relational and equity dimensions of the context (and continuing decision making process) should be taken into account;
- The 'multi-modal' systems thinking approaches propose between 14 and 17 modalities that should be considered: credal/pistic, ethical, juridical, aesthetic, economic, operational, social, epistemic, informatory/communicative, historical, logical/analytic, sensitive/psychic, biotic, physical, kinematic, spatial and numerical/quantitative (Dooyeweerd, 1958; De Raadt, 1997; Basden, 2002; Khisty, 2006; Lombardi and Brandon, 2007).

Although potentially considered as sub-sections or more general classes of some of these dimensions of the context, temporal, behavioural and abiotic considerations could also be added to the list.

Observing all of these potential contextual elements, it can be considered that the 'problem situation' of any decision is a personal construction – or social construction if there is more than one person interested in the 'problem' or 'decision' – that covers three main domains, as defined by Habermas (1984): the personal world, i.e. including cognitive and normative dimensions; the social or inter-personal world, i.e. including relational, political, institutional, economic and equity dimensions; and the material world, i.e. including the physical, biotic, abiotic, technological, infrastructural, spatial and temporal dimensions, and interrelations or 'interbeing' among them (Morin, 1990; Low, 2002). As Edgar Morin (1990) explains, this is the type of relation that occurs when 'the whole is in the part, which is in the whole'. These interrelations could include the operational, behavioural, kinematic and numerical dimensions of the context.

A problem situation can also be considered as a cognitive artefact – a representation or construction – of the decision-aiding process. This specific meaning is attributed to it in the Tsoukiàs (2007) decision-aiding model presented in Figure 3.5.

However, to decide what type of decision-aiding or water planning and management approach may be required in a certain situation, a number of other parameters of a context may first be analysed, such as the general uncertainty, complexity and degree of conflict related to all of the contextual dimensions listed here.

B.3.1 Complexity

The concept of complexity is most commonly defined in relation to the study of 'systems'. *Systems* and the science associated with them are thought to have appeared as a result of studies emerging in the mid-nineteenth century (Matthews, 2004). This included work by Mill and Lewes in the domain of philosophy, the 'equilibrium theory' proposed by Koehler and the early work of biologists associated with von Bertalanffy, as well as the studies of 'wholes' (related to the organisation of living systems) by Angyal (1941) and Feibleman and Friend (1945), and early work on structuralism (see Durand (1979) for details). Two large branches of 'systems theory' then developed (Simon, 1996; Ison *et al.*, 1997): 'general system theory' (e.g. von Bertalanffy, 1950) and 'cybernetics' (e.g. Ashby, 1956), as well as a range of other variants, including organisational theory (e.g. Selznick, 1948; Ackoff, 1960; Katz and Kahn, 1966), chaos theory (e.g. Lorenz, 1963), synergetics (e.g. Haken, 1981) and autopoiesis (e.g. Varela and Maturana, 1973). These were followed by a series of more general 'systems approaches' in a range of disciplines, including 'resilience theory', 'complex adaptive systems', 'management complexity' theories and 'post-normal science' (refer to Matthews (2004) and Allison and Hobbs (2006) for more information).

From this large body of scholarship, a variety of definitions of 'systems' can be developed (Emery, 1969; Durand, 1979; Matthews, 2004), including that a system is:

- A complex whole, formed by a number of heterogeneous and interrelated components;
- An organised global entity of interrelations between elements, actions or individuals; and
- A set of dynamically interacting components, organised in space and time relative to a functional goal.

The *complexity* of a system can then be attributed, based on a number of properties (Le Moigne, 1977; Durand, 1979; Morin, 1990; Allison and Hobbs, 2006), including:

- Those inherent in the composition of the system, including the number and characteristics of its components and the interrelations between them; and
- Those linked to uncertainties, randomness, ambiguity and seemingly chaotic properties of their own internal composition and external environment.

As a part of complexity, *uncertainty* can be defined as something that is not entirely known. It can be considered a human-perceived condition of having limited knowledge about a future outcome, or an existing state of something. However, there are also many other definitions and views of the concept (Myšiak and Brown, 2006; Bammer and Smithson, 2008). The concept of *risk* is closely related to uncertainty; yet it tends to be used to explain the likelihood of occurrence of potential event consequences that are undesired (Standards Australia, 2004). A number of types of uncertainty may have an impact on water planning and management, including those related to:

Known knowns Uncertainties most often related to cross-perceptual issues, with one person questioning the beliefs and representations of another when both believe the subject matter to be 'known' in their own minds. Such representations could include competing scientific (or non-scientific) theories.

Unknown knowns Uncertainties that occur when one person does not have knowledge of something, but there is someone else whom they do not know who does possess such knowledge. One of the ideas of participatory processes is to attempt to avoid such uncertainties.

Known unknowns The most common type of uncertainty, which is recognised as such, and thus tends to be the only type of uncertainty taken into account in risk-management approaches. For water management problems, these uncertainties often include issues of climate, hydrological responses, infrastructure capacity, future land-use changes, human behaviour (especially stakeholders related to the problem under analysis), political environment and the economic climate. However, the list of potential uncertainties could be extended to include the list of contextual dimensions at the beginning of this section. During the problem analysis stage, such uncertainties can also be joined by uncertainties related to data i.e. quantity and quality, and the models or processes used i.e. methods of calibration and validation, modeller or analyst capacity, data match to needs and valid usage of data sets (Maier and Ascough II, 2006; Brugnach *et al.*, 2007).

Unknown unknowns Uncertainties that no one knows anything about – they could be considered as 'surprises'. Despite the lack of knowledge surrounding such uncertainties, some water planners and managers are trying to develop systems that are the most adaptable or resilient possible so that they may potentially adapt or recover in the face of an unknown unknown (Pahl-Wostl *et al.*, 2008).

Considering these two principal aspects of complexity (i.e. the number of interrelations and uncertainties), water 'systems' and their associated planning and management systems can be observed to have become more complex through history, as outlined in Chapter 2 and Appendix A. Present-day complexity in the water sector related to interrelations and uncertainties will be further outlined here, along with some of their consequences for the future. Sections on 'conflict' will then follow separately, even though it is also an intimately intertwined part of complexity.

INCREASING WATER SECTOR COMPLEXITY DUE TO INTERRELATIONS

Water planning and management are becoming increasingly complex due to the interrelations between the water sector and almost every other sector in the rapidly globalising world. The 'information' revolution, including the growth of the Internet and mobile telecommunication networks, has aided the speed of knowledge transfer and communication. Population growth, environmental degradation, resource depletion and international economic markets are inducing strain and competition for local resources in almost every region of the world; and political decisions or individual actions in one region of the world are likely to have follow-on effects and unpredictable ramifications in other regions, owing to a multiplicity of scale dynamics and network connections (Vlachos, 1998; Buchanan, 2002; Lankford, 2008; Lebel and Garden, 2008). Because of this increase in interrelations, defining appropriate scales for water planning and management is also becoming an increasingly difficult task when water basins or administrative boundaries no longer bound the problem situations involved.

This is especially true in relation to a number of key sectors, such as land management and agricultural produce, health (through biological systems), ecological systems and energy systems, where there is a strong need for integrated planning and management. Some of the other interrelations between water and agricultural produce and water and health have already been highlighted in Sections A.1.2 and A.2.1. As an extra example, the issue of water and energy interrelations will be discussed here, as it is a more recent example of a strong human-created interdependence outside the traditional natural water–energy cycles.

In modern societies, water use and management are intimately entwined with energy production. Water is used to create and use energy, and energy is required for the sourcing, treatment and transport of water in many areas. A fifth of the world's total energy is produced through hydropower (Pearce, 2004), and cooling of other types of power stations, such as nuclear and coal, requires large amounts of water. Consequently, it is thought that these important relations, if not effectively considered and managed, will lead to increased conflicts, as both water and energy become more valuable and less easily obtainable resources. For example, in countries with Mediterranean or temperate climates, it can already be observed that summer peak energy periods often occur in drought years when temperatures are at their most extreme and in times when there is increased water scarcity, high evaporation and strong competition for the resource. At such a time, dam releases for hydropower, nuclear and coal power station cooling, and energy for desalination to produce more water, are all likely to create additional strains on limited water resources, causing conflicts with other water users. The new issue of desalination also brings the interdependence between water and energy ever closer, along with a wide range of other interrelations, including: land use, as competition for coastal land is becoming increasingly intense because of population growth and urbanisation; and ecological systems, as discharges of the salty brine in sensitive coastal regions are likely to cause environmental degradation, which includes biodiversity loss.

INCREASING WATER SECTOR COMPLEXITY DUE TO UNCERTAINTIES

Predicting the impacts of planned water management actions over a variety of scales and interrelated sectors, or the impacts of outside phenomena, such as climate change, on a local region's water and environment, is also becoming an increasingly complex task. In the past, such tasks were left to scientists and engineers, who formulated their predictions based on traditional water resource models or tacit expert advice. However, with the increasing list of contextual uncertainties and new system interrelations (highlighted in this section), such traditional decision-aiding techniques and modelling for planning and management of water-related actions have fallen under heavy criticism (KNAW, 2003). Although some impacts of actions may still have predicable outcomes, others are likely to produce non-linear or chaotic outcomes and still others will result from unforeseen or surprise events (Gleick, 1998). Moreover, the scientific uncertainties over assumptions used in traditional water resources models are likely to cause further problems and prove invalid for a variety of reasons, leading to the necessity to develop new methods (Borgman *et al.*, 1970). For example, the assumptions of homogeneity and stationarity (which often involves ignoring outlying statistical points) used in statistical models for hydrology rarely appear to hold true (Alexander *et al.*, 1970; Vogel *et al.*, 1998; Daniell and Daniell, 2006; Kuczera, 2008; Milly *et al.*, 2008), and there are similar debates on the assumptions of human behaviour translated into economic and social models related to water management.

One of the major issues in water resources and hydrological modelling is how to account for and deal with changes that have occurred over the life of the collection of the data set, such as land-use changes, changing withdrawals, decision behaviour, and dam and infrastructure construction over the catchment's life. There is now also mounting evidence of climate change outside normal variability patterns affecting some of the world's catchments (Gleick, 1998), which is likely to invalidate the models based on long-term data sets in these areas, even if they have not been subject to the other changes listed previously. Perth in Western Australia is a prime example of where the majority of experts agree that climate has undergone changes in two step-intervals in the past 30 years. From an average 10–20% drop in rainfall over the last 30 years in Perth's catchment, there has been an average 40–60% (or more) drop in streamflow entering the city's dams (Wentworth Group, 2006), as shown in Figure B.2.

These unpredictable effects have had a major impact on the water management systems of the city. These types of change are also now being found in Australia's other cities, such as Sydney (Wentworth Group, 2006) and Melbourne (Tan and Rhodes, 2008), and are likely to be found soon in many other regions around the world. In light of such changes, relying on future prediction based on past data records to aid future water planning and management decisions may prove dangerous or very costly in the future (Gleick *et al.*, 2004). Such issues highlight the new complexities that need to be given attention in today's and tomorrow's water planning and management practices.

B.3.2 Dealing with complexity

Traditional scientific methods (i.e. reductionist or positivist) have focused for many years on controlling or reducing complexity by isolating elements of problem situations and attempting to understand their internal and external function

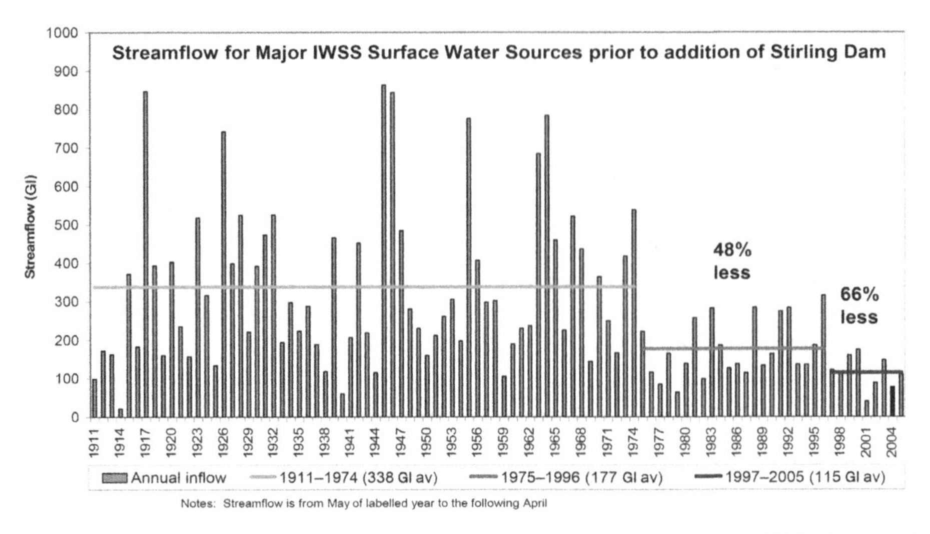

Figure B.2 Climatic step-wise changes – Perth dam inflows 1911–2005. Source: WA Water Corporation (2006); Wentworth Group (2006).

relative to other elements, often via causal mechanisms (Kuhn, 1962). Experimental methods with control samples or groups of people to determine system or element behaviour in an attempt to control or reduce complexity are typical of such approaches (Tharenou *et al.*, 2007). Considering the common incapacity to control or reduce complexity sufficiently for successful management, many methods have been developed that consider complex systems in a more embracing manner to aid decision making, as partially described in Section 3.1.2. Those methods still considering individual elements and the causal interrelations of the system's elements include causal loop diagrams and system dynamics models (Forrester, 1961; Sterman, 1989; Zwaan and Radvansky, 1998; Lasut, 2005; Forrester, 2007). Such models may be constructed as part of the problem situation, formulation or evaluation model stages of decision-aiding, depending on the typology of their components and purpose.

Many other types of modelling technique have been developed to embrace and live with both complexity and uncertainty, some of which do not attempt much to understand the complex mechanisms, and others that are more focused on understanding how the system complexity is manifested. Statistical methods, such as artificial neural networks (Kohonen, 1988; Daniell, 1991), allow the 'black box' (the complex system and its behaviour) to remain largely closed, with the results able to be used after validation with real-world data. Such methods can be appropriate when good quality quantitative data are available on a range of system properties, but are less useful where high uncertainties or gaps in data exist, i.e. climate change or land-use changes, altering the underlying data patterns (Daniell and Daniell, 2006) They are also unhelpful for developing a 'causal' understanding of the system. A number of heuristic techniques for finding optimal or near-optimal solutions in complex systems and for multi-objective decision making include: simulated annealing (Kirkpatrick *et al.*, 1983; Dougherty and Marryott, 1991), tabu search (Glover and Laguna, 1997; Tung and Chou, 2002), genetic algorithms (Goldberg, 1989; Simpson *et al.*, 1993), ant colony optimisation (Maier *et al.*, 2003; Dorigo and Stützle, 2004) and particle swarm optimisation (Kennedy and Eberhart, 1995; Chau, 2004). Other complex decision making processes may be aided by artificial intelligence (AI) applications, such as 'expert systems', which are based on pre-established decision rules of reasoning derived from expert advice (Rowe and Davis, 1996; Pomerol, 1997; Letcher, 2002).

In some water management applications, typical objectives of attempting to control or reduce uncertainty and cope with system complexity include trying to increase the *reliability* or *stability*, as well as the *robustness*, of systems and to reduce system *vulnerability* or *susceptibility* (Hashimoto *et al.*, 1982a; 1982b; Howell, 1989). Methods used for decision-aiding with these objectives have included developing statistical and probabilistic models (i.e. for calculating climatic variability and flood risks) and using safety factors above model estimates to further reduce uncertainties of design failure (i.e. for dam construction). Monte-Carlo simulation (Kuczera and Parent, 1998; Rousseau *et al.*, 2001) is one common method. To pursue system robustness, techniques that involve developing scenarios of possible system futures and then using decision tree analyses or simulation models to test for system failure or inadequacy are common (e.g. for designing and testing water supply systems) and sometimes result in building *redundancy* into systems (Dandy *et al.*, 2007). The majority of these methods have typically focused on uncertainties related to elements of the 'evaluation model', rather than other stages in the decision-aiding process. Some of the most common methods for aiding decision making under uncertainty and risk include expected-utility methods (Schoemaker, 1982) and those based on prospect theory (Kahneman and Tversky, 1979). However, some forms of robustness analysis also address issues in the 'problem situation and formulation stages' (e.g. Rosenhead, 2001) and some recent standards and models for aiding risk management processes encourage a more complete treatment of uncertainties from the problem situation to the final recommendations (Standards Australia, 2004).

A large range of multiple criteria decision analysis (MCDA) tools and approaches have also been developed to help structure the complexity of decision problems and make subjectivity explicit (Belton and Stewart, 2002), especially linked to decision makers' values and preferences. This very broad family of methods, which helps decision makers to take multiple criteria explicitly into account in their decision problems, includes: multiple objective linear (or mathematical) programming (MOLP and MOMP) methods, such as goal programming (Charnes *et al.*, 1955); compromise programming (Zeleny, 1973); multi-attribute utility (and value) methods, i.e. MAUT, MAVT (Keeney and Raiffa, 1976; Edwards and Newman, 1982; Bunn, 1984); the ELECTRE (Roy, 1985) and PROMETHEE (Brans and Vincke, 1985) methods, which are based on 'outranking' problem management alternatives over one another; and the analytic hierarchy process (AHP) (Saaty, 1980) and SMART methods (Edwards, 1977) based on additive-value models. More example methods and useful descriptions of differences in underlying backgrounds and assumptions of these methods can be found in Guitouni and Martel (1998), Bouyssou *et al.* (2000), Dodgson *et al.* (2000) and Belton and Stewart (2002). Many of these methods can also be 'fuzzified' to further cope with uncertainties or impreciseness, with example applications in the water sector including those of Fleming and Daniell (1996), Bender and Simonovic (2000) and Srdjevic and Medeiros (2007). Such multi-criteria methods, although allowing for complexity to be embraced, may also prove too complex in their underlying mathematical structures and associated meanings for decision makers to understand them adequately, thus leading to the likelihood that decision makers will have less confidence in them (Hajkowicz *et al.*, 2000; Belton and Stewart, 2002).

Modelling methods that place a greater emphasis on dynamic complex system understanding and analysis include many from the systems theory family outlined in Section B.3.1, which include cybernetics, system dynamics and multi-agent systems. Multi-agent systems, in particular, can be used to concentrate on analysing emergence effects (Weiss, 1999; Janssen, 2002; Perez and Batten, 2006), and system dynamics models on understanding feedback effects, which are often counter-intuitive (Sterman, 1989; Forrester, 1992a). In terms of further exposing and understanding complexity, a number of qualitative or semi-qualitative methods have been developed, including various forms of cognitive mapping, some which can also be 'fuzzified', to cope with a range of uncertainties (Kosko, 1986; Hobbs *et al.*, 2002; Giordano *et al.*, 2005) or are used as part of a large range of 'problem-structuring' methods (Rosenhead and Mingers, 2001b). Some of these problem-structuring methods have already been noted in Sections 3.1, 3.3 and B.2.1 and will be further investigated in Section C.3.3. A large range of models designed principally for improving understanding of spatial and temporal complexity, e.g. by using geographical information systems (GIS) and simulation models, which can be used in the development of the problem situation or for use as evaluation models, are outlined in Agarwal *et al.* (2002) and Parker *et al.* (2003).

To embrace or live with uncertainty, emphasis is placed on increasing system *flexibility* and *adaptability* (or *adaptivity or adaptiveness*), as well as on learning to work with ambiguity. Flexibility can be associated with the idea of increasing *responsivity* to change and adaptivity to the ideas of *resilience* of systems and their *recoverability* and *transformability* (Holling, 1973; Howell, 1989; Walker *et al.*, 2004). Methods of dealing with uncertainty in this manner aim to promote system *diversity*, in order to keep the maximum number of future paths open, develop alternatives that maximise *reversibility* potential (i.e. the precautionary principle), build scenarios and adaptation strategies in anticipation of change, and build human capacity to cope with and react to change and uncertainty. Examples of practical approaches for achieving these aims are mostly participatory-based exercises of prospective analyses (Bouleau, 2003b) and other forms of 'what-if' analyses using interactive simulation models and role-playing games (Pomerol, 1997; Bousquet *et al.*, 1999; Barreteau *et al.*, 2001; Barreteau, 2003b). Working in and with ambiguity has required the development of alternative methods, many of which focus on mathematical methods of expressing and dealing with uncertain scientific knowledge (Hipel and Ben-Haim, 1999), including: fuzzy sets, fuzzy logic and possibility theory (Zadeh, 1978; Fleming, 1999; Bender and Simonovic, 2000; Srdjevic and Medeiros, 2007); rough set theory (Pawlak, 1991); interval analysis, grey systems theory and grey programming (Moore, 1979; Huang *et al.*, 1992; Bass *et al.*, 1997; Chang *et al.*, 1999); qualitative physics (Faltings and Struss, 1992); Bayesian or probabilistic networks (Batchelor and Cain, 1999; Jensen, 2001; Ticehurst *et al.*, 2007); and Markov chains (Meyn and Tweedie, 1993; Stewart, 1994). Most of these methods concentrate on the evaluation model stage of the decision-aiding process.

If uncertainty is to be further exposed and understood before the evaluation model stage, there is a need to outline: limits and gaps in knowledge; model-related assumptions and performance in all phases of the decision-aiding process; and representations of the world, beliefs, perceptions, values and preferences. This includes assumptions about the past state of the world – what happened through history and what it means; the current state – including dynamic understanding; and the future states – what is possible, probable, plausible and desired. To expose and understand these future-related uncertainties, visioning, future building or scenario building and analysis are the most commonly employed methods. Miser and Quade (1988) consider that a 'good' scenario should possess a number of attributes – consistency, plausibility, credibility, rationality, relevance, utility and probability – while others see visioning and scenario building as a more creative and exploratory process that can drive innovation. For other knowledge and human-related assumptions, many elicitation methods are used, including: cognitive mapping; brain-storming; interviews; facilitated discussions or dialogue; and other reporting or mapping techniques (Axelrod, 1976; Buzan and Buzan, 1993; Eden, 2004; Creighton, 2005; Tharenou *et al.*, 2007). For example, explicit mapping of various uncertainty types – environmental, values and choices – on an 'uncertainty graph', as part of a strategic choice approach, has been used to develop greater participant understanding (Friend and Hickling, 1987; Hickling, 2001). These types of methods are commonly used in the problem situation and problem formulation stages of the decision-aiding process, as well as in the exploratory part of evaluation model use. A large number of other methods designed to understand the evaluation model uncertainties, i.e. input data, framing, structure, parameter values and output, also exist, including sensitivity analyses and uncertainty analyses (Pomerol, 1997; van der Sluijs *et al.*, 2004; Jakeman *et al.*, 2006; Maier and Ascough II, 2006; Myšiak and Brown, 2006; Brugnach *et al.*, 2007).

Other complexities and uncertainties that need to be coped with in water management are driven by increasing human interest and conflicts over priorities for water use.

B.3.3 Conflict

It is generally considered that the scarcities created by unequal geographical distribution and interrelations of water systems with human populations and development needs (sometimes causing physical, economic and social scarcities) have fuelled conflicts between various water uses, i.e. industrial, rural, urban and pure subsistence (Ohlsson, 2000). However, this is a simplified view of water-related conflict, the underlying story of which will be examined in a little more detail here.

Conflict can be considered a human-perceived state of discord. Such a state may result from differences in a range of factors, including values (normative positions); beliefs and representations (cognitive positions); goals; needs; rights; interests; priorities; and actions or power relationships. Conflicts can occur at a number of levels, including: intra-personal; inter-personal or intra-group; and inter-group (where 'group' is considered to have the largest possible definition, i.e. a number of individuals with something in common, a society, a state, or a real or virtual ideological programme or movement). Conflicts may therefore also occur at a variety of spatial or virtual scales, including: in an individual's mind; locally; intra-nationally; and internationally. When a conflict has been expressed in the public domain it can be considered 'manifest' and if it remains hidden it can be defined as 'latent' (Rinaudo and Garin, 2003).

In the case of water and its management, 'conflict' has been further defined as '*a social situation in which at least two actors try to, at the same time, gain access to the same set of resources*' (Thomasson, 2004) or a situation where there are '*two or more entities, one or more of which perceives a goal as being blocked by another entity, and power being exerted to overcome the perceived blockage*' (Frey, 1993). However, it is noted that these definitions are likely to be too closed to explain the full complexity of water-related conflicts, considering the multiplicity of conflict causes previously listed in the general definition. As mentioned in Section A.2.1, water systems and resources can also be considered as sources, instruments or targets of such conflict (Gleick, 1998). Examples of water being used as an 'instrument' or 'target' of conflict have been given in Section A.2.1 and more are outlined in Gleick (2006).

When water is the 'source' of a conflict, the situation can be attributed to one or more of quantity, quality and timing (Wolf *et al.*, 2005), underlying the more fundamental causes, i.e. value, belief, interest and power differences. One of the major challenges for water planners and managers in the future will be to be able to successfully work and deal with, as well as transform, conflicts (Delli Priscoli and Wolf, 2009). This involves working through a number of stages in the life of conflicts, including: identifying both manifest and latent conflicts; effectively managing the conflicts; and resolving or neutralising them (Vlachos, 1998).

Conflict identification and analysis can occur in anticipation of a latent conflict manifesting itself or once a manifested conflict has been noted. A number of analysis tools have been created to help to identify different aspects of conflicts, including: their sources or causes (listed previously); their participants (concerned parties); their impacts on the problem situation, i.e. increasing the apparent overall level of conflict, complexity or uncertainties; and their relations to other conflicts, whether they be water-related or otherwise (e.g. Yoffe *et al.*, 2001; Rinaudo and Garin, 2003). Carrying out such activities early during the definition of a 'problem situation' may help to better anticipate, diagnose or prevent conflicts manifesting themselves (Vlachos, 1998), all of which will prove valuable in the following stages of water planning and management.

Conflict management is a process designed to work with manifested conflicts. It has been suggested that strategies of conflict management could include: avoidance; accommodation; competition; compromise; and collaboration (Thomas, 1976), depending on the relations and the intent of satisfying the party's own and others' interests (Thomas, 1992), as presented in Figure 4.2. Much literature exists on the mechanisms of mediation, negotiation, trust-building or arbitration, and can be used to manage conflicts to attempt to obtain the best possible solutions for both parties, often considered to be 'collaboration', which will help to achieve long-term 'win–win' outcomes (Fisher and Ury, 1981; Bellenger, 1984; Pruitt and Rubin, 1986; Priscoli, 1990; Wolf, 2002b; Nandalal and Simonovic, 2003; Zeitoun and Warner, 2006). Some available decision-aiding methods for conflict management will be outlined in the next section.

Conflict resolution is the final stage of the conflict management process, where an agreement is made that is accepted by the parties in conflict. Such a conclusion or neutralisation of a conflict could occur through collaboration, consensus building or other arbitration processes (e.g. White *et al.*, 2007), as previously noted.

Water conflicts can be particularly complex, owing to the sheer magnitude of possible participants, organisations and government departments involved from different sectors of interest (i.e. indigenous and recent cultures, agriculture, water-supply companies, industry, environmental groups), that are all likely to manifest a large range of concerns, different value sets, representations of the world, power levels and capacities to act. Many such conflicts are likely to be associated with other conflicts involving even more participants and interests over a variety of interrelated scales, rendering analysis and management of such conflicts complex and full of uncertainties.

B.3.4 Dealing with conflict

Decision-aiding methods to deal with conflict can be grouped according to categories, including approaches to:

- *Control* or *reduce* conflict;
- *Embrace* and *live with* conflict; and
- *Expose* and *understand* conflict.

To control or reduce conflict, many decision-aiding approaches exist, including those associated with 'conflict management', references for which were briefly outlined in the previous section. There are also a number of 'group decision support systems' (GDSS) (DeSanctis and Gallupe, 1987) or interactive computer or internet-based tools, which have been developed as a basis for aiding these negotiation, consensus building or compromise processes. These include group multi-criteria decision support or negotiation systems, such as those from the *Decisionarium* tools website (SLA, 2007), including 'WEB-HIPRE' (Mustajoki *et al.*, 2004) and 'joint-gains' (Hämäläinen *et al.*, 2001) software, which have been used for water management decision-aiding purposes, and the MULINO decision support system (mDSS) (Myšiak *et al.*, 2005). Further examples of such tools are given in Bruen (2007). However, depending on the extent of conflicts, such tools may not prove sufficiently transparent and their underlying assumptions (i.e. their mathematical structures) may be disputed in some problem situations (Holz *et al.*, 2004).

Other forms of conflict management attempt to embrace and live with conflict rather than to try to control or reduce it (Hardy and Phillips, 1998). For example, it has long been considered that debate and diversity of opinion are important for innovation and effective political governance. Most methods of embracing and living with complexity are necessarily participatory. Many political methods fall into this category of decision-aiding approaches and include mediated debates, discursive struggles, deliberation and dialogue (Forester, 1999), policy analysis models (Mayer *et al.*, 2004) and other communicative tools (Bots *et al.*, 2005), and theoretical aids, such as Habermas' (1984) communicative action theory or critical systems thinking (Ulrich, 1991; Midgley, 2000). This is not surprising, considering that '*If there are no conflicts over meaning, the issue is not political, by definition*' (Fischer, 2003). Decision-aiding approaches to aid such conflicts include a range of interactive 'information and communication tools' (ICT) and group decision support systems, which have been designed for certain stages of the decision-aiding process and, more generally, aiding 'social learning processes' (Bandura, 1977; Pahl-Wostl and Hare, 2004; HarmoniCOP, 2005b).

As a basis for either type of conflict management, a preliminary phase of exposing and understanding conflict, also referred to as 'conflict identification and analysis', can occur in anticipation of a latent conflict manifesting itself or after a manifested conflict has been noted. A number of analysis tools have been created to help to identify different aspects of conflicts, including their sources or causes, participants, impacts on the problem situation, and their relations to other conflicts. For example, Keeney (1992) suggests 'value-focused thinking' as a base decision making process that can aid the early identification and differentiation of conflicting viewpoints on consequences and desirable consequences of alternatives. Some ICT tools are also well suited to encouraging open definition and conflict identification in the problem situation stage, including actor or stakeholder analysis methods, cognitive mapping, facilitated discussions and questionnaires (HarmoniCOP, 2005b). Other common methods of mapping out different stakeholder positions and working on issues of conflict include some multi-objective analysis methods, such as the Delphi method or the nominal group technique (Goicoechea *et al.*, 1982; Fleming, 1999).

Of course, another eventual way of dealing with conflict is simply to circumscribe it, as Machiavelli suggests.

B.3.5 Messes

From the previous parts of this section, Chapter 2 and Section 3.1, it can be seen that water planning and management are commonly dominated by complex, uncertain and conflict-ridden problem situations; in other words, messes.

As noted in Figure 3.1, *messes* can be defined as dynamic situations that consist of complex systems of interacting and changing problems (Ackoff, 1979). Specifically related to natural resources planning and management, Lachapelle *et al.* (2003) describe that:

> *Wicked problems and messy situations are typified by multiple and competing goals, little scientific agreement on cause–effect relationships, limited time and resources, lack of information, and structural inequities in access to information and the distribution of political power*

From Sections B.3.1 to B.3.4, it can be seen that the complexity of a problem situation is largely driven by the number of uncertainties and the number of interrelations, which include the complexities of the human system (i.e. the number and level of conflicts present). Although all of these categories are heavily interrelated, it is useful to divide them arbitrarily, to create a conceptual model of problem situations, as shown in Figure B.3. Such a model can be used to understand the different types of problem situation that may be observed, from 'simple' problems to messes. Problem situations that have few interrelations, are not ridden with uncertainties and exhibit low levels of conflict, can be considered 'relatively simple' and, as such, may be able to be 'solved' using traditional scientific investigations or by engineering problem solving. Problem situations that exhibit higher levels of conflict but still have few uncertainties and interrelations may lend themselves to management using typical conflict resolution or political methods. However, those with high levels of conflict, uncertainties and interrelations can be thought of as good quality messes that will require inspiration, at the least, to manage them.

Following on from a similar line of reasoning about the incapacity of technical experts to deal with certain types of messy or 'wicked' problem situations, Funtowicz and Ravetz (1993) have suggested that methods of 'post-normal science' are required in contexts where facts are uncertain, values are in dispute, the stakes are high and decisions must be made urgently, as shown in Figure B.4.

This categorisation can generally be translated onto the surface between the uncertainties and conflicts axes of Figure B.3. When stakes and uncertainties (i.e. uncertainties of an epistemological nature) are high, 'post-normal' science implies a need to 'democratise' scientific practice, as the 'normal' scientific methods (Kuhn, 1962) are no longer considered relevant or practicable in such situations (Funtowicz and Ravetz, 1993). It is suggested that planning and management for the messiest situations should be predominantly 'politics with science advisors' (Burchfield, 1998), where public participation becomes an intrinsic part of the planning process, for example through 'interactive planning' (Ackoff, 1999) and the role of science and its relationships to policy will be explored (Lachapelle *et al.*, 2003). The other categories of 'professional consultancy' and 'applied science' could be matched to the types of professional judgements made by engineers and the use of typical scientific methods in Figure B.3, corroborating a general typology of strategies for managing messy and simpler problem situations.

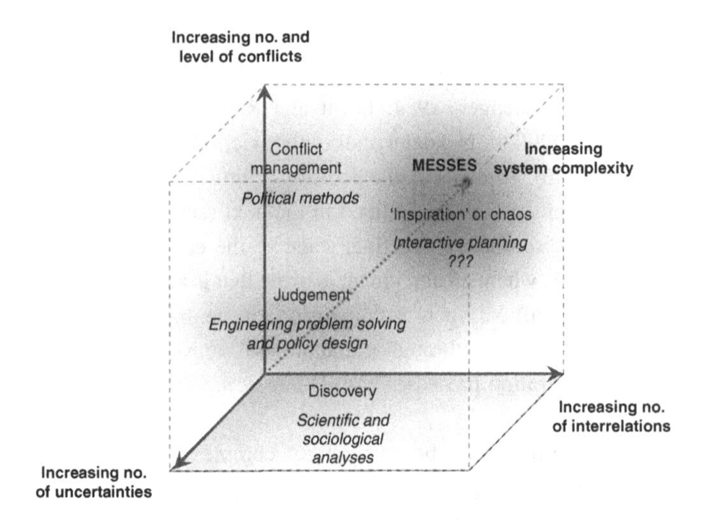

Figure B.3 Problem situations in terms of uncertainties, interrelations and conflicts with potential management methods.

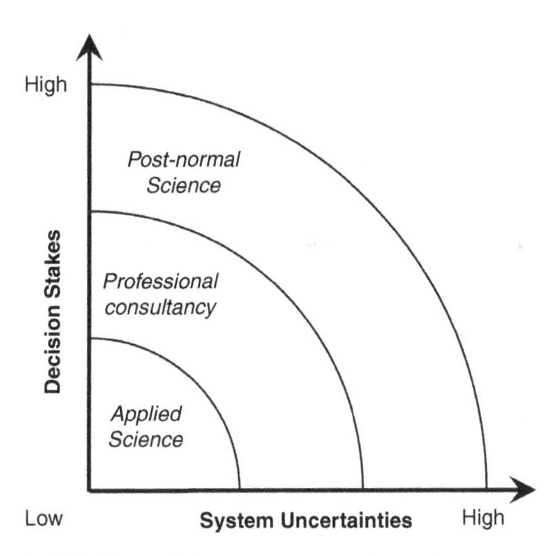

Figure B.4 Problem-solving strategies according to Funtowicz and Ravetz (1993).

Although there have always been issues of uncertainty, complexity and conflict in past societies, civilisations and their water systems, these have never been felt to the same degree as in today's world. In the past, some seemingly sustainable solutions to certain problems appeared to have been found; yet whether they will meet the needs and desires of today's communities is another matter. Although some problem situations in the water sector are still relatively simple – e.g. determining the amount of treatment chemicals to be applied to stored water to bring it within drinking water regulations, or designing a pressurised water network within a number of flow and budgetary constraints – and can be performed using a range of advanced engineering techniques, there are many more which are much more messy. Whenever larger-scale issues are investigated, or more people are interested in the water issues, the problems are likely to be significantly messier. A number of potential methods for working in messy situations have already been highlighted in Chapter 3, so only the issue of inspiration noted in Figure B.3 will be looked at in the final section of this appendix.

B.3.6 A need for new ways of thinking for decision-aiding in the water sector?

> *There is an old Zen story about a small fish that went on a long journey to find a Zen master. When at last he found the old wise master, he asked him: 'What is that thing called water that they talk about?' How could a fish not know about water? The reason the concept of 'water' presented such a challenge to the little fish was because he was so thoroughly immersed in it.*
>
> Beaumont W. Vance (2007)

Although there have been recent changes in the underlying principles of water management practices in recent times, many of the on-the-ground approaches and practices used in today's world (especially in the West) are still largely impregnated with rational Cartesian thinking. This is not surprising, considering the current educational programmes in school and universities, which have a tendency to present the traditional views of science as an objective pursuit of knowledge, and engineering as 'problem solving' first, and then present 'alternative' views, such as 'problem structuring' (Rosenhead and Mingers, 2001b), much later (if there is time in the curriculum!). Inhabitants of the Western world are largely programmed to think in a particular way; of differentiating between objects and subjects, and consequently what is objective or subjective. This is even influenced by our Western languages, which have been constructed in relation to these 'subjects' and 'objects'. However, like the little fish in the Zen story, this way of seeing and understanding the world limits our comprehension of other possible viewpoints that might exist. Einstein is often quoted as having noted many years ago that, *'The problems that exist in the world today cannot be solved by the level of thinking that created them.'* This means that we should be searching for alternative paradigms of thinking, which may prove more effective in helping us to aid the management of today's and tomorrow's complex issues. Embracing a broader range of ways of thinking may ensure that decision-aiders can appreciate and help to co-construct a greater range of perspectives for their potential clients to consider in their decision problems; or find more adapted decision-aiding methods to use with clients who hold different general types of world views.

Many alternative paradigms of thinking and understanding the world have been developed throughout history, a number of which have already been lightly touched upon in the previous sections of this appendix. A paradigm can be thought of as a model for understanding reality (Allison and Hobbs, 2006), which is made up of sets of assumptions about how the world works, and language and concepts (Kuhn, 1962). These include assumptions about determining: *ontologies* – 'what exists?' and what constitutes this reality; *epistemologies* – 'what is knowledge?' and how knowledge (or different types of it) is constructed, the relationship between the knowledge-possessor and the known, including the knowledge-possessor's reality of existence and the value of this knowledge; and *methodologies* – 'how can new understandings be created?' and what systems of study or processes can be used for understanding (Kuhn, 1962; Le Moigne, 1999; Berthelot, 2001; Allison and Hobbs, 2006). To realise just some of the potential paradigms or existing theories and general philosophies that could help us to think differently about the world and aid water planners and managers to address today's water issues in another way, Table B.4 presents a small but eclectic collection of recent and not so recent approaches for consideration. It is noted that each does not necessarily constitute an entire paradigm or even theories that have become stabilised, but each of them should start to provide a little food for thought for curious and creative decision analysts looking to broaden their own world views and inspire their decision-aiding practice.

Apart from the approaches presented here, there are, of course, many others that have been unfortunately neglected, some of which are already demonstrating interesting insights into the water sector, including Chinese ecological philosophies (Wang and Li, 2008) and network theories, for example, social network theory and small-world networks (Buchanan, 2002).

From Table B.4, it can be seen that a number of common themes are emerging in these alternative approaches, including self-organisation; learning and evolution; integrating existing dualisms (living with paradoxes); intuition; creativity; the recognition that nothing can be considered as fixed or stationary (systems, time, space); and the realisation that everything depends on perspective.

A number of these approaches were used to inspire the design and implementation of the Montpellier Pilot Trial decision-aiding intervention outlined in Chapter 6 and Appendix D: Japanese learning and innovation theory was used as a basis for the selection and placement of methods through the process design, in an attempt to incite individual and collective learning and innovation; spiral dynamics inspired a mix of different work methods to suit participants' underlying value-systems and working preferences (based on questionnaire responses in the first workshop); and the panarchy concept inspired the three inter linked-scales version of the oval mapping technique for investigating decisions at a personal, neighbourhood and regional (water basin) level.

Table B.4 *Potential approaches for inspiring decision-aiding in water management*

Approach and example references	Basis, aims and assumptions	Opportunities and challenges
Spiral dynamics and mimetics (Beck and Cowan, 1996; Dawkins, 1976; Blackmore, 1999; Rixon *et al.*, 2006)	*Basis:* Cultural psychology, information theory, development and evolution theory. *Aims:* To give a conceptual basis to understand and manage cultural value systems and their development, information transfers and adoption behaviours. *Assumes:* The existence of memes (or ^{v}memes) – units of cultural, imitation behaviour (or values) that can be transferred and develop through human or societal development (mimetics – as opposed to biological genes and genetics).	*Opportunities:* Presents a useful basis for understanding and embracing cultural development and value-based concerns of individuals and groups, as well as to better study adoption behaviours. *Challenges:* For spiral dynamics – difficulty in analysing and 'labelling' groups as predominantly exhibiting particular value systems (that may be considered as higher or lower in the development chain) – risk of theoretical rejection in a participatory setting – theory does not consider physical processes or ecological evolution; for mimetics, the transferral mechanics proposed in the literature are harshly criticised.
Integral approaches (László, 2004; Wilber, 2000; Gidley, 2007; Roy, 2006). The integral review: http://integral-review.org/	*Basis:* Philosophy, psychology, physics, biology – integration of predominant paradigms and consideration of all aspects of human–environment systems: integration of dualisms – subjective–objective divide, and individual–collective – through development and evolutionary phases in the cosmos. *Aims:* To encourage a holistic understanding of the world from the material to the spiritual. *Assumes:* The evolution of consciousness and that all individuals are part of wholes in organised holarchies.	*Opportunities:* May provide a useful basis for common understanding between epistemologies – that there are many complementary ways of viewing the one world. *Challenges:* A number of the integral approaches exhibit inconsistencies in the presentations of their theories; understanding the bases of these theories is likely to involve intensive background reading, thought and interpretation.
Panarchy and complex adaptive systems (CAS) (Forrester, 1961; Gunderson and Holling, 2002; Matthews, 2004; Perez and Batten, 2006; Pahl-Wostl *et al.*, 2008)	*Basis:* Systems theory, ecology, human development and evolution – considers the evolutionary and self-organising nature of complex adaptive systems that are nested one within the other and interrelated across space and time scales. *Aims:* To provide a theoretical basis for understanding transformations in human and natural systems. *Assumes:* Four main stages in a 'panarchy' adaptive cycle – exploitation, conservation, release and reorganisation – and inter-scale interconnectedness dynamics; other CAS assume different process mechanics for social learning and self-organisation, emergence and transition behaviours.	*Opportunities:* Relatively simple frameworks and concepts for describing and understanding complex phenomena; comprehensive body of background theory to complex adaptive systems and techniques that can be used to model them; already gaining widespread support in the water sector as an alternative management paradigm. *Challenges:* The panarchy model of CAS still remains controversial and requires more empirical studies to provide supporting evidence; the approach remains a largely objectivist vision of the world with less focus on the integration of subjective viewpoints and the place of the researcher in the world.
Constructivist epistemologies (Piaget, 1967a; 1967b; Bateson, 1972; Watzlawick *et al.*, 1974; von Glasersfeld, 1989; Morin, 1990; Astolfi *et al.*, 1997; Le Moigne, 1999)	*Basis:* Philosophy of science, alternative learning theories, mathematics, physics, biology, cybernetics. *Aims:* To give an alternative view on how knowledge is created (from the realist and positivist epistemologies).	*Opportunities:* Provides an alternative epistemological basis for understanding knowledge creation and its relation to social systems – already gained wide support in the social and management sciences and as the basis of studying collective action.

Table B.4 (*cont.*)

Approach and example references	Basis, aims and assumptions	Opportunities and challenges
	Assumes: All knowledge is constructed (cognitively or socially through learning). In the more extreme forms, knowledge is not a 'representation' of the real world, but rather a collection of conceptual structures that are adapted or viable within the knower's field of experience.	*Challenges:* Objective 'facts' do not exist. All knowledge is contingent on the constructions of the knower.
Japanese innovation and learning theory (Nonaka and Takeuchi, 1995; Nonaka and Toyama, 2003; Gourlay, 2006)	*Basis:* Japanese philosophy, psychology, organisational theories. *Aims:* To foster innovation and learning. *Assumes:* Oneness of humanity and nature, oneness of mind and body, oneness of self and other; knowledge is considered as a dynamic human process of justifying personal belief towards the truth (it is about intention, action and meaning); innovation is created through processes of knowledge 'conversion' between and within tacit and explicit knowledge.	*Opportunities:* To learn how to foster creativity and create innovation and learning on the individual, group, organisational and inter-organisational levels; a relatively easy-to-understand theory. *Challenges:* Heavily criticised by some authors, owing to a lack of comparative grounding with other learning and organisational theories, as well as lack of empirical evidence.
Indigenous Australian philosophy (Yunkaporta, 2006, 2007a, 2007b; Jackson *et al.*, 2005; Rose, 1996)	*Basis:* Ancient philosophy. *Aim:* Basis of understanding life in the universe. *Assumes:* Rich variety of complementary knowledge systems (holistic, synergistic, communal, ancestral, logic 'webs' of CAS (biomimicry), circular logic, pluralism (between language and cultures) and deep narrative; connections and integration of 'country', 'the dreaming' and their places in the universe where everything is alive and intelligent.	*Opportunities:* Presents a cohesive understanding of the world that has allowed the sustainable management of life in the Australian environment for thousands of years. *Challenges:* Extreme differences from the Western way of life and the representation of their place in the world and the universe, which may limit comprehension capacity for many and increase the likelihood of misunderstandings.

Appendix C Understanding participatory modelling

This appendix supplements Chapter 3 by extending the brief definitions and outlines of participatory modelling previously provided.

C.1 THEORETICAL BASES OF PARTICIPATORY MODELLING

Some basic definitions of 'participation' and 'modelling' will first be presented to outline the differences between 'traditional' and 'participatory' modelling and their potential domains of use. Following this brief review, the subject of how participatory modelling can, as well as why it should, be used for water resources management and planning will be treated, and potential participants will be identified.

C.1.1 Participation

> *Appropriate policy in a democracy is determined through a process of political debate. The right course of action is always a matter of choice, never of fact.*
>
> Davidoff (1965)

The debate on what constitutes true 'participation' and why it is important, especially in the political decision making sense linked to theories of democracy, has intensified since the 1960s, with many publications treating the topic explicitly (e.g. Davidoff, 1965; Arnstein, 1969; Pateman, 1970; Borton and Warner, 1971; Dryzek, 1990; Fischer, 1990; Beierle and Cayford, 2002). This issue is intimately related to some of the hypotheses behind the use of, or need for, participatory modelling in the production of models used to aid collective decisions. A number of these have already been presented in Chapters 2 and 3. This section aims to outline a number of classifications of participation that stem from work on 'public', 'citizen' or 'stakeholder' participation, principally from the urban or rural planning and environmental and risk policy and management literature bodies, in order to better understand what is meant by 'participation' and inform the creation of a classification of 'participatory modelling'. One commonly cited definition (e.g. Evan and Manion, 2002; Linnerooth-Bayer *et al.*, 2005; Mazri, 2007) considers that 'public participation' is constituted of '*forums for exchange that are organised for the purpose of facilitating communication between government, citizens, stakeholders and interest groups, and businesses regarding a specific decision or problem*' (Renn *et al.*, 1995). This leaves the types of 'exchanges' to be specified.

One of the key assumptions behind the theoretical musings, practical research and implementation in much of the literature on participation in planning and decision making is that some traditional forms of representative democracy in Western countries are proving inadequate to meet the needs and interests of citizens directly impacted by government decisions, which leads to the need for different forms of 'direct' or 'participative' democracy. For example, the development of 'advocacy planning' (Davidoff, 1965) was '*an attempt to increase the power of deprived or suppressed citizen groups by fighting apathy, guiding their complaints, and formulating their ideas to the bureaucracy*' (Khisty, 2000). Similarly, 'transactive planning' (Friedmann, 1973) and 'radical planning' (Grabow and Heskin, 1973) attempted to promote decentralisation and community democratisation in different ways, to empower individuals and communities and promote social learning (Friedmann, 1993).

Related to this perceived need for a transfer of power in urban planning from centralised administrations to deprived citizens, Arnstein (1969) developed her 'ladder of citizen participation'. The scale (represented in Figure C.1) is based on power distribution, in the general sense outlined in Dahl (1961), between traditional decision makers, such as government authorities, and the generally affected public, whom Arnstein refers to as the 'have-nots'. The three general tendencies of decision makers' use of public participation in planning programmes are represented as 'non-participation', 'tokenism' and 'citizen power'. A translation of the equivalencies of these levels in the French system is presented in Martin (2003).

Many other power-based classifications of public, citizen or stakeholder participation in decision making processes have since been proposed or adapted to specific domains, although

Table C.1 *Five-step classification of public participation in water management. Adapted from Mostert (2003b)*

Level of participation	1. Information gathering, dissemination	2. Consultation and hearings	3. Discussion	4. Co-decision making	5. Decision making
Outcomes	The public is provided with or has access to information	The views of the public are sought	Real interaction takes place between the public and government	The public shares decision making powers with the government	The public performs public tasks independently
Approach	Leaflets, brochures, mailings, briefings, use of media, Internet, etc.	Reply forms, opportunity to comment in writing, hearings, meetings, interviews, opinion polls, stakeholder analysis, internet discussions	Small or large group meetings, workshops, roundtables, brainstorming sessions, internet discussions	Negotiation, e.g. resulting voluntary agreement, stakeholders represented in government bodies, small or large group meetings	Water use association and other NGOs performing public functions, popular initiative

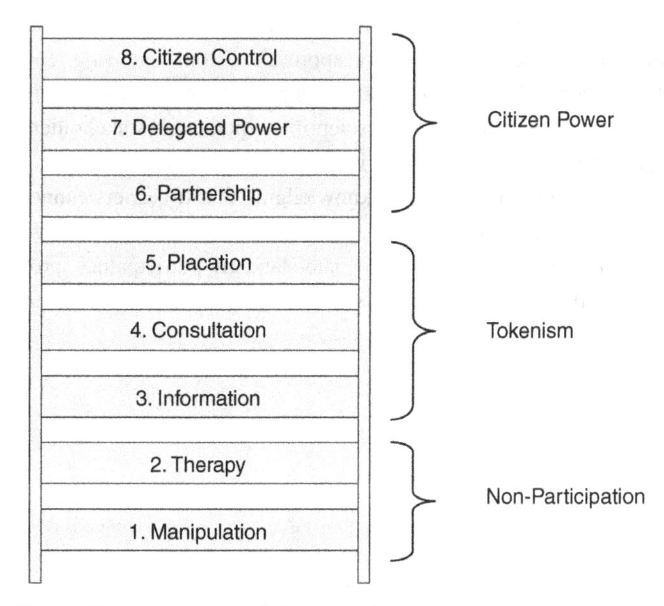

Figure C.1 The ladder of citizen participation. Adapted from Arnstein (1969).

other types of power, such as in 'non-decision making' (Bachrach and Baratz, 1962) are not considered. For example, the five-step version of Mostert's (2003a; 2003b) classifications for water planning and management is outlined in Table C.1, and, in operational research, Mazri (2007) proposes a four-step classification for decision-aiding processes for risk management, where: level 0 is 'information'; level 1 is 'invited response to information'; level 2 is 'consultation'; and level 3 is 'implication in decision making'. Treating power in a different, more 'experiential' manner, Rocha (1997) developed a 'ladder of empowerment', where empowerment is treated as a social action process that may occur at a number of levels, from the individual to the community. Cornwell's (1996) six-step classification is more similar to this second version of power through participation, as co-learning and collective action are included: 1. co-option; 2. cooperation; 3. consultation; 4. collaboration; 5. co-learning; and 6. collective action. Further insightful review and analysis on power and learning in participation, plus a number of other participation classifications, can be found in Kelly (2001) and Ker Rault (2008). As well as these classifications, other disciplines have treated the underlying democratic and citizen participation issues in modern technocratic societies. For example, in the political sciences a succinct overview of the history of development of the need for more participative forms of democracy in environmental and risk management decision making is given in Fischer (2000). Similarly, a brief review of public involvement in science and technology policy from the evaluation sciences perspective is found in Rowe and Frewer (2000).

Despite power being an important means of classifying participation, it is not the only factor influencing 'quality' participation. Other authors have chosen to broaden the definition. For example, Pateman (1970) classifies participation types based on combinations of 'levels of interaction' and 'political power', as outlined in Figure C.2.

Fung (2006) has gone further, describing various 'approaches' to participation through his 'democracy cube', which is based on axes of: authority and power; types of participant; and communication and decision mode. He suggests that it could be used to inform institutional design choices for public participation planning initiatives. His use of the cube to represent the difference between government or private agency work and public hearings is shown in Figure C.3.

Similarly, to inform institutional design and participation processes, it is important to determine the objectives, the resources available to support the participation process and the phase of the decision making process in which the methods are to be applied. A number of these are outlined in Table C.2.

Depending on the objectives, resources available, and decision making phase targeted for the participation process, different methods may be selected. A number of methods, ranging from less interactive to more interactive, are outlined in Table C.3, along with what they might ideally be used for, as well as the key challenges associated with each method, its potential cost and how many people can participate. Further discussion of these methods and many more are available in a range of publications (e.g. Forester, 1999; Chambers, 2002; Aslin and Brown, 2004; Creighton, 2005).

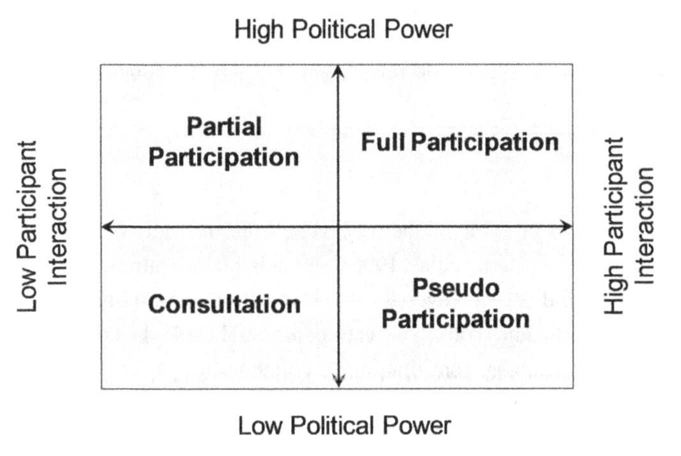

Figure C.2 Participation classification based on interaction level and political power. Adapted from Pateman (1970).

Some methods in Table C.3 allow for participants to be chosen at will (e.g. workshops, Delphi, mail-outs), some are typically open to all (e.g. broadcasts, town hall meetings, some online gaming or forums) and others have specific methodologies for the selection of participants (e.g. citizens' juries or consensus conferences). Most methods require careful design, implementation and monitoring to ensure that they have the best possible chance of meeting their planned objectives. It is very common for a suite of different methods to be employed with different participants for separate stages of the decision making process. A range of documents (e.g. Aslin and Brown, 2004; Tan *et al.*, 2008; von Korff *et al.*, 2010) explains how this might be carried out, along with the case study examples provided in this book. A range of expertise is available internationally in water management, community development and business that can be mustered for these processes, including facilitators, mediators, decision analysts, communications experts and participatory process management specialists. Nevertheless, despite the best intentions, some barriers can prevent effective stakeholder participation and collaborative approaches from occurring, and these need to be understood and managed. Examples of barriers for both decision makers and scientific experts, and stakeholders and the public, are presented in Table C.4.

At this point it is worth acknowledging that if barriers cannot be overcome, it may end up being better not to try to engage stakeholders than to convene sub-standard participatory processes that are likely to disappoint stakeholders or lead to

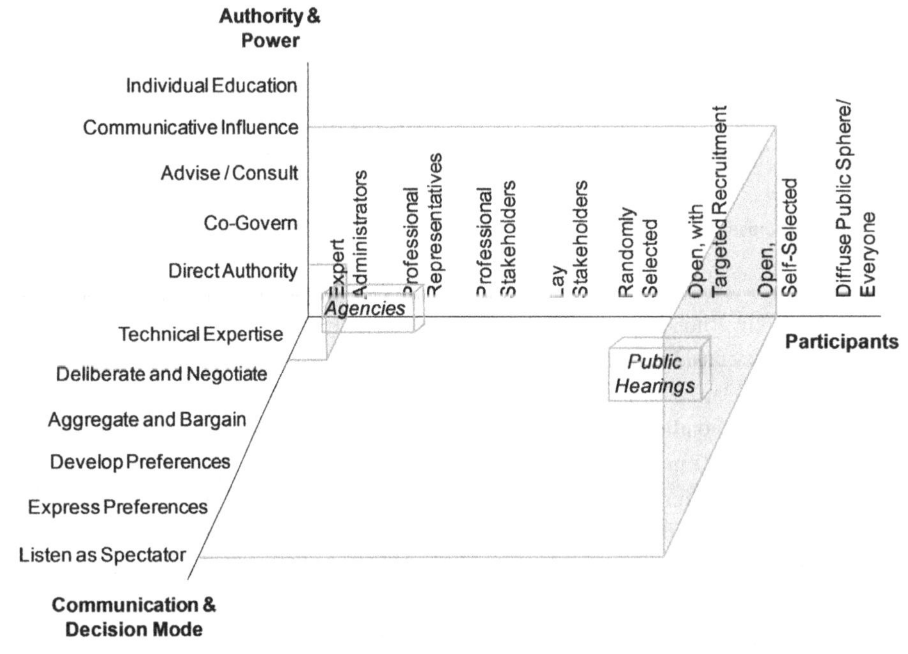

Figure C.3 The democracy cube for informing institutional design. Source Fung (2006).

Table C.2 *Potential objectives, resources and decision making phases that can influence appropriate choice of methods and participants. Source: Daniell (2011)*

Objectives	Resources	Decision making phases
• Information provision • Education • Improving two-way communication • Social learning • Enhancing legitimacy of decisions • Enabling democratic governance • Conflict resolution • Legal or organisational requirements • Building personal relations and social capacity • Achieving an improved water management outcome	• Time and finance • Skills in designing and using methods • Organisational will and leadership • Existing trust levels and relationships • Power to make and implement decisions • Knowledge of the policy area • Stakeholder interest and capacity in engaging	• Identifying and structuring issues and values • Situation analysis • Visioning for the future • Eliciting preferences • Developing and assessing management options • Negotiating choices • Implementation planning • Monitoring and evaluation

'over-consultation', as more damage than good is likely to result (see Barreteau *et al.* (2010) for further discussion on this point).

This book will tend to refer to 'participation' in the case where more than just 'consultation' is occurring; where there are quality two-way interactions and knowledge transfers. The question of 'pseudo-participation', which could also be matched to certain forms of non-participation or tokenism on Arnstein's scale, has been further discussed in this book when looking at ethics in participatory research (Section 9.2.2).

Examination and definitions of 'modelling' in its general and participatory forms will now be addressed.

C.1.2 Modelling

Modelling can be defined as the process of developing and providing an abstraction, representation, 'meta-object' or model of reality. Costanza and Ruth (1998) claim that the building of models is an essential prerequisite for human understanding and for making a choice between a number of options or alternative actions. For an individual, these models often constitute 'mental models' that are formed through extracting the relevant elements of previous observations and experience, as well as their interrelations (Johnson-Laird, 1983). A mental model relating to a particular question or problem can then be analysed through these relations to elicit an answer or to choose a relevant solution or course of action (Chown, 1999). The majority of mental modelling occurs qualitatively, with relationships and causalities between variables, space and time being simplified so that rapid analysis of alternative options is possible (Zwaan and Radvansky, 1998). In situations where quantitative values become important to decision making and system understanding, or the relationships between elements become more complex and variance over spatial and temporal scales needs to be analysed, mental models may be insufficient to underpin adequate analysis or decisions (Forrester, 1992a). To provide useful abstractions of these more complex realities, other types of modelling are used. Varieties of modelling methods may range from making these mental or 'cognitive' maps explicit, using words, diagrams, concrete structures or mathematical equations that can be used to communicate ideas with other people, to create reusable theories or to aid understanding and more complex decision making. If basic understanding, scientific theories and mental models of certain system subsystems are insufficient to create more explicit models in other forms, such techniques as statistical modelling using series of observed data may be able to reveal some of the missing links. Some modelling methods have already been briefly outlined in Section 3.1.2 and Section B.3.

As all models are only different abstractions of reality, certain hypotheses and assumptions are always present in their construction. Models can be analysed by adjusting or changing these hypotheses, assumptions or various initial parameters. Because all models exhibit underlying assumptions or hypotheses, they can be challenged or rendered illegitimate by someone who does not agree with or accept them (Korfmacher, 2001). This property of models is of extreme importance when they are to be used by a third party or a number of parties to aid decision making. If the model is deemed unsatisfactory by these parties, then the decisions informed by the model are also open to challenge. Landry *et al.* (1996) pointed out that models may serve different purposes concurrently and can be thought of as instruments or tools, which, throughout their lifetime, permit various modes of human–model interaction, as depicted in Figure C.4.

From Figure C.4 it can be seen how models can be used as enabling or constraining devices and in a number of modes. They

Table C.3 *Common 'less interactive' and 'more interactive' methods for stakeholder engagement with some of their properties. Source: Daniell (2011)*

	Methods	Ideal use	Key challenges	Cost	Participant numbers
Less interactive	Mail-outs, press articles, broadcasts	Broad scale information distribution, awareness raising	Tailoring information to audience, finding attractive format	$–$$$$	Few to very large numbers
	Information stands, road shows	Providing overview information, providing people to explain information	Making information easily understandable to people – finding knowledgeable people able to answer questions	$–$$$$ (depending on length or size of road show)	Potentially large numbers
	Town hall, public meetings	Providing overview information, providing people to explain information	Can heighten conflict if information is contentious or disputed, difficult to hear many voices	$–$$	Dependent on size of meeting hall. Typically <2000
	Public presentations, Q&A sessions	Providing information of interest and encouraging some debate	Requires a good facilitator to maintain a positive Q&A session	$–$$$	Dependent on room size unless televised. Typically <200
	Mail, phone and in-person surveys	Eliciting information from a targeted population	Obtaining expertise to develop and administer a useful and well-constructed survey	$$–$$$$	Dependent on survey design and resources to carry it out. Potentially large numbers
	Delphi analysis (typically experts)	Developing a structured expert view on an issue	Facilitation of method use and choice of experts	$–$$$	Three to many (especially web-based Delphi).
	Consultation by written submission	Eliciting feedback with a view to considering new information and differing opinions in decision making	Not being a superficial process, synthesis and treatment of submissions	$–$$$$ (depending on synthesis work)	Potentially large numbers
More interactive	Citizens' juries, consensus conferences	Developing judgements on controversial or little publicly examined topics	Organisation of the events, having political buy-in to considering decisions, recommendations	$$–$$$$	Approx. 10–150
	Participatory modelling	Developing shared representations as a basis for joint investigations and informing decisions	Managing organising team and participant dynamics, effectively structuring complex information	$–$$$	Approx. 5–50
	Facilitated workshops, focus groups	Encouraging dialogue and collaborative work, including making trade-offs through use of techniques such as multi-criteria decision analysis	Establishing agreed workshop aims and finding effective facilitators who can work with participants to achieve them	$–$$	Approx. 10–30 per workshop. More can be handled in parallel by multiple facilitators

Games (role-playing, simulation, online. . .)	Developing understanding of a specific situation and impacts of actions	Finding resources for game development, having access to appropriate props, technology	$–$$$$ (depending on game development costs)	Variable depending on game and platform
Problem-structuring methods	Aiding decision making in complex, uncertain and conflict-ridden situations	Finding facilitators with a working knowledge of the required methods	$–$$$	Approx. 5–50
Visioning, scenario building, search conference	Developing and assessing potential futures	Finding facilitators with a working knowledge of the required methods	$–$$$$	Approx. 5–150
Participatory planning, GIS	Jointly developing action plans and spatialising information provided by participants	Finding effective facilitators and having access to GIS technology, maps or spatial models	$–$$$$	Variable depending on scope of planning process – large numbers can participate through online GIS systems
Participatory evaluation	Encouraging participant reflection and learning with a view to applying lessons in the future	Finding facilitators and evaluation specialists with a working knowledge of the required methods	$–$$$$	Variable depending on scope of evaluation
Discussion forums (online, TV, in person. . .)	Encouraging dialogue, debate and mutual learning	Finding discussion facilitators and mediators, developing appropriate platforms for interaction if online or on TV	$–$$$$	Variable depending on media – effective interaction likely with relatively small numbers
Multi-lateral negotiations	Developing joint agreements or treaties	Finding effective mediators, chairpersons and information synthesisers or drafters	$–$$$$$	Variable. Typically 3 to 200
World cafés	Developing conversations and collective understanding of multiple interconnected issues	Organisation of the event, continuing the conversations after the end of the event and acting on knowledge	$$–$$$$	Variable depending on aims. Can range from 12 to hundreds or thousands
Advisory panels	Developing and synthesising knowledge and opinions to inform decision making processes	Selecting a broad range of advisors who will be able to work effectively together and add value to decision making	$–$$$$	Variable depending on aims. Can range from 5 to hundreds or thousands

Table C.4 *Barriers to stakeholder participation and collaborative approaches. Source: Daniell (2011)*

For decision makers and scientific experts	For stakeholders and the public
Lack of will to involve others in decision and science processes	Lack of interest or time to become involved in such processes (especially in an 'out of crisis' period)
Lack of organisational support and leadership	Previous bad stakeholder engagement experiences
Inadequate resources, including time, finance, knowledge and a lack of training as facilitators and with participatory methods	Lack of other resources to participate (e.g. knowledge, technology, financial support)
Lack of personnel continuity (difficulties building and maintaining relationships and trust)	Mistrust in coordinators
Inability to manage stakeholder expectations and conflict	Scepticism that participation will make a difference
Legal, security or other institutional constraints	Nothing obvious in it for them

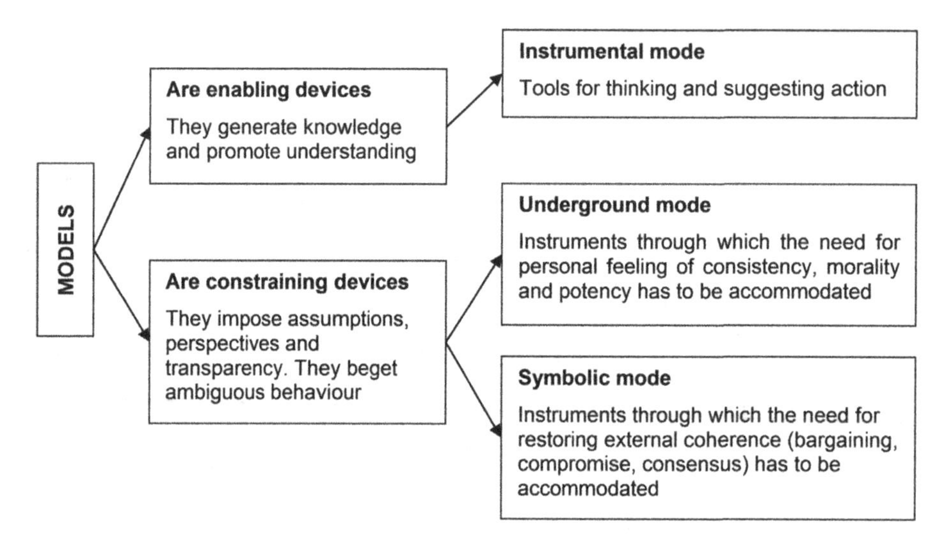

Figure C.4 Different uses of models. Source: Landry *et al.* (1996).

may be used to allow open reflection or suggestions of options for action in the instrumental mode, or they could be used in modes requiring accommodation of modellers' and decision makers' (model users') (uncommonly) observable traits of feelings and values in the underground mode and the more observable interests of each party, through methods such as negotiation or consensus building in the symbolic mode. Model legitimacy is likely to be based on the perceptions of the model in all three modes described in Figure C.4, and not just in the instrumental mode. Traditional technical models are mostly produced with the maximum attention paid to the instrumental mode of use. The legitimacy of models in the public decision making or policy analysis domain is sometimes difficult to achieve, as the models have not been specifically built to allow the accommodation of various interests, values and feelings of all who will use them to guide their decisions (Mintzberg, 1979a). Increased attention to the underground and symbolic modes of model use is thus required.

For example, a lack of attention in determining the 'real' or 'right' problems to study, right from the beginning of the problem identification stage of a modelling process for decision-aiding, may also have been the concern hidden beneath the question, 'Why are so many models built and so few used?' (Lilien, 1975). Here it is the user of the model who must be made explicit, as usually at least the model's designer will use it for something. Behind this question lies another reason why many modellers, both in water management and planning applications and in other domains, are currently looking at the possibilities of participatory modelling as a way to increase the explicit and meaningful use of their models by others. However, several warnings have been issued over the need merely to involve people so that a model can be improved or created; 'a model for model's sake', rather than focusing on the entire process of problem solving and decision-aiding from the earliest problem identification stages (Nancarrow, 2005).

Participatory modelling is different from traditional modelling, in that the source of information, assumptions and hypotheses in the model do not stem from just one point of view, representation or set of values. It is a process of collective creation of a common model by a number of people or stakeholders with varying interests and world views. Participative modelling requires the management of conflicting points of view, as well as the assembly and coordination of views, differing value systems and various understandings, without necessarily integrating them. The methods, processes and needs required to achieve such socially orientated objectives are commonly different from those required in unique viewpoint modelling methods. Owing to the inherent involvement of a variety of stakeholders in the modelling process, this type of modelling is more likely to exhibit increased levels of underground and symbolic modes of usage, outlined in Figure C.4. Loucks (1992) noted that although models produced in such settings are more apt for the problems addressed, they must also be technically valid, not only from the perspective of modelling process participants, but also by peers or experts in the field of analysis. A balance must therefore be achieved between ensuring model legitimacy and model validity in participatory modelling exercises.

Although the theoretical basis of 'participatory' modelling, defined as the creation of a common model taking into account more than one individual's point of view, is relatively clear, how this process can be achieved, and to what degree it is really 'participatory' is relatively less so. For example, the amount of participation required to 'take a view into account' can vary dramatically from brief verbal consultation at any period during initial modelling stages with a stakeholder of the model, to the model's stakeholders co-constructing and being physically involved in every stage of model development from problem definition to decision implementation.

In its purest form, the participatory modelling process could be considered as a continuous spiral of collective decision cycles, related to every section of the modelling. Each aspect of the problem to be analysed is agreed upon collectively, along with the objectives of process, and each model hypothesis is agreed upon, as well as how the model is to be used to aid final collective decisions related to the objectives. If at any stage individual decisions occur that are not discussed and accepted collectively and that have an impact on changing the process or model form, then the participatory nature of the process in its purest form will not have been maintained. In reality, the purest form of participatory modelling remains almost impossible to achieve, owing to a large number of factors, including undiscussed facilitator interventions and time constraints. For example, any analyst, facilitator or modeller involved in a participatory modelling process has his or her own values and perceptions of what is occurring, and therefore cannot be considered neutral. Unless the facilitator's (or analyst's or modeller's) objectives, decisions and proposed processes are discussed and accepted by the participants, some individual decisions are highly likely to impact the process. In terms of time constraints, participatory modelling is notorious for the large amount of time required to discuss all details of problems or hypotheses of the model, as many conflicts may occur or the number of model variables may become excessive. Total convergence on all such details is rare within a set timeframe, unless a model is particularly small and simple. It is usual then only to wait until acceptance is achieved on a set number of major issues, with other minor issues often decided at the discretion of certain participants, such as the analyst, the facilitator or the modeller. In some cases, this may have some negative impacts on the process, such as reduced participant comprehension and acceptance of the model. In other cases, there may be benefits, such as decreasing the repetitiveness and length of sessions to encourage sustained participant interest and attention levels. It is noted that an excessively time-consuming process or ineffective participatory process is likely to involve less and less motivated participants. On the other hand, if participation is effective then the process can be considered really participatory.

C.1.3 Why apply participatory modelling to water resources planning and management?

The most fundamental flaw in contemporary water policy is that many value questions in which ordinary citizens have a great interest are being framed as technical questions.

Ingram and Schneider (1999)

Unlike traditional modelling carried out by one person or institution, which may or may not include information from other stakeholders and which is used for decision-aiding, 'participatory modelling' allows a number of different points of view to be explicitly represented and collectively reflected upon by a group of stakeholders before a collective decision is made (Ferrand, 1997). Such an approach to modelling is thought to be suited to dealing with the increasingly messy situations facing the water sector, which have proven difficult to manage with traditional, technical, scientific methods, largely because stakeholders' knowledge, perceptions, preferences and values become too important not to be taken into account. Under the conditions of uncertainty, complexity and conflict in the water sector, traditional water resources models designed to aid decision makers only focused on evaluating management alternatives – e.g. to optimise a certain number of objective functions, to choose the best 'economic' option, or even to give the decision maker a range of options corresponding to different risk levels – rather than supporting the full decision process from the definition of problems. Such isolated evaluation models are commonly overlooked or shelved, as they are incapable of taking into account a multitude of social and environmental factors that can commonly carry more weight in a political debate than 'the best technical solution', from just one or a limited number of points of view of the modellers. In particular, technically or economically optimised solutions are not likely to be able to adequately treat the messy kinds of water-related 'problem situations' outlined in Chapter 2, Chapter 3 and Section B.3.

Messy water planning and management problem situations instead require approaches to decision-aiding that can allow the representation of various points of view, mutually agreed classifications of the problems and the development and implementation of collectively acceptable or 'reasonable' solutions (ADVISOR, 2004; HarmoniCOP, 2005b). As models are most commonly used to aid water management decisions, a logical advance towards trying to meet these new demands is to involve stakeholders with conflicting views in the process of building these models. This process of the co-construction of models by multiple stakeholders, or participatory modelling, has now been applied in a variety of different forms to a reasonable number of water resources problems around the world, with varying perceived levels of success. Examples will be outlined at the end of this section.

One of the reasons why the 'success' of participatory modelling for water resources management and planning applications is so difficult to gauge is that behind the use of participatory modelling there commonly lies a multiplicity of assumptions and objectives. For any particular situation, these objectives could include but are in no way limited to:

- Gaining a common understanding of problems;
- Solving the 'right' problems;
- Developing a platform or tool to aid communication, negotiation or consensus building;
- Explicating tacit knowledge, preferences and values;
- Improving the legitimacy of a model;
- Reducing conflict;
- Education;
- Enhancing both individual and social learning;
- Promoting creativity and innovation;
- Building social capacity;
- Investigating the effects of group behaviours or scenarios on a 'micro-world' system without having to use the real world;
- Informing decisions; and
- Instigating action or improving the adoption of chosen problem solutions or management options.

Although these are stated as objectives of participatory modelling, the degree to which participatory modelling may help to achieve such objectives, especially when compared with other modelling and non-modelling methods for water resources management and planning, remains to be systematically evaluated. The form of the process and participation is also likely to vary markedly, as will be discussed briefly in Section C.2.

C.2 PARTICIPATORY MODELLING CLASSIFICATION

Despite steadily increasing volumes of literature on participatory modelling, few people have attempted to define the concept explicitly, in terms of who is participating and in which phases of the water planning, management and modelling process the participation occurs (Daniell *et al.*, 2006b). A large quantity of work exists on what 'participation' means in decision making terms, as discussed in Section C.1.1. In most examples, the emphasis is on who has the balance of power for final decision making, i.e. the 'choice' phase of a decision process (Simon, 1977), but other issues of process are not specifically mentioned. These participation classifications, although useful in a very general sense for the question of 'participatory modelling', do not explicitly treat the issue of the place of a modeller (i.e. scientist, engineer or analyst) or expert knowledge. Other disciplines, such as operational research, have spent much more effort on focusing on the place of models and expert knowledge in decision making processes, although less effort has been paid to an analysis of 'participatory modelling' as a concept, especially in the inter-organisational, multi-stakeholder group context that is commonly encountered in water resources and planning.

C.2.1 Defining participants

In the policy and sociology domains, the place of technical knowledge, management and the public or 'citizens' has been well analysed (Dewey, 1927; Fischer, 1990; Mermet, 1992), especially with respect to who is involved in a decision making process (as distinct from a decision-'aiding' process). For example, in public policy and management decision making processes related to complex water management and planning problems, it is suggested that three main groups of participants could be represented: scientists, external experts or researchers; governments, policy makers and managers; and community stakeholders and the general public (Thomas, 2004). Similarly, Dietz *et al.* (2003) consider participants as 'scientists', 'officials' and 'interested parties'. The category of 'officials' or 'policy makers and managers' is most problematic, as it could be easily divided into three further groups on the basis of their legitimisation: elected officials with power over final decision making; employed managers with high decision making power (public and private); and government-employed bureaucrats or policy advisors who may have significant political influence but little real decision making power. In some cases, these bureaucrats could also be shifted to the 'scientists' category, depending on their expertise. Other much more complex categorisations, including important actors, such as the media, could also be envisaged (Althaus *et al.*, 2007), and classifications on choice of participants will probably need to form part of the discussions

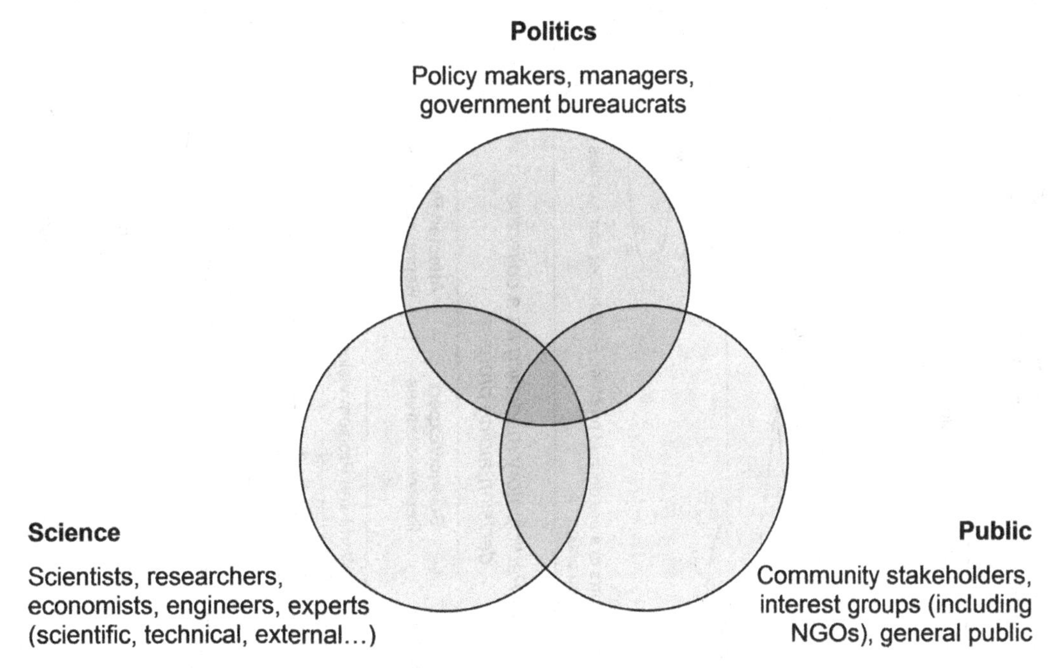

Figure C.5 Possible interactions for water management. Adapted from Thomas (2004).

on participatory process design (Mazri, 2007). As an extremely simple and idealised scheme in terms of 'politics', 'science' and the 'public', visualisation of the potential interactions of some of these parties that could occur at any stage in these processes is given in Figure C.5.

Continuing to consider the potential interactions of the three sets of highly idealised parties in Figure C.5, Thomas (2004) presents a critique of water management and planning scenarios under either individual, or combined, actions by the three parties. He maintains that any set acting entirely on its own is likely to be more ineffective in significantly improving water management than when sets work together. Each combination of sets working in pairs may have certain advantages and disadvantages: for example, a collaboration between science and politics is likely to be more efficient and will ensure that decisions are made based on sound scientific knowledge but does carry the risk that public backlash could occur if the decisions are deemed unacceptable by the wider population. Collaborations between science and the public are likely to improve the knowledge of both sectors, which can potentially drive changes to management and policy if lobbying takes place, although if unsuccessful, the lack of decision making power will prove a downfall. For collaborations between only the public and politics, policy is likely to be acceptable to the public but lacking in any scientific basis, which could result in negative impacts, such as environmental degradation and poor or technically infeasible solutions. Combinations of all three parties at some stage throughout the water management and decision making process, although potentially more time-consuming, are likely to produce the best outcomes, especially for complex and uncertain water problems.

Generally, no well-trialled procedures exist for finding the 'right' set of participants for a participatory modelling exercise, especially for the case where such a process is being used as a collective decision-aiding process. However, it is thought that a diverse selection of participants from a well-performed stakeholder analysis related to the perceived problem situation under question is likely to be a good starting point (Hare and Pahl-Wostl, 2002). In ensuring that science, politics and the public are represented in water planning and management inter-organisational working groups, it is also hypothetically likely that the potential for knowledge transfer and integration to help a decision-aiding process could be maximised, although this claim remains largely unevaluated. Other suggestions on current 'best practice' and potential risks to avoid, when considering participant choice, are outlined in Allen *et al.* (2002), Hare *et al.* (2003) and van Asselt *et al.* (2001).

C.2.2 Defining a four-stage classification

From the literature briefly reviewed in the previous sections, a four-stage classification of increasing 'participation' of actors in a decision making process is presented in Figure C.6.

The classification is based on a five-stage process, similar to the typical 'phase' decision processes outlined in Table B.2: world and problem vision (i.e. Simon's (1960) 'intelligence'); model design; solution design; choice; and action, with three

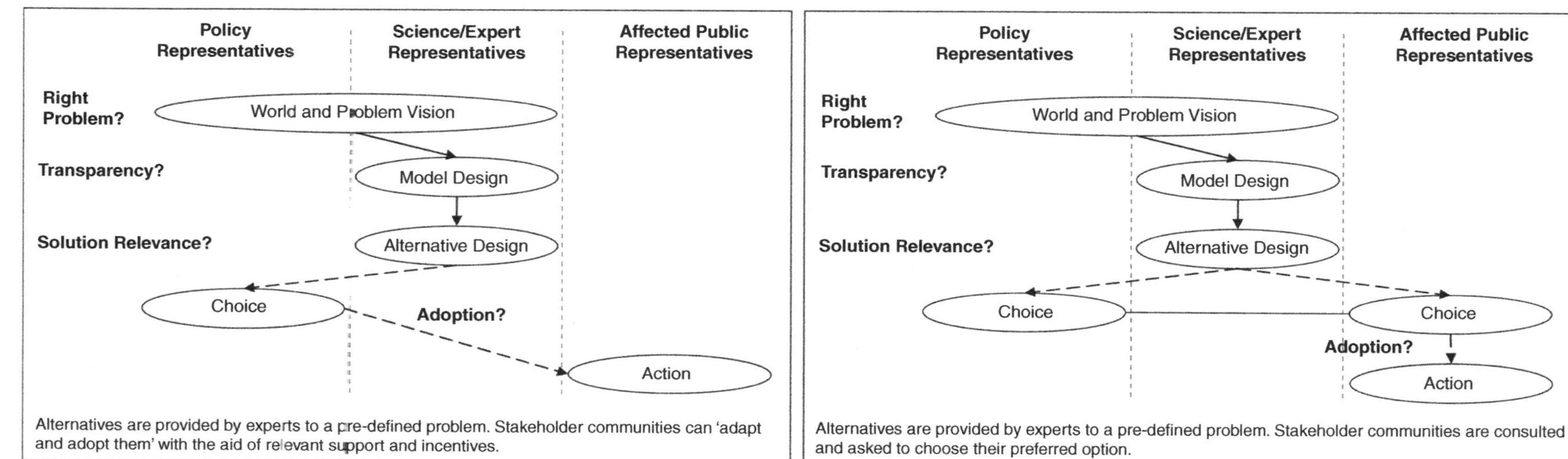

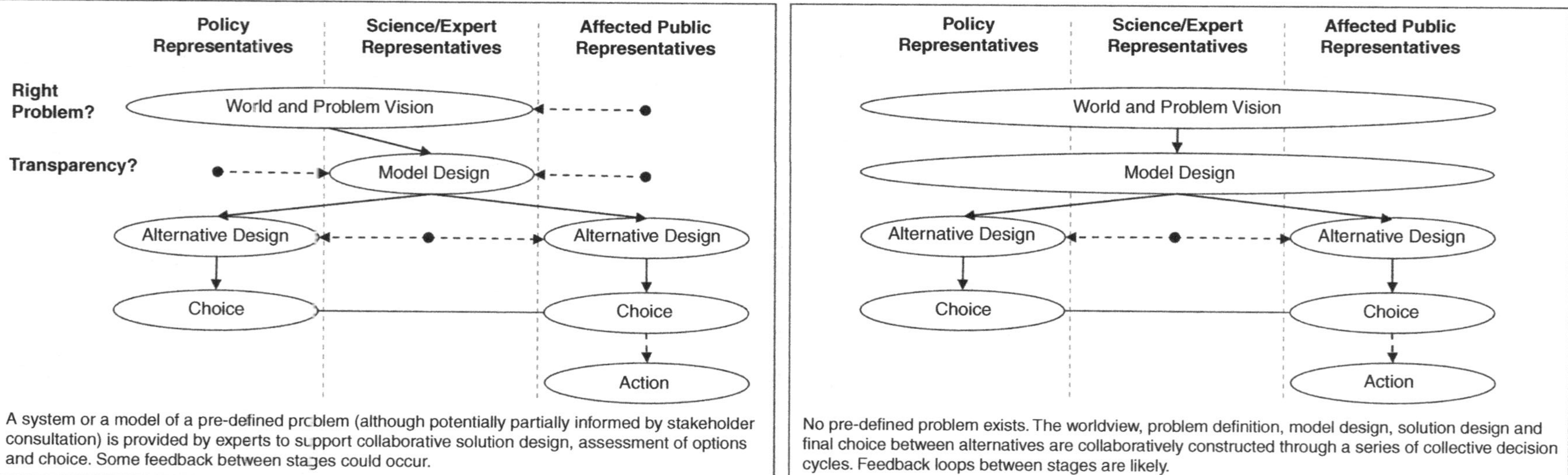

Figure C.6 Four-level modelling classification, with increasing levels of stakeholder engagement.

main groups of actors who could be involved in a participatory modelling exercise: policy makers and managers; technical experts; and stakeholder communities. Figure C.5 provides a more complete list of the types of actors that each of these generalised categories could include. These are not the only possible classifications that could be imagined, as: some actors may fall into more than one category; the process could be divided into more or fewer stages; there could be feedback loops between the process stages; or, as a result, more intermediate levels of classification could exist. These four levels of classification, presented and further explained here, have been specifically chosen because they match some of the more commonly observed modelling processes in water management and planning.

LEVEL 1: TRADITIONAL EXPERT MODELLING

The first level of the modelling process classification is in fact not at all participatory in nature. It corresponds to traditional forms of modelling in water management and planning, where the managers tell the modellers or engineers (technical experts) what problem needs to be solved. It is noted that engineers or technical experts in a government bureaucracy may determine the problem themselves (owing to their combined management-expert role). The technical experts then analyse this pre-defined problem and develop any model(s) they require to develop a set of solutions (or range of options), which they can present to the managers. At this stage, either based on the experts' recommendations or on some other criteria, such as the budget or fear of public backlash, the managers and policy makers will choose their preferred option or solution, which will then be implemented. This phase of implementation commonly means that the affected public or 'stakeholder communities' should make some kind of change or carry out an 'action', for example, to meet the requirements of a new piece of legislation, or prepare for a new measure of water supply and demand management, such as a new dam, water-use restrictions or recycled water systems.

When dealing with complex disputed problems, often with high levels of uncertainty, this type of modelling process has been extensively criticised (Rosenhead and Mingers, 2001) and has often failed to reach a reasonable level of adoption of, and thus solution to, the problem. There are a number of reasons why this approach may fail to obtain the expected results, which include: actual or perceived irrelevance of the proposed or chosen solution alternatives, either from the points of view of the stakeholder communities or the policy makers and managers; social inertia and various forms of social opposition from the stakeholder communities who have not been included in the modelling or decision making process; lack of integration between the stakeholder communities' visions and the other actors' visions, i.e. the 'right' problem may not be addressed; lack of model transparency; and inadequate support from the policy makers and managers for implementation, i.e. adoption incentives may be unrealistic or real education needs are not covered.

LEVEL 2: TRADITIONAL EXPERT MODELLING WITH LIMITED STAKEHOLDER COMMUNITY INVOLVEMENT

The second level of the classification could be included under the umbrella of multi-criteria decision aid and, more generally, of classical decision support systems: options are proposed and a choice is offered. The model itself is not given or manipulated directly by the decision makers, who in this case include the stakeholder communities.

The main criticisms of such methods follow many of those from the first level, which include the fact that sound options exist outside the initial set of those proposed, as well as the relative difficulty of weighting options and finding a fair and acceptable social choice when the design may not cope with localities' specifications and idiosyncrasies. It is, however, the usual choice for public decision making, as 'public consultation' is often a legislated obligation, and scientific and technical experts' knowledge can be efficiently drawn upon.

LEVEL 3: INTEGRATED EXPERT MODELLING FOR COLLECTIVE DECISION-AIDING

The third classification level is less common (although it is currently gaining support through 'integrated water resources management approaches'), as it assumes an open scenario approach based on a given model of the situation. Scientific or technical experts provide an 'integrated' model (often taking into account information from stakeholder communities and policy makers and managers), which is used by decision makers and other stakeholders to explore and select scenarios or actions. The model is 'external' from this point of view.

The advantage of this type of modelling process is that the scope of actions and the solution design can be more controlled and appropriated by decision makers. However, it still assumes an a priori agreement on the model and its accuracy for the local situation, which leads to the question of whether the 'right' problems were addressed from the affected public's point of view and whether the model is sufficiently transparent to be understood and regarded as 'legitimate' by the public and the policy makers and managers.

LEVEL 4: PARTICIPATORY MODELLING AS A COLLECTIVE DECISION-AIDING PROCESS

The fourth and final classification level is what will be defined in this book as 'participative modelling', where the role of the scientists, engineers and technical experts is to provide support and requested knowledge for the local stakeholders and other

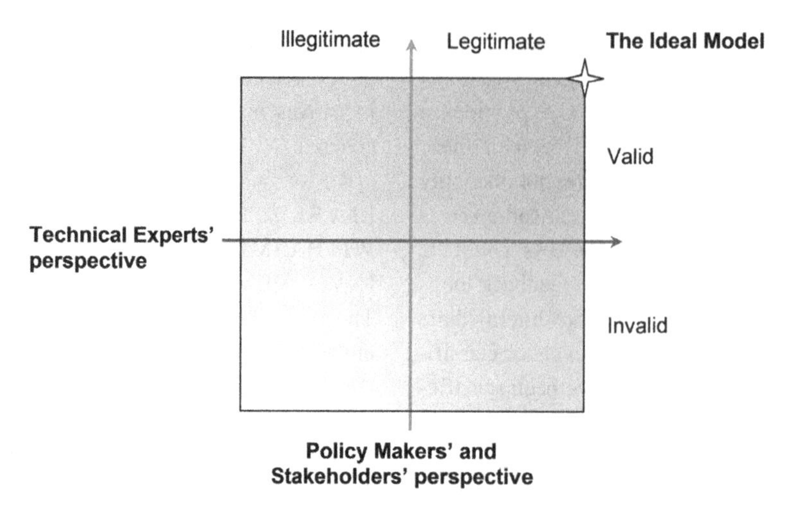

Figure C.7 Seeking the ideal model – stakeholder legitimated and specialist validated. Adapted from Landry *et al.* (1996).

decision makers to formulate, structure, build and use a model or several models (i.e. the 'meta-objects' outlined in Chapter 3) of their messy problem situation. The participatory modelling process constitutes a series of collective decision cycles of problem identification, problem definition, model structure and hypotheses, potential solutions, scenarios and their evaluation, culminating in final collective decisions on planned actions for implementation.

Unlike the other previous classification levels, in 'participatory modelling' the stakeholder communities and the policy makers and managers are involved in the collective 'world and problem vision', 'model design' and 'alternative design' phases before making their final collective 'choices'. It is hypothesised that participatory modelling:

1. Helps to examine the 'real' underlying problems;
2. Increases trust, appropriation and understanding of the models created, as assumptions and uncertainties are more likely to be explicitly identified and discussed;
3. Generates greater creativity and innovation;
4. Leads to an improved ability to respond to change through enhancing social capacity, adaptability, flexibility and resilience;
5. Leads to greater individual and social learning;
6. Produces richer and more realistic action plans; and
7. Provides a greater chance of adoption or implementation of problem management alternatives.

It is also noted that participatory modelling may have a number of drawbacks, including being more costly in time, money and personnel in the short term than the other classification levels (Korfmacher, 2001), although this is not always the case, as discussed in Section 7.5.3.

Another aspect to consider is that, although participatory modelling can potentially be used to avoid many pitfalls of expert-centric modelling processes, a balance between both participant discourse and scientific rigour should be encouraged. As models built in participatory modelling processes are attempting to incorporate previously lacking stakeholders' perspectives on the world and problems to be modelled, they tend to have a risk of over-focusing on these aspects in order to obtain the stamp of approval or legitimacy on the model from the participants, often to the detriment of scientific rigour and validation surrounding model structure, inputs and outputs. In theory, the design of a participatory modelling process should support the construction of a model that is concurrently viewed as legitimate by stakeholders and considered valid by technical standards, as represented in Figure C.7.

However, in real life decision-aiding, as applied in operational research, models constructed may often not meet the required standards for academic peer review, as they have not been created as models, but rather as analysis tools in larger decision processes. It is therefore suggested that, along with seeking stakeholder legitimisation of the participatory modelling process, two extra tests should be considered: whether the model and modelling process are theoretically sound and meaningful; and whether they are operationally complete and useful for participants.

C.3 PARTICIPATORY MODELLING METHODS USED FOR WATER MANAGEMENT

To expand upon the previous theoretically orientated descriptions and discussions of participatory modelling methods, a more practical look at two of the most commonly used processes for participatory modelling in water resources management and planning applications will be briefly outlined and discussed: those from the system dynamics group of methods, specifically the 'shared vision modelling' promoted by the US Army Corps of Engineers in the Institute for Water Resources and a derivative termed 'mediated modelling' (van den Belt, 2004); and those based on a multi-agent systems approach, in particular Barreteau's 'companion modelling approach' (Barreteau, 2003b). Another large group of methods that can be used in participatory modelling processes, 'problem structuring approaches' (Rosenhead and Mingers, 2001b) (mentioned in Sections 3.3.3 and B.3), some of which have been used in the water domain, have also been outlined in Section 3.3.4 to aid their comparison with the other two approaches.

C.3.1 System dynamics and shared vision modelling

Models based on Forrester's (1961) systems dynamics have been used in water resources management since the theory's conception in the USA (Yeh, 1985), although using them in a participatory setting by many stakeholders is much more recent. The first test of the 'shared vision planning' methodology, which includes the building of 'shared vision models', occurred between 1989 and 1993 in the 'National Drought Study' (Palmer *et al.*, 1993). Some of the study test applications were considered successful for the learning and agreements drafted and others less so, often because of final political vetoing or lack of adoption of model results (Werick and Whipple Jr., 1994). These methods have since been reused in many other areas of the USA and more recently in other countries. The US Army Corps of Engineers states:

> *Shared vision models are computer simulation models of water systems built, reviewed, and tested collaboratively with all stakeholders. The models represent not only the water infrastructure and operation, but also the most important effects of that system on society and the environment.*
>
> (Werick and Whipple Jr., 1994)

It is noted here that the final 'model' in this form of participatory modelling is considered a technical computer simulation model, unlike other participatory modelling methods, which may be used to build other 'non-computer' models. At the base of the shared vision model development the two principal questions cited as being the most important are, 'Who will use the model?' and 'How will the model be used?' (IWR, 2007). These pre-model considerations and the later model development take place as part of the shared vision planning structure with a group of stakeholders, which is based on the traditional US Army Corps of Engineers planning principles (Werick and Whipple Jr., 1994):

1. *Build a team and identify problems*;
2. *Develop objectives and metrics for evaluation*;
3. *Describe the status quo; what will the future look like if we do nothing?*;
4. *Formulate alternatives to the status quo*;
5. *Evaluate alternatives and develop study team recommendations*;
6. *Institutionalise the project or plan*; *and*
7. *Exercise and update the project or plan (adaptive management).*

In this planning process, the shared vision model provides a basis for formulating alternatives, measuring their potential effects and providing information to help evaluate alternatives. Throughout the process, in which the stakeholders are expected to participate from the team building and identifying problems stage, 'collaborative negotiation' and a 'deliberative decision process' takes place between the various stakeholder groups (Stephenson, 2003). The process is usually carried out during a number of workshops, with modellers and their colleagues working behind the scenes to finish the data collection for quantification of model parameters and linking functions, as time restrictions often prevent the model from being completed in workshop time. The common format of tools or modelling methods used is a communally built causal loop diagram – a particular form of cognitive mapping designed to elicit 'if–then' dynamic statements between variables – which is then translated into a quantitative 'stock and flow' type dynamic model, using platforms such as STELLA (High Performance Systems, 1992) and VENSIM (Ventana Systems, 1995), which allow simulations to be carried out with the results visualised on the computer. Human behavioural effects or decision patterns tend not to be explicitly considered in this kind of modelling, except when they can be statistically modelled in terms of water use or other easily quantifiable variables.

An offshoot process of this original systems dynamics inspired participatory process is 'mediated modelling' (van den Belt, 2004), which formalises these two steps of qualitative and quantitative model building, as well as the idea of 'working behind the scenes' in the process steps of model building:

1. *Agreement on software, ground rules, issues.*
2. *Development of a qualitative model in a series of workshops.*
3. *Modeller between workshops works on model.*
4. *Development of a quantitative model for 'What if?' scenario testing.*

The author also goes on to state what she believes such a method can achieve in terms of improving the degree of understanding of the system dynamics of the problems being studied and in increasing the degree of consensus among stakeholders, as shown in Figure C.8.

Although the potential advantages of mediated modelling over other techniques are presented in Figure C.8, the common pitfalls of the approach are not made so evident. One element of the

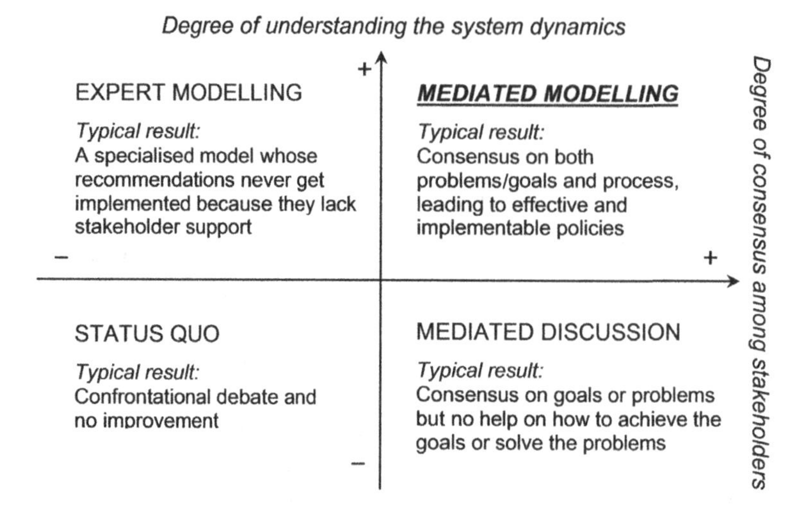

Figure C.8 Process results from mediated modelling. Adapted from van den Belt (2004).

system dynamics approaches that sometimes comes under criticism is the fact that it is often still a modeller, usually a technical expert, who is required to help guide participants through the modelling to a consensus. Apart from potential difficulties of bringing a group to a range of consensus decisions, the fact that more often than not the modeller is required to perform a large amount of work behind the scenes has, in many case studies, rendered the final model less transparent or unbiased than originally intended. Moreover, the assertion that many of these processes lead to 'implementable policies' remains highly contentious (Scholl, 2004). There is currently very little knowledge on what these modellers (and facilitators) do behind the scenes and little theory on how their contributions may be organised or supervised both in participatory workshops and outside them.

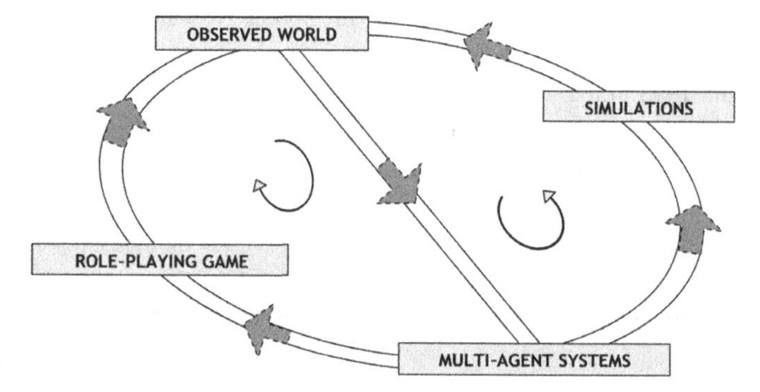

Figure C.9 The companion modelling approach. Adapted from Barreteau (2003b).

C.3.2 Multi-agent systems and companion modelling

Multi-agent systems, despite having many similarities with certain aspects of system dynamics, are based on the body of 'complexity' theory, relating to 'chaos' theory and self-organising systems (Ison *et al.*, 1997), which evolved quite separately from the system dynamics body of theory. While system dynamics attempts to characterise the behaviour of whole systems through their feedback structures (a 'top-down' approach), multi-agent systems focus on individual agents – representations of people or objects – with in-built rules. These agents interact with other agents, permitting visualisation of emergent macro system behaviour (a 'bottom-up' approach) (Scholl, 2001). Multi-agent modelling thus allows the explicit inclusion of behavioural modelling or qualitative information; two aspects that are commonly considered as shortfalls of system dynamics modelling processes.

Although the theory of multi-agent systems has been under development for at least 30 years, it was not until the end of the 1990s that it started to be used in participatory settings for water resources management applications. One of the most well-known approaches from these first experiments is now known as the 'companion modelling approach', so called because it can be used as an instrument in mediation processes that can co-evolve with the temporal and adaptive dimensions of social processes (Barreteau, 2003b).

'Companion modelling' differs somewhat from other forms of participatory modelling processes in that various types of social simulation tool, such as collectively constructed simulation models and role-playing games, are used with stakeholders to help negotiation and collective decision making over how to manage common resources (D'Aquino *et al.*, 2002a). A primary objective of this approach is also collective learning (Bousquet *et al.*, 2002), which is, in theory, aided by integrating different stakeholders' points of view through the process of development and use of common artefacts (Barreteau, 2003a), the equivalent of the 'meta-objects' introduced in Sections 3.2.3 and 3.2.4.

As presented by Barreteau (2003b), and shown in Figure C.9, the companion modelling process is cyclic, and is comprised of the following main stages which can be repeated as many times as required:

1. Field studies and bibliography, which supply information and hypotheses for modelling and raise questions to be resolved using the model;
2. Modelling, i.e. converting current knowledge into a formal tool to be used as a simulator; and
3. Simulations, which, conducted according to an experimental protocol (computer model or role-playing game), include participants, challenge former understanding of the system, raise new questions for a new batch of field studies, and generate participation behaviour in modelled outcomes.

The social interactions and behaviours of the participants are of particular interest in this approach. These are observed through the role-playing games, serve the purpose of helping to validate the multi-agent models scientifically from the researchers' point of view, and may also aid the participants to determine the legitimacy of the model.

Current multi-agent systems applications to water resources management and planning remain largely initiated by researchers on problems of smaller spatial scales. Such applications are typically carried out with small groups of participants, unlike the system dynamics approaches that were originally initiated by planners and engineers for large-scale planning processes with greater numbers of stakeholders. More recently, multi-agent systems-based participatory modelling and related role-playing games have been used for larger-scale problems with the inclusion of water policy makers (i.e. Dray *et al.*, 2005). However, these approaches remain highly experimental, with a strong focus on learning for future adaptive management rather than on aiding collective decision making. An overview of some recent experiments related to the companion modelling approach can be found in D'Aquino *et al.* (2002b) and Etienne (2010).

Despite the relatively few concrete examples of participatory modelling using multi-agent systems in water resources planning and management to date, multi-agent modelling combined with increased participant interaction is being heralded by many researchers as the way of the future (Parker *et al.*, 2003). This is especially because of its capacity to be linked to spatial information, such as GIS maps, and to include both qualitative and quantitative information in simulation tools (Daniell *et al.*, 2004). Following the power of multi-agent systems technology used in the artificial intelligence field and in computer gaming for planning, strategy and learning, such as in computer simulation games including SimCity™ and Civilisation™, the race has commenced to create similar level platforms to aid water resources management and planning (Poujol, 2004). One recent example is the Dutch SimCity-like game, 'Splash!', which allows players to manage river basin activities in order to attempt to satisfy the various urban, rural and industrial constituencies (Wien *et al.*, 2003). The task of opening up these games for easy re-personalisation of the environments and stakeholders' behaviours and values will require more time, as will embedding such tools in multi-stakeholder group participatory decision-aiding processes for water planning and management.

C.3.3 Problem structuring methods

Unlike participatory modelling methods based on systems dynamics or multi-agent systems modelling, problem structuring methods typically use more qualitatively based tools, in some cases with supporting computer software. Most of the methods were developed in the field of 'soft' operational research, as introduced in Section 3.1.1, the most common of which include:

- Soft systems methodology (Checkland, 1981);
- Strategic choice approach (Friend and Hickling, 1987);
- SODA: 'strategic options development and analysis' (Eden, 1989);
- Robustness analysis (Rosenhead, 1980);
- Drama theory (Bennett and Howard, 1996) and its precedents, including hypergame modelling (Bennett and Dando, 1979; Bennett *et al.*, 1989) and metagame analysis (Hipel *et al.*, 1974; Howard, 1989); and
- The viable systems method (Beer, 1984).

Such methods typically aim to express and structure underlying assumptions, as well as to develop the adaptive capacity of participants to deal with problems exhibiting high levels of complexity, uncertainty and conflict. This section will only provide brief details on the strategic choice approach and SODA. Rosenhead and Mingers (2001b) and Heyer (2004) provide further information on these and other problem structuring methods.

THE STRATEGIC CHOICE APPROACH

The strategic choice approach (Friend and Hickling, 1987) is one of the problem structuring methods most commonly used in the inter-organisational context. Its developers consider that it has a number of distinguishing features, including that it:

> *Places more emphasis on: 'structuring communication' than on 'reinforcing expertise'; 'facilitating discussions' than on 'exploring systems'; 'managing uncertainty' than on 'assembling information'; 'sustaining progress' than on 'producing plans'; 'forming connections' than on 'exercising control'*
>
> (Friend, 2001).

One of its principal goals is to elicit information about and understand different types of uncertainty, including:

- Environmental uncertainties related to external factors, such as climate and the economy or a lack of information on variables required for management;

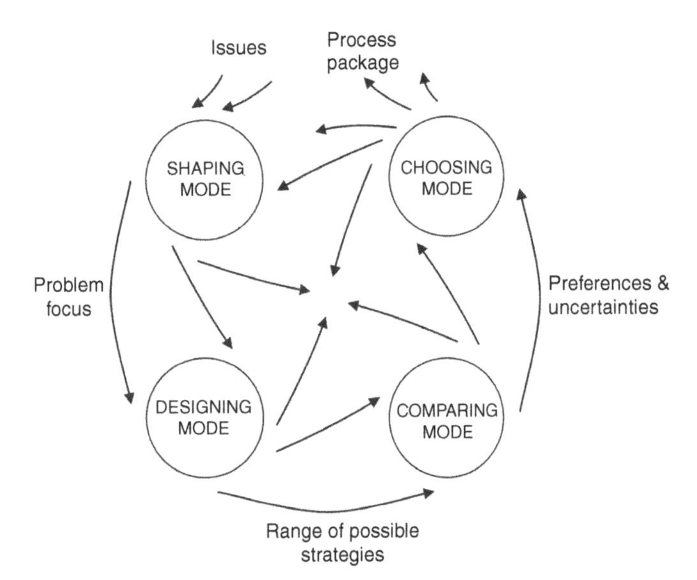

Figure C.10 The strategic choice approach. Source: Friend (2001).

- Value uncertainties related to the underlying political goals and values of the participants and related organisations that render strategy analysis and planning difficult; and
- Choice uncertainties related to whether other management alternatives exist under the current problem formulation or if the problem situation could be framed differently, to uncover other choice alternatives.

It has also been developed with the underlying hypothesis (Friend and Hickling, 1987) that in a time-constrained environment, decision makers do not tend to follow a 'phase-like' process (see Section B.2), but rather appear to switch haphazardly between various focal points and different associated types of reasoning. In the methodology of the strategic choice approach, Friend (2001) has therefore preferred to outline and support four 'modes' of working with decision makers in a decision-aiding process, between which the participants of the process may switch at will, as presented in Figure C.10.

For each of these four modes, Friend (2001) outlines a range of activities and thought processes that the participants of the decision-aiding groups are likely to use, as well as a number of specific methods, some based on computerised supports such as STRAD software (Stradspan Limited, 2007), which may be used in the implementation of the strategic choice approach; a few of which have been summarised in Table C.5.

Example applications of the use of the approach in the inter-organisational water sector include: de Jong (1996); de Carvalho and Magrini (2006); and Malmqvist *et al.* (2006). Although positive results were generally reported, issues of difficulties in finding common ground between stakeholders and agreeing on the definition of system boundaries were encountered (Malmqvist *et al.*, 2006). Strategies for negotiating common decisions are not overly evident in the approach and may therefore warrant further attention.

STRATEGIC OPTIONS DEVELOPMENT AND ANALYSIS

The problem structuring method, 'strategic options development and analysis' or 'SODA' (Eden, 1989), and its enveloping methodology, entitled 'journey making' (*jo*intly *u*nderstanding, *r*eflecting, and *ne*gotiating strateg*y*) (Eden and Ackermann, 1998), is another well-known approach, which stems from a very different theoretical background to the strategic choice approach. SODA uses the technique of 'cognitive mapping', the formal basis of which has been derived from Kelly's (1955) personal construct theory. In adopting this theoretical position, 'individuals' are considered the primary generators of meaning and they cognitively construct their own realities principally based on activities perceived to be related to making decisions for action (Midgley, 2000). By taking this subjectivist ontological stance, it is implied that there is not one 'objective reality', and that valid practice revolves around exploring individually and subjectively perceived alternative options for action, and not on the generation and testing of hypotheses (Midgley, 2000). However, it must be noted that this type of cognitive mapping is not the only type that is referred to in the literature. Other variants and their underlying theoretical and philosophical bases are outlined in Bonner (1993), Spicer (1998), Eden and Spender (1998) and Lauriol (1998).

Focusing on the use of cognitive mapping in group settings, SODA and 'journey making' have the stated underlying aims of facilitating: consensus building among process participants; development of emotional and cognitive commitments to action; and the management of '*political feasibility*' (Eden and Ackermann, 2001). The proposed process of 'journey making', as it is presented in Eden and Ackermann (1998), is made up of a cycle of seven steps: jointly understanding, reflecting and negotiating strategy; confirming and (re-)designing strategy; strategically managing stakeholders; strategically managing the environment; managing continuity and strategic change; exploiting planned and emerging strategic opportunities; and detecting, emerging, strategising. This cyclic process of creating strategy has been specifically designed to pass through a number of stages in a particular order, unlike the strategic choice approach, in order to achieve the above-stated aims. This constraining of direction to help turn strategy making into '*a conscious and purposeful activity*' has been designed in a deliberate manner to allow members of an organisation to understand and reflect on current organisational strategies before moving on to negotiating and re-designing new strategic directions and the possible implications of these directions for stakeholders, the environment and managing change (Eden and Ackermann, 1998).

A number of methods have been specifically developed and are typically used in a process like SODA and its internal elements of cognitive mapping and computer-supported analyses. The process of cognitive mapping may either be carried out through

Table C.5 *Suggested elements of the strategic choice approach. Adapted from Friend (2001)*

Mode of functioning	Focus	Typical activities
Shaping	Identifying and formulating the decision problems: expanding or narrowing the focus of decision problems	1) Identifying 'decision areas': areas or issues of concern, with choices to be made between alternative courses of action 2) Identifying 'uncertainty areas': areas where there is a lack of information or areas over which the participants have little influence or control 3) Constructing a 'decision graph' (including highlighting doubts or disagreements) of the identified 'decision areas' connected by 'decision links' where relations exist between them 4) Choose a 'problem focus' by outlining a boundary of specific interest on the 'decision graph'
Designing	Identification of alternative options (courses of action) to address the decision problems	1) Develop a set of alternative options for each of the identified 'decision areas' 2) Analyse the options from different 'decision areas' using pair-wise comparison and combine them if it is feasible to do so: 'Analysis of Interconnected Decision Areas (AIDA)' (Luckman, 1967) 3) Determine whether the alternative options are as 'mutually exclusive' and 'representative' as possible of the range of potential choices for each 'decision area', and adjust the option list if required 4) Examine the 'compatibility' of the options between decision areas using an 'option graph' or 'compatibility grids' (matrices of pair-wise option combinations with the possibilities of compatible, doubtful and incompatible) or 'option tree' (available in the STRAND software)
Comparing	Analysis of implications and effects of alternative courses of action	1) Define a number of 'comparison areas', which could be specific 'criteria' or more general domains where 'intuitive' judgements can be used 2) Work using a 'cyclic approach' to comparison (comparing pairs and excluding options from the 'short-list') by oscillating through: increasing and decreasing the 'range of choice' (option list to compare); restricting and widening the range of 'comparison areas'; bringing 'uncertainty areas' to the 'foreground' or keeping them in the 'background' 3) Comparisons can be noted using simple scoring measures or by using a 'comparison grid' to look at pair-wise 'comparative advantage' of options
Choosing	Making decisions and finding agreements on current commitments and future management processes	1) Reassess analyses created in the other three modes, paying particular attention to the potential effects of uncertainties, where an 'uncertainty graph' (Hickling, 2001) could be used, and to levels of urgency to make decisions for each of the decision areas 2) Develop a 'process package' (or 'commitment package') outlining the key elements examined during the decision-aiding process in a structured form and choose alternatives, i.e. outlining current 'decision areas' with the actions to be taken and the required explorations of uncertainty areas, as well as those still to be examined and decided upon in the future, including who is responsible for implementation, and when and how it should occur

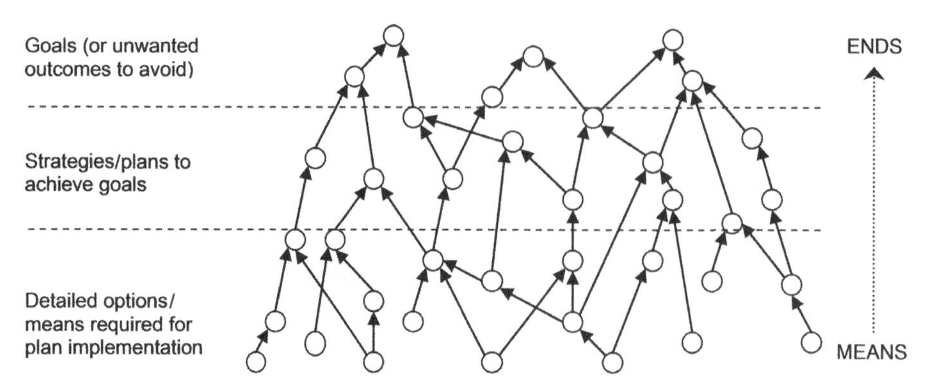

Figure C.11 SODA cognitive map structure. Adapted from Ackermann and Eden (2001) and Eden (2004).

individual interviews, from which composite group maps can be constructed by the analyst, or directly carried out in groups using a technique such as the 'oval mapping technique' of Ackermann and Eden (2001; Eden and Ackermann, 2001). As considered by Eden (cited in Clarke and Mackaness (2001) and Heyer (2004)), such maps can be thought of as models of 'action-orientated thinking' where the directional links represent some type of causal argument. The general structure of the cognitive maps that is sought through these techniques in SODA is shown in Figure C.11.

In a group situation, the SODA method is usually carried out in small groups using simple paper and pen supports and is often accompanied by a computerised version using the DecisionExplorer® software (Banxia Software, 2008). Maps developed through the facilitated participatory process may then be analysed in terms of both content and structure, using many available qualitative and quantitative techniques (Eden *et al.*, 1992; Langfield-Smith and Wirth, 1992; Daniels *et al.*, 1994; Mohammed *et al.*, 2000; Eden, 2004); one aim of which can be to inform better decisions on future strategies to pursue.

Despite its increasing levels of adoption, one of the potential challenges of using this problem structuring method in water applications is in its capacity to be adapted for use in the inter-organisational domain. This challenge may soon be overcome, as recently there have been increasing numbers of articles treating such issues (Franco, 2008). For example, the possible expansion of SODA and the 'journey making' methodology into the public policy domain was proposed with an example practical application in 2004 by its originators (Eden and Ackermann, 2004). About the same time, other experiments of using SODA in a couple of different inter-organisational settings appeared, including its application in participatory forestry management (Hjortsø, 2004). Finally, there have also been some very recent attempts to apply SODA to water sector applications (i.e. Giordano *et al.*, 2007). From this last experimental water conflict management initiative, SODA was used for problem identification of the interests and preferences of the different stakeholders. It was found to be particularly effective for the stakeholders to build and share their perspectives with each other, although the vagueness of natural language used in the maps was a challenge for synthesis efforts (Giordano *et al.*, 2007). SODA also only formed one module of an integrated decision support system for consensus achievement (IDSS-C) (Giordano *et al.*, 2007). This highlights that the method can probably not be used by itself to easily aid group decision making but should be coupled or integrated with other methods for evaluating or choosing options.

Appendix D Montpellier pilot trial, France

Following the definition of the research framework in Chapter 6, this appendix will further present the pilot intervention case: a trial in Montpellier of a pre-designed conceptual methodology for the participatory modelling of water resources problems and of the evaluation protocol presented in Section 5.5. The appendix commences by outlining the context of the intervention. This is followed by some descriptive elements of the co-engineering process, including the original design of the process' participatory modelling methodology and evaluation protocol, as well as a brief outline of the participatory modelling implementation process. Results obtained through the evaluation procedures will then be highlighted, followed by a discussion of these results and of the differences between the pre-designed methodology and the methodology applied in practice. The discussion will focus on a number of preliminary insights gained from the intervention; a critique of the methods used; suggestions for improvement for future applications; and questions remaining to be examined.

D.1 LOCAL CONTEXT AND OBJECTIVES

As explained in Section 6.2.1, the Montpellier intervention was chosen because of a number of constraints. The intervention context did not meet the ideals of a 'real world' messy problem situation or of a participatory modelling group with representation from policy, science expert and an affected public (refer to Section C.2). It is still of value, as it allowed experimental validation of a theoretically developed participatory modelling methodology and the evaluation protocol. Moreover, it provided a relatively risk-free environment of 'learning by doing', in which practical experience in designing and mixing methods, organising, facilitating, modelling and simultaneous evaluation of participatory processes was obtained. Implementing and learning from such an intervention was therefore still useful, and in retrospect was seen to be a necessary step for intervening in future real-world, more politically risky, cases. This section will outline the local context and general objectives of the intervention in more detail.

D.1.1 Background and role of the Montpellier methodology trial

The background context of this intervention is largely linked to the challenges of integrating 'non-traditional' forms of public participation mechanisms into France's water planning and management practices. France has a strong tradition of representative democracy and organised public political action to protest against injustices seen to exist in the centralised state decision making processes, such as the well-known student–worker rebellion of 1968 (Arnstein, 1969). France already has organised stakeholder groups and government management agencies that work together in legally formalised participative structures, such as the regional water basin committees (comités de *bassin*) and local water committees (CLE) for catchment planning (Bouleau, 2003a). However, there is little possibility for marginalised or unorganised individuals to participate actively in the decision making processes that affect them. The European Union's *Water Framework Directive* (EU, 2000), as outlined in Section 2.1.2, encourages '*the active involvement of all interested parties in the implementation of this Directive*'. This raises the question of whether alternative public participation mechanisms could be introduced into France to complement, aid and inform the work of the existing stakeholder structures to achieve the 'good statuses' of all water bodies by 2015.

The Montpellier intervention was created with the aim of contributing to this debate. The region surrounding and incorporating Montpellier, which is included in the water catchment management structure (*SAGE*) of '*Lez-Mosson-Etangs Palavasiens*', faces a range of water management challenges, including: rapid population growth and seasonal influxes (22% more in summer); high climatic variability leading to floods and water scarcity issues; water pollution, principally from urban runoff; and biodiversity protection (SAGE LMEP, 1999). The region has a number of principal land uses, including: urban and viticulture, predominantly on the littoral plain and on the low alluvial plain; wetlands on the littoral plain; and native 'garrigue' vegetation and forests, from the alluvial plain to the higher plateau, as shown in Figure D.1.

Based on these regional challenges and characteristics, where water management is characterised by high complexity,

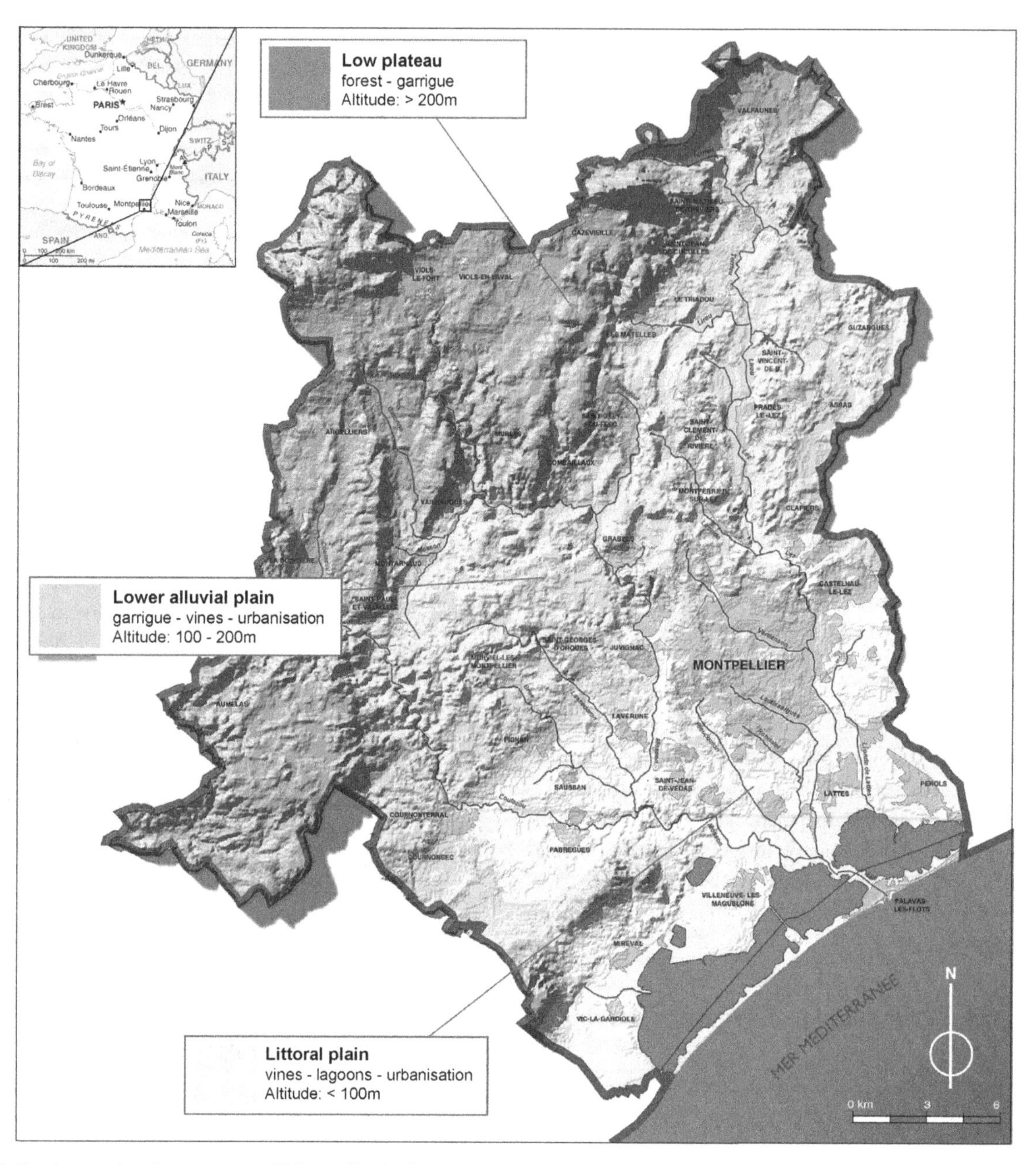

Figure D.1 Catchment planning area around Montpellier in France. Adapted from SAGE LMEP (1999).

uncertainty and conflict, a preliminary proposal to work with interested members of the public on developing water management scenarios that could help to meet the 2015 objectives was developed. The proposal, which was based on participatory modelling used in a 'reverse-engineering' scenario planning process (Fleming, 2005), was sent for internal institutional approval in early September 2005. However, because of the level of conflict over water management in the region, the intervention research project was thought to be too politically and institutionally risky. Suggestions were then made on how these risks could be reduced: these included only inviting students to work on an 'abstracted version' of the region's water issues.

D.1.2 Intervention questions of interest and objectives

The research questions to which responses could be sought through the intervention were consequently changed from their original form and adapted to the constraints of the situation. These principal questions of interest included:

- To what extent can participatory modelling aid learning and changes in water-related perceptions and practices of a selected subset of the general public?
- How and to what extent can methods be combined through the various stages of a participatory modelling process to incite collective learning, creativity and change?

- What are some of the potential unforeseen advantages and constraints that can be uncovered through the intervention research process?

The objectives of the Montpellier methodology trial were therefore to:

1. Test the relevant hypotheses of the impacts of participatory modelling in this abstract case (outlined in Section C.2.2) through extensive evaluation procedures that include the 'ENCORE' model components.
2. 'Learn by doing' about method selection, evaluation procedures and other process constraints and opportunities in a relatively risk-free research environment.
3. Gain a greater understanding of the 'general public' perspective on water issues.

The next two sections will outline how these questions were investigated and introduce how a diversity of objectives and roles of members of the project team impacted the intervention design and implementation processes.

D.2 THE CO-ENGINEERING PROCESS AND INTERVENTION DESIGN

Because of the institutional context of the two principal researchers who drove the development, design and implementation of this intervention (my director and me), it is rather difficult to define when the intervention was 'initiated', considering the historical evolution of defining the research project questions, and the case selection and case constraining process. The point of initiation will be taken as the moment that the 'abstract' nature of the Montpellier case was defined, in order to delineate one external boundary of the research object of interest: 'the co-engineering process'. This choice does disadvantage analysis of the 'real' water issues of the Montpellier region that were mentioned previously, although it also has the advantage of presenting a coherent case intervention, which is based on the adopted research questions and objectives in Section D.1.2. Therefore, the next section will start description and evaluation of the co-engineering process from this point.

D.2.1 Initiation process

As alluded to in the previous sections, the intervention was initiated by three principal actors: my director and me, and our institutional research chief, whose role was to develop the intervention in a manner that would ensure a low institutional risk; the completion of a research initiative that would be of benefit to the institution; and reasonable budgetary costs. Apart from preliminary scoping of the intervention, which included the caveat of the avoidance of regional water politics, the institutional research chief was only active in an advisory and budget operator role throughout the process.

The research director had originally proposed my project topic, and so he maintained an active interest in my completing the research investigation along the guidelines proposed. This included the use of the ENCORE model, outlined in Section 5.5, to evaluate the impacts of a participatory modelling process implemented through the intervention, and the stipulation that the intervention would be completed within the required time constraints. From the research director's point of view, my role in the process was to act as an evaluator. The research director considered that it was the institution's and his role to provide the protocol for the participatory modelling process. This meant that the initial intervention proposition also required other researchers in the institution to be found who were willing to aid with the design and implementation of the participatory modelling process, including someone willing to undertake the modelling work for a calculable computer model and to facilitate the process.

I had the objective of completing my research work to the best of my capabilities, which would involve, from my point of view, carrying out the evaluator's role of designing and implementing the evaluation procedures for hypothesis testing, as well as assisting with the participatory modelling process design. This was predominantly because of my general interest in participatory modelling processes and the theoretical and literature review work on the subject that I had completed prior to the intervention. Although such an objective seemed to be in opposition to the research director's idea of the roles, it was not really a large issue, as the participatory modelling process for the intervention had in fact not already been designed; nor was any modelling protocol capable of being adopted without first adapting it to the context. However, any mixing of roles might have affected the 'objectivity' of the evaluator, but this was not thought to be the most crucial element of the process, considering the 'intervention research' stance that I adopted, which was to be the principal framework for developing research insights. I also wanted to make the intervention as 'real' as possible for those involved, and especially to try to ensure that they could understand their stakes in, and impacts on, the water management process. This was thought to be of importance so that the evaluation results might help to create some useful insights for other water management and participatory modelling interventions, which would still be relevant to the central questions of this book.

There was a reasonable level of uncertainty and potential complexity in the roles to be undertaken, as well as the process to be designed and implemented. The stakes for each of the actors were also quite high, especially for the director and me, as we would have to attempt to maximise the research benefits from carrying out this 'abstract' water management exercise.

In terms of who was to be chosen to be involved in the methodology design phase of the intervention, only my director and I were confirmed participants in the design team at this stage of the process.

The objectives and resources of each of these actors appeared at this point to be complementary for carrying out the intervention, rather than in conflict.

D.2.2 Design process

For the design process, the research director and I also possessed our own objectives, stakes and resources. The research director had a great deal of experience in simulation modelling, multi-criteria analysis and working in a participative manner to create models of various kinds, such as cognitive maps or those that could be used as a basis for role-playing games. He also wanted to compare a process based on 'participatory modelling' with one on 'integrated expert modelling', where two 'test' groups would need to be chosen. One group would then complete a full modelling process from defining the problem to using the model to form recommendations, and the other would be provided with a model and could use it to analyse options or scenarios in order to formulate its recommendations with the aid of a multi-criteria analysis. In other words, the director had an underlying objective of creating a simulation model through the first participatory process of the 'abstract' water-related system, which could then be used by the second group.

I had some experience with simulation modelling, although not much in creating such models in a participatory fashion with a group of non-specialists. I also had a much stronger interest in planning and procedural concerns for group work, which had been developed through my literature review, as well as through previous management studies and group project work in Australia. I therefore had the objective of attempting to leave as much of the modelling work as possible to someone else and to concentrate my efforts on procedural design, process evaluation and continuous improvement and adaptation en route, based on evaluation insights through the methodology implementation. Through my literature review, I had already found a number of methods, models and theories that I was interested in using or exploring through the participatory modelling exercise and evaluation. These included: Ackermann and Eden's (2001) 'oval mapping technique'; the MERF process for reverse engineering the planning cycle in Figure A.5 (Fleming, 2005); the elements of Tsoukiàs' decision-aiding process model described in Section 5.1.3 (Tsoukiàs, 2007); Nonaka and Takeuchi's (1995) innovation and organisational learning cycle; the four quadrants or fields from integral theories and learning theories (Schumacher, 1977; Wilber, 2000); spiral dynamics (Beck and Cowan, 1996); and the panarchy concept (Gunderson and Holling, 2002) (see Table B.4).

During the design process, an administrative and logistical assistant was also found from the institution to help with the participant selection process, budgetary procedures and other institutional administrative issues for the project.

D.2.3 Selection of participants

In the two weeks prior to the commencement of the test, publicity flyers were distributed throughout the university campuses of Montpellier by the administrative assistant, another colleague and me. The flyers gave very little detail on the content of the exercise, except that it would include 'discussions, games and interactive activities on the theme of water'. Interested students were invited to contact the research institute by either phone or email. It was also noted that the participants would receive a small amount of money to cover their expenses, almost half of which would be paid at the end of the exercise to ensure that the participants would continue to participate at all of the planned sessions between the 19th of October and the 23rd of November.

Response to the flyers was reasonably rapid, with nine participants being selected a week before the start of the exercise. Despite working only with students, the director and I had the aim of finding a group of nine students exhibiting a high level of diversity. Nine students were chosen so that the group could be broken up into threes for some exercises. For choosing the participants, a series of answers to 'pre-selection' questions (contact details, age, studies, number of years living in Montpellier, place of birth and two questions relating to their knowledge of the main uses and problems related to water resources) were elicited from participants during a preliminary phone call by the administrative assistant. The final group, selected from the 15 interested candidates, included four men and five women from four countries (France, Morocco, Tunisia and Luxembourg), with ages ranging from 18 to 35 (average of 25), levels of study ranging from a completed technical high school diploma to those with almost completed PhDs, and areas of study that included economics, psychology, accounting, philosophy, law, information and communication sciences, languages, history, geography, water engineering and agricultural engineering.

D.2.4 Problem scale boundary choices

To involve the group of participants in the process to the maximum level possible, some basic choices regarding the problem scales and boundary of the study to be explicitly treated were taken before designing the rest of the methodology. Based on the assumption that students have a stake and responsibility over the management of their own lives and local surroundings, including the way they use and alter their available water resources, three nested scales of analysis relative to the situations of these participants were outlined by the research director and me:

- Life and home;
- Neighbourhood or village; and
- River basin or region.

The rationale behind the choice of these scales was to allow the students to examine the possible impacts of their actions at a larger scale, and the impacts that regional or local decision makers or exogenous factors could have on their lives. The nested scales mimicked the panarchy concept, which I was interested in exploring. Final decisions in the process could therefore be concrete at an individual level, or suggestions for management at the higher spatial levels. This scale choice had a universal character of being able to be applied to many water systems around the world, but the disadvantage of not explicitly allowing the effects of national or international policy or other drivers to be examined. However, owing to the objective of building a simulation model during the process, spatial limits around a water basin were thought to be an important constraint.

D.2.5 Process methodology design

The process methodology was co-designed by the director and me, based on my preliminary work, and then collectively redefined. The first version that I proposed followed the four stages of the Tsoukiàs (2007) model and the planning stages of the MERF model (Fleming, 2005). However, the research director insisted that not enough emphasis had been placed on the modelling section of the methodology and that this section should be re-examined. He suggested designing a role-playing game as part of the methodology, such as SelfCORMAS (D'Aquino *et al.*, 2003) or JustGame (Ferrand *et al.*, 2005). Such a game and its accompanying model could be used by the students to 'play out' their roles, which is similar to the case that was explained in the 'integrated methodology proposal for participatory modelling' in Section 3.3.5. These proposals were then collaboratively expanded to produce a planned methodology incorporating six participatory workshops, as shown in Figure D.2.

At this stage of methodology design, a number of methods had been considered for use, such as: cognitive mapping for the problem situation models and the oval mapping technique for the problem formulation phase, which would be supported by DecisionExplorer® software; UML modelling to conceptualise the simulation model; and a multi-criteria analysis matrix for the decision or choice of planning options phase. Modelling platforms and the form of the game to carry out the scenario analysis through role analysis were still undecided, as a modeller had not yet been found. More general methods, such as individual brainstorming on cards, work in small groups and collective group discussions were planned.

The second part of the planned trial to compare the 'integrated expert modelling' level of participation was to involve a second group of students in the 'introduction', 'scenario exploration through role analysis' and 'decision or choice of planning options'.

D.2.6 Evaluation protocol adaptation

I used the 'participatory modelling process' from the evaluation protocol given in Table 5.2 as the basis for developing the evaluation methods that were to be employed during the methodology implementation. The procedure planned for use included *ex-ante* and *ex-post* questionnaires with closed and open questions during each session, some group oral debriefing at the end of each session, and external process observation with the aid of audio and video recordings. *Ex-post* interviews with the participants would be carried out if time permitted and participants were willing to be contacted. Thirteen questionnaires were planned in total, as an extra *ex-ante* questionnaire was to be administered during the first session after the presentation of the methodology. The majority of the questions to be asked were to be in closed response form on a five-point Likert scale. The five-point scale was thought to be more appropriate to this study than the four-point 'forced-choice response' format, which has no neutral or 'neither agree nor disagree' category. The five-point scale was preferred to remove bias from the answers. Questions were developed to take the context, objectives, process, content and outcomes of the participatory modelling intervention into account.

Most of the questions to be posed in the methodology had been defined in the months leading up to the planned intervention by further specifying questions under the seven hypotheses for testing (Section C.2.2). Between 7 and 25 potential questions were outlined for each hypothesis, as well as for the categories of 'interactive', 'procedural' and 'distributive' justice (Marks, 2004). These were largely formed from reflection using the ENCORE model and previous questions collaboratively developed with a colleague at the French research institute (Finaud-Guyot, 2005), as well as a range of other references, including Marsh *et al.* (2001), Beck and Cowan (1996) and Loubier and Rinaudo (2003).

The 'preliminary questionnaire' had the aim of eliciting information on participants' value systems and work-related preferences; the *ex-ante* questionnaire of the first session was to elicit a representation of the participants' cognitive state, preferences and practices relative to water issues, as well as some socio-economic information on personal and family history. The following *ex-ante* questionnaires, at the beginning of each session, were to gain participants' impressions of the previous session. The *ex-post* questionnaires of each workshop were to assess changes, relationships between the participants and their

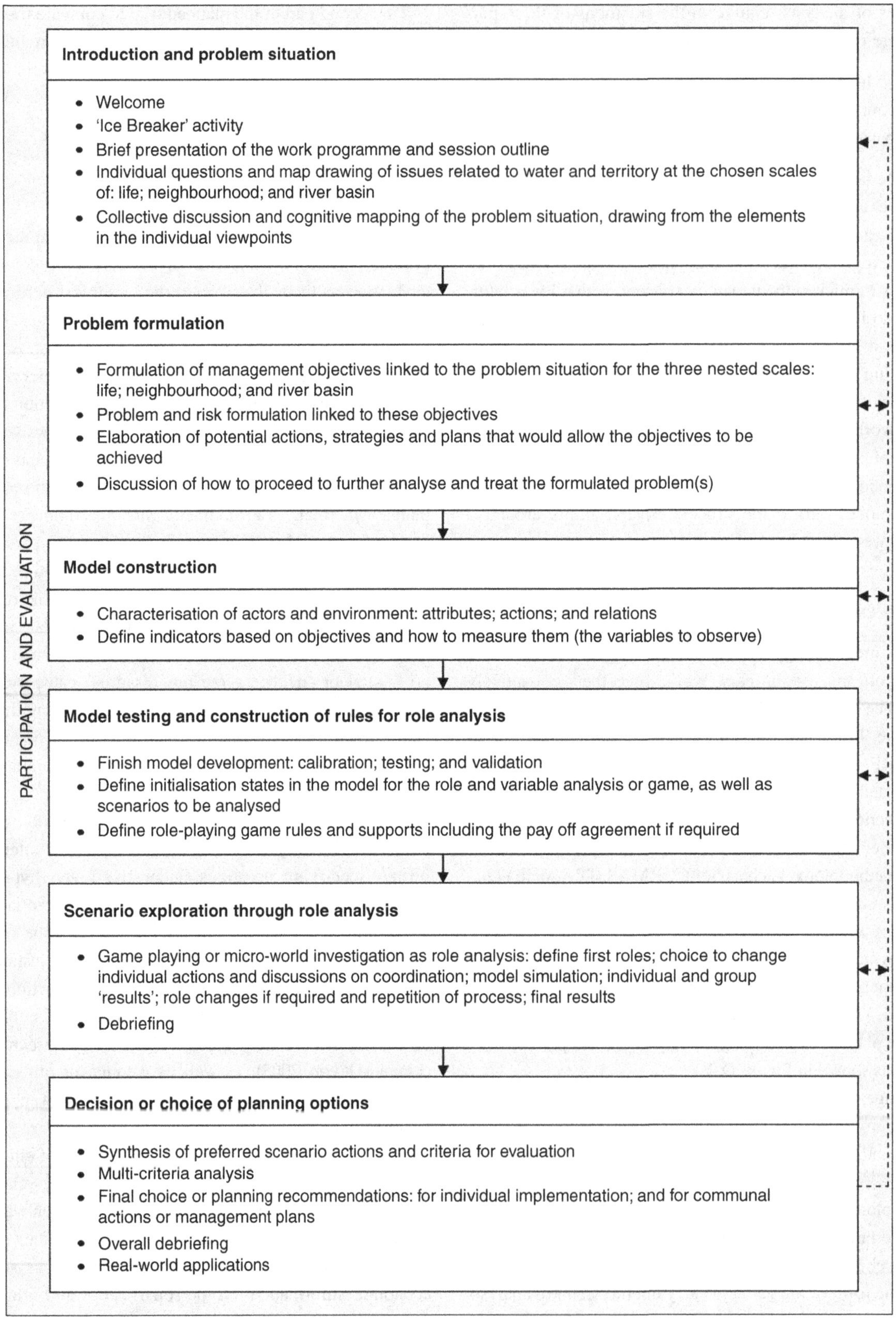

PARTICIPATION AND EVALUATION

Introduction and problem situation

- Welcome
- 'Ice Breaker' activity
- Brief presentation of the work programme and session outline
- Individual questions and map drawing of issues related to water and territory at the chosen scales of: life; neighbourhood; and river basin
- Collective discussion and cognitive mapping of the problem situation, drawing from the elements in the individual viewpoints

Problem formulation

- Formulation of management objectives linked to the problem situation for the three nested scales: life; neighbourhood; and river basin
- Problem and risk formulation linked to these objectives
- Elaboration of potential actions, strategies and plans that would allow the objectives to be achieved
- Discussion of how to proceed to further analyse and treat the formulated problem(s)

Model construction

- Characterisation of actors and environment: attributes; actions; and relations
- Define indicators based on objectives and how to measure them (the variables to observe)

Model testing and construction of rules for role analysis

- Finish model development: calibration; testing; and validation
- Define initialisation states in the model for the role and variable analysis or game, as well as scenarios to be analysed
- Define role-playing game rules and supports including the pay off agreement if required

Scenario exploration through role analysis

- Game playing or micro-world investigation as role analysis: define first roles; choice to change individual actions and discussions on coordination; model simulation; individual and group 'results'; role changes if required and repetition of process; final results
- Debriefing

Decision or choice of planning options

- Synthesis of preferred scenario actions and criteria for evaluation
- Multi-criteria analysis
- Final choice or planning recommendations: for individual implementation; and for communal actions or management plans
- Overall debriefing
- Real-world applications

Figure D.2 Conceptual methodology for the participatory modelling process.

impressions of the methods used and process undertaken. The final *ex-post* questionnaire at the end of the whole participatory modelling process was to elicit a second representation of the participants' cognitive state, preferences and practices relative to water issues for comparison, as well as information on the participants' perception of the overall process.

The questionnaires were not all prepared prior to the methodologies' implementation, but rather developed in an adaptive fashion by me and discussed with my director just prior to each workshop session. A number of examples of the Montpellier trial questionnaires are provided in Daniell (2008).

D.2.7 Logistical considerations

A set of six three-hour long workshops was planned for the 19th, 24th and 26th of October and the 16th, 21st and 23rd of November, 2005. A space of one month was purposely left between Workshops 3 and 4, to provide some time to process the information from the first sessions and to create the computer model from the participants' conceptual modelling, so that it could be discussed and tested in the following session.

To allow participants adequate space to carry out the participatory modelling activities in a convivial atmosphere, a wine bar with a large outdoor area was booked for the sessions. Food and drink were also provided to participants during the sessions. In addition, participants were to be paid a small amount of money to cover their costs and to ensure their continuous participation. Ten euros per participant per workshop was paid by the administrative assistant at the end of the exercise, plus a portion of 500 euros, which was set aside for the role-playing game payoff.

D.3 PARTICIPATORY MODELLING PROCESS IMPLEMENTATION

The participatory modelling process was implemented in the form of seven consecutive three-hour long workshops (WS1–WS7), with a range of activities in each one, as shown in Figure 6.2, between the 19th of October and the 23rd of November, 2005. One extra workshop had been added to the planned six, the reason for which will be explained in Section D.3.1.

Throughout the process, the nine students appeared to participate actively in a convivial atmosphere. The principal methods used during the participative process, which are outlined in more detail in Daniell (2008), included: two types of cognitive mapping (current situation and strategy mapping); UML modelling; a dynamic interaction specification game; payoff construction; brainstorming; a role-playing game supported by a Microsoft Excel model; and open debate.

The next sections will present some aspects of the coordination of the project team members involved in the implementation process and the differences between the proposed and implemented methodologies. This will be followed by a brief content analysis of the process using the Tsoukiàs (2007) decision-aiding model and a number of other evaluation results from the process.

D.3.1 Implementation process roles

A number of actors were involved in aiding the implementation of the participatory modelling process. The director and I took the principal management roles. A facilitator was found from the institution to help facilitate the workshops, although he was only present for the first workshop, as will be explained later in this section. A newly arrived modeller in the institution was seconded to work on the project from the 13th to the 20th of November and attended WS4 to present the preliminary stages of the computer model to the participants. In addition, a friend who had previously supported a participatory process at the institution volunteered to assist the implementation by video recording the sessions and providing additional technical and logistical assistance when required. Finally, one external observer from the institution was also present for WS5 and WS7.

WORKSHOP DESIGN AND LOGISTICS

At the beginning of the process, the director and I reached the understanding that the very short time limit for the process and between the workshops meant that we would have to be adaptive in the roles that we would play through the process, as we could not necessarily rely on much outside help. Our common objective appeared to be to implement the planned process as effectively as possible in the time available.

The process for designing each workshop was that I prepared the evaluation forms and a preliminary presentation for the meeting, incorporating any necessary information or feedback of results from the previous sessions. I would then meet with my director on the day of the workshop to check through and make modifications to the evaluation forms, as well as to work collaboratively on the design and agenda of the workshop. This included carefully designing the methods to be used and documenting them on a PowerPoint presentation. This presentation would then be printed for the evening workshop session, along with the required questionnaires and any other necessary supports. The design process continued through the implementation process and was carried out in an adaptive manner, which dealt with ongoing needs, constraints and arising issues.

The logistics of implementation for each session then involved at least a couple of the project team members arriving at the venue in the institution's van at least half an hour before the session to set up the space with chairs, tables, pinboards, visual supports, computer, projector, video and audio recorder and the paper and pens to be used by the participants. The friend who volunteered to help with these logistics worked effectively with the other team members to carry out this set-up process and to ensure that the equipment was kept in effective working order throughout the sessions.

FACILITATION

The facilitator was an internationally professionally accredited facilitator whose objective was to carry out his role according to suggested best practice rules. The rules that he subscribed to, and suggested that the project team respect, were that the facilitator should remain neutral to the content provided and not provide his own opinions on the topic. His role was to moderate the discussions, ask for clarification of ideas and provide general principles of interaction and other ground rules for the session, such as that workshop start and end times should be respected and that mobile phones should be switched off.

During the first workshop, the introductory presentations and facilitation of discussions were shared between the research director, the facilitator and me. Regarding the purpose of facilitation in this process, the research director had a differing opinion from that of the facilitator, considering that facilitator neutrality was not an important requirement in this participatory modelling and research exercise; and that knowledge and content of the research objectives would be needed to adapt the facilitation needs effectively. There also appeared to be a confusion of the roles that each actor would play in the process. For example, would the facilitator explain the exercise and rules to the students, or would the research director, who had a much better understanding of the methods and their objectives, explain and moderate the activity? These differences of opinion and challenges of each other's perceived rules and roles created a conflict at the conclusion of the session, which led to the facilitator not participating in any further sessions. This meant that the role of principal facilitator was taken on by the research director and I acted as a secondary facilitator when required.

MODELLING

After the first three workshops, a modeller who had previously worked on multi-agent models to support role-playing games arrived at the institution and, when asked by the research director, agreed to create the computer model for this exercise. Owing to the time constraints of the exercise, his work started on a Sunday afternoon one week prior to the model needing to be ready for the role-playing game. I worked as the main consultant for the modeller, who transferred the existing UML conceptual modelling that the students had carried out at the end of WS3 to CORMAS (Bousquet *et al.*, 1998), a multi-agent modelling platform with which the modeller was very well acquainted, as he had helped to design it. This first stage of modelling worked well, with the modeller attending WS4 to present the first form of the model to the students. However, he also explained to them that there was not yet enough specification, or relationships between the model's objects, dynamics, or quantification, to make it ready for the simulations required in the role-playing game. These aspects were worked on in WS4 and WS5 by the students and in meetings in the institution between the workshops by the research director, the modeller and me. Following these meetings, I ended up creating the role-playing game cards and writing the necessary equations and data to populate the 'abstract', yet realistic, hydrological part of the model, while the research director wrote the population dynamic equations in MSExcel, as the modeller was not an expert in these fields. However, transferral of these modelling principles during the weekend before the workshop proved a major challenge. It was also the first time that I had worked collaboratively with a modeller who was not an expert in the field that the model was created for, which presented a range of conceptual and understanding transferral issues.

After discussions between the research director, the modeller and me on the morning of the day that the role-playing game and model were supposed to be ready, it was decided that we would have more chance of having a workable model and game if we forgot about the CORMAS model and transferred everything to MSExcel. In terms of model performance, there would be few disadvantages, except for the spatial layout interface, as the CORMAS model did not include any behavioural rules. This decision meant that for the last few days of the participatory process, the modeller would no longer be involved and the modelling role would revert entirely to the director and me. The consequences of this decision have been outlined in Section 6.6, and will be further discussed later in Section D.5. The research director was responsible for running the model with the participants in the workshop, and I was responsible for distributing and organising the rounds of the role-playing game in the final session.

PARTICIPANTS AND PROJECT TEAM

At the beginning of the participatory modelling process, the idea that this process was one of 'participatory research' was presented to the students and reiterated by the research director: '*We are here to learn from you and you are here to learn from us.*' Although the project team maintained its organisational roles throughout the process, a certain amount of liberty was accorded to the participants to decide whether they wanted to spend more time on certain exercises and discussions, or were happy to move on. This included giving them the option of what start times would suit them for each workshop, and whether and when to hold extra workshops or work after the planned finish times. The process was also adjusted to include aspects that they suggested. In this way, the students occasionally played a part in directing the project. However, the 'participatory' nature of the two-way

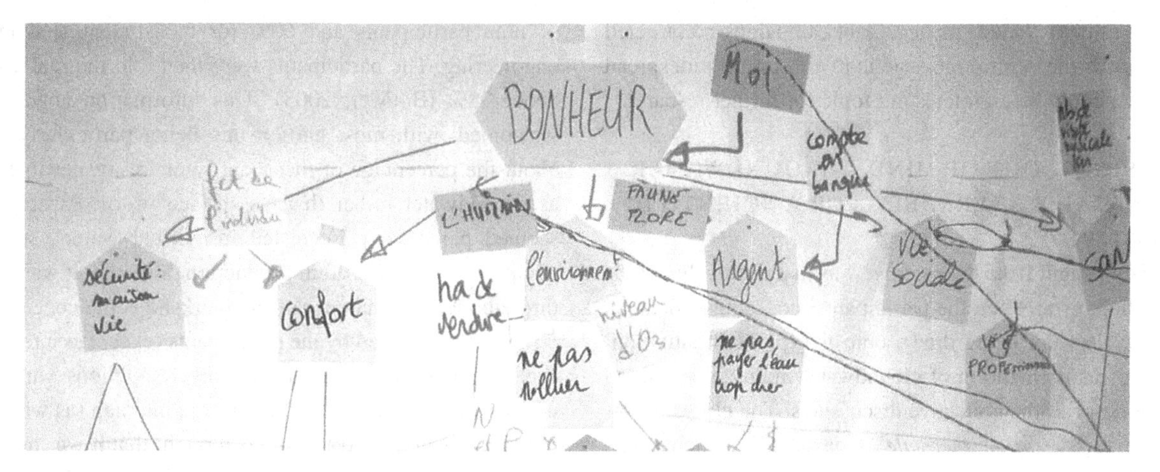

Figure D.3 Happiness as the major objective for water management at the individual level.

exchanges still generally appeared to be 'working for each other' rather than working 'collaboratively with each other'.

WORKSHOP TIMING AND STUDENT PARTICIPATION: METHODOLOGY ADJUSTMENT

One of the largest differences between the planned and implemented methodology was the length of time needed to complete the planned activities. The programme was to consist of six workshops of three hours duration on the evenings of the 19th, 24th and 26th of October and the 16th, 21st and 23rd of November. However, it was established during the first three sessions that the programme had been designed with not quite enough time to complete the required work at a comfortable rate, even though participants worked for 3.5 hours in the first and second workshops. This meant that the participants were asked if a seventh workshop could be added, which they agreed to hold on the 14th of November. Despite the addition of this workshop, which permitted the third stage of the methodology to be completed as required, the length of time between the last few workshops was too short to be able to complete the required modelling and game creation activities successfully, as mentioned previously. This led to the process programme being at least one session too short to complete the designed collective decision supported by multi-criteria analysis. Unfortunately, owing to the unavailability of the participants and the organisers at a later date, a supplementary workshop could not be added, so the planned methodological stages were never entirely completed.

D.3.2 Content-based results and procedural issues

The students involved in the participatory modelling process produced a very large quantity of elements and insights related to water and its management. Hundreds of map and modelling elements were produced, as well as hours of rich discussion and group interactions. From this work, just a few of the more surprising results and aspects of the process content will be outlined here, as they stimulated my learning to a greater extent than some of the other findings.

WATER FOR HAPPINESS

One of the first content results of the participatory modelling process was that effective or 'good' water management was equated to the overarching objective of achieving personal happiness (*bonheur*). This statement appeared in the collective problem formulation map, as illustrated in Figure D.3.

Although most people understand that water is vital for life, it is rarely considered at a public level that happiness is a result of, or can be linked to, good water management, especially in developed countries. Effective water management tends to be linked, rather, to other surrogate indicators of development, such as economic growth as measured by the 'gross domestic product', economic costs caused by water-related risks, such as floods, droughts or the number of water supply interruptions, or environmental indicators, such as chemical concentrations or biota counts for assessing the state of water quality. At the end of the problem formulation map exercise, the students considered that they had never really thought about just how important water was for them to live happy lives. This being the case, they wondered why 'happiness' of people in society could not be taken as an objective for effective water management. To make this a more concrete proposition, they then classified the sub-objectives of happiness at an individual level that were linked in some way to water. These included security (life and house), comfort, the environment, money, social life, professional life and health, as shown in Figure D.3. One student stated simply in the discussion, '*Without water we don't have health.*' Other links were made to show that to have 'security' it was necessary to choose a house that was not in a flood zone, or to create an evacuation plan to help protect this security aspect, which was linked to water issues and thus happiness. Considering this idea, could more effective water management campaigns be targeted at individuals if the central message is to encourage widespread happiness? It could be an interesting topic for further research.

OUT OF SIGHT, OUT OF MIND – GROUNDWATER AND VIRTUAL WATER: NEED FOR A SCIENTIFIC PERSPECTIVE?

The next key insight from the exercise was that during the whole first workshop, even when the participants' conceptual ideas on the water cycle were being drawn onto the 'problem situation' maps, there was no mention of groundwater in any of the small groups' work, or in the collective discussions. The closest references were to a '*station thermale*' (spa resort), which could potentially be considered as being supplemented by groundwater (from hot springs), and '*pompage*' (pumping) for agriculture, although it was not specified whether this meant ground- or river-water pumping. After this major omission when thinking about water management was noticed, a question was included in the questionnaire at the beginning of the next workshop. Discussion on this point then referred once to obtaining mineral water from 'springs' and to pollution seeping into groundwater aquifers. One indicator, 'groundwater level', linked to avoiding water scarcity for agriculture, then appeared in one group's map in WS3.

During the session on the participants' objectives, as well as later throughout the other workshop exercises and in feedback through the evaluation questions of WS1 and WS2, it had been made evident to the director and me that the group wanted to learn more about an 'expert' or 'scientific' vision of water-related problems. Although not originally planned for in the process, it was thought that some exterior thoughts could be voiced without bias to the original problem visions of the participants. Introduction of a scientific vision (constructed by the research director and me) was achieved in WS3 through the following means:

- A set of questions regarding quantities of fresh and accessible water on the planet was put to the participants, with responses and discussion following the individual responses of the group members;
- An introduction to the concept of 'embodied' or 'virtual' water, with some data on how much water is required to grow or manufacture certain products, as well as a breakdown of average 'embodied' water use in a Sydney (Australia) household; and
- A fourth problem situation vision created using the same rules as those used by the participants was presented and discussed.

One of the questions related to what percentage of fresh liquid water on the planet was found as groundwater. The participants' estimates for this quantity ranged from 1% to 30% for eight out of nine participants and 60% for the student studying water engineering. The participants were then told the real estimate is about 98% (Bouwer, 2003). This information appeared to be welcomed, with most participants being particularly surprised about the percentage of the quantity of freshwater that is found as groundwater rather than as surface water. Participants also seemed particularly interested in 'virtual water', such as the quantities of water used in production, although it was noted that throughout the remainder of the workshops the concept was not directly re-discussed by the participants (except as a reason to eat organic food and no meat). However, opinions varied on the utility of the 'scientific' problem situation map (as with the ones they had created). Most students thought that it was too complex to be of use; yet the water engineering student believed it presented a more 'correct' or 'realistic' vision of the 'real problems' than her group's own.

A FEW HOURS FOR AN INTEGRATED VISION OF WATER ISSUES?

After just an hour and a half of work, the students had filled three boards with an image of objectives of water management, plus a large range of strategies and actions for treating them. These boards corresponded to the three interlinked scales for water management shown in Figure D.4.

One of the participants commented after completing the exercise that it was quite complex to understand how all the elements on the different scales interacted. However, on closer examination with DecisionExplorer®, using a 'hierarchical cluster analysis' to look at groups of linked concepts, it appeared that much was quite coherent. With a target size of 30 concepts and a minimum size of eight concepts, seven clusters were created across scales, the central ideas of which were:

- Personal and community security and well-being development;
- Water use reduction;
- Integrated urban water resources planning and management;
- Building social capacity and direct democracy to encourage respectful water governance and reduce vulnerability to natural disasters;
- Mitigation of water scarcity (quality and quantity related);
- Sustainable energy systems; and
- Preserve water and the environment.

Throughout the following stages of the modelling process and during the role-playing game, many of the ideas in these sectors were re-elicited, with the exception of the energy system. The oval mapping technique appeared to provide a useful and efficient method of eliciting and synthesising information in a manner that allowed its reuse and further investigation. Moreover, the mapping exercise using the three scales helped to make the problem areas more 'real' to the students,

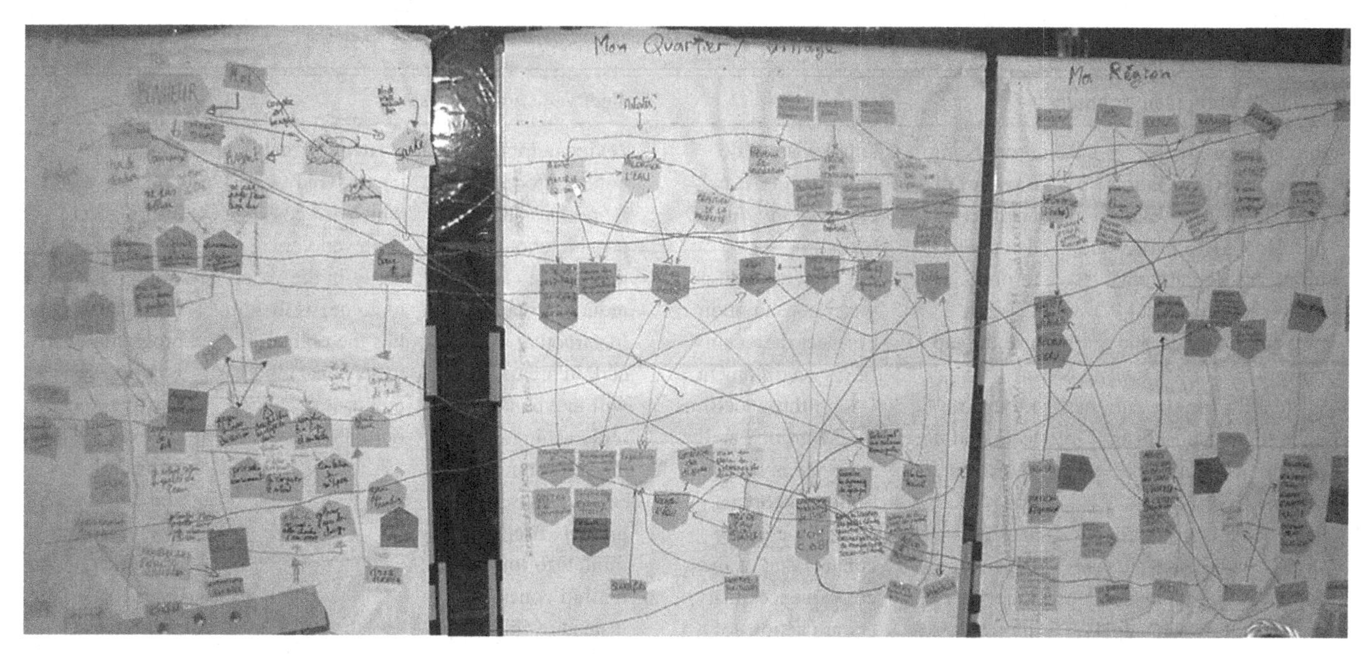

Figure D.4 Collective problem formulation across three scales: the life or private home scale (left); the neighbourhood or village scale (middle); and the water basin or region as a whole (right).

as they could see the possible impacts of many actions on their own lives (and happiness). Many of the problem areas that the students had chosen to focus on, such as development of well-being, capacity building, water use reduction and preserving the environment, were shown to be areas where the students could influence water management as a whole. This exercise helped to demonstrate to the students that they had 'real stakes' in the water management issues of the world around them.

FINAL ACTION PLANS – APATHY, EXTREMISM OR A HAPPY MEDIUM?

By the end of the workshop series, more diversity was evident in the opinions and stated positions of the participants, as can be seen from the individual action plans and suggested collective actions presented in Table D.1.

During the final debriefing of the process, one of the participants, whose action plan is shown in the first box of Table D.1 mentioned that she was disappointed in the people and society around her. In particular, she noted that it was incredible that these bright young students in the room did not appear to have any real courage or interest in the future. However, despite this disappointment manifested by one participant, the others did not seem to share this opinion. From participant observation and the evaluation questionnaires, it appeared that value systems of the participants became more polarised over the last two workshops. Conflict had been growing between a few of the participants over the direction that individuals should take in helping to improve water management. Some believed that individual action and sensitising close networks were all that were required, whereas others believed that more major political involvement was needed. Taking another direction entirely, one student preferred centralised management by technical government or elected officials working in the public interest. Finally, another participant even suggested to '*take the maximum profit*', after some other comments, which were more in line with generally conserving water and sensitising the population! It was interesting to see that a number of the participants had seemingly reverted to their original value systems in the final sessions, when comparing the evaluation questionnaire responses from the beginning of the first workshop with their positions in the final workshop. This was after having flirted with a variety of new ideas and critically appreciated others' views in the first series of debates.

Some of these issues on changes through the workshops and responses from the questionnaires will now be presented and discussed.

D.4 EVALUATION OUTCOMES AND DISCUSSION

This section will concentrate on the evaluation results related to some of the hypotheses and dimensions of ENCORE. The principal quantitative and qualitative results concerning learning, changes in awareness and acceptance and comprehension of model hypotheses and uncertainty will be presented here, along with a number of overall process outcomes.

Table D.1 *Three individual and three collective examples from the final recommendations*

Individual action plan examples			Collective action suggestions		
– Join an awareness movement (association type) related to water problems → to participate in increasing individual awareness and to put pressure on the public powers – Write a text on the current aberrations (subsidising private interests!) and attempt to diffuse it – In the long term, to provide myself with better means for collecting rainwater and treating water (algal basin type) – Not use any polluting detergents even when 'I am in a hurry' – Try to accept 'the reality' and actual state of the level of consciousness to not therefore fall into the hate of and disgust of my human sisters and brothers and for myself → become extremist → isolate myself and cut myself off from others	For me, being a keeper of one part of this universal resource, I can: – turn the tap off when I brush my teeth and do the washing up – not throw out left-over water – inform those around me of the stakes or issues and try to sensitise them to the problem – catch rainwater – drink wine ☺	I do not wish to do anything more and conserve my line of action = – I pay attention to my water consumption – I respect the quality of my environment = I don't throw detritus on the ground and I recycle my waste – I inform people around me of my conduct and of the consequences that certain actions of ours can have – Despite that, I still sometimes take a bath! → do not fall into extremism – Be conscious of the problem and act as best as I can – I am open to any new action propositions that I can implement at my scale	– Review agricultural policy: producing more does still not provide enough to live → we can reduce irrigation (destructive in certain areas). This will maybe allow replenishing of the groundwater stocks – Finance 'deviation' projects from high rainfall to low rainfall zones instead of financing useless roads – Stop giving construction permits in floodable zones	Pay attention and try to avoid wars, which are possibly and even very likely in the years to come, as therein lies the obligation to sensitise others and try to change our habits. . .	– Awareness campaigns and information – Use the media – Water management programmes in schools –'creating good reflexes' – Restrictions and enforcement – Take water out of rental charges – Fines for pollution

D.4.1 Quantitative participant perceptions

The first sets of results shown in this section have all been created in a similar fashion. For each of the closed questions in the questionnaires on the five-point Likert-type scales, the responses were coded from 1 to 5 (i.e. strongly disagree = 1, mostly disagree = 2, neither agree nor disagree = 3, mostly agree = 4, strongly agree = 5) and then the average of the participants' responses taken (n = 9 for WS1, WS4, WS5, WS6, WS7; n = 8 for WS2, WS3). The average for a repeatedly asked question from each workshop is then plotted with a line joining each response to show evolution trends throughout the participatory modelling process. To interpret the following graphs with more background information, the reader is referred to Figure 6.2 for a summary of the methods used in each workshop. It is also noted that the y-axis scales of each graph have not been kept constant over the set, to maintain visual clarity of some of the data sets. The Likert-scale label tags available on each graph are given to support result interpretations. Only a small set of results is displayed and discussed here. Further results are available in Daniell (2008).

As outlined in Section D.2.6, an analysis of the change in cognitive state of the participants, the 'cognitive' dimension of ENCORE, was attempted by posing the same questions in the first and last questionnaires of the workshop series. The results of these participant self-evaluations of their general 'understanding of issues' and 'awareness of the effects of their actions and practices' in a number of domains are represented in the spider graphs in Figure D.5.

From both of these evaluations, what appears to be most startling is not the apparent increase in understanding and awareness related to the water and politics aspects, which could be expected, considering the topic of the workshops, but rather the lack of change or small regressions in most of the other domains. There may be a number of reasons for these results. The first hypothesis is that the cognitive learning that occurred through the workshops was about the participants learning to doubt their existing knowledge. Another potential hypothesis was that the

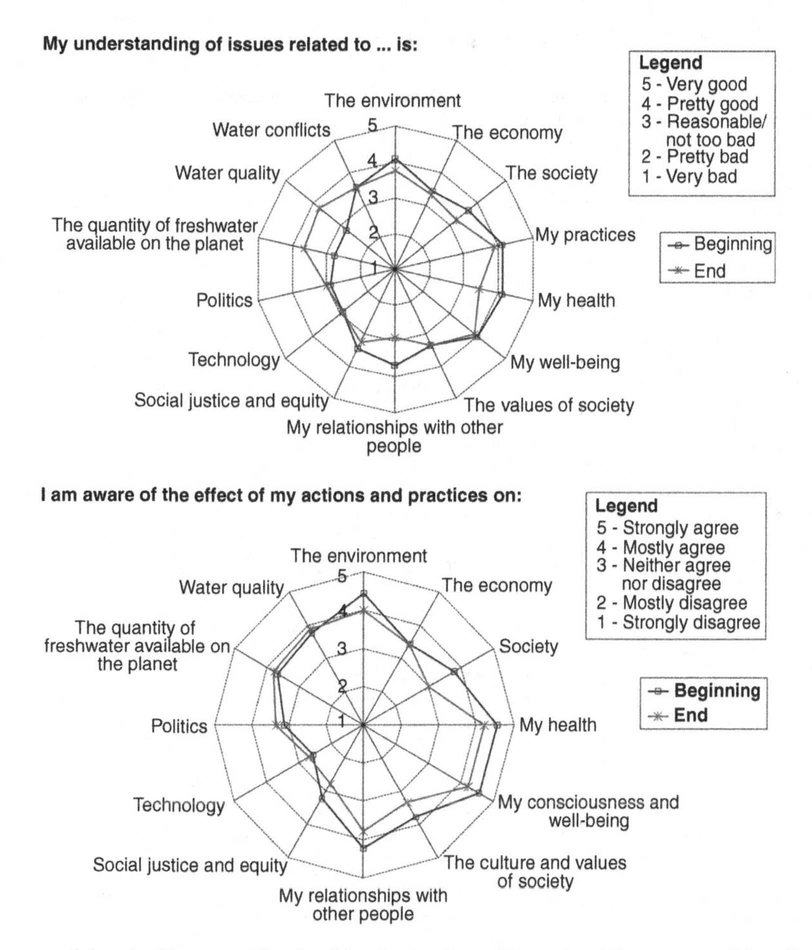

Figure D.5 Changes in average participant self-assessed levels of 'understanding of issues' and 'awareness of the effects of their actions and practices'.

participants had learnt to fill the questionnaires out differently through the workshops. One participant had mentioned during the sessions that he doubted the coherence of his own responses from workshop to workshop. This may mean that the evaluation methods should be reassessed.

As well as trying to assess cognitive states, attempts were made to determine the levels of participant learning in a number of other ways. Firstly, participants were asked directly at the end of each workshop how much they thought they had learnt on a range of aspects. A selection of these results is given in Figure D.6 (top). To gain yet another perspective and to attempt to triangulate these results, participants' impressions of their learning experiences in the previous workshop were also elicited in an *ex-ante* questionnaire in WS2–WS7. Some of the responses to the *ex-ante* workshop questionnaires are outlined in Figure D.6 (bottom). Note that the results of WS2 are the perceived learning that occurred in WS1 etc.

From the depth of learning results in Figure D.6 (top), it appears that learning in a number of areas continued through the workshop sessions, even if it was just 'a bit' during each session. Perhaps what is more interesting is that the greatest depths of learning appeared to occur during the sessions with more open debate (WS6 was almost entirely a large-group discussion and WS3 had about 1.5 hours of large-group discussion). The more structured activities on the model and game building did not appear so conducive to participant learning. The learning in the first workshop also seemed to be based more on relationship building than on other types of learning (the 'relational' component of ENCORE). The learning also dropped slightly for the less creative UML modelling and game-building workshops (WS4 and WS5).

The results of previous workshop learning experiences in Figure D.6 (bottom) are quite encouraging in that, especially where similar questions were posed (i.e. on learning about other participants), they confirmed the main findings of the *ex-post* workshop questionnaires on learning experiences. The impression of learning or retention of knowledge appears to remain relatively constant over time, even over the two-week break between WS3 and WS4. This finding is important, as it infers that if less time is available for evaluation procedures, the above question could potentially be omitted from the *ex-ante* questionnaires, without fear of losing too much information.

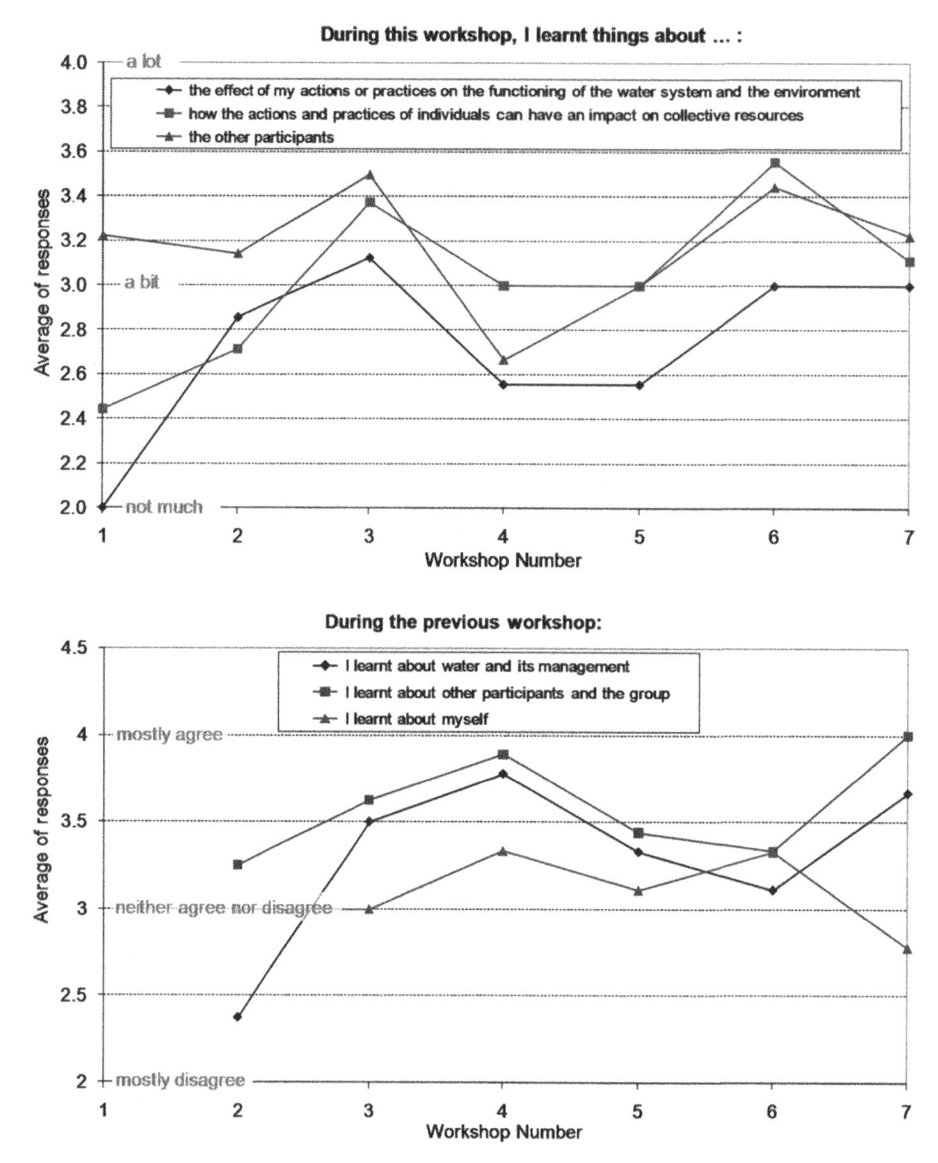

Figure D.6 Mean participant self-assessed depths of learning in workshops (top) and impressions of previous learning experiences (bottom).

The next results, shown in Figure D.7, concentrate more on the process' impact on participants' thoughts and practices, once again in a retrospective manner, using the *ex-ante* questionnaires.

From Figure D.7, it appears that at least some of the workshops had an effect on the thought and action patterns of participants in their lives outside the interaction space (the 'operational' dimension of ENCORE). However, the exact magnitude of that effect is more difficult to gauge, as the participants' level of thought and action related to water (the baseline level) that was present before the workshops is not measured. As a minimal hypothesis, it could be assumed that the lower average values seen in the *ex-ante* WS5 responses (i.e. on the effect of WS4) are close to the base participant characteristics. For future studies, if this question is of interest, it would be valuable to try to assess the baseline characteristics at the beginning of the first workshop by asking the questions in a more general form (e.g. '*Considering the problems associated with water, I think about them ...*').

The results in Figure D.7 indicate that there was a drop in effect after WS5. The model-building activities of WS4 were not as heavily focused on the objectives and actions of individuals, such as students, but on a broader range of roles. Alternatively, the more technical and convergent thought processes of the conceptual modelling process for the simulation model may not have been of particular interest to the participants, nor did it instil the required types of divergent creative thinking that may be more likely to lead to changes in practice. Such hypotheses obviously require further investigation.

Finally, to determine how well the participants understood the simulation model and role-playing game they created, a number of questions were posed based on the hypotheses and uncertainties of the model. The results of these questions are shown in Figure D.8.

From Figure D.8, it can be observed that, as the modelling process progressed, at least until the last workshop, there were general improvements in the participants' awareness of the

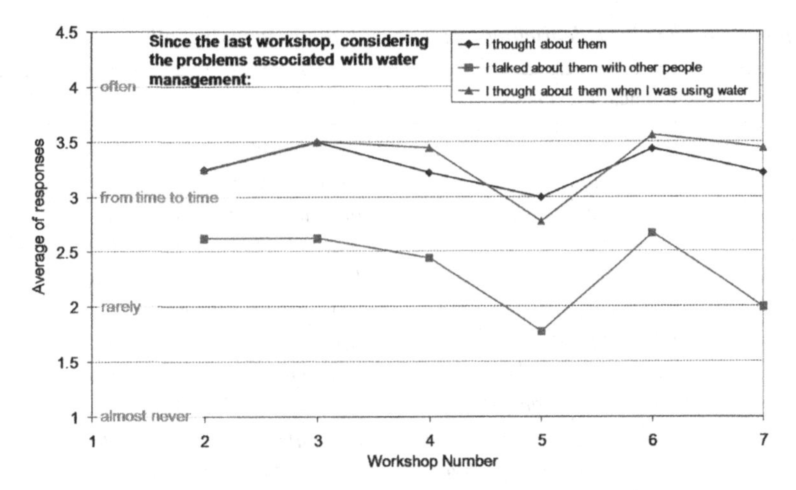

Figure D.7 Mean participant water management related thoughts and behaviours. Note that the results of WS2 are the perceived learning that occurred in WS1 etc.

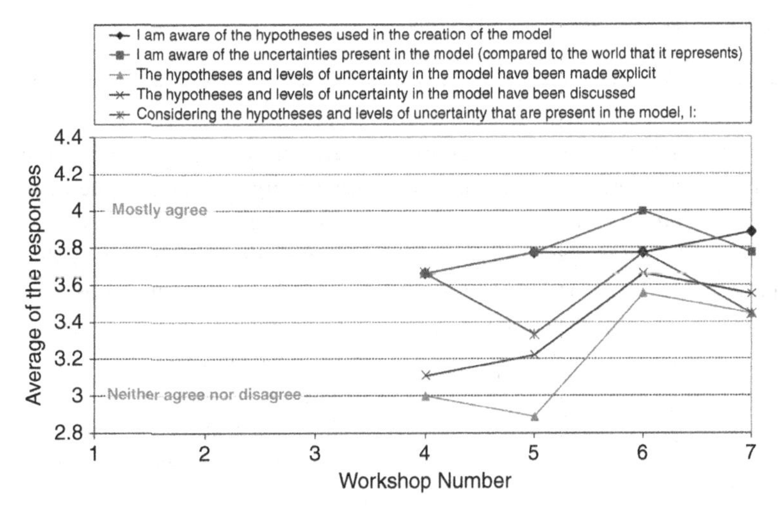

Figure D.8 Participants' perceptions of model hypotheses and uncertainties.

model-related hypotheses and uncertainties. In particular, it appears that taking the time to debate and examine the model and some of its hypotheses collectively in WS6 aided the process of model understanding. These results help to support the second half of the hypothesis that, '*Participatory modelling increases trust, appropriation and understanding of the models created, as assumptions are more likely to be explicitly identified and discussed.*' However, further examination of participants' 'trust and appropriation' of the model was required. This proved to be a challenge, owing to the unfinished nature of the model. With the unfinished model, the hypothesis could potentially even be refuted for this exercise considering one of the participant's final thoughts on the outcomes of the participatory modelling process as noted in the final questionnaire:

> *I think that a great job of awareness raising has been done, and that is already a lot (from my point of view the first step and the most important). But considering the results that can be drawn from the model and my actions during the game, I remain more sceptical.*

Such a hypothesis obviously requires further testing.

D.4.2 Insights from participant discussions, debriefing and observation

Although creating mean values from crosses in boxes is a useful exercise for helping to gain an understanding of general trends that were present in the participatory modelling process, it is difficult to create sufficient choice in questions to gain a contextual understanding of 'why' some boxes were crossed rather than others. Also, the more emotional rather than logical reactions to the process may be more difficult to assess from closed-response questionnaires, as filling out the form in itself required some form of directed and logical effort. For these reasons, qualitative forms of evaluation were also undertaken in this study to build a

Table D.2 *Participants' impressions from WS1 debriefing matched with their study discipline*

Psychology	Communication studies	Water engineering	Philosophy
Have I learnt a lot on water? I don't think so... I liked the ways of animating the work group, the situation and the supports... For a first session, the group dynamic has installed itself, the cohesion as well, which allows our little group to go quickly... I can't wait; I'm impatient to have the feedback from the boards. I would like to see what the others have done.	*Participation, collaboration, a holistic vision, different people – we manage to reach agreement in a group, on a vision, which we have related to water. I like the context very much. Even if it's a little cool, even if it rains, it doesn't matter, it's a warm atmosphere. There are people who have come here because we have a certain vision, which is somehow similar. So we find ourselves here and I find that very nice. I learnt to see how others conceive water as well, so that's already a lot.*	*I learnt more on the way of working in a group and then to face up to certain ideas of others... There are people here that don't have any links to water, who have a vision which is maybe less specialist than ours [referring to her training and the last participant's background in engineering]. But it's true that afterwards, the exercise we did with the PostIts®, it allowed us to share different ideas.*	*I am in a state of confusion. There's a bit of chaos in my mind – there are all these PostIts® on the boards. I'm enjoying myself a lot. For everything water-related, I don't think I've learnt much. I think it will come later... I don't know where I am, but I am well.*

richer picture of the intervention, including participant group debriefings, participant observation and analysis of audio and video recordings of the sessions and process artefacts.

At the first debriefing session, this point on the challenges of filling in the closed question responses was confirmed by one participant when she was asked her opinion of the questionnaire:

It's not easy to put a cross on something, on a feeling in fact – maybe it will become easier by the end of the sessions ... through the training in replying to the questionnaires.

However, there are advantages to the questionnaires, such as aiding reflection, as another participant noted:

The fact of having a questionnaire... it's good because it allows us to have a feedback on what we've done.

A number of insights drawn from the qualitative analyses of the process have already been discussed elsewhere in this chapter, including at the end of Sections D.3.2 and D.4.1, so this section will just provide several insights on some additional participants' feelings and thoughts at different moments through the process.

Considering the participants' first impressions of the process, a number of comments given in the first debriefing session are outlined in Table D.2.

These comments helped to gauge the participants' general feelings about the exercise and what they thought they had learnt. As can be seen from Table D.2, it appears that a number of the participants learnt more about the other group members, their views and methods of working together, rather than anything specifically water-related. These views are backed up by the quantitative results from WS1, shown in Figure D.6. It is also interesting to note how the participants' personality traits and disciplinary backgrounds are transferred in the responses, especially related to comfort levels with the exercise and levels of mental clarity experienced by the end. The participant with the psychology background was interested in analysing logically how the group had formed; the participant from an arts (communications) degree had a more emotional response, explaining her feelings and senses about the workshop; one of the participants with a technical background logically noted the differences in techniques of analysing and treating problems from what she was used to; and finally the participant who was found to be the 'deepest thinker' (an older unemployed part-time student with an interest in philosophy) had a rather mixed emotional and logical response, showing that saturation levels of information processing appeared to have been reached.

Throughout the process, the types of response given by the different participants remained fairly constant and true to their disciplinary backgrounds or the personality traits that they had originally shown. For example, the 'deepest thinker' summed up his experiences at the end of the last workshop as:

I have the feeling of never having been rational at any moment.

A number of other comments from the end of the process are outlined in Section D.4.3. Perhaps the only general exception to this trend was the psychology student who by the end of the workshop was no longer 'impatient' to see the results or interested in being part of the group, but rather 'disappointed' in her fellow human beings and wanting to exclude herself from others, as previously outlined in Section D.3.2. In other words, there appear to be relatively few major normative shifts in the participants through the process or adopting of new value systems (the 'normative' dimension of ENCORE).

When asked about the model and its usefulness, such underlying personality traits and preferences for processing information were also evident. For example, the communications student stated that:

> *If it [the model] is very scientific, I don't understand it any more. It is a loss of humanity. I prefer words to numbers.*

She also equally noted that she thought that it was too difficult to quantify elements such as 'success', 'employment' and 'health' meaningfully in a model. However, from the more technically minded individuals, this type of quantification and model received less resistance, with one participant stating that:

> *It [the model] can help us to make decisions, like a tool.*

The question of what type of model, if any, is required to aid groups to make collective decisions, and by whom and how it should be constructed, still remains unresolved. More discussion on this point can be found in Section 6.5.1.

D.4.3 Overall intervention outcomes

A number of more general results concerning the effectiveness, efficacy and efficiency of the participatory modelling process implementation will now be examined, followed by a number of innovations that occurred through the process.

EFFECTIVENESS

Was taking a participatory modelling approach to helping the students' collective decision-aiding on water management problems the right thing to do? This question requires thought about the legitimisation of the approach from both the students' points of view and validation from external perspectives.

From the students' perspectives, one of the signs of legitimisation was that all the students actually attended almost all workshops in the series and were even willing to attend an originally unplanned one. In other words, the process appeared useful or interesting to them. From the final qualitative responses in the questionnaires on whether the students had any other general comments on the process, there were a number that appeared to legitimise the process. As an example on the process being of interest, one of the students wrote that the workshops were:

> *Very interesting and enriching both on the problem of water and from a human perspective. And fun, interactive and interesting in the process.*

Another student mentioned how the process was personally 'encouraging' for her:

> *It is encouraging as an approach. It stimulates our intellect, our reflective capacity and our sense of responsibility.*

In terms of 'why' the process seemed to be encouraging, one student explained in the last workshops that:

> *Afterwards, with these sessions, obviously I will pay more attention when I see water, speak about water and speak about water with others... We've created a type of interaction between ourselves, where we are able to raise problems that we have not necessarily brought up before, and that we've done between nine people... That will recreate itself, we'll talk about it with people around us. I won't say that it could become like an epidemic, but it could happen like that... even if there aren't any solutions at the end of these discussions, there will still be a raising of awareness of those who feel somehow concerned.*

There were also responses that legitimised the approach by suggesting to the organisers that the participants would like to participate in a similar exercise again. For example, one of the students wrote:

> *Thank you to you [the organisers] and maybe see you soon. Don't hesitate to contact us again for another similar experience of even something else.*

This idea of being willing to repeat a similar experience was shared by most students, with only one being sure of not wanting to participate in such a process again.

From an external perspective, it appears that creating an abstract setting for the students was also a reasonable way of not giving the students false hopes of governance actions on their lives, but rather worked on building the idea of taking responsible actions at their own level to impact the larger system. This idea of 'responsibility' was also noted by a student in one of the previous quotes.

Despite the process not being entirely completed as planned, the methodological basis of the process could be considered to have been given external *ex-post* validation, as similar processes based on this one were accepted for implementation in the Australian and Bulgarian interventions outlined in this book. In other words, carrying out this particular process and the results obtained were considered sufficient to give the two principal organisers sufficient knowledge and confidence to intervene in other water management processes.

EFFICACY

Determining the intervention's efficacy requires investigating to what extent the means used in this intervention allowed the objectives of the process to be met. As outlined at the beginning of this appendix, there were three collectively stated objectives for this process, the progress towards which will be outlined separately here.

Firstly, the objective to '*Test the relevant hypotheses of the impacts of participatory modelling in this abstract case (outlined in Section C.2.2) through extensive evaluation procedures, which include the 'ENCORE' model components*' was only partially achieved, owing to the difficulty of trying to test hypotheses related to real-world situations with high stakes in a 'low-risk and abstract' environment. However, using elements of the ENCORE model and the hypotheses as the basis of questions in the evaluation procedures did provide a number of interesting preliminary results. These included that:

- The participants were slightly in agreement that they were aware of and accepted the hypotheses and uncertainties in the model (including that these had been explicated and discussed); yet there appeared to be a potential partial refutation of the 'trust'-building part of the hypothesis in this intervention: '*Participatory modelling increases trust, appropriation and understanding of the models created, as assumptions are more likely to be explicitly identified and discussed,*' as outlined in Section D.4.1.
- The problem situation and formulation sessions and those with longer group debates induced the greatest creativity and learning for the participants (about water, others in the group and themselves), which provides partial support to the hypotheses that '*Participatory modelling generates greater creativity and innovation, and leads to individual and social learning.*'
- There appeared to be a slight process impact on participant thoughts and actions, as after most of the workshops the participants had been thinking about the water problems, especially when they had been using water, and occasionally talking about them with others.
- Changes occurred in the cognitive frame of participants between the start and end of the workshops, with the average understanding of problems and knowledge about the impacts of their practices and actions regressing in most sectors, except for those directly related to the quality and quantity of water, leading to another hypothesis that increased knowledge has led to the realisation of having much more still to learn about many things.

Next, the objective to '*"Learn by doing" about method selection, evaluation procedures and other process constraints and opportunities in a relatively risk-free research environment*' was mostly achieved, apart from not being able to trial a multi-criteria evaluation method, owing to time constraints in the workshop programming, as outlined in Section 6.6.

Finally, the objective to '*gain a greater understanding of the "general public" perspective on water issues*' could only be assumed to be partially achieved, owing to the subset of the 'general public' population who were involved in the test. The group did exhibit a good range of diversity of views, despite all being students, and they appeared very open in outlining their vision of water and its management. One result of note was that the group believed the ultimate goal for water management on a personal level should be 'happiness' – whether such an idea is more broadly shared in the general public will require further research. However, other insights that seem more likely to occur in the general public included the participants' lack of knowledge about how much freshwater on the planet occurs as groundwater, and a general lack of consciousness about the water 'embodied' in food and manufactured products.

EFFICIENCY

Determining the efficiency of the process is related to asking whether a minimum of resources was used for the achieved outputs. It is difficult to estimate comparatively the extent to which the process was efficient. However, a few indices were obtained from other sources. Firstly, the participants' opinions on whether they thought the process was 'cost-effective' or whether they thought '*the results obtained could not have been obtained in a more cost-effective manner*' elicited three 'mostly agree' responses to two 'mostly disagree' responses (plus four 'neither agree nor disagree'), which indicates little, apart from the fact that they do not have strong opinions on the question. From the organisers' perspective, the process seemed very efficient from a time resource perspective, considering the quantity of raw information produced for research analysis in the space of just over a month and the rapid learning they experienced. From a monetary perspective, the efficiency was perhaps not optimal, as participants were paid for their participation, which would perhaps be unnecessary in a 'real' situation.

INNOVATIONS

Determining the innovations of this intervention involves investigating to what extent new forms of collective action and knowledge emerged through the process.

From a perspective external to the group involved in the intervention, a certain amount of knowledge constructed through and after the process was made explicit on the full trial of a participatory modelling methodology with a diverse group of university students. Furthermore, the participatory process involved a previously untried combination of adapted problem structuring methods, conceptual modelling for computer simulation models, role-playing game design and payoff construction for a multi-scale, multi-role analysis activity of water management scenarios, at the individual and collective level.

Considering to what extent a new form of collective action was created through the process, a number of propositions can be made. By examining the process from an operational perspective, a number of aspects of knowledge creation from the participants' and organisers' perspectives have already been investigated in previous sections, demonstrating that the form of collective action that took place during the participatory modelling process did encourage some knowledge creation on both an individual cognitive level and a group relational level. In other words, a 'collective multi-disciplinary student–researcher learning group' was created through this process. In this form of collective action, each individual was able to reflectively construct his or her own knowledge as a result of method-supported interactions; both with the 'tools' or 'artefacts' used and created through the process, and inter-personally within the interaction space.

From this insight, it could be imagined that such a form of collective action could be recreated as an advantageous method of providing university-level education linked to water management and participatory processes. Running a participatory modelling process as a type of 'non-traditional' education programme could be envisaged, especially for research-oriented universities or international Masters programmes. In such a programme, it could be considered that there are no real 'teachers' but rather only 'mutual learners', with different existing bodies of knowledge to be investigated, exchanged and constructed. Many improvements and adaptations for this purpose could be imagined, including that the students could design part of their own payoff (marks for the programme) and play a larger role in evaluating the process, perhaps through a journal of observations and final process analysis. Depending on the range of students and their background disciplinary skills, it could be envisaged that a few students in the group take the role of building a computer model, if one is required.

D.5 REMARKS ON THE CO-ENGINEERING PROCESS

From this intervention, it appears that the 'engineering' of the participatory modelling process by a cohesive team becomes 'co-engineering' when the 'team' members have divergent objectives. Divergent objectives may be manifested as cognitive, normative, relational or operational conflicts, or remain latent. In the latter case, there is a distinct possibility that one party will 'drop' an objective if he or she is not interested in entering into a negotiation or situation of conflict, in order to attempt to achieve it. Another possibility is that one actor could try to silently work towards an objective that he or she is aware is not shared, in the hope that the actors with divergent objectives will forgo their own. Manifest conflicts, depending on their importance and impacts on the activities that the project team is to perform, will tend to require some form of management to allow the project work towards central objectives to continue.

From the co-engineering process of this intervention, just a few examples to demonstrate these theoretical insights on conflicts that create the 'co-engineering process' will be provided.

One manifested cognitive conflict in the co-engineering process, which had an impact on the design and implementation of the participatory modelling process, related to my underlying epistemological beliefs and those of my director. I was sceptical about taking a positivist approach to hypothesis testing in a small-scale social situation, for which no sufficiently comparable 'control' could be found. I instead believed that taking a pluralist epistemological approach suited to what was to be encountered in the intervention, and drawing upon constructivist, action or positivist epistemological positions, would be more beneficial to my research objectives. It is partly for this reason that the systematic testing of the hypotheses was not entirely completed, as I did not value highly enough the comparative testing of the hypotheses in an initially proposed separate 'integrated expert modelling' case. Rather, the time not dedicated to hypothesis testing was spent analysing and theorising the type of collective action in which I had been involved and in determining what knowledge had been constructed through this action, and later in initiating my involvement in other interventions, instead of completing the unfinished simulation model for the second planned section of the trial. Such a cognitive conflict is loosely linked to the question of whether the research is to be carried out *on*, *about* or *with* others (Heron, 1996), which then links it to normative and ethical considerations. Whether this type of cognitive conflict can be resolved, or would even be of value to resolve, is still under debate. I found that this cognitive difference was a useful source of creative tension and a vital driver for self-reflection and epistemological analysis throughout the intervention research process.

One example of a more normative conflict in the co-engineering process, which has already been highlighted in Section D.3.1, occurred between the original facilitator and the research director. The lack of appreciation of each other's 'rules' or value systems related to how the facilitation should be carried out ended in a conflict, which resulted in the facilitator leaving.

In terms of operational conflicts related to project team practices, one of the manifested conflicts occurred between the original modeller, my director and me, as outlined in Section D.3.1. Each actor exhibited different competencies and had different previous practical experience in the domain of modelling. The objectives of how we were each to work with the others also appeared to be divergent and could not be converged in the short period of time required.

These last two conflicts also appeared to exhibit relational aspects. It could be considered that, perhaps because of a lack of prior work on relationship building before working together in the resource-stressed intervention environment, levels of trust between the actors involved were potentially not at sufficient levels to work through the normative and operational conflicts in more constructive ways, which resulted in the consequence of a project team member leaving the process.

As a result of analysing such conflicts from this process, it could be suggested that team-building exercises between members of a project team who have never worked together before could potentially be a beneficial use of time to avoid extreme conflict management solutions. If such time is not available, it is suggested that co-engineering a model with design team participants who have never previously worked or modelled together before is not advisable for time-limited projects. Likewise, designer-facilitated processes are suggested when there is not sufficient time allocated for transferring understanding and skills to the facilitator, so that he or she is able to facilitate the required activities effectively.

D.6 DECISION-AIDING MODEL USE

Analysing the intervention process in a more formal manner, Table D.3 presents the elements of the Tsoukiàs (2007) decision-aiding process model and how members of these sets evolved through the process.

The model was of particular use for understanding the limitations and future needs of improving a participatory modelling methodology used in a complex and time-limited setting. Some of these issues will be further discussed in the next section.

D.7 PROCEDURAL DISCUSSION

From the Montpellier methodology trial process, several points worthy of further thought and discussion have emerged through a preliminary phase of reflection, which include: how time management can be improved; finding optimum levels of procedural and model complexity; and determining an adequate balance between participant world views and information submitted for group analysis by outside 'experts'.

D.7.1 Time management

The first and most important lesson that must be stressed about participatory processes, such as the one undertaken in this test, is that they take time. As the first phases of the proposed methodology are supposed to allow relationship building (in terms of trust, confidence and respect for one another), as well as creativity and innovation, to take place, they cannot be easily rushed without negatively affecting these aspects. In the Montpellier case, the first test of the methodology, the idea of 'adaptive participatory management' was adopted as the foundation of decision making concerning the timing or flexible reprogramming of the sessions. This choice was made to give more power of decision to the participants on the running of the workshops and the progress of their own work, with the eventual aim of aiding their integration into the process and personal empowerment. Although the outcomes of these aims were not necessarily achieved or even assessable, the co-decision making between the participants and organisers of the workshops did appear to encourage participants to work at their natural pace and allow them to complete the problem situation and problem formulation stages of the methodology more comprehensively than would have been possible under the original programme schedule. The adaptive management led to an extra workshop being proposed and accepted by the participants after the first two stages of the methodology, in order to attempt to increase the time available for the remaining stages. Owing to participant and organiser unavailability late in October, the extra workshop took place two days before the last three workshops planned for mid-November. This led to time difficulties for the computer modelling and aspects of the game creation work, which took place outside of the workshop (theoretically one week in total), building on from the participants' model and game definition work. This one week proved to be too short a time to produce the complex model that the participants had outlined.

Possibilities for overcoming such time issues in the future could take a range of forms, including:

- Changing the workshop objectives – time management may be improved by more explicitly defining what the group expects the final outcomes of the workshop to be, rather than by collectively redefining these objectives and the methods used to achieve them based on the amount of time remaining. This may help the group to focus more on the achievement of objectives within a time limit, as well as result in the effective usage of workshop time. However, such a goal-based focus may lead to other process elements being neglected, such as relationship building, time for debate, consensual decision making and individual reflection time.
- Pre-defining a maximum model complexity to ensure a sufficiently simplified (but adequate) model can be constructed – a pre-defined idea on the complexity of a required model could allow easier calculation of the time required for the modelling process. However, exactly how to determine the required complexity of a model that will be useful for the application to be defined by a group of stakeholders is far from an easy task, and just by itself could take a whole series of workshops unless it is only arbitrarily chosen by an outside expert (which is very likely to go against the philosophy of the participatory process and may not meet the requirements of the stakeholder group).

Table D.3 *Decision-aiding process model element elicitation and evolution*

	Manifestation of model elements	Evolution through process
Representation of the problem situation: $\mathbf{P} = \langle \mathbf{A},\mathbf{O},\mathbf{R} \rangle$	**A** – *set of actors*: $\mathbf{A} = \mathbf{C} \cup \mathbf{K}$, $\mathbf{T} \subseteq \mathbf{A}$	
	C – *subset of core-participants*: nine members of the set defined during the 'design process'	Little evolution through workshops – one member fewer in WS2 and WS3 (see Section D.3.2)
	K – *subset of associated stakeholders*: elicited as a subset of 'factors' and 'management' in WS1 and 'actors' in WS3 (as part of the problem formulation exercise and in the 'scientific' problem situation map)	Approximately: 13 discernible actors or actor groups identified in the problem situation maps; 21 identified in the problem formulation map; 49 identified in the 'scientific' problem situation map (not all of them common to the list). The seven 'most important' actor types were then chosen for UML and dynamics modelling in WS4 and then split into nine actors in WS5 for the game roles. Many of the last nine did not explicitly appear in the first problem situation maps but could be related to the 'uses' outlined
	T – *subset of project team members*: found within institutional network	Membership different for initiation, design and implementation process. Two implementation members only participated for one workshop each (see Section D.3.1)
	O – *set of objects*: elicited as 'stakes or objectives', 'problems' and a subset of 'factors' in WS1	The objects or stakes elicited in the problem situation map were carried by a group representative into the problem formulation mapping exercise, where many of the objects or stakes were listed in a more ordered form as the 'objectives'. A number of the other system objects, such as the elements of the hydrological and societal systems were further outlined and defined in the UML modelling and simulation model. When linked together with relations and processes, these objects formed the 'working micro-world'
	R – *set of resources*: not explicitly elicited in WS1, but some 'stakes or objectives' could be classed as resources; some more elicited through the 'means' category in the problem formulation map	Although a few resource elements were made explicit, such as '€ (money)', 'expertise', 'courage', 'social links' and 'wisdom', this category was treated implicitly through much of the process. During the role-playing game, the 'players' would have to use their own intrinsic resources (i.e. ingenuity, logic, communication skills) and those attributed to them in the game (i.e. a budget) to act. Some of these were evaluated through the questionnaires
Formulation of the problem and objectives: $\boldsymbol{\Gamma} = \langle \boldsymbol{\Pi}, \boldsymbol{A}, \mathbf{V} \rangle$	***Π*** – *set of problem statements*: the essence of these problem statements was elicited as the 'objectives' in the problem formulation map	The problem statements were linked to the objectives that were made explicit on the objectives map. These objectives were discussed as a group to determine the mode for important 'problems' to be treated or 'objectives' to achieve. From a hierarchical cluster analysis using DecisionExplorer®, the map could be broken down into seven main problem areas (see Section D.3.2)
	A – *set of potential actions*: these were elicited as 'actions' in the problem formulation map. Some other actions which could contribute to managing these problems were found in the 'plans or strategies' category of the problem formulation map elements, as well as in a number of categories: 'management'; 'factors'; 'uses'; and 'stakes' of the problem situation maps	The potential actions that could be taken in the system were formulated generally within the problem situation maps. More were then specified relative to certain actors in the problem formulation map, UML and dynamic interactions modelling

Table D.3 (*cont.*)

	Manifestation of model elements	Evolution through process
	V – *set of potential points of view*: a number of these were defined as the indicators of the objectives or sub-objectives in the problem formulation map	Through the problem formulation mapping, a number of points of view were made explicit on what creates personal 'happiness', the principal objective on the personal level. A number of these 'sub-objectives' and their corresponding indicators then further evolved through the modelling and role-playing game creation to form the most important elements in the 'set of dimensions' and the 'set of preference criteria' in the 'model exploration and options evaluation' meta-object. Other indicators also formed potential points of view from which other actors such as the 'government' could observe the action sets, such as the 'unemployment rate' or 'groundwater levels'
Model exploration and options evaluation: $\mathbf{M} = \langle A^*, \{\mathbf{D}, \mathbf{E}\}, H, U, F \rangle$	A^* – *set of alternative sets of actions*: elicited in WS4 during the UML modelling as the 'processes' on the 'actor' cards; and in WS5 during the role-playing game design	The alternative action set or 'options' for evaluation were first specifically defined principally in WS4 during the UML modelling as the 'processes' on the 'actor' cards; and during the role-playing game design, scenarios, role-playing game cards in WS5; and in other actions during the game and scenario analysis in WS6
	D – *set of dimensions*: elicited as 'attributes' during the UML modelling in WS4	Some of the defined attributes in WS4 had stemmed from the 'indicators' on the problem formulation maps. The final set of dimensions explicitly used in the model and role-playing game were selected by the modellers, based on time and capacity constraints. Other dimensions held implicitly by participants were sometimes voiced or remained implicit in the discussions on 'good' management in WS6
	E – *set of corresponding scales*: developed behind the scenes by the modellers; some hypotheses on trends checked with participants in WS6	Owing to the 'abstract' nature of the problem situation, data showing the trends seen in the Montpellier region were found or created by the modellers to the extent possible in the time constraints. For example, approximate patterns of rainfall for three regions within the basin were found for equivalent-type Mediterranean basins to create the hydrological part of the integrated simulation model. Other assumptions where the modellers had less expertise were discussed with the participants (see Section D.3.2)
	H – *set of preference criteria*: developed as models of satisfaction in WS5 for each actor to be included in the role-playing game	Criteria or preferences on what is important for each actor were developed in WS5 with reference to the work that had already been completed in the UML modelling and indicator definitions on the problem formulation maps. Other implicit preference criteria were evident in the group discussions throughout the workshops. Some of the participants' preferences were also elicited through the evaluation questionnaires and examined from observation
	U – *an uncertainty structure*: the principal uncertainty structure in the model was linked to the rainfall distribution and was decided upon by the modellers	The question of uncertainty and its structure was not treated in a systematic manner through the workshops. Some thought was given to it through the scenario analyses and discussions on system complexity. Uncertainty in the role-playing game was principally driven by actor

Table D.3 (*cont.*)

	Manifestation of model elements	Evolution through process
		choices (outside of the model and set choices given on the game cards), including the scenario choices of the organisers
	F – *set of operators*: a number of processes and relations were defined in the UML and dynamic interactions game. The remainder were developed by the modellers	The explicit, numeric operators for linking the elements in the above sets were mostly defined by the modellers. However, especially due to the lack of simulation model use in much of the scenario exploration exercises, many implicit operators and arguments were used by the participants to decide on and defend their actions
Final recommendations: $\boldsymbol{\Phi}$	$\boldsymbol{\Phi}$ – *set of final recommendations*: partially provided by participants in WS7	Owing to lack of time to analyse further the preferred alternative sets of actions through a multi-criteria analysis, the participants gave their personal 'action plan' as their set of final recommendations for their own 'real' lives. A number of suggestions for collective management, which they felt most important or appropriate, based on their own value systems and implicit analyses, were also developed. The planned idea of creating a time-dependent action plan (now, 5, 10, 20 years) was not completed

- Increasing the time between the model construction participant workshops, model testing and model exploration and role analysis (game) workshops – if time were to be increased between these three workshops, there is more chance that there will be sufficient time for the computer model to be built and then adequately tested before being discussed and approved by the participants. Although the model may reach a more complete form because of the extra internal time, it is not guaranteed that the model assumptions will be any better explicated than before, as the short workshop times do not easily permit extensive debate. For this to occur, more 'in-workshop time' would be needed, with participants interacting more directly with the model.
- Reducing time spent in the workshop on filling in questionnaires and in administrative activities – to dedicate more time to the workshop activities in the methodology, it may be possible to reduce the time required for the filling in of questionnaires and administrative activities significantly, by confining them to 'home time'. In other words, all the explanation of contract details and background information on the research could be distributed by mail to the participants before the commencement of the workshops, along with the first questionnaire, and later questionnaires for just after or before sessions could also be taken home, filled in and brought back to the next workshop. Although leading to significant 'in-workshop' time benefits (a potential 20% augmentation), this approach may lead to participants feeling bombarded with information and tasks, and would lead to participants spending more overall time on the process. There are also risks in the 'control' of how and when the questionnaires are filled in which could significantly bias or entirely discredit results drawn from them.

This short range of possibilities demonstrates that there is no easy way to improve time management significantly, although increasing the time between workshops is likely to have the most immediate benefits on the worst of the problems encountered in the Montpellier test, especially for the workshop organisers. If extra time is not possible, then not using a quantitative simulation model and, instead, sticking to qualitative methods might be the best option.

D.7.2 Complexity

The question of how to deal with complexity was one of the recurring themes throughout the series of workshops; both methodological process complexity and complexity in terms of the real-world water problems and their relation to society (and how they can be represented or collectively modelled). In particular, the questions that appear the most pertinent include, 'What level of complexity is required?' and 'How can this level be determined?' This includes not only which environmental, economic and societal phenomena and interactions should be taken into account, but also on which scale or range of scales (i.e. microscopic, individual human level, regional, global...) the problem should be analysed. Unlike the imposed set of scales for the Montpellier trial, it is proposed that these scale levels should be determined by the principal group of stakeholders involved in the process at the very beginning. To determine what complexity 'needs to be' or 'ought to be' represented, a practical approach based on time constraints, accessible information and process objectives is proposed. It may be more feasible to represent some subsystems, where reasonable data are available, in a completely quantitative manner and to leave other more difficult-to-quantify subsystems and their interactions to be only qualitatively analysed if they are required in problem solution option evaluation. For example, if a multi-criteria analysis of options is undertaken in the last stage of the methodology (as was proposed for the Montpellier test, but did not occur owing to time constraints), difficult-to-quantify measures that could include in-house water quality, happiness or aesthetics could be comparatively assessed on a qualitative scale where 'very negative' to 'very positive' effects could be expected when compared with the current situation (i.e. – –, –, 0, +, ++ or a scale in words). This may reduce computer model complexity and therefore modelling time requirements. This sort of analysis may also prove more comprehensible for stakeholders who have difficulty assimilating the meaning of numbers, reducing the barriers between those who are more quantitatively or qualitatively oriented.

Although this analysis comes nowhere near to answering the questions of what an 'adequate' level of complexity is or how it can be found for a certain problem, it is thought that by using socially acceptable decision making procedures for the context of the problem to determine a preferred solution to the problem for the current point in time (even if the decision is based on partial information), the effects of the decision can be monitored over time and the situation adaptively managed as new information, scientific research results or 'best practice' considerations come to light. A hypothesis of the participatory methodology outlined in this appendix is that by involving stakeholders in a decision making process related to water resources management, social capacity will be built and learning will have occurred that will allow the stakeholder communities to deal similarly with future problems. It is the process of bringing people together to work on common issues and problems, and to come to collective decisions that can be successfully implemented, that is considered to be the most important, rather than just the production of a one-off solution.

D.7.3 Outside 'expert' information vs. stakeholders' world views

Throughout the Montpellier trial, an important question emerged regarding if, when and how much of the facilitators' (the research director's and my) own 'expert' knowledge, or accessible information, should be included in the process. Having originally been designed to accommodate a range of stakeholder knowledge from different disciplines or interest groups, the proposed methodology did not explicitly treat the question of how information or knowledge 'outside' the group sphere could be included. It was initially proposed that each stakeholder individually create his or her own point of view on the problem situation before stakeholders collectively exchange points of view and further formulate the problem together. This process would allow each individual to question and portray his or her own understanding and perspective of the problem, before being influenced by other people's perspectives, which, apart from being considered a useful starting point for collective problem analysis, can produce richer insights from a social research point of view.

For example, in the test no one represented groundwater as being in any way linked to the water cycle and its associated problems before it was mentioned in one of the questionnaires. However, after the first two workshops, the participants asked the facilitators whether the problem situation diagrams were going to be 'corrected', and when they were going to be given the facilitators' 'view' of the problem. In the pre-test discussions over what could be done in the case of a lack of information in the stakeholder group, it was suggested that an exterior 'expert' view could be given, but only on the request of the participants (as could happen in a real case study when the group of stakeholders believes it does not have the necessary or correct knowledge to treat a certain aspect of a problem, and so would usually set the task of asking one of the group members to find it, either by consulting external information or by asking an expert). For this reason, when the 'facilitators' view' was asked for, it was provided in the same problem situation form as that which the participants had constructed, with the addition of some generally accepted scientific facts related to world water quantities. What the group then did with this information was left to their discretion.

The main question, which came right towards the end of the workshops, was whether it would have been more efficient to have provided more general information about water and its related problems at the beginning of the workshops, or even to have just provided a model that the participants could have improved. In terms of efficiency, it may have proved advantageous. However, the participants had varying ideas on the issue, with one participant stating that she was frustrated not to have had more information provided at the start, whereas another student stated that:

> *I would not have come back [after the first workshop] if you [the organisers] had not given us a chance to formulate our own visions.*

Therefore, no clear conclusions can currently be drawn on when or whether 'expert' opinions or views should be included in a process of participatory problem structuring and modelling for decision support applications. However, if 'expert' views are asked for by all of the stakeholder group members, then it is most probably more reasonable to supply them at some stage before decisions must be made, rather than refuse to do so at all.

Further information on the Montpellier pilot trial process, including the evaluation questionnaires and detailed implementation summary, is obtainable on request.

Appendix E Supplementary information on the LHEMP Process, Australia

This appendix supplements Chapter 7 by extending the brief methodology design and implementation descriptions previously provided.

E.1 DEVELOPMENT OF A PLANNING PROCESS

This section presents extracts of the conceptual design of the participatory modelling process, which I proposed and presented to the Hornsby Shire Council and its estuary manager before the intervention. The full details were provided in the internal working document, *Preliminary Project Proposal – A Process of Knowledge Integration for the Berowra Creek Estuary Management Plan 2010* in May 2006. As can be observed from the title, this proposition was based on the renewal of the Berowra Creek Estuary Management plan, rather than the Lower Hawkesbury Estuary Management Plan (LHEMP).

E.2 PRELIMINARY PLANNING NEEDS DESIGN

During a meeting with me on the 7th April 2006, the Hornsby Shire Council's estuary manager outlined his first thoughts on what kind of planning needs would need to be written into the tender for the Berowra Creek Estuary Management Plan (BCEMP). He composed his plan and requirements of the process in four main stages, with a number of sub-questions or actions, as shown in Table E.1.

Even from this first range of needs highlighted in Table E.1, the estuary manager and I thought that a number of these needs seemed to lend themselves to the application of a participatory modelling process (shaded in grey), as part of an overall planning process. Questions related to how scientific tools such as the Bayesian network estuary model (** in Table E.1) could best be constructed and used to support the estuary planning and management processes, and could also be addressed by a wider group of the estuary's stakeholders through a participatory modelling and planning process.

E.2.1 The first proposal: four stage participatory modelling and planning process design

Considering the needs of the Berowra Creek Estuary Management Plan revision process, I developed a process proposition, which is outlined next. This initial proposal included that the final processes were to be further designed in collaboration with the Hornsby Shire Council, consultants, the Berowra Creek Estuary Management Committee and any other necessary stakeholders and community members. The process also needed to be adapted according to resource constraints.

The proposed process was one of participatory modelling, which took the form of a series of workshops with a group of stakeholders from the general spheres of politics and management, technical experts and stakeholder communities. Determining who should be involved would be one of the first steps in the process and could simply involve the members of the existing Berowra Creek Estuary Management Committee if this was deemed sufficient. Once a general form for this plan revision process had been decided upon, the tender for external consultants to aid in the process would need to be sent out. Further discussions and redefinition could then be undertaken with the consultants selected and other stakeholders to decide on the specificities of the process.

I believed that, as originally envisaged by the estuary manager, it would be useful for the consultants first to undertake an external review of how the implementation of the current Berowra Creek Estuary Management Plan had been carried out and what issues remained. Once this information was available, a series of participatory modelling workshops to establish the other stakeholders' viewpoints on these issues and to comment on the consultants' findings would provide a good basis to begin collectively examining the range of issues related to the process opportunities outlined above. The four-stage participatory modelling process design for this application is outlined in Figure E.1.

Table E.1 *Preliminary planning needs structure*

1. Assessment: actions	What has been completed from the 2002 BCEMP? Were these actions adequate (based on what indicators)?
2. Assessment: issues	Which issues from the 2002 BCEMP are current or ongoing? Which issues are redundant? Are there any new issues? What could be the future scenarios? What indicators are being used to assess these issues?
3. Knowledge: to be integrated into the planning process	1995 Estuary processes study Prototype Bayesian network model**
4. Action plan	Community and expert opinions Goals Actions Costing Gaining support for plan from governments, communities, private sector

The process design, shown in Figure E.1, was based on an adaptation of a six-stage participatory modelling framework, which I had previously designed (outlined in Appendix D). The main reason for adapting the process was that it was originally also designed to allow for the construction of a quantitative computer simulation model as a basis for further process examination and decision making. With such a model not being specifically required for this particular phase of plan revision, a model testing and calibration phase was omitted. The process also mimicked the structure in an adapted format of some 'decision-aiding processes' (i.e. Tsoukiàs, 2007). It was noted that some of the decisions sought in the final stage might take place in earlier stages of the process and could simply be re-summarised in this phase, as the whole participatory modelling process is formed from a series of collective decision cycles. Throughout the process, it was also envisaged to have a small set of internal evaluation questions or activities for a number of reasons: to monitor progress towards objectives; to drive collaborative improvements (either for this process or for future processes); to resolve any problems or conflicts more quickly; and to aid personal and collective learning. An 'external' evaluation, separate from this internal evaluation, could also prove advantageous.

E.2.2 External process evaluation

At this stage of the project initiation, there had already been interest in the preliminary proposal stages of this project (on applying participatory modelling methods to aid knowledge integration for estuary planning) from researchers from around the world using similar methods to try to improve natural resources management. As the concept of 'participatory modelling' was still relatively new in the context of aiding the sustainable development of natural resources, and the communities that rely upon them, an international evaluation project funded by the French National Research Agency under their 'Agriculture and Sustainable Development' programme had been created to compare applications of these methods in over 20 case studies in a number of countries. One of my supervisors was managing part of the project entitled *Companion Modelling: A Research Approach Supporting Sustainable Development* (ADD-COMMOD), as well as the Australian case studies. This supervisor was enthusiastic about this project proposition with the Hornsby Shire Council and he and his colleagues wanted the opportunity to evaluate the process that would be undertaken and its later effects, to be able to compare and contrast it with other such participatory initiatives around the world. If the Hornsby Shire Council were willing to participate in this comparative evaluation project, the costs of an external evaluator working on the project would be covered through a combined grant from the French National Research Agency and the Sustainable Ecosystems group of the CSIRO. The evaluation would follow a standard procedure, which would require a number of elements, including a description of the project context, process observation and an *ex-post* summary (all compiled from the point of view of the external evaluator); which would be based on interviews, questionnaires and document analysis. Some costs in the form of time to reply to questionnaires or interview questions would probably be incurred through the evaluation process for the participants in the participatory planning process, although it was thought that the overall opportunities and benefits of being a part of such an international project could outweigh these costs. As well as giving the participants of this evaluation a chance to reflect individually on the process and what it achieved, a thorough *ex-post* analysis of this 'participatory knowledge integration planning' method would allow all participants and other outside individuals and organisations to discover how the process was

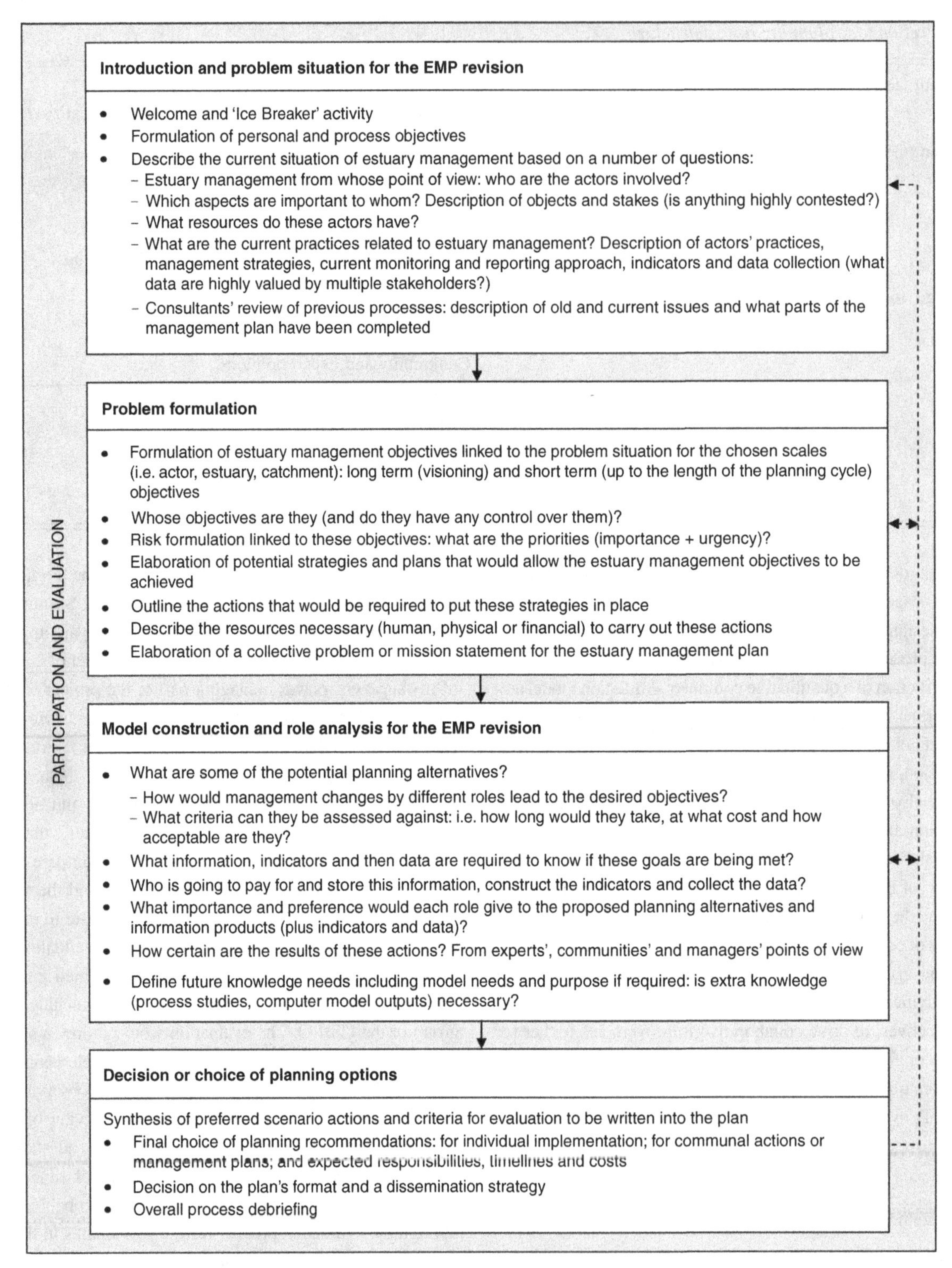

Figure E.1 Proposed Berowra Creek Estuary Management Plan revision process.

implemented and access more general feedback on the process and its outcomes. Such an evaluation could therefore help the transparency of the process and make it easier for other groups to repeat, improve or adapt their own projects in the future, as good records would be available. Another potential benefit would be the national and international exposure that the estuary and its management team and communities could receive (in the form of publications and presentations) to highlight their commitment to pursuing new initiatives to achieve sustainable development.

Table E.2 *Proposed timeline of plan revision process*

Month	Proposed actions
June 2006	– Tender process
July 2006	– Consultants selected
August 2006	– Plan revision process organised and publicised
	– Participants (stakeholders) found
	– Research administration completed
	– Planning review: consultants to analyse previous processes and determine what has been done or not (in consultation with council and stakeholders where required)
	– *External evaluation: process commenced*
September 2006	**– 1st participatory workshop (half-day or one-day) stage 1**
	– *External evaluation: context evaluation finished, process observation started*
October 2006	**– 2nd participatory workshop (half-day or one-day) stage 2**
November 2006	**– 3rd participatory workshop (half-day or one-day) stage 3**
December 2006 or January 2007	**– 4th participant workshop (half-day or one-day) stage 4**
	– Public consultation and opinions about plan sought
January 2007 or February 2007	**– 5th participant workshop (only if required)**
	– Plan completed and sent for approval
	– *External evaluation: process observation completed*

E.2.3 Proposed timeline and required resources

During the meeting of 7th April 2006, and in subsequent communications, the estuary manager stated that he would like the plan revision process to be finished by the end of 2006 or early 2007. I also explained that this timeframe would suit me, as I planned to return to France in March 2007 to continue work on other aspects of my research project. I would also be in Europe for about a month from 9th June 2006. It was suggested that the exact timeline of the project stages be organised with the Hornsby Shire Council and then adapted to meet the consultants' preferences. From my previous experience, I knew that a reasonable amount of time was required between participatory workshop sessions to allow the organisers to treat the information and work received, in order to organise the following session adequately, so this was factored into the timeline.

It was therefore envisaged that the process stages outlined in Figure E.1 could be carried out over a series of four or five half- or whole-day workshops spread over a number of months and integrated into a larger planning process, as highlighted in the timeline found in Table E.2.

In Table E.2, the process steps to be undertaken by the external evaluator (which in most cases would require the participation of those involved in the planning process) are given in italic text, and the set of participatory workshops to be conducted are shown in bold text, for easier differentiation from the other general tasks of the planning process. 'Research administration procedures' refer to the necessity for any researcher from the Australian National University to adhere to the university's ethics standards related to human research, whereby any participants taking part in the research project would be asked to sign a formal agreement acknowledging that they agreed to take part in the workshops and evaluation procedures (questionnaires, interviews and process observation). Such an agreement would also be required from the Hornsby Shire Council, where the ethics principles relating to confidentiality, storage of data related to the project and a number of other issues would be outlined.

A number of resources (human, physical and financial) would be required to conduct a participatory planning project successfully. Firstly, smoothly running workshops require pre-planning and thus adequate human resourcing. This pre-planning includes determining and inviting participants, workshop programme structure and logistical organisation. During the workshop, facilitators and potentially an evaluator would also be required. Other stages of the planning process, such as analysis of the existing plan and communication strategies for distributing the plan for public comments, would also require human resources. From discussion with the estuary manager, it appeared that the consultants who would be tendered for could take care of some of these responsibilities. I would also make myself available at no cost to the Hornsby Shire Council to help with the organisation and facilitation of the planning process workshops. If the Hornsby Shire Council agreed to be a part of the international evaluation project cited above, the project manager would provide an externally funded evaluator.

For effective participatory workshops, a number of support materials are often useful for encouraging open participant interaction and creativity. Suggested physical resources to be made available for the series of workshops would include:

- A venue (with chairs, not too many tables and sufficient space for participants to interact without physical barriers);
- Food and drink for participants;
- Computer, projector and projector screen (plus the required cables and power boards);
- Facilitation materials: butchers' paper or large paper equivalent; pinboards, pins; hexagons (coloured paper shapes); string; adhesive tape; coloured marker pens; pens; sticky dots; PostIts® etc.;
- Recording equipment: camera; audio; video (aids evaluation and reporting, increasing process transparency); and
- Information sources related to the planning process: maps; models of the estuary and catchment; previous plans; other related documents etc.

The costs and the question of who would provide these physical resources needed to be discussed in more detail with the Hornsby Shire Council before the project commenced.

E.2.4 Proposition conclusions

The proposal document had been written to outline a potential participatory process for the Berowra Creek Estuary Management Plan revision. Apart from being based on innovative research methods that would potentially bring more positive publicity to the region, this collaborative project was expected to have a number of benefits for the Hornsby Shire Council's environment and communities. Firstly, the project of knowledge integration was a process that would add value to the large volume of existing high quality research, work, monitoring programmes and planning that had recently been undertaken related to the estuary. This work could include, but not be limited to, the previous Berowra Creek process studies, the draft NSW Oyster Industry Sustainable Aquaculture Strategy (DPI, 2006), the Hawkesbury Lower Nepean Catchment Blueprint (DLWC, 2003), the Bayesian network model of the Berowra Creek catchment, the Community Sustainability Indicator Program and Report Cards (HSC, 2004a), the Shire's Management Plan (HSC, 2005b), the 2004/05 State of the Environment Report (HSC, 2005a), the Water Management Plan (HSC, 2004b), the Sustainable Total Water Cycle Management Strategy (UTS/SKM, 2005), the Water Quality Monitoring Program (HSC, 2005c), and other plans and research related to urban development, tourism, boating and ecological processes of the estuary.

An approach for how such knowledge integration could take place throughout this plan revision was outlined as a four-stage participatory modelling process. Such participatory processes are considered to help improve management, planning and monitoring processes through improving: communication and understanding between stakeholders; education; individual and collective learning (knowledge production); relationships and building social capacity to aid the sustainability of the environment and society; and the adoption of more sustainable behaviours. The proposed process, which was to be further developed and decided upon in collaboration with the Hornsby Shire Council, consultants and other stakeholders, drew upon a number of current 'best practice' concepts in evaluation and monitoring, water resources participatory planning, integration and decision-aiding theories.

Part of this process could also be used to discover the current knowledge gaps or uncertainties surrounding estuary management and how such initiatives as model building (e.g. the proposed Bayesian network Berowra Creek Estuary model) could be used to address such gaps. The need for model building for specific strategic outcomes could thus be properly integrated into the estuary's planning and management process. A proposal for the project to be evaluated externally and comparatively with other participatory modelling natural resources management case studies around the world has also been described.

As has been detailed in Sections 7.1 and 7.2, this initial project proposal was then expanded to support the creation of a new regional estuary management plan for the Lower Hawkesbury River, with three day-long participatory workshops. The next section provides further in-depth information to support the co-design and co-implementation summary presented in Section 7.3.

E.3 DETAILED CO-DESIGN AND CO-IMPLEMENTATION

The principal elements of the co-design and co-implementation phases, including both relational and operational aspects, are outlined in this section. A brief description of the methods used throughout the LHEMP process follows. Further information on the process can be found in Coad *et al.* (2007) and BMT WBM (2008).

E.3.1 Workshop 1: co-design

The detailed co-design phase for the participatory workshop process was commenced when the project manager and an environmental scientist from the engineering consulting firm, the estuary manager and I met in late October 2006 to design the contents of the first workshop. It appeared that the estuary manager, the project manager and I had amicable, mutually respectful and good working relationships. The environmental scientist was supportive of her boss and they appeared to have a good working relationship. This was only the second time that the project manager, and the first time that the environmental scientist, had met me. It was also the first time that all four of us were required to work closely in a group, so the relationships were still in the process of being formed, with each participant trying to understand the others' personalities, objectives and skills they had to offer the project.

Although the meeting was constructive and the work objectives for the meeting were achieved, clear-cut roles for each participant in the process design and implementation were not formally established, as the project manager, estuary manager and I all had an interest in how the workshop process was run and did not yet know the capacities of the others to meet these objectives. We also exhibited a number of similar leadership or managerial qualities, which led to an interesting dynamic in the design team through the process; we all had technical education and research backgrounds in environmental science and engineering at a high level (the project manager had finished his PhD in 2004 and both the estuary manager and I were working on PhDs), as well as interests and different levels of experience in planning and management processes. As we all held legitimate positions in the project to be aiding the design and management of the participatory process, we agreed a priori to continue to design and develop it collectively. This led to all three of us proposing process suggestions, designing and commenting on one another's work and negotiating final supports and outcomes to be presented and used in the workshops. Who would facilitate or present each activity outlined in the agenda was negotiated at the time of writing the agenda. A brief outline of the methods used in each workshop is given in Section 7.4.1 and more information on the process and activities in and between workshops is given in Section E.4.

Participant selection, although discussed by the project team, was principally decided upon by the estuary manager and the invitations were sent out by the project manager. A good cross-section of the local stakeholders was maintained by involving the estuary management committee members from planning zones on both the northern and southern sides of the estuary. Many appropriate representatives from the state and regional agencies with responsibilities over sectors of estuary management or funding of estuary management programmes had to be more specifically tracked down, as some of them were not commonly involved in local-level estuary management. This selection meant that many of the local stakeholders already knew each other and had previously worked together on estuary management issues, but many of the agency staff did not know each other or the local stakeholders. Some of the consequences of this selection and how it was adapted through the workshops will be outlined in the following sections.

E.3.2 Workshop 1: co-implementation

For the first workshop held in November 2006, it turned out that I designed the first sets of methods to be used (and their supports) for the morning session. The project manager facilitated the large group sessions and decided how the last discussion of the day would be run. Four engineering and planning consultants facilitated the small group work. One facilitator had a rather large group of participants with strong personalities that she found challenging to run, especially as she was having a problem that day with her hearing. Considering these difficulties, I asked if the estuary manager would be able to support this facilitator in her role, which he was happy to do. Working together, they were able to guide the group to complete a good amount of work. The external evaluators were critical of the project manager's 'gate-keeping' behaviour of only selectively writing down some of the participants' suggestions in the large group discussion he was facilitating at the end of the workshop. The slow speed of working in the large group also meant that the final exercise of the workshop was not completed as planned, which left the synthesis work to be carried out and justified by the consultants and me before the next workshop. Despite these small issues, the workshop was generally perceived by the project team members and the participants to have been a success and reasonable working relationships had been established. It was noted after the session that the estuary manager's boss thought that it would have been appropriate for the estuary manager to have played a larger role in presenting the project, rather than the consultant, so this was taken into consideration for the distribution of roles in the second workshop.

E.3.3 Workshop 2: co-design

The co-design and co-implementation of the second workshop was an illuminative example of how objectives of those in the project team and stakeholder group can affect process outcomes. At the second meeting between the estuary manager, project manager, environmental scientist and me for the design of the second and third workshops in mid-January, the project manager was interested in renegotiating how the remaining work on the project would be run, as has been examined through the negotiation outlined in Section 7.3.2. The only supporting information that will be presented here on this co-engineering event is a summary of the arguments put forward for having either a larger agency-only (government, industry and commercial representatives) workshop or no programme change. These arguments, which are presented in Table E.3, are important, as at the end of this negotiation everyone in the meeting was aware of the potential risks and benefits of the choice of the agency-only workshop, many of which occurred, as outlined, during the rest of the process.

During this working meeting, the estuary manager also presented the consultants and me with a process diagram based on the Australian and New Zealand Risk Management Standard, which he wanted to see used in the remaining workshops. He also mentioned his preference for changing the vocabulary used in the process to fit the risk-management framework and current

Table E.3 *Key arguments for and against the major programme change*

Arguments put forward for changing the programme to an agency-only risk assessment workshop	Arguments put forward for not changing from the original programme
It is often difficult to get agency representatives to participate in large participatory workshops with community representatives for a number of reasons. Firstly, they sometimes feel obliged to represent only the 'public image' of their role and the current political lines of their institutions, rather than their true feelings on management possibilities. Next, large workshops can often be rather confrontational, with agency representatives being 'attacked' by some community representatives on gripes they have with the agency's policies, which they feel they have little control over. Finally, many agency representatives have large jurisdictions of management and limited time to participate in all the planning and management processes that take place in their territory, so they are required to prioritise their actions and often only participate in the most important or personally interesting processes.	Not inviting the participants to the second or third workshops after telling them in the first session that they would be part of a participatory process and responsible for making many of the planning decisions (partly because of time and budgetary constraints!) would be seen as bad form and could produce a 'backlash' against the process and the future success of the process. It would be better to run the agency workshop if it were required as an extra one to keep the original programme of three mixed stakeholder workshops, if time and budget would allow for this eventuality.
Agency support and funding is required for the successful support and implementation of this plan. There is more chance of getting this support (especially from agencies that do not usually participate in our programmes) if there is an 'agency-only' workshop. It may be seen as something unusual and thus worth attending, less confrontational and a good opportunity to discuss management issues from a purely management point of view. The 'risk-assessment' session may also be seen as an 'appropriate' agency task that can tap their expertise.	Risk assessment, even if it attempts to make explicit uncertainties, is an inherently subjective process, which is especially true in this broad estuarine context. Therefore, the interest in using it is to get stakeholders to better understand the nature of risks though developing a common values-based assessment of them and then to use this method as a basis for 'calculating' priorities for treatment. Performing an agency-only risk assessment would lead to the need to 'sell' the results to the other participants in the following workshop and so the risk prioritisation could be refused and the process compromised.
Community involvement is very important to the success of this plan, but so are the agencies, as without them it will be next to impossible to fund and implement the plan. If the changes to the programme are sufficiently well explained, the community representatives will understand why they took place, even if they are initially disappointed. They have already participated well in the first workshop to develop the lists of assets and risks that will be assessed, and will also have the opportunity to create strategies and actions for these risks, so in the overall process they will not have lost much of the directional power.	As risk assessment is subjective, all stakeholders have just as much potential to contribute to it (especially as some of the assets the risks were to be assessed against were not particularly technical, such as 'scenic amenity'), and many of the 'community' representatives have more in-depth knowledge or scientific expertise on the estuarine system, industries and community values than some of the agency staff external to the estuary.

managerial language, in particular to use the word 'assets' for the 'values' that had been elicited. I had started to look at the standard a couple of days before the meeting but the consultants had not yet had a chance to do so. As none of the project team members had actually used the standard before, although they had some background understanding of the approach, they could not finalise the design of methods and content of the next meetings, owing to a lack of information. No one had very strong preferences over vocabulary use as long as the change was outlined to participants with reasoning, so 'asset' was to be adopted for the next workshops.

Over the next month after the meeting, the consultants finalised the synthesis report and documented the list of risks to be assessed in the next workshop, the estuary manager decided on a list of participants and I worked on developing the method to be used for the assessment, including preparing a number of risk tables and a calculation model upon which the assessment could be based. These roles were all fairly well self-defined and the completed work was checked by the others. Working relationships through this time were limited to email correspondence and phone conversations, with a few more lively debates taking place between the project manager and me. The short timing of the project and the consultants' workload meant that they were a little too snowed under to take the time to understand and critique the risk assessment approach that I had developed, which led to them preferring that I take on the facilitation of that part of the workshop.

E.3.4 Workshop 2: co-implementation

For the second workshop, held in mid February 2007, the roles were distributed differently from the first workshop, with the estuary manager presenting the project, the project manager giving updates on the workshop outcomes and the synthesis report results and facilitating the introductory sessions with stakeholders, me presenting and facilitating the large group sessions related to the risk assessment and four engineering and planning consultants facilitating the small group work. During the second workshop with the agency staff, a different aspect of co-design or co-implementation was demonstrated, as well as the advantages and disadvantages it produced. While discussing an example risk as a whole group before breaking into smaller groups to analyse the other risks, the participants requested a last-minute change to the workshop programme, to run through the 'water quality' risk as a whole group, as they thought it was one of the most in need of discussion, rather than completing it in the small groups as the cross-validation risk. Although this change was discussed briefly between the project manager, me and the rest of the group – the new co-design group – I accepted it in my role as the facilitator so as not to go against the participants' collective wishes. I put a proviso on this change saying that this should be all right as long as in their small groups they each completed the rest of the half-finished risk, which was being worked on for validation purposes. However, this decision did have a couple of ramifications on the risk assessment process, and particularly on its validation. The little extra time spent collectively on the water quality risk, although productive, meant that there would be less time available for the small groups to work their way through the other required risks. In light of this problem, the project manager and I found a solution, which was to break the large group down into much smaller groups than originally planned (i.e. pairs or threes rather than groups of four or five), so that all risks could be completed in the available workshop time.

This solution did achieve its original objective of finishing the risks, but there was no time to complete the remaining half risk for validation purposes. This left the question of whether the results of the risk assessment could be scientifically validated, as different groups might have different tendencies of rating behaviour. However, legitimisation of the results could still take place if the participants believed sufficiently in the process or the capacities of the other participants to accept their judgements. Such an agreement to support the results, despite their potential weaknesses, is in some way what occurred at the end of the workshop. The participants discussed a couple of surprising results and accepted without too many complaints that all of the risks had been prioritised as in need of treatment in the next phase of the process. It is likely that the rating of all risks as 'requiring a response' helped the lack of opposition to the risk assessment process, both from the participants who took part in it, and those in the third workshop. In essence, this meant that the second workshop did not have as much of an impact on changing the content of the LHEMP process as could have been the case if the prioritisation had changed the scope of interest for strategy building. There was thus less opposition and reaction to it and the majority of participant evaluations of the workshop were positive.

Through this workshop, there was positive relationship growth between me and a number of the participants, with these participants feeling able to recommend constructive adaptations to the process and suggestions for the next workshops, which I took on board positively, further discussed privately with them in the next workshop and then passed on to the project manager and estuary manager. It appears that the all-agency environment did provide a safe place for open discussion, as had been speculated by the estuary manager. A levelling of power between the project team and stakeholders present was also perceived, a situation that was not so evident in the mixed workshops, where some of the agency representatives relied on the project team to use methods that would protect them from sustained community criticism and direct the work in a more constructive manner.

E.3.5 Workshop 3: co-design

Following the risk assessment workshop, both the project manager and I went about trying to validate or invalidate the results of the risk assessment in our own ways, as there were some obvious limits to scientific validity of the results, which might not be accepted by stakeholders external to the process. This resulted in the project manager asking the workshop attendees to give their overall priorities or feelings on the risk importance from their own points of view, as he thought the risk assessment had returned illogical results for which he would be ultimately responsible. It also resulted in my performing and documenting sensitivity analyses on the risk model to check the levels of uncertainties and potential effects of preference ratings (on asset importance) that had not had time to be obtained, in order to mitigate criticisms of the model. In the end, apart from effects on the process, the negotiations over the form and content and results gained from the second workshop did not appear to have particularly constructive effects on the relationship between the project manager and me. The external evaluator noticed a certain amount of unsupportive body language displayed by the project manager in the second workshop towards me and I felt some tensions. However, in a review of the video recordings for the workshop to analyse this element, the lack of support did not appear too evident and our verbal exchanges remained pleasant as we attempted to work constructively together.

As there were only two weeks between the second and third workshops, no face-to-face meeting was held, in part because of the geographical distances between the project team members, who lived in three different Australian cities. Rather, communication was confined to email and the occasional phone call. I predominantly designed the final workshop methods, wrote up the explanation of my proposal into a PowerPoint presentation and clarified a number of points with the project manager. I also wrote the first version of the agenda, which was slightly altered by the project manager following our last phone conversation before the workshop. The consultants also prepared the previous process results for presentation and an 'info-pack' for participants, which included scientific and relevant planning information for each of the risks. Working relationships between the project manager and me appeared to improve in this last phase. I also began to understand better some of the constraints that the project manager was under from all his other projects.

E.3.6 Workshop 3: co-implementation

For the third workshop in early March 2007, the roles were distributed in a similar manner to those in the second workshop, with the estuary manager opening the workshop and the project manager updating and discussing process results to date, presenting the day's agenda and leading the brainstorming session on individual strategy and action development related to the 16 risks. I described the strategy mapping method to be used during the day, the four engineering and planning consultants facilitated the small group sessions and the evaluator managed both the questionnaire implementation and video recording. Relations appeared predominantly amicable in this workshop between the project team members, as well as between most of the project team members and the participants.

In this workshop, as in the first workshop, the methods developed by me and implemented with the aid of the consultants did achieve their goal of considering conflicts in a constructive manner and not spending too much time on them to the detriment of other issues, as seen by the positive participant and external evaluation results. This management capacity of the project team helped to build trust with some of the key stakeholders. From this relationship building, I now maintained a friendly and collaborative relationship with one of these stakeholders, whom I had not met prior to the process. However, the exclusion of a number of community members from the second workshop also led to at least one of them voicing her disappointment and how she had felt disempowered, even though the third workshop was constructive and she understood why the decision of the programme choice had been made.

E.4 DECISION-AIDING MODEL USE

Analysing the intervention process in a more formal manner, Table E.4 presents the elements of the Tsoukiàs (2007) decision-aiding process model and how members of these sets evolved through the process.

E.5 PARTICIPATORY PROCESS DESCRIPTION

This section provides supplementary information to Chapter 7, focusing on summaries of workshop activities from the two research reports on the participatory workshops submitted to the Hornsby Shire Council and BMT WBM as part of the creation of the Lower Hawkesbury Estuary Management Plan (Daniell, 2007a; 2007b). For additional reference material, including the agenda, worksheets, questionnaires and result summaries on these workshops, readers are referred to these two reports.

E.5.1 Workshop 1 process overview

The activities undertaken during the first workshop included a general welcome and project introduction, which were given before starting the personal introductions and then writing down individual values and issues related to the estuary. In the second session of the morning, the participants broke into small groups to discuss their collective values and issues. In the afternoon, a large group discussion was held to combine and contrast the findings from the four small groups and to distil a set of common values and goals for the estuary. This process therefore constituted starting with individual voices and opinions at the beginning of the day and slowly melding them into a collective vision by the end of the workshop.

SESSION 1: INDIVIDUAL VIEWS

As part of the personal introductions, participants were asked to introduce themselves to the group by giving their name, where they were from and, in not more than ten words, one thing they liked and disliked about the estuary. Although the 'ten words' rule was not commonly adhered to, this first activity of 17 minutes duration provided a succinct overview of many of the issues and values that were to be discussed throughout the remainder of the day. Likewise, the diversity of stakeholders and views represented was clearly defined for all the participants in the room.

The participant introductions were followed by some individual reflection time, where participants were asked to identify from their own perspective a number of values and issues related to the estuary. Ten minutes were given for writing values on green cards (one value per card) and ten minutes for writing issues on yellow cards. Each participant was asked to hold on to his or her cards and take them to the small groups after the morning break.

Table E.4 *Decision-aiding process model element elicitation and evolution (the researcher is, of course, the author)*

	Manifestation of model elements	Evolution through process
Representation of the problem situation: $\mathbf{P} = \langle \mathbf{A}, \mathbf{O}, \mathbf{R} \rangle$	**A** – *set of actors*: $\mathbf{A} = \mathbf{C} \cup \mathbf{K}, \mathbf{T} \subseteq \mathbf{A}$	
	C – *subset of core-participants*: 38 members in the set over the whole process	Many more participants were invited in the ongoing co-design phase than actually came to the workshops (over 60) – invitations decided or to be negotiated with the project team. Evolution of core-participants was: 30 in WS1; 17 in WS2 (agency participants only); and 19 in WS3
	K – *subset of associated stakeholders*: elicited in project team meeting discussions; elicited in WS1 in the issues and value sheets and linked to management responsibilities, interests and resources; elicited in evaluations as 'anyone else who should have participated in these workshops'. Defined in WS3 in the 'responsibilities' on the strategy maps	Approximately 30 discernible actors or actor groups identified in the issues and value sheets and evaluations (i.e. state government departments, local governments, private businesses, research groups, community associations and user groups). Approximately 20 of these actor groups were invited or represented in the core-participants – the actor networks linked to each of these representatives is thought to be large but was not specified entirely as part of the project. Main evolution included increasing awareness that more representatives from other local governments were required in the core-participants. Groups re-signalled in WS2 and WS3 were indigenous groups, the federal government, upstream actor groups, such as farmers and tourists, and more scientific experts, such as hydrologists
	T – *subset of project team members*: 15 discernible members from five institutions – water managers, scientists and planners from two local governments; engineering consultants; planning consultants; university researchers	Membership different for development, design and implementation process. Thirteen present in some part of the development, seven present in some part of the co-design and ten present in some part of the co-implementation. Four actors played the largest key roles through the whole process: the estuary manager (local government), the researcher (privately funded PhD scholar), the project manager (private engineering consultant) and the external evaluator (university researcher funded by an international project and CSIRO)
	O – *set of objects*: elicited as issues, values, goals, visions and threats in WS1; as risks, along with their sources, causes and potential impacts in WS2; and these objects and their relations also elicited and presented in the synthesis report by the project team	One hundred and forty eight individually elicited values reduced to eight by the participants in WS1 (aided by the facilitators), and then increased to nine in WS2 after one was added by the project team. Sixty individual goals and visions elicited in WS1 reduced to three by consultants and slightly adapted in collaboration with participants in WS2. One hundred and seventy two individual issues and the threats on the issue sheets from WS1 were used in the construction of 15 risks and their descriptions given in the synthesis report: one more risk and the risks' potential causes and effects (over 100) given in WS2 evolved into 16 risks and 104 causes and effects by the end of WS3 (from the synthesis report, WS2 and a few WS3 additions)
	R – *set of resources*: elicited in WS1 in the issue and value sheets as authority, information held and available data; some also elicited as	Approximately 35 different data resources outlined in WS1 on the issues and value sheets. Many other resources were treated

Table E.4 (*cont.*)

	Manifestation of model elements	Evolution through process
	'values' (e.g. water of good quality is seen to be a resource or asset); resources to be obtained were outlined in WS1 and WS3 in the responsibilities and some of the actions	implicitly through much of the process or just in oral discussion, such as the funding needs and what is available from various sources, as well as resources such as knowledge, trust, leadership and authority. Others were mixed into other categories, as some resources were closely associated with objects or actors (e.g. boating, community and tourist facilities, car parking). Evolution of resources made explicit predominantly occurred after the completion of WS3, where required resources and responsibilities for each proposed action in the plan were outlined. Resources that actors could potentially mobilise to carry out their priority actions were estimated by the researcher in the actor–action–resource matrix as an example use of the tool
Formulation of the problem and objectives: $\boldsymbol{\Gamma} = \langle \boldsymbol{\Pi}, \boldsymbol{A}, \mathbf{V} \rangle$	$\boldsymbol{\Pi}$ – *set of problem statements*: 16 prioritised tolerable and intolerable risks requiring treatment which were developed from the object subset	The problem statements were considered the risks requiring treatment. This ended up being the 15 that the consultants had synthesised from the relevant objects and other information obtained through the document review, plus one more obtained from WS2
	$\boldsymbol{A}$ – *set of potential actions*: these were elicited as 'actions' in the strategy map. Some other actions that could contribute to managing these problems were found in the 'strategies' category of the strategy map's elements, as well as in some of the issues and values cards	Actions were brainstormed individually in WS3 although some had already been elicited in WS1 in some of the issue and value cards (i.e. potential actions given with issues such as 'education needed', 'need for tanks to be installed in gardens' or in any card that had a 'lack of' something, where the something was often a potential action, such as 'lack of education programmes for visitors' where the potential action that was added to the set could be considered as 'provide education programmes for visitors'). In total on the strategy maps there were about 620 potential actions or strategies and more had been signalled by participants as existing in other reports or plans
	$\mathbf{V}$ – *set of potential points of view*: a number of these were defined as the indicators on the strategy maps linked directly to specific actions or strategies. A set of potential points of view were also taken as the values of each of the stakeholders elicited in WS1 and some were presented in the issue and value sheets linked to available or required data	The values elicited in WS1 evolved as outlined in the 'objects' set, to become a set of nine points of view (asset categories) under which potential actions could be evaluated and compared. Other indicators also formed potential points of view from which other actors could analyse the effects of certain actions – approximately 70 on the strategy maps. These were checked and more were outlined by the consultants and estuary manager in the draft risk response plan
Model exploration and options evaluation: $\mathbf{M} = \langle \boldsymbol{A}^*, \{\mathbf{D}, \mathbf{E}\}, \boldsymbol{H}, \boldsymbol{U}, \boldsymbol{F} \rangle$	$\boldsymbol{A}^*$ – *set of alternative sets of actions*: derived from full set of potential actions by the project team	Alternative actions to be evaluated were derived from the full set of potential actions described above, including those on the strategy maps and those from a literature review of existing plans and potential actions. The 620 actions were reduced to 317 by the researcher, and the consultants further reduced this to 149 strategies to be evaluated
	$\mathbf{D}$ – *set of dimensions*: derived from AS/NZS 4360:2004 and associated documents by researcher between WS1 and WS2, others	Five prime dimensions used in the risk assessment model were considered relative to each of the nine assets and included: consequence, likelihood, risk level, knowledge uncertainty and

	derived from planning needs voiced in the workshops, such as financial cost and time to implement actions	management effectiveness. Other dimensions used in the plan included 'initial capital cost' and 'indicative ongoing cost'
	E – *set of corresponding scales*: derived from examples in AS/NZS 4360:2004 and associated documents by the researcher between WS1 and WS2. Plan-related scales set after the final workshop by the consultants and estuary manager	Five prime dimensions measured on numerical scales of 1 to 5, with 5 corresponding indicative qualitative descriptions for the 5 points along the scale. Derived from information in AS/NZS 4360:2004, existing documents and the synthesis report by the researcher and checked by the other project team members. Scales for costs used in the draft plan were in Australian dollars, either a once-off payment or an annual cost
	H – *set of preference criteria*: participant risk priorities obtained after WS2 and individual distribution of participant preferences on actions obtained during WS3. Some other preferences were voiced in discussions or could be observed during the workshops	Risk priorities given by 14 participants in the categories, 'high', 'medium' and 'low'. Sixteen preferences were distributed on the actions and strategies on the strategy maps. Other implicit preference criteria were evident in the group discussions throughout the workshops. Some of the participants' preferences were also elicited through the evaluation questionnaires and examined from observation. For example, ecosystem health was considered to take precedence over economic and social considerations for a number of participants. No weightings of assets were obtained from participants due to lack of workshop time and participant interest in giving them, although some were tested in a sensitivity analysis of the risk prioritisation model
	U – *an uncertainty structure*: uncertainty structure considered explicitly in the categories 'likelihood' and 'knowledge uncertainty'. Model uncertainty explicated in part through sensitivity analyses	Physical system uncertainty considered explicitly by the participants using the likelihood category (relative to consequences of a risk's impact on each asset). Uncertainties due to human perceptions and available knowledge relative to consequence and likelihood estimations also given for each asset. Model sensitivity analyses were carried out by the researcher after WS3 to demonstrate mathematical choice uncertainties (operators). Uncertainties or sensitivities in the tolerability bounds discussed in the project team
	F – *set of operators*: operators for calculating the risk level in the prioritisation model were from Wild River and Healy (2006). All others developed ad-hoc by the project team, or carried out intuitively by participants or project team members	The explicit, numeric operators for linking the elements in the above sets were mostly defined by the researcher for the first prioritisation model, then by the consultant for the model to assess the potential risk reduction capacity of the set of alternative actions (labelled as strategies). Mostly very simple operators and visualisation. Few operators set by participants except in the case of implicit operators and arguments used by the participants to synthesise, present or defend their views
Final recommendations: Φ	Φ – *set of final recommendations*: given as 32 short-listed strategies in the draft action plan. Participants' final recommendations had been previously given by their distribution of 16 preferences on actions	Thirty-two short-listed strategies were given in the draft action plan after the consultants' and estuary manager's secondary risk assessments. One hundred and thirty-two actions or strategies had been prioritised by the participants in WS3. Most of the highest-ranked actions by participants were present in a similar form in the final 32 strategies

Table E.5 *Value and issue sheet questions*

Value sheet	Issue sheet
What is the value?	What is the issue?
Who or what holds this value?	For whom or what is this an issue?
Why is it of value?	Why is it an issue?
Where is the value applicable?	Where is the issue applicable?
How is this value currently preserved?	How is this issue currently managed or mitigated?
Who is responsible for preserving this value?	Who is responsible for managing this issue?
What existing information and data can be used to describe this value and who holds it?	What existing information and data can be used to describe this issue and who holds it?
What additional information and data would be necessary to describe the value?	What additional information and data would be necessary to describe this issue?
What is threatening this value?	What values is this issue threatening?

SESSION 2: SMALL GROUP WORK

Groups were selected randomly according to coloured dots that had been stuck on the chairs: red, yellow, blue and green. Some groups were a little uneven as there were a few spare chairs in the room, so a few participants were redistributed before commencing the small group activities. Once the small groups had convened in the adjacent room, the participants, along with a group facilitator, were given a number of activities to complete. Firstly, the participants were encouraged to present their value and issue cards to the other group members and to start sorting them into general categories. It was suggested that the groups attempt to distil the values and issues into about six to ten categories each. These could then be prioritised by the group members and written onto a 'values–issues' matrix as a summary of the group's main issues and values. As well as a matrix for participants to fill in with their distilled values and issues, participants were encouraged to localise their issues on a large map of the estuary, in order to facilitate greater discussion and understanding amongst group members of the broad range of issues that were put forward. It was evident that some groups favoured working with this visual medium of expression, whereas others preferred to stick to lists of written material and discussion around them.

Once the groups had reviewed, selected and written their values and issues on the matrices, they were asked to determine whether there were any interactions between them (which issues impact negatively upon the values), and if so, to mark the corresponding box. This allowed groups to see the most important issues to resolve that had the greatest impacts on a large number of values. As part of the small group sessions, participants were also asked (with the aid of their facilitator) to fill in a number of 'value' and 'issue' sheets for their most important values and issues, using the general category headings from the matrices. The questions from the values and issues sheets are outlined in Table E.5. A summary of all of the information gathered from these sheets is available in Daniell (2007a).

During the two hours dedicated to the small group activities, each group completed seven or eight sheets, to varying degrees, as well as their matrices and issues maps. At the end of the small group session, the participants were offered lunch. This included a kind donation of oysters by local oyster farmers. Both the new Pacific Oysters and the new disease-resistant Sydney rock oysters from the estuary were available for tasting and greatly appreciated by the workshop attendees.

SESSION 3: LARGE GROUP DISCUSSION

The afternoon session involved a large group discussion, where the values from the four groups' matrices were compared and contrasted. This process was used to develop a common list of core values for the estuary, resulting in a final list of eight values (refer to Table 7.2). Using this list of values, participants were then asked to reflect individually on how they are related to 'goals' for the estuary. The participants were each given a few cards on which to write either goals related to these values for the estuary or overall vision statements related to all or a number of the goals. They were then given the opportunity to share their goals and visions with the rest of the group, and a number of these goals and visions were written on the board and further debated. The large group session was then drawn to a close, with the facilitator describing the following steps of the project process and thanking the group for their participation and hard work. An overview of the day's activities is provided in Figure E.2.

For further information on the results of these activities, please refer to *Summary Report: Community Workshop 1 for the Lower Hawkesbury Estuary Management Plan* (Daniell, 2007a; 2007b).

PRELIMINARY OUTCOMES AND PREPARATION FOR WORKSHOP 2

From Workshop 1 and the external document review performed by the consultants, a list of nine estuarine values (asset

Figure E.2 Workshop 1 activities.

categories) was developed for use in the following stages of the LHEMP development process, as presented in Table 7.2. A list of 15 risks to the estuary (derived from the Workshop 1 issues and external document review) was also developed, as shown in Table 7.3.

Prior to the second workshop, I used the set of nine values, the Risk Management Standard (AS/NZS 4360:2004) and a number of other references (ANZECC, 2000; WHO, 2003; Fletcher *et al.*, 2004; Standards Australia, 2004; Billington, 2005; Everingham, 2005; SP AusNet, 2006; Standards Australia, 2006; UEC, 2006) to develop a series of 'risk tables'. These tables (Tables E.6 to E.10) were first distributed to participants as Appendix B of the 'Lower Hawkesbury Estuary Synthesis Report' (BMT WBM, 2007). The use of these tables will be explained in the next section.

E.5.2 Second workshop process overview

The second workshop commenced with a general welcome, project background update and a presentation of the day's agenda, followed by a brief session of personal introductions. Prior to the morning break, a session was run to obtain the participants' confirmation on the estuarine goals, assets and risks. The risk analysis method to be used during the day's activities was also presented. In the second morning session, two risks were discussed and assessed as a whole group; then in the afternoon session, the group was broken into pairs to assess the remaining risks. Once the assessments were completed, the priority categories were computed and the results discussed as a whole group.

INTRODUCTIONS AND CONFIRMATION OF GOALS, ASSETS AND RISKS

As part of the personal introductions, participants were asked to introduce themselves to the group giving their name, which agency they represented and, in a few words, the biggest risk that they believed the estuary faced. This session provided a good overview of many of the issues identified in the first workshop, including:

- Storm-water discharges;
- Risks outside catchment and development within it;
- Overdevelopment and overuse (twice);
- Catchment impacts (urban) on water quality;
- Drier forecast for the river basin and effect on other risks;
- Impacts of water quality;
- Lack of freshwater inflows into the river;
- Developments in the catchment, and impacts on water quality and quantity;
- Pollution from residential lots;
- Development in the catchment and associated water use;
- Pollution and overuse;
- Cumulative effects of all the different impacts all flowing into the one area;

Table E.6 *Consequences risk table*

Asset category	Consequence level description				
	1. Insignificant	2. Minor	3. Moderate	4. Major	5. Catastrophic
Scenic amenity and national significance	Little to no impact, or short-term (reversible) impacts, on scenic amenity	Minor or medium-term impacts on scenic amenity (some reversible)	Moderate or long-term impacts on scenic amenity (mostly irreversible)	Major and permanent long-term impacts on scenic amenity	Extreme and permanent long-term impacts on scenic amenity
	Impacts have little to no community significance	Impacts have low community significance for the region and nation	Impacts have some community significance for the region but little nationally	Impacts have high community significance for the region and some nationally	Impacts have high regional and national community significance
Functional and sustainable ecosystems	Little to no discernible effects on aquatic or terrestrial ecosystems or impact is so small as to be considered trivial	Aquatic or terrestrial ecosystem health temporarily compromised over a localised area	Aquatic or terrestrial ecosystem health compromised in a localised area for a long time period or temporarily over a wider area	Aquatic or terrestrial ecosystem health compromised over a wide area for a moderate term	Aquatic or terrestrial ecosystem health severely compromised over a wide area and for a long term
		Possible minor changes in species abundance and community structure but these could be mistaken for being due to seasonal changes or natural variation. Recovery would probably occur within a short timeframe	May result in significant changes in native species abundance and community structure or major habitat loss or triggering of algal or nuisance species growth. Recovery may take several years	May result in major changes in native species abundance and community structure or major habitat loss or triggering of algal or nuisance species growth. Recovery may take many years	May result in extensive losses of organisms and habitat with the potential for whole ecosystem destruction. Recovery may occur in the very long term or not at all
Largely undeveloped natural catchments and surrounding lands	Little to no impact of development, or short-term (reversible) impacts on land-use patterns	Minor or medium-term impacts of development on land-use patterns (some reversible)	Moderate or long-term impacts of development on land-use patterns (mostly irreversible)	Major and permanent long-term impacts of development on land-use patterns	Extreme and permanent long-term impacts of development on land-use patterns
	The quality and quantity of runoff remains unchanged (relative to normal variability patterns)	Possible minor changes to runoff quality or quantity outside normal variability	Significant changes to runoff quality or quantity outside normal variability	Major changes to runoff quality or quantity outside normal variability	Extreme changes to runoff quality or quantity outside normal variability
Recreational opportunities	Little or no impact on recreational opportunities	Minor or medium-term impacts on some recreational opportunities, most activities remain unaffected	Moderate or long-term impacts on some recreational opportunities or minor impacts on most activities	Major and permanent long-term impacts on some recreational opportunities or moderate impacts on most activities	Severe and permanent damage to a large number of recreational opportunities

Sustainable economic industries	Little or no impact on resources, industries and activities of economic significance	Minor impacts on some resources, industries and activities of economic significance	Moderate or long-term impacts on some resources, industries and activities of regional economic significance	Major impacts on some resources, industries and activities of regional and national economic significance	Severe and permanent impacts on some resources, industries and activities of high national economic significance
		Possible short-term losses of employment or financial hardship	Loss of employment or sustained financial hardship in some industries (potentially recoverable in the medium term)	Widespread employment losses or high industry financial losses (potentially recoverable in the long term)	Widespread employment losses or extreme financial losses (not recoverable in the long term) or total collapse of some industries
Culture and heritage	Little or no impact on areas or items of cultural significance and traditional ways of life	Minor permanent impacts to some areas or items of local cultural significance or minor unwanted impacts on traditional ways of life	Permanent damage to some areas or items of local cultural significance or moderate unwanted impacts on traditional ways of life	Permanent damage to areas or items of local and national cultural significance or major unwanted impacts on traditional ways of life	Widespread permanent damage to areas or items of national cultural significance or total destruction of traditional ways of life
Improving water quality that supports multiple uses	Insignificant impact on water quality and flora, fauna and habitat. Insignificant impacts on optical properties, temperature, dissolved oxygen, nutrient levels and salinity outside natural variability. Presence of toxins and undesirable species (heavy metals, pesticides, bacteria, algae etc.), which do not exceed water quality guidelines (i.e. ANZECC, WHO) anywhere in the estuary	Minor localised effects on water quality but without long-term impacts on aquatic ecosystems. Minor localised impacts on optical properties, temperature, dissolved oxygen, nutrient levels and salinity outside natural variability. Presence of toxins and undesirable species (heavy metals, pesticides, bacteria, algae etc.) exceed water quality guidelines (i.e. ANZECC, WHO) in a few areas (such as at discharge points) but does not limit most estuary uses (fishing, oyster farming, recreation) in other areas	Significant localised effects but without longer-term impact on aquatic ecosystems, and short-term and localised effects on water quality that impacts some estuarine uses. Significant localised impacts on optical properties, temperature, dissolved oxygen, nutrient levels and salinity outside natural variability. Presence of toxins and undesirable species (heavy metals, pesticides, bacteria, algae etc.), which exceed water quality guidelines (i.e. ANZECC, WHO) in a few areas that have short-term impacts on some estuary uses (fishing, oyster farming, recreation)	Damage to a moderate portion of the aquatic ecosystem resulting in moderate impacts on aquatic populations and habitats and long-term impact on water quality that impacts some estuarine uses. Significant widespread impacts on optical properties, temperature, dissolved oxygen, nutrient levels and salinity outside natural variability. Presence of toxins and undesirable species (heavy metals, pesticides, bacteria, algae etc.), which exceed water quality guidelines (i.e. ANZECC, WHO) in most of the estuary that have major impacts or long-term effects on some estuary uses (fishing, oyster farming, recreation)	Damage to an extensive portion of aquatic ecosystem resulting in severe impacts on aquatic populations and habitats and long-term impacts on water quality and most estuarine uses. Extreme widespread impacts on optical properties, temperature, dissolved oxygen, nutrient levels and salinity outside natural variability. Presence of toxins and undesirable species (heavy metals, pesticides, bacteria, algae etc.) which exceed water quality guidelines (i.e. ANZECC, WHO) in most of the estuary that have devastating long-term impacts on some estuary uses (fishing, oyster farming, recreation)

Table E.6 (*cont.*)

Asset category	Consequence level description				
	1. Insignificant	2. Minor	3. Moderate	4. Major	5. Catastrophic
Community value	Little to no impact on local communities and their well-being, health, social equity, access to services and participation levels (in local activities, governance processes etc.)	Minor long-term or moderate short-term impacts (mostly repairable) on local communities and their well-being, health, social equity, access to services and participation levels (in local activities, governance processes etc.)	Significant long-term or major short-term (mostly repairable) impacts on local communities and their well-being, health, social equity, access to services and participation levels (in local activities, governance processes etc.)	Major long-term or devastating short-term (some repairable) impacts on local communities and their well-being, health, social equity, access to services and participation levels (in local activities, governance processes etc.)	Extreme and widespread devastating long-term impacts on all local communities and their well-being, health, social equity, access to services and participation levels (in local activities, governance processes etc.)
Governance, legal and media	Little or no impact on existing governance structures	Minor impacts on existing governance structures (minor changes required for improvement or small disagreements between governing agencies)	Moderate impacts on existing governance structures (significant changes required or disagreement between governing agencies)	Major impacts on existing governance structures (major changes required or major disputes between governing agencies)	Extreme impacts on existing governance structures (total breakdown of existing structures or irreconcilable disputes between governing agencies)
	Low-level legal and regulatory issues	Minor legal issues, non-compliances and breaches of regulations	Serious breaches of regulations with possible investigation, report to authority with prosecution or moderate fine possible	Major breaches of regulations. Major litigation likely	Significant prosecution and fines. Very serious litigation including class action
	Public concern limited to local complaints	Minor, adverse local public or media attention and complaints	Significant adverse local public and media attention. Possible limited criticism from outside groups (NGOs, national media)	Significant adverse national media, public and NGO attention	Serious international public and media outcry

Table E.7 *Likelihood risk table*

	Likelihood level description				
	1. Rare	2. Unlikely	3. Possible	4. Likely	5. Almost certain
*Likelihood of risk **impacts** occurring*	Occur only in exceptional circumstances	Could occur but not expected	Could occur	Will probably occur in most circumstances	Is expected to occur in most circumstances

Table E.8 *Risk level matrix*

Likelihood level description	Consequence level description				
	Insignificant	Minor	Moderate	Major	Catastrophic
Almost certain	H	H	V	E	E
Likely	M	H	H	V	E
Possible	L	M	H	V	V
Unlikely	L	L	M	H	V
Rare	L	L	M	H	H

Table E.9 *Knowledge uncertainty table*

	Description of knowledge certainty				
Level of confidence	1	2	3	4	5
	Perception only, no information to support opinion	Perception based, some information on processes but not directly relevant to local region, or information at a regional level has significant limitations	Limited information, information could relate to cause or effect, expert knowledge would lead to this outcome – may be some differences in opinion	Information available and could relate to cause or effect, process has been described and documented at a regional level, experts can verify this position	Information is available and represents the process, and relates to cause and effect, process has been described and documented at a regional level, experts readily agree on this position

Table E.10 *Current risk management effectiveness table*

Rating	Guide to effectiveness
Excellent	Systems and processes exist to manage the risk, and management accountability is assigned. The systems are well documented and the system is effective in mitigating the risk.
Good	Systems and processes exist to manage the risk. Some improvement opportunities have been identified but not yet put into action.
Satisfactory	Systems and processes exist to manage the risk. Recent changes in operations require confirmation that accountabilities are in place and understood and that the risk is being actively managed.
Poor	The system and process for managing the risk has been subject to major change or is in the process of being implemented and its effectiveness cannot be confirmed.
Unsatisfactory	No system exists or no process exists to manage the risk.

- Inappropriate development and catchment-based pollution;
- Upper catchment influences and compliance: development not complying with legislation, onsite maintenance and non-compliance with standards, policing of use on the waterways, overuse; and
- People.

These introductions were followed by the consultant's presentation of the goals, asset categories and list of risks. Firstly, three amalgamated goals were discussed, specifically to allow agency members who did not attend the first workshop to voice their opinions on their formulation:

> **Goal 1** Recognise and respect the unique and diverse scenic and natural environment of the estuary through the integrated and holistic management of human and environmental interests.
>
> **Goal 2** Maintain sustainable economic, recreational and social uses without compromising the high quality and functional ecosystems upon which they rely.
>
> **Goal 3** Preserve and foster the sense of belonging, culture and respect for the estuary among existing and new residents and other users.

The discussion about the formulation of these goals predominantly focused on the lack of specific emphasis on the preservation and enhancement of 'sustainable ecosystems' that support many of the other assets. It was suggested that this ecosystem importance be more strongly highlighted in the first goal (for example by using more active words – *preserve and enhance*, instead of *recognise and respect*) and equally in the second goal. Suggestions for changes to the second goal included switching the order of the phrase to avoid the use of 'without compromising': '[*Conserve, protect and enhance*] or [*Maintain*] *functional and sustainable estuarine ecosystems upon which economic, recreational and social uses rely*', so that the idea of 'quality' ecosystems supporting the social, recreational and economic uses was strongly supported. For the third goal, there was discussion as to whether the word, '*heritage*', should be added, or whether it can be considered as part of the term 'culture'. It was also strongly suggested that '*managers*' be added to the stakeholders in the third goal: '*Preserve and foster the sense of belonging, culture*[, *heritage*] *and respect for the estuary among existing and new residents, users and managers*'.

During this discussion, a number of agency representatives expressed the opinion that they did not want to take the responsibility for 'signing-off' on these goals, and that prior to the LHEMP's acceptance they should be resubmitted equally to other community and commercial stakeholders for their comments. Support and respect for the community and commercial stakeholders was also voiced more generally, including the value that they contributed to the creation of the EMP and the need for their continued inclusion in the process to ensure the success of the plan and its impacts on the estuarine region.

Table E.11 *Questions for risk identification*

Risk sheet questions
What is the risk?
What are the sources or causes of this risk?
What are the main potential impacts of this risk?
Where, or to whom, will these impacts occur?
What are the current strategies used to manage this risk?

Following on from this discussion, the list of asset categories developed predominantly during the first workshop promoted no further discussion and so was taken as accepted for the following phases of the planning process. The list of risks was accepted in a similar fashion, although it was noted that '*treated sewerage*' should be added to the sources of 'regulated freshwater inflows', documented under risk no. 3 in the synthesis report (BMT WBM, 2007). The question of 'inadequate management' was also raised as a further risk that should be considered. The discussion highlighted how inconsistencies between different council planning practices and state legislation could lead to the possibility of local plans and objectives being overridden by state or federal governments. Difficulties in integrating plans and objectives over spatial and administrative scales were also seen as drivers of potential management failures. Based on these views, a 16th risk, '*inadequate or dysfunctional management mechanisms*', was proposed as an addition to the 15 previously defined risks (refer to Table 7.3).

RISK ANALYSIS METHOD

I presented the method to be used to analyse the list of risks in the morning and afternoon sessions to the participants, using the diagram similar to that in Figure E.3 as a basis for explanation.

A 'risk sheet' was developed as a guide to aid participants through the 'risk identification' and 'risk analysis' stages of the Australian Risk Management Standard AS/NZS 4630:2004 (refer to Figure 7.2). The questions on the 'risk sheet' used to complete the risk identification, following on from the risk descriptions in the Synthesis Report (BMT WBM, 2007), are given in Table E.11.

An example was then given to demonstrate how to use the 'risk tables' to fill out the table for assessment of the risk against each of the asset categories. This assessment included identifying pairs of consequences and likelihoods of risk impacts on each of the estuarine assets, and then finding an associated 'risk level' i.e. a function of the consequence and likelihood as outlined in the 'risk level matrix' shown in Table E.8.

A level of 'knowledge uncertainty' related to these assessments and the current level of 'management effectiveness' (for mitigating a risk's impacts relative to an asset category) was also to be assigned. Once the table was filled out, the risk could then

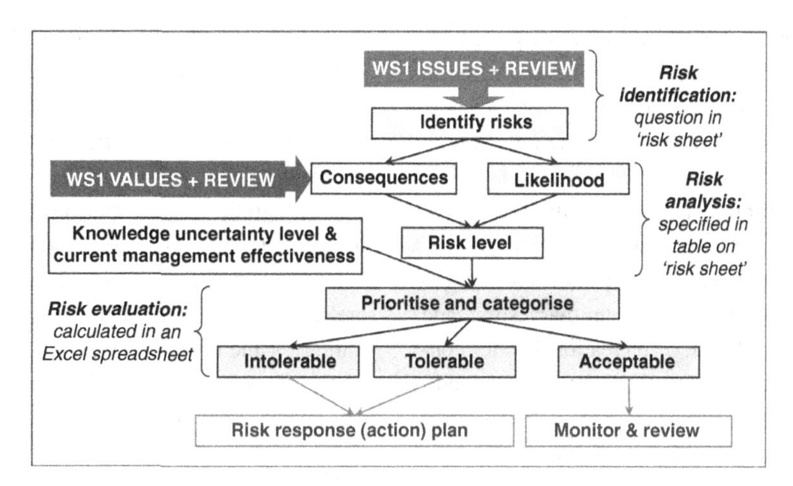

Figure E.3 Risk assessment outline in Workshop 2.

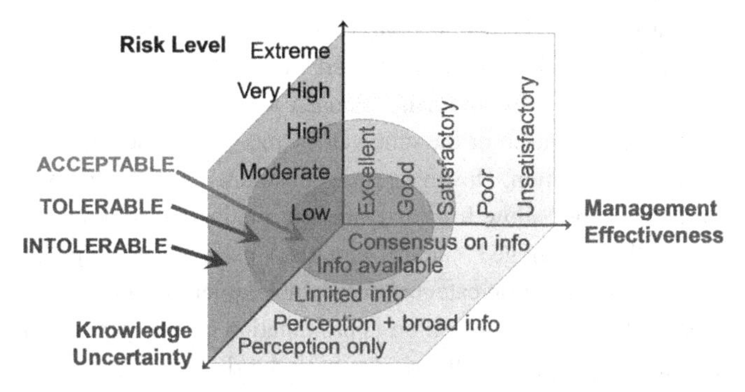

Figure E.4 Risk prioritisation: intolerable, tolerable and acceptable.

be evaluated into one of three categories: intolerable; tolerable; or acceptable, as a function of the risk's 'risk level', 'knowledge uncertainty' and 'management effectiveness', as conceptualised in Figure E.4.

Those risks categorised as either tolerable or intolerable would then be treated in the third stakeholder workshop as part of the risk response (or action) plan for the estuary. No further action on risks found to be 'acceptable' would be undertaken, except for routine monitoring and review at a later date (to determine whether the risks' status had changed and required management attention).

FACILITATED STAKEHOLDER RISK ANALYSIS

Following the morning break, I facilitated the first risk to be treated, the 'risk of excessive sedimentation', as a whole group session. The group worked through and discussed each of the questions on the 'risk sheet' (Table E.11), and the facilitator wrote the responses to these questions on flip chart sheets. Once the group had discussed each question, they moved on to filling out together the risk sheet's table using the 'risk tables' (Table E.6 to Table E.10). This included determining the 'consequences' and 'likelihoods' of risk impacts on nine previously defined estuarine assets, as well as an associated 'risk level', the uncertainties related to these classifications and the level of current management effectiveness of the risk related to each asset. Discussion on the values in each of these boxes was relatively rapid with, on average, three or four different opinions being solicited and discussed before decisions were recorded and the facilitator moved the conversation on to the next box. During these discussions, participants noted that there was a real need to document assumptions when they were going through the assessments, so that the results could be later traced back to specific thinking. The 'notes' section on the risk sheet's table was dedicated to this purpose and the participants were encouraged to discuss their assumptions or thoughts, with the main issues being noted by the facilitator.

Half-way through the sedimentation example, after 40 minutes of work, one of the participants suggested that the large group should also go through the assessment procedure for the water quality risk together, as a large range of opinions was likely to be presented and the participants wanted to hear the others' views. It was originally planned that each group would treat this risk in their small groups as the validation case for checking the results of the other risks. However, the consensus in the room appeared to be that if the participants in the room were broken up into small groups, some of the knowledge, information and useful exchanges related to this important issue would not be possible. It was therefore decided that the water quality risk would be treated by the whole group and the remaining section of the sedimentation risk would be completed in each small group for the purposes of validation.

Discussion on the 'risk of water quality and sediment quality not meeting relevant environmental and human health standards' was livelier than for the excessive sedimentation risk and almost all group members participated in the discussions. The specific definition of the risk and the 'standards' to which the estuary's water quality should adhere incited a particularly lively discussion. It was suggested that guidelines for the direct harvesting areas outlined in the NSW Oyster Industry Sustainable Aquaculture Strategy (DPI, 2006) could be taken as an 'aspirational goal' for all estuary waters. However, despite some agreement with this aspiration for human uses of the waterway, it was noted that for other objectives, such as ecosystem health, the guideline levels of faecal coliforms may potentially be too low, as other estuarine species use them as food sources. Therefore, adhering to the ANZECC guidelines for the 'protection of aquatic ecosystems', with the exception of oyster harvesting and recreational zones (where the ANZECC 'recreational water contact guidelines (primary and secondary)' are to be followed), would also be of benefit to the estuarine ecosystems. It was noted that this approach would additionally be consistent with the Healthy Rivers Commission's water quality objectives for the Hawkesbury–Nepean Region.

Emerging issues not yet covered under the current water quality guidelines, such as the impacts of new medications (e.g. anti-cholesterol drugs) and hormones, were also discussed. It was believed that a current lack of information about the impacts of such substances on the estuary is an issue that should be treated as part of this risk and spur evolution of the guidelines related to it. However, it was outlined that sewage treatment plant (STP) systems are being continuously updated to attempt to treat potentially damaging new substances and each STP discharge in the estuary area is currently toxicology-tested for, and must meet the guidelines for, about 115 substances. The other water monitoring programmes carried out in the estuary by the Metropolitan Water Authority, HSC and the NSW Food Authority (including the oyster growers) were also mentioned at this stage. Finally, the issue of discharges emanating from areas of acid sulphate soils and the difficulty in defining indicators to measure their effect on the estuarine assets was raised but no conclusions drawn.

Overall, the discussion of the five questions on the risk sheet for the water quality risk took about 15 minutes, and the table of consequences, likelihoods, risk level, knowledge uncertainties and management effectiveness took about 45 minutes to complete. After the definition of the risk, the other questions were relatively quickly dealt with, as the participants were simply asked to add to or comment on whether they agreed with the risk summaries in the Estuary Synthesis Report (BMT WBM, 2007). The results from these discussions were transcribed from the flip chart and white board table onto a risk sheet to facilitate data entry into the computer.

During the lunch break, the facilitators discussed how the afternoon session could be changed to maintain the objective of completing all 16 risks by the end of the day, considering the time that had been reallocated to completing the water quality risk as a large group. It was decided that the remaining participants (a few had to leave at lunchtime) would treat a couple of the remaining risks in groups of two or three. Following lunch, the participants were therefore split up into small groups and allocated a couple of risks each. The groups (or pairs) then worked on their allocated risks with the aid of the risk tables and Synthesis Report (BMT WBM, 2007). Facilitators were also on hand to answer any queries and to retrieve the completed risk sheets from the groups to speed up the computer entry of results. The small groups or pairs finished their tasks after approximately an hour and a half. They were then invited to complete the day's evaluation form, observe the entry of the results into an MSExcel spreadsheet and to help themselves to a cup of tea or coffee while waiting for the completed risk analysis outcomes.

From the values in the risk sheets' tables, the individual asset risk levels were numerated into the pre-prepared MSExcel spreadsheet using a logarithmic scale (base 2), as utilised in the CERAM method of environmental risk analysis (Wild River and Healy, 2006). The overall risk level for each risk was then calculated as the average of the nine asset risk level values. Similarly, the overall knowledge uncertainty and management effectiveness values were calculated as an average of the nine asset values. It was originally planned to determine weighting preferences for each of the nine values and then use a weighted average calculation. However, time constraints in the workshop process did not allow for this extra information to be obtained. These three averaged values were then used to look up the relevant prioritisation category (acceptable; tolerable; or intolerable), which was based on an approximation of the diagram in Figure E.4. Once all of the results from the risk sheet tables had been entered into the computer, the final list of risk levels and tolerability indices were presented to the participants and some time was allocated to discussing and interpreting the results (see screenshot in Figure E.5).

WHOLE-GROUP DISCUSSION OF RESULTS

Using the MSExcel spreadsheet calculations, all of the risks were classified as requiring treatment (either tolerable or intolerable). Before examining these results in more detail, the participants were asked whether they had any difficulties in filling in the risk sheets. The two principal comments related to this question were that for one of the risks, 'risk of inadequate monitoring to measure effectiveness of the EMP', the group that treated it did not assess the risk's impacts against the majority of the assets, owing to a perceived lack of relevance. It was thought that this risk was in some ways a 'second-order' risk and rather difficult to assess in the same way as the others. Other participants noted that the risks, as outlined in the synthesis report, were very broad. This meant that, in some cases, concrete examples within a risk had been taken as a starting point for the assessment. Such specific examples e.g. looking at fishing and oyster industries within the 'sustainable economic industries' asset category, were noted with qualifying comments on the risk sheets. A few participants remarked that they still felt defining high-level risks to be a useful activity, especially to help to

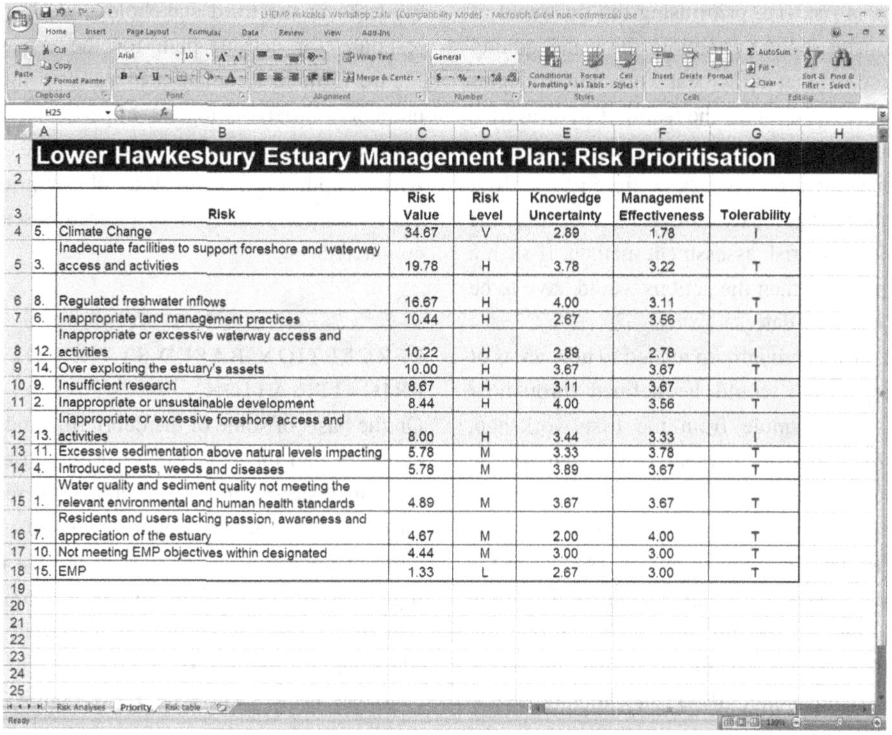

Lower Hawkesbury Estuary Management Plan: Risk Prioritisation

	Risk	Risk Value	Risk Level	Knowledge Uncertainty	Management Effectiveness	Tolerability
5.	Climate Change	34.67	V	2.89	1.78	I
3.	Inadequate facilities to support foreshore and waterway access and activities	19.78	H	3.78	3.22	T
8.	Regulated freshwater inflows	16.67	H	4.00	3.11	T
6.	Inappropriate land management practices	10.44	H	2.67	3.56	I
12.	Inappropriate or excessive waterway access and activities	10.22	H	2.89	2.78	I
14.	Over exploiting the estuary's assets	10.00	H	3.67	3.67	T
9.	Insufficient research	8.67	H	3.11	3.67	I
2.	Inappropriate or unsustainable development	8.44	H	4.00	3.56	T
13.	Inappropriate or excessive foreshore access and activities	8.00	H	3.44	3.33	I
11.	Excessive sedimentation above natural levels impacting	5.78	M	3.33	3.78	T
4.	Introduced pests, weeds and diseases	5.78	M	3.89	3.67	T
1.	Water quality and sediment quality not meeting the relevant environmental and human health standards	4.89	M	3.67	3.67	T
7.	Residents and users lacking passion, awareness and appreciation of the estuary	4.67	M	2.00	4.00	T
10.	Not meeting EMP objectives within designated	4.44	M	3.00	3.00	T
15.	EMP	1.33	L	2.67	3.00	T

Figure E.5 Risk prioritisation results.

decide whether more time should be spent in the following stages of the planning process in defining the sources and causes of the risks, as well as strategies and actions to treat these more specific areas of the risks.

At the beginning of the discussion of results, it was noted that some risks were pushed into the intolerable area not only by their risk level, but also by their high knowledge uncertainty (i.e. 'risk of inappropriate land management practices'), their low score of management effectiveness (i.e. 'risk of inappropriate or excessive waterway access and activities') or all three factors combined (i.e. 'risk of climate change'). This type of information could be useful for helping to develop strategies to treat the risks in the next workshop effectively. For example, risks with a high level of knowledge uncertainty may be suited to being treated with research-based solutions, or other similar methods of reducing this knowledge uncertainty.

When looking at the prioritisation of the risks, one participant mentioned that the water quality risk was not as high as could have been expected, and that the result was rather 'counter-intuitive', based on the major concerns highlighted in the introductions earlier in the day, such as pollution and stormwater runoff. The meaning behind this result was then discussed among the other participants. Theoretically, this risk had not been classed as intolerable, owing to its moderate risk level, relatively low knowledge uncertainty and high management effectiveness. It was thought that this result might have been not as highly prioritised, despite its perceived importance, as every risk had also been assessed against 'water quality' as an asset. Therefore, the importance of maintaining or improving water quality to support the estuarine uses and ecosystems was also an inherent part of the assessment of all risks, and some risks that have larger potential impacts on water quality were highly prioritised (e.g. inappropriate land management practices: for example, land clearing can increase erosion and sediment levels in the estuary, as well as allow more polluted runoff to reach the estuary). In an attempt to justify these differences of perceived importance of risks and the assessment procedure, one participant mentioned that if they had been asked, '*What risk will have the largest impacts on the estuary but does not receive enough management attention*,' the list of initial 'biggest risks' might have been rather different. This view was backed up by one of the participants who had treated the 'risk of climate change', saying that it was a good example as the risk's impacts are likely to occur with significant impacts; yet the management regime is not there and the knowledge uncertainty is high.

PARTICIPANT WORKSHOP DEBRIEFING AND EVALUATION QUESTIONNAIRES

Following the discussion of prioritisation results, the participants were asked whether they had any other comments or questions related to the workshop or the LHEMP process. The first question from a participant was a general process question directed at the project management team and related to whether further risk assessment at a sub-risk level would be carried out to determine the internal priorities of a risk (i.e. prioritising the treatment of boat discharges over onsite septic systems in the water quality risk). The consultants replied that in the next workshop strategies and actions would be developed for all the 'sub-risks', or to treat the various 'causes and effects' of the risks, and then prioritised by participants. However, owing to a lack of time and budget, each sub-risk would not be rigorously assessed in, or after, the next workshop using the same risk assessment method. If such a level of detail were aspired to, then the actions would have to be assessed individually at a later date.

The next comment that was brought up related to how areas of responsibility for each agency could be defined throughout the planning process. An example from the first workshop, '*Who is responsible for removing a dead cow found on the estuary foreshore?*' was used to illustrate the point that there is a lot of overlap between management agencies of the estuary and foreshores. This management overlap was seen as one of the reasons why, '*Issues sometimes get bumped from local to state government, then between departments, and often nothing gets done.*' Reactions to these comments included that the recent 'Waterways Review' (SJB Planning, 2005) had started to review management responsibilities and these were laid out in the 'governance table' (SJB Planning, Attachment 9, 2005). However, this table was not yet sufficiently specific to provide illumination for a definitive answer to issues such as the 'dead cow'. It was noted that responsibilities for certain actions would be investigated in the next workshop and further defined in the plan writing stages of the LHEMP process, although defining workable management responsibilities would require ongoing planning and cooperation between all agencies.

Finally, one participant commented that, compared with the last workshop, this workshop was very effective, as, '*If you can keep emotion out of it you can move forward a lot more effectively*'. When prompted by the facilitator to expand on this comment, the participant responded that, '*Sometimes emotion can polarise debate around certain issues that don't pose a 'real risk'.*' Another participant commented that this was sometimes difficult, as emotion surrounding important issues is natural. The point was made that, as the estuary is so big, focusing on just a couple of issues may not be a very effective way of moving forward. The facilitator replied that the methods used in the next workshop would attempt to reduce this problem and encourage participants to focus on developing strategies and actions for the whole range of intolerable and tolerable risks.

E.5.3 Preliminary outcomes and preparation for Workshop 3

Considering the risk evaluation results from the second workshop that had defined all risks as being either 'tolerable' or 'intolerable', the decision was taken to continue to study and treat all of the risks in the third stakeholder workshop as part of the risk response (or action) plan for the estuary (as represented in Figure E.3). In an attempt to validate these findings, despite a number of methodological imperfections that occurred in the workshop, owing to last-minute changes to meet stakeholders' wishes, both a follow-up study of risk priority preferences and a brief sensitivity analysis of the risk assessment outcomes were conducted.

PERCEPTION-BASED STAKEHOLDER RISK PRIORITISATION

On the basis of some of the debriefing and evaluation comments, as well as an interest in comparing and cross-checking the risk analysis results with the participants' perceptions of risk importance, the consultants sent a follow-up email to the workshop participants asking them to rank the list of 16 risks as either a high, medium or low priority, in their opinion. A 50% response rate from the participants was achieved, and the responses are illustrated in Figure E.6.

It can be quickly seen from Figure E.6, that the 'risk of residents and users lacking passion, awareness and appreciation of the estuary' is the lowest prioritised risk. Apart from this risk, the overall ranking of the other risks was dependent on the method of statistical analysis used. For example, if the risk with the largest number of 'high priorities' were to be ranked first, then that would be the 'risk of inappropriate land management processes'. However, if the numbers of high, then medium, priorities (or a number of other methods) were to be considered for achieving the top rank, then the 'risk of water and sediment quality not meeting relevant environmental and human health standards' would be ranked first. A limited number of statistical analyses were performed on these stakeholder risk priorities to determine their rankings, including: attributing different numerical values to the three priority categories and taking averages; sorting based on numbers of quantities of risks in each priority category; and calculating medians. From these analysis options, the rank ranges for the perception of each risk's priority are given in Table E.12.

Observation of the risk rank range levels in Table E.12 shows a number of differences between the stakeholder perceptions of the risk and the multi-asset based risk assessment. Specifically, the water quality risk was perceived to be the most important, as described previously in the risk evaluation session debriefing. The other major difference was the particularly low comparative ranking of the 'risk of inadequate facilities to support foreshore and waterway access and activities'. This could potentially result from a number of factors, including that these values only represent the 'agency' perspective and that for these agencies' management domains, this factor is not immediately thought of as a priority.

Table E.12 *Rank ranges for stakeholder-defined risk priorities*

Risk name	Rank range
Water quality and sediment quality not meeting relevant environmental and human health standards	1–2
Inappropriate land management practices	1–3
Over-exploiting the estuary's assets	1–3
Inappropriate or unsustainable development	1–4
Climate change	5–6
Introduced pests, weeds and disease	5–6
Insufficient research	5–7
Inappropriate or excessive waterway access and activities	5–7
Inadequate monitoring to measure effectiveness of EMP	5–9
Regulated freshwater inflows	5–11
Excessive sedimentation	5–12
Inappropriate or excessive foreshore access and activities	5–14
Inadequate or dysfunctional management mechanisms	5–15
Not meeting EMP objectives within designated timeframes	11–14
Inadequate facilities to support foreshore and waterway access and activities	13–15
Residents and users lacking passion, awareness and appreciation of the estuary	14–16

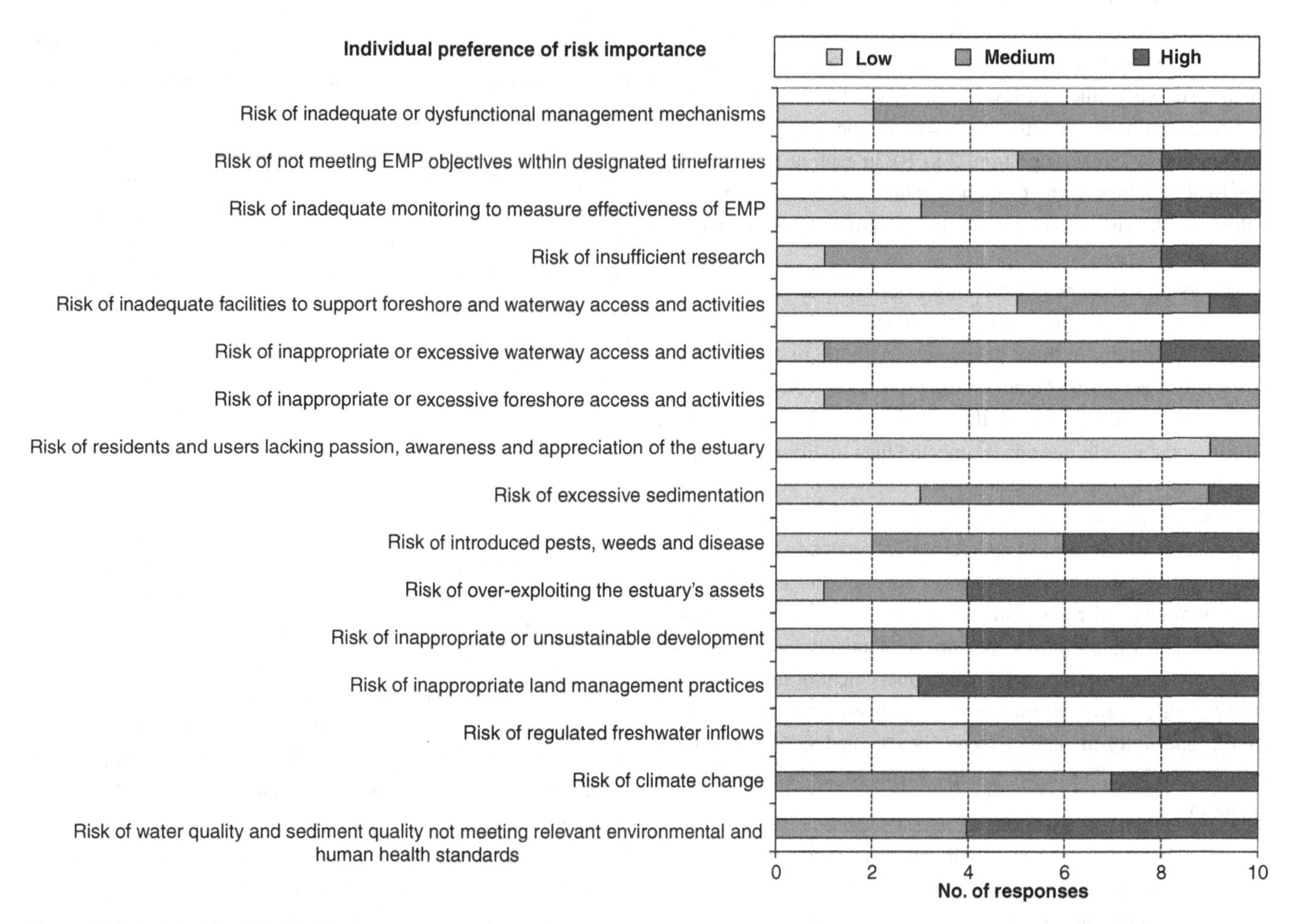

Figure E.6 Stakeholder risk priorities.

RISK ASSESSMENT SENSITIVITY ANALYSIS

To understand which aspects of the risk assessment most influenced the numerical outcomes, a sensitivity analysis of a number of different factors was made. In all of these analyses, the original values given by the participants were conserved. It is noted that it is likely that some of the small groups had different styles of interpreting the risk tables from others, and thus, owing to a lack of validation to check these differences, the results of the risk assessment can only be taken as a general guide. This aspect of the risk assessment process is described in more detail in Section E.6.1. However, a number of different aspects related to the mathematical model of calculating risk levels, asset weightings and ranking based on different factors can still be analysed.

In total, the effect of 33 different combinations of parameters was utilised to test the sensitivity of the model and risk rankings. Three base mathematical models were used as a part of these combinations for calculating the risk levels based on the consequence and likelihood ratings: firstly, the original logarithmic scale (base 2), as previously described and employed in the CERAM method of environmental risk analysis (Wild River and Healy, 2006), where the risk levels vary from 0 to 128; secondly, the traditional model of: risk level = consequence × likelihood, where the risk level ratings vary between 1 and 25; and finally, a model of: risk level = consequence + likelihood, where the risk levels range from 2 to 10. In each of these cases, the final rankings were examined when sorted based on the values of: the risk levels; knowledge uncertainty; and management effectiveness.

The asset weightings (which were all equal in the original model) on the rankings were also modified to examine the model's sensitivity to these parameters, with the risks only then being sorted according to the resulting risk level. To consider this in a more meaningful way than randomly or systematically adjusting the weightings one by one and then in different groups, which, although an interesting exercise, would be extremely time-consuming, the choice was made to define four scenarios of preferred weightings. In simpler terms, groups of assets were chosen to be 'preferred' over others according to a preference for: environmental enhancement and maintenance of ecosystem services [ENV]; an active and an economically and socially viable community [CE]; successful management of the estuary's water quality to support multiple uses [MAN]; and maintenance of the undeveloped nature, scenic beauty and heritage of the estuary [HIST]. The asset groups favoured in each of these scenarios are:

[ENV] Functional and sustainable ecosystems; water quality for multiple uses;

[CE] Community value; sustainable economic industries; recreational opportunities;

[MAN] Effective governance; water quality for multiple uses;

[HIST] Scenic amenity and national significance; largely undeveloped surrounding lands; culture and heritage.

In each of these cases, the relevant asset weightings were increased from 1 to 5, and then to 10, to look at the impact of more extreme preferences relative to the other assets. From these 33 different parameter changes, the maximum and minimum rankings of each of the risks are displayed in Table E.13. The full summary table is presented in Daniell (2007b). The ranks are first given according to the risk levels and then according to the knowledge uncertainty and management effectiveness. Risks of the 'intolerable' kind are noted in bold text and those considered as 'acceptable' in italic text. All risks noted in normal text were calculated as 'tolerable' risks.

Table E.13 shows that, based on the participants' input and the sensitivity of the evaluation model, a number of the risks were more sensitive than others in terms of their relative rankings of risk levels: in particular, the 'risk of regulated freshwater inflows', which was extremely sensitive to the asset weightings. This was the only risk to range from 'intolerable' to 'acceptable' over the analyses. Under all equal weightings, this risk was 'tolerable'. However, under all three mathematical models, when the weightings were changed on the [ENV], [MAN] and [CE] scenarios, this risk became 'intolerable', and under the [HIST] scenario became 'acceptable'. Interestingly, regulated freshwater inflows were the subject of much discussion during the first and second workshops, with opinions ranging widely on how much of a risk they actually posed. This range of opinions is also backed up in the risk perceptions in Figure E.6. It therefore appears that if there had been time to collect stakeholder preferences on the importance of different assets in relation to one another, a more specific picture of this risk (relative to the concerned stakeholders) could have been calculated using this risk assessment model.

Other elements of this sensitivity analysis worth noting were that the tolerability indices tended to follow the relative ranks of the knowledge uncertainties, as well as the level of management ineffectiveness (second ranking column in Table E.13), as should be expected. The water quality risk became 'acceptable' under one scenario ([MAN]), where the weightings advantaged the positive perception of management systems in place to manage this risk, as discussed earlier; otherwise, it remained largely insensitive to other parameter changes and largely differently ranked when compared with the risk priority preferences in Figure E.6. Although not looked at in great detail here, the selection of tolerability index boundaries is another aspect of the model used in this workshop that could be easily debated with participants and tested for their sensitivity in the final results.

If such a risk assessment process were to be repeated, a suggestion to better analyse and aid understanding of these risks

Table E.13 *Rank ranges in risk model sensitivity analyses*

Risk name	Risk level rank range	Knowledge uncertainty and management ineffectiveness rank range
Climate change	**1**–9	**1–4**
Regulated freshwater inflows	**1**–*14*	5–14
Inadequate facilities to support foreshore and waterway access and activities	**1**–4[a]	6–12
Inappropriate land management practices	**4**–9	**2**–8
Inappropriate or excessive waterway access and activities	**3–6**[b]	**2–5**
Over-exploiting the estuary's assets	1–8	10–11
Insufficient research	**5**–9	**7**–10
Inappropriate or unsustainable development	2–10	9–15
Inappropriate or excessive foreshore access and activities	4–10[c]	**7**–9
Excessive sedimentation above natural levels impacting the environment	9–12	8–14
Introduced pests, weeds and diseases	9–*12*	13
Water quality and sediment quality not meeting the relevant environmental and human health standards	10–13[d]	11–12
Residents and users lacking passion, awareness and appreciation of the estuary	10–14	1–15
Not meeting EMP objectives within designated timeframes	7–14	3–6
Inadequate monitoring to measure effectiveness of EMP	15[e]	3–4

[a] Also 1 (tolerable) found and the likelihoods are likely to be somewhat overrated compared with other risks.
[b] Also 5 (tolerable) found.
[c] Also **9** (intolerable) found.
[d] Also *12* (acceptable) found.
[e] This risk assessment was not entirely completed during the workshop.

would be to define specifically the 'inherent' part of each risk (when there is no management or the management systems in place for this risk fail), as well as the 'residual' risk (the risk posed when the management systems are in place and working as they should). In the case of this workshop process, the 'management effectiveness' was used as a surrogate for the difference between these two risks levels, but it might become clearer from the participants' points of view if these two different risks were made explicit. It is most likely that the inherent (unmanaged) water quality risk would be ranked extremely highly, extrapolating from the priorities, and that the 'residual' risk, as it is currently managed, would not pose an excessive problem compared with other less well-managed risks.

Other potential variants on the process used for carrying out these risk prioritisations could include breaking the risks down into 'sub-risks' or risks concentrating on 'sub-areas' of the whole estuary. If such analyses were to be conducted, more time should be dedicated to the task. In the future, other multi-criteria methods of analysis could also be used for asset preference elicitation and the ranking procedures used in the model, rather than the simple weighted-average method utilised in this workshop. However, in a participatory setting, these choices of method should be carefully made, as some of the mathematical models may be more difficult to understand and may not be as readily accepted by participants as the basis for the already subjective task of risk ranking.

E.5.4 Workshop No. 3

The third workshop commenced with a general welcome, presentation of the day's agenda and a session of personal introductions. This was followed by a short project background update and presentation of the strategy mapping technique to be used for the day's activities. The first session of individual brainstorming was run to determine potential strategies and actions for each of the 16 risks. The second session focused on the strategy mapping exercise, which was undertaken in small groups. Once the small groups had finished discussing their allotted risks, they could then add to the strategy maps of the other groups' risks. In the afternoon session, responsibilities and monitoring needs were added to the strategy maps and the participants distributed their preferences on strategies or actions for each of the risks. The workshop ended with the final participant evaluation questionnaire.

INTRODUCTIONS AND RECONFIRMATION OF GOALS, ASSETS AND RISKS

Following the general welcome and agenda for the workshop, a round of personal introductions was begun, with everyone being asked to present themselves to the group giving their name, where they were from, and to describe, in no more than 10 words, 'an innovative or radical strategy' to address one of the estuarine risks. The session of introductions lasted approximately 15 minutes, with the responses drawn from the strategy question ranging in 'radicalism'. It was interesting to note that the perception of 'radical' relied not only on new or 'out-there' ideas, but on what could actually occur. Thus strategies seen to be utopian or near impossible to achieve, even if they were already established goals for the management of the estuary, were seen as 'innovative or radical'. One strategy of this kind put forward by a participant, '*Work together to put this plan in place and see the outcomes*,' elicited laughter from the rest of the participants, potentially because of the perceived 'innovative' nature of such a proposal.

Other strategies to manage the estuarine risks put forward by the participants included:

- More monitoring throughout the catchment system to know what is going in and coming out;
- Focus attention on underlying factors (i.e. population growth; ecosystems being primary);
- A big police operation against illegal development;
- Assess risks and people's different perceptions of them to supplement knowledge of what the risks actually are and their magnitudes;
- Undertake an environmental and economic impact study of the whole estuary for different risks (e.g. a study of boating risk impacts now compared with 15 years ago to help identify 'real' estuarine risks);
- Conversion of all sewerage treatment plants to recycled water plants to reduce nutrients and flows to the river;
- Zone estuary for different waterway uses and put speed limits on vessels;
- Create an inventory for the estuary and catchment so we know what we are managing (i.e. fish stocks; maps of sea grasses);
- The need to encourage 'responsible use of the river' through education and making facilities for responsible use available;
- Investigate triggers for collapse of assets (i.e. fish populations);
- 'Zap' sedimentation: need to first establish where it is coming from;
- Educate in more 'user-friendly' ways (i.e. develop brochures and booklets related to estuarine issues as 'people don't read lengthy documents'; approach schools);
- Establish risks and the priority they pose to the estuary and surrounding land so that management efforts can be prioritised appropriately;
- Guarantee environmental flows and periodic flooding (i.e. go back to original system, productivity and values);
- Use less water in Sydney and recycle effluent for drinking;
- Determine what values and services the estuary provides, then make people who use it pay to look after it;
- Develop a strategic approach to managing the ecological footprint of the estuary: break the estuary down into development areas and business areas; determine the impacts of global warming on these areas and their footprints; and
- Develop the collective vision and collaboration to put the LHEMP in place.

After this round of introductions, the consultants presented the background information to this third workshop in a five-minute summary, which incorporated a brief recap of the activities in the first and second workshops, including how the 'issues' developed in Workshop 1 became 'risks' and the 'values' became estuarine 'assets'. The updated goals for the estuary were then presented:

Goal 1 Preserve and enhance the unique and diverse scenic and natural environment and functional ecosystem of the estuary through integrated and holistic management.

Goal 2 Maintain sustainable economic, recreational and social uses without compromising the high quality and functional ecosystems on which they rely.

Goal 3 Preserve and foster the sense of belonging, culture and respect for the estuary among existing and new residents, users and managers.

These were followed by the lists of previously defined nine estuarine assets and 16 risks.

This presentation was not met with too much vocal comment or criticism. One participant commented that where 'functionality' or 'sustainability' of ecosystems is mentioned, 'biodiversity' should also be attached to these terms, as, '*Functionality could occur with a fraction of the species.*' On the risk prioritisations, a small objection was raised by a few participants about the 'medium' ranking of the 'dysfunctional management' risk, as they thought that it should have been assigned a higher risk. Another participant also described the risks and underlying causes as being confused (i.e. the 'residents lacking awareness' and 'dysfunctional management'), as some were causes of others. The facilitator replied that many of the risks were in fact interrelated and could be considered as risks or causes in certain circumstances but that they still all needed to be examined for potential treatment.

STRATEGY MAPPING METHOD

After the consultants' project background update, I presented the method to be used to create input to the Lower Hawkesbury Estuary Action or 'Risk Response' Plan in this workshop. The 'strategy mapping' technique had been specifically adapted for the risk management treatment process of the estuary after being

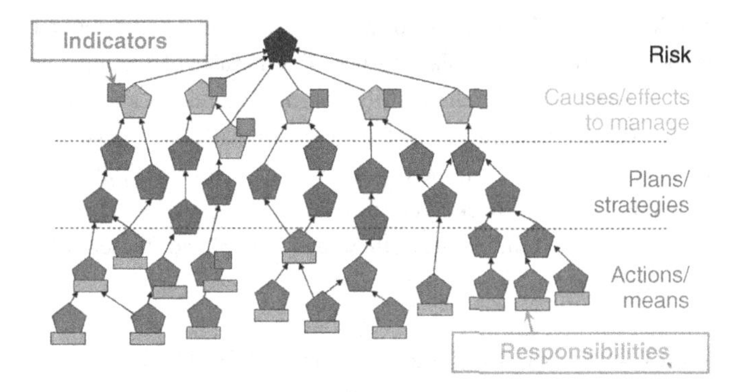

Figure E.7 Elements of the strategy mapping technique.

originally adapted from Ackermann and Eden's (2001) 'oval mapping technique' and its use in part of the Montpellier pilot trial, outlined in Chapter 6 and Appendix D.

The strategy mapping technique was to be used to: develop strategies and actions to treat the risks; define who would be responsible for carrying out the actions; and determine how these strategies could be monitored (i.e. required indicators, information for management and data). More specifically, the objectives of using the technique in this workshop were to:

- Encourage creativity and active participation;
- Visually structure information and allow everyone to see and add to the information produced: 'piggy-back' brainstorming.

The construction of the strategy maps was explained using the image in Figure E.7 as a basis.

The risks and their causes and effects that had already been identified in the first two workshops became the top rows of the strategy maps, as shown in Figure E.7. The strategies (or plans) and actions (or means) for each risk were first to be generated on coloured cards, through a session of individual brainstorming. Two minutes was to be allocated for each risk. The idea was to try to write as many cards as possible (one card per action or strategy) in the two minutes and to be creative. The feasibility of actions could be commented on at a later stage of the process, as the creative ideas might feed others' ideas in the meantime – thinking 'outside the square'.

The individual contributions would then be sorted by risk, and preliminary structuring of the cards would take place in small groups. The strategies and actions were to be added by the participants under the associated causes or effects of the risk. They could be joined in a hierarchical fashion to show the dependencies of each strategy and action on the rest of the possible management system. Once all of the existing cards had been structured on the maps, other strategies and actions could be added by the groups to fill the maps out (i.e. 'piggy-backing' and further formalisation of ideas). If some causes and effects had been left unmanaged, these areas should be especially targeted. After the small groups had finished with their assigned risks, the participants would be free to look at and add to the other groups' strategy maps. They could add strategy or action cards, or comment on what was already there (if they were against or in agreement with the strategies and actions), but were asked not to remove anything.

Following the completion of the structure of the strategy maps, the participants would then be asked to define stakeholder responsibilities (i.e. who would (or could) be responsible for carrying out the actions), and to determine how the effects of the actions could be monitored (i.e. required indicators and targets, information for management and data), as shown in Figure E.7.

At the end of the mapping exercise, participants would then be asked to distribute their preliminary preferences on the strategies or actions, so that these preferences could be considered by the consultant team in the plan write-up. To visualise these preferences, each participant would be given 16 sticky dots to distribute on the strategies and actions of the 16 risks. They were asked first to initial them and then to place them on their preferred actions on each of the maps (no more than one dot per action).

IMPLEMENTATION OF THE PARTICIPANT STRATEGY MAPPING

To start the strategy mapping exercise, the consultants led the session of individual brainstorming for each risk. All of the participants in the room were handed a pile of orange and pink cards by the facilitators and asked to write on the top of each card the number of the risk they were writing the cards for, as well as their initials for tracking and confirmation purposes. The two minutes of brainstorming for each risk were carefully timed by one of the facilitators, with a signal being given at 1min 30 s for each risk, to allow the participants to finish off the risk they were working on and to prepare for the next one. During the two minutes for each risk, the session facilitator would introduce the risk with its previously identified causes and effects with the aid of a pre-prepared PowerPoint slide, which would remain on the screen for the participants to refer to during their brainstorming.

The participants worked very productively and conscientiously through this 32-minute period of brainstorming, producing over 700 separate strategy and action cards between them. They were then invited to take a well-earned morning break while the facilitators sorted the cards into piles for each risk and checked that all the risk maps had been correctly pre-prepared with the risk name and the yellow cause-and-effect cards.

In the next session, the participants were asked to split up into four groups and to try to avoid having too many similar interests in each group (i.e. no two participants with the same affiliation and a mix of community and agency representatives). This self-selection process appeared to work reasonably well, although one group did not have an agency representative. Each group of four to five participants was then allocated half a room each, with four pre-prepared risk strategy maps, the corresponding piles of strategy and action cards from the brainstorming session, and a facilitator. In most groups, the participants, either individually or in a pair, decided to start the process by organising the cards on the maps for each risk. This choice sped up the strategy mapping process quite remarkably and allowed much more time for the participants to discuss, alter and add to the maps after the initial structuring. It was interesting to note the different techniques of participants for attempting this task. Some participants sorted all the cards into sources, causes and hierarchies before sticking them to the map, whereas others worked with one card at a time, sticking it to the map and drawing in its interdependencies.

In this structuring phase, the participants were told that they could add cause or effect cards to the maps, if these were required. Once the maps were largely constructed, the group members could then work together and discuss options to manage their risks, adding more cards to the maps. When the groups were satisfied with their four maps, they were then encouraged to view and add to the other 12 risk strategy maps. These phases of the strategy mapping were completed more quickly than had been planned for in the workshop agenda, with most groups being satisfied with their risk strategy maps and the other groups' maps before the lunch break. This meant that they had a reasonable amount of time to talk with other participants about particular strategies or other more general topics.

In the afternoon session, participants were asked to move on to identifying indicators, targets and monitoring strategies to manage the risks and implementation of actions, as well as potential stakeholder responsibilities for actions. A few participants noted at this stage that they could not yet definitively place management responsibilities on the actions (unless it was already part of a planned programme), as they would have to confirm them with their superiors. To be able to do this at a later date, the participants asked that the consultants provide them with a copy of the proposed action tables as soon as they were written, so that they would be able to seek out the required confirmations of management responsibility. Despite this potential issue, most strategy maps were marked with quite a number of stakeholder responsibility and monitoring PostIts®. Defining 'concrete' indicators and monitoring strategies to measure the impacts of actions in reducing or mitigating the risks was found to be a challenge by some participants. For example, 'erosion' was marked as one indicator for monitoring. Erosion could be considered as an indicator, but for this indicator to be useable, a target state of erosion needs to be established spatially and temporally, and data would have to be collected using special techniques over various spatial scales and at different time intervals. Monitoring schemes to this scale of detail were rarely noted, except where reference was made to existing monitoring schemes in the estuary region. This issue is further discussed in Section E.6.2. Partway through the defining of stakeholder responsibilities and monitoring needs, a number of participants had to leave early, so the distribution of preferences and workshop evaluation were brought forward from the time that had been originally programmed in the agenda. The remaining participants then went back to defining monitoring needs and stakeholder responsibilities after they had completed these other activities.

PARTICIPANT PREFERENCE DISTRIBUTION

Considering the risk strategy maps, the participants were asked to think about which strategies or actions they would prefer to see put in place through this planning process. They were then each handed 16 sticky yellow dots to distribute on their preferred strategies and actions over the 16 risk strategy maps. The participants were asked to mark their initials on their dots and only to leave one dot per strategy or action card. This activity was useful as it allowed the participants more time to read and absorb the content of the 16 risk strategy maps before making their choices. It was also mentioned to the participants that if they did not find enough actions that they wanted to support, they still had the opportunity to add more strategy or action cards. Figure E.8 (bottom right corner) shows an excerpt of one map with a couple of highly prioritised strategies, as well as some of the other activities from the second and third workshops. These priorities were then to be considered by the consultants in the plan-writing phase after the strategy maps had been condensed into a useable format.

E.5.5 Preliminary outcomes and preparation for plan writing

As some participants mentioned in the evaluation questionnaires, the challenge after the strategy formulation workshop was to turn the content of the results into something useful for the rest of the LHEMP process. The final outcomes from the third workshop were the 16 risk strategy maps, which collectively included around 900 cards, PostIts® and comments. For this information to be used effectively by the consultants in the plan-writing process, I first treated it in a number of ways.

The first step undertaken was to convert all of the maps' information into an electronic format. This process was performed using the software DecisionExplorer®, a program specifically designed for cognitive and strategy mapping of complex problems. Each of the colours and categories of cards, PostIts® and comments was conserved in DecisionExplorer® to aid with the analysis of these concepts. An example of the 'water quality' risk's conversion from paper to electronic format was shown in Figure 7.5. From DecisionExplorer®, the elements written on the various categories of cards (risks, causes and effects, strategies, actions, responsibilities, monitoring needs and comments) were

Figure E.8 LHEMP Workshop 2 and 3 activities.

then extracted to Microsoft Excel for further treatment. The hierarchies between actions and strategies found in the original strategy maps were checked for consistency and some rearrangements made where necessary (some strategies were found to be actions of other strategies, as a couple of participants had pointed out in the workshop evaluations). Monitoring needs, responsibilities, priorities and other comments associated with particular actions or strategies were transferred directly into tabular format. To make the information more accessible for the use of plan writing and the development of an 'action' or 'risk response' table, a number of other operations were performed:

- Repeated actions or strategies under the same risks were merged into one when discovered.
- Some of the more 'radical' strategies or actions were checked for feasibility within the bounds of this planning jurisdiction. Those found to lie outside the jurisdiction were omitted (i.e. federal government responsibility).
- Where actions were similar to those proposed in existing plans covering the estuarine study area (either written down by the participants or discovered during the plan analysis), references to these existing actions and their proposed timeframes were noted.
- Where actions or strategies were marked for treating more than one risk, the reference to the other risk(s) was noted.
- A preliminary coherency check was undertaken between strategies and actions to examine compatibility. Those thought to be incoherent (i.e. in terms of time for implementation, or opposite system impacts) were marked as needing more analysis before the plan was written.

This risk–response table based on the participant contributions comprised 317 actions distributed over the 16 risks, an average of just under 20 actions per risk. This table was then sent to the consultants, who further modified it based on their review of existing management strategies, planning documents and legislation. Final priorities were developed through a secondary risk assessment process outlined in BMT WBM (2008), along with additional consultation with key stakeholders.

E.6 LHEMP FINAL RECOMMENDATIONS

In the draft LHEMP, 149 strategies were outlined and assessed for treating the 16 risks. Of these strategies, 32 were outlined as short-listed strategies, which were suggested as having high implementation priority in terms of risk reduction potential (BMT WBM, 2008). These strategies are outlined in priority order in Table E.14.

Since the plan's finalisation and acceptance by the local governments, many of these prioritised strategies have led to projects being financed and implemented, as summarised in the Estuary Management annual reports (HSC, 2009; 2010).

Table E.14 *Prioritised strategies with high estuarine risk reduction potential*

No.	Strategy description
1	Conduct assessments to determine the carrying capacity of land areas (based on water, air, land capabilities) and limits for sustainable development within the entire catchment.
2	Develop a strategy for sustainable recreation across the Lower Hawkesbury Estuary, which states the sustainability of locations, facilities and access based on recreational survey and other data.
3	Collect information to inform amendments to planning controls based on the assessment of land capability, estuary carrying capacity (future population and development within the catchment), ecological assessments and HSC Housing Strategy.
4	Ensure planning instruments incorporate best practice, including: sediment, erosion and stormwater controls (e.g. construction controls, plans and WSUD); use of water reduction devices and maximum permeable surfaces, landscaped area calculations; protection of native vegetation; sewage management (e.g. low risk OSSMs); restriction of landscaping and gardens to endemic species; energy efficient design and ESD.
5	Determine sustainable limits for recreational activities (types, numbers and locations) and the requirements for existing or new facilities and access to achieve sustainable limits on foreshores and waterways of the estuary (i.e. suitable locations, unsustainable locations requiring removal, locations requiring restoration, new sustainable locations).
6	Employ a river keeper for the Lower Hawkesbury Estuary, to assist in compliance, education and on-ground works (e.g. boat speeds and zones, seagrass protection, effluent discharges, littering, fishing, foreshore habitat protection, foreshore and waterway activities).
7	Incorporate climate change strategy to mitigate local climate change impacts into planning instruments, management plans or strategy activities (i.e. with tools such as vulnerability maps).
8	Minimise clearing of vegetation on privately owned land, via new LEP template (e.g. Clause 34) and existing biodiversity strategy.
9	Submit the EMP to the appropriate minister for gazettal by the NSW Government.
10	Develop an estuary processes and issues checklist (EPIC) and integrate the checklist into councils' planning controls. (The checklist needs to be completed and submitted with DA documentation. The checklist will require applicants and council planners to assess the likely impacts of DAs upon the natural processes, estuary values and sustainability of the Lower Hawkesbury Estuary.)
11	Liaise with relevant state agencies to ensure integration of EMP actions into their relevant management plans or strategy activities (e.g. HNCMA's Catchment Action Plan, DPI Fisheries Sustainable Oyster Aquaculture Strategy etc.).
12	Establish a Lower Hawkesbury Estuary Management Committee to be facilitated by HNCMA, which incorporates Pittwater, Gosford and Hornsby Councils for a coordinated approach to estuary management.
13	Undertake remote and real-time environmental monitoring for the Lower Hawkesbury Estuary (e.g. chlorophyll-A probes, wind speed probes, salinity, flow meters, satellite data), and make data available to the public.
14	During the review of plans of management for all parks and reserves (national and council), ensure estuary assets are preserved (including habitat values for native animals, animals listed under the TSC Act 1995, prescribed burning and bushfire suppression undertaken according to park/reserve fire management plan etc.).
15	Provide an annual progress report that gives a review of monitoring data, progress in implementing EMP actions and outlines the status of estuarine health.
16	Undertake an independent review and update of the EMP every three years to improve performance continually in meeting the EMP objectives and protecting estuarine health.
17	HNCMA to appoint an estuary manager for the entire Lower Hawkesbury Estuary, to administer and update existing management plans and access state, federal and private industry funding sources, and develop a Hawkesbury Estuary Management Plan.
18	Improve the understanding of local impacts that may arise from climate change (e.g. produce vulnerability maps) and the management responses to such impacts (changes to infrastructure, planning provisions etc.).
19	Ensure adequate waste disposal facilities for people aboard boats and recreational fishers on land. This includes installation/provision of approved bins on hire boats, commercial fishing boats, moored boats and boats with trailers, and supporting waste services on land.
20	Establish a regular monitoring programme to monitor the impacts of recreation at various locations and times of year (such as peak periods), to ensure ongoing sustainability of such locations.
21	Consider a 'resident's pack', which outlines the estuary values, regional significance, ways to preserve such values, and includes existing brochures (from councils, DPI Fisheries, NSW Maritime, NPWS etc.) on stormwater, endemic plantings, bush care, boating maps, seagrass maps, aquatic weeds etc.
22	Undertake an audit of planning compliance to review the effectiveness of development conditions to protect estuary assets and achieve sustainability (e.g. an audit of the types of development being approved for consistency with sustainable growth limits and estuary asset protection goals).
23	Encourage vigilance in reporting non-compliance with regulations and environmental conditions, degradation (e.g. sediment erosion controls, OSSMs, vegetation removal or destruction, stormwater control and maintenance, recreational activities etc.) and pollution incidents (e.g. algal blooms, oils spills, chemical spills etc.) to appropriate authorities (e.g. 'river hood watch programme').

Table E.14 (*cont.*)

No.	Strategy description
24	Continue to lobby for reuse of water from STPs, to reduce freshwater demands in the catchment.
25	Develop a set of biological indicators (e.g. food chain or structural biota), which will assist in measuring climate change impacts.
26	Provide a forum for discussion about issues relating to the estuary and EMP progress.
27	Enhance weed management programmes across catchment, particularly in estuarine vegetation.
28	Enhance existing pest eradication programmes, particularly in estuarine habitats.
29	Define and map minimum buffer widths for riparian or foreshore vegetation in relevant planning documents (LEPs, DCPs etc.) to protect estuary assets and account for landward migration of habitat due to sea level rise.
30	Ensure suitable controls are contained within planning instruments for the design of foreshore development, including recreational facilities to maintain the estuary shoreline in as natural state as possible and minimised potential for bank erosion.
31	Riparian zones in agricultural areas fenced to prevent access of livestock to estuary. Protect and encourage rehabilitation of riparian vegetation.
32	Educate recreational users and general visitors about estuary values and the estuarine system, recreational impacts, and actions they may take to reduce impacts on priority areas (seagrass, harvest areas, recreational swimming) in the estuary (e.g. signage, boating stickers, brochures etc.).

E.7 FURTHER DISCUSSION POINTS

This section expands the discussion of Section 7.6, focusing in particular on workshop-related issues, including: the effects of last-minute programme changes in participatory processes; and monitoring, evaluation and management cycles.

E.7.1 Effects of last-minute programme changes

Apart from the effects of the major last-minute change in having an 'agency-only' workshop for the risk assessment stage of the participatory process, as discussed in Section 7.3.2, there were a couple of other last-minute changes that also had certain impacts on the project, but this time that occurred within or separate to the workshops.

In the first workshop, the whole-group discussion method used in the final session was too time-consuming to reach the desired outcomes of a list of synthesised goals, assets and risks. In the end, the asset list was fully completed but the goals and risk lists were synthesised by the consultants for the synthesis report (BMT WBM, 2007). Although not severely impacting the rest of the process, this led to a little confusion over 'ownership' of the goals and risks, as in the second workshop some agency representatives did not want to comment on the goals as they were developed directly by the whole stakeholder group, or to be seen to be changing the goals around without the input of the community representatives. A similar type of confusion resulted in the third workshop, because some participants did not feel as if they had been included in the creation of the risks, as they did not realise that the input of their 'issues' in the first workshop had been synthesised into the 'risk' list. This misunderstanding might have been avoided if the synthesis report had been sent directly to all participants in the process, rather than just the participants of the second workshop (with the other participants specifically having to 'request' a copy). For future workshops in the preliminary phase of problem identification and goal setting, it might be worth reworking the methods or increasing the workshop length to be able to complete the planned synthesis activities as a whole group in the available time. Likewise, sending the synthesis report to all participants might help to reduce confusion over process and outcomes.

In the second workshop, a last-minute change to the workshop programme was requested by the participants to run through the 'water quality' risk as a whole group. Although this change was accepted by the facilitator so as not to go against the participants' collective wishes, it did have a couple of ramifications on the risk assessment process, and particularly its validation. The extra time spent collectively on the water quality risk, although productive, meant that there would be less time available for the small groups to work their way through the other required risks. In light of this problem, the solution found was to break the large group down into much smaller groups than originally planned (i.e. pairs or threes rather than groups of four or five), so that all risks could be completed in the available workshop time. This solution did achieve its original objective to finish the risks, but time to complete another risk for validation purposes, such as the end of the sedimentation example, did not eventuate. This leaves the question of whether the results of the risk assessment can be validated, as different groups may have different tendencies of rating behaviour. Theoretically, it is now very difficult to validate these results, although their sensitivity can be further examined, as was shown in Section E.4.3. However, 'legitimisation' of the results can still take place if the participants believe sufficiently in the process or the capacities of the other participants to accept their judgements. Such an agreement to support the results, despite their potential weaknesses, is in some ways what occurred. The participants 'accepted' without too many complaints that all of the risks had been prioritised as either tolerable or intolerable and they were willing to 'treat' them all in the next phase of the process. It is likely that the rating of all risks as 'requiring a response' helped the lack of opposition to the risk assessment process, both from the participants who took part in it and those in the third workshop. In essence, this meant that the second workshop did not have as much of an impact on changing the content of the LHEMP process as could have been the case. There was thus less opposition and reaction to it.

Finally in the third workshop, a couple of last minute programme changes were made; firstly, because the strategy mapping exercise did not take as long as planned, and secondly, because some participants had to leave early. The changes were made to allow participants who needed to leave early to prioritise their actions and to fill out their evaluation questionnaires. Despite the fact that other participants did not have to leave, they also wanted to follow suit and ended up also assigning their action priorities and filling out their evaluation questionnaires before going back to the other activity that had been planned (i.e. adding responsibilities and monitoring needs to the strategy maps). This programme change seemed to prompt more participants to leave when they were satisfied by their contributions to the strategy maps, rather than wait around for the end of the planned workshop time and the planned events of sharing certain strategies of the risk strategy maps and discussing the next phases of the workshop. In the end, there were so few participants left working on the strategy maps that even some of the facilitation team took the advantage of leaving early. This rather interesting exit phenomenon, which not many of the participants witnessed (as they had also already left), presented some obvious difficulties in officially closing the session, and so a final official close never really occurred. Most participants were thanked individually and asked about their visions and hopes for the rest of the process before leaving, but a collective strategy for the next steps of the process was not officially presented. In retrospect, to avoid this problem, the workshop probably should have been officially closed early, just after the questionnaires were returned, and then the participants who did not have to leave invited to continue to work after the workshop's close.

As a summary to this discussion on the effects of last-minute programme changes, it is worth specifying that change is a natural part of participatory processes. However, this change and the need to accommodate flexibility in participatory processes presents some interesting challenges to researchers, consultants and project managers who work with them, as the outcomes are often unpredictable for a variety of good reasons, only three of which are mentioned here.

Firstly, the power base of decision making and process or project content often shifts in the direction of the participants and their interests, which can be difficult for the project instigators to deal with, as often their personal objectives for the process and outcomes will not be entirely achieved. The validation issue of the risk assessment process is a good example of this, where the project team was aiming for a 'validation' of a more scientific and robust kind, but instead had to live with a 'legitimisation' of results, something potentially more important for most of the workshop participants.

Next, there is the question of the uncertainty of reaching outcomes (or especially those specifically planned by the project team), as inviting participation has a tendency to 'open the box', define problems differently and create innovative ways of approaching and managing them. Not knowing exactly where a participatory process is going to lead to at the end, even if there are some excellent unforeseen outcomes, will sometimes require a 'leap-of-faith' from the project managers at the beginning of the process, which, when considering their responsibilities, they are sometimes not willing to make.

Finally, learning and changes in social relations and conflicts can occur as a result of (or lack of) participatory processes, both of which were observed through this LHEMP workshop process. Decisions to instate or stop participatory approaches to management are both likely to change the state of informal learning, stakeholder capacity building, social relations between people (both inside and outside the stakeholder communities) and conflicts, so project managers are often rather cautious about changing the status quo of management operations.

All this means that change resulting from participatory processes is probably inevitable, but with good management and careful design of projects, taking into account known constraints, this change can be of the positive kind and actively encouraged using well-chosen methods. Flexibility and the ability to develop effective contingency plans in the event of unexpected changes, and having enthusiastic and experienced facilitators, can also help to improve the chance of success of participatory processes and their outcomes, as well as reduce the more negative impacts of last-minute programme changes.

E.7.2 Monitoring, evaluation and management cycles

Estuary planning and management is a continuous process that requires ongoing monitoring and evaluation to determine whether management objectives are being reached, as well as dissemination of the correct monitoring information so that management strategies can be adapted when required (often through a new cycle of planning). It is suggested that monitoring and evaluation can be most useful for managing resources when a system for carrying out the process is designed as part of the overall planning stage, rather than tacked onto the end of a management process (SKM, 2004). This LHEMP process therefore had as its goal the incorporation of the analysis, synthesis and creation of monitoring and evaluation strategies (including objectives, information needs, indicators and data) throughout the workshop process and adjacent review.

In the first workshop, goals for the estuarine system were developed, which were principally summarised as the 'preservation and further enhancement' of the estuarine values or assets. During this first workshop, when the 'values' and 'issues' were defined, the participants were also asked to answer the following monitoring related questions: '*What existing information or data can be used to describe this value or issue and who holds it?*' and, '*What additional information and data would be necessary to describe this value or issue?*' This first phase of collecting existing and required knowledge on information and data sources yielded a large number of responses, with the summarised results presented in the first stakeholder workshop report (Daniell, 2007a).

However, as with the difficulties in carrying out a review of the existing and already proposed management strategies and actions before the second two workshops (discussed in Section 7.6.2), time and budgetary constraints did not permit the expert review of the information and data sources outlined by the participants to be carried out before the second and third workshops. This, therefore, had the same impacts on the third workshop, with stakeholders not knowing whether they were adding onto existing monitoring systems and proposing indicators when they were writing the 'monitoring needs and indicators' for the strategy maps. It was also interesting to note that during this activity, some participants appeared to find the definition of specific indicators or data collection programmes for monitoring quite a challenge. More specifically, a few participants found it difficult to focus on how to 'measure' work towards objectives and targets. For example, there was a large discussion over water quality objectives, where 'water quality' written on one of the strategy maps as a monitoring need had to be further broken down into specific indicators for a variety of uses, such as 'faecal coliforms' for primary contact recreation activities (i.e. swimming) and oyster harvesting or 'salinity levels' for certain estuarine flora and fauna (i.e. sea grass and oysters). Although it would have been useful, indicators or data were rarely more specifically defined by stakeholders to incorporate when and where data measurements would be taken, how the indicator would be constructed from data sources and how the information products from the indicators could be best constructed and disseminated to aid managers and other stakeholders (refer to Fleming (2005) for a more in-depth discussion on how effective monitoring and evaluation strategies can be constructed). Perhaps this situation could have been aided by a longer explanation and example of what kind of 'monitoring needs' description could most aid the estuarine management processes.

During the final workshops, a number of needs and issues were highlighted that could be addressed in the 'integrated monitoring and reporting strategy'. One such need was for good information dissemination strategies that would: provide simple systems for information disposal and retrieval; provide managers and stakeholders with relevant and easily understandable information (i.e. simple maps with indicator values, rather than lengthy reports); and underpin required stakeholder or general public education needs. One issue highlighted was that occupational health and safety regulations currently prevent some stakeholders from aiding in monitoring and evaluation processes (and other projects that could be beneficial for the estuary). This issue had specifically been encountered at a local-government level in council-managed zones, where the local government is required to take out insurance for community volunteers on projects, such as 'clean-up' days. Such costs unfortunately currently limit the number of good-willed or altruistic community stakeholder-aided initiatives that can be carried out, including estuarine monitoring programmes. Finding alternative solutions to this type of issue as part of the 'integrated monitoring and reporting strategy' may prove very beneficial for effective management of the estuary in the long term.

Process monitoring of the planning process is another very important part of an effective monitoring and evaluation strategy, a practice that has been embraced during the development of the LHEMP and discussed in Sections 5.5 and 7.5, with further information being provided in Section E.7. External process monitoring carried out by the researchers and participant evaluations provides valuable knowledge about benefits of, and potential problems or issues related to, the planning process before any such problems or issues become unmanageable. If such evaluations and participant comments are taken seriously, much can be learnt and processes and management continuously adapted and improved. These discussion sections have largely benefited from and been illuminated by the evaluation results of the LHEMP process. It is hoped that others may also learn from the implementation description and evaluation results of this Australian process.

E.8 FURTHER INFORMATION ON THE EVALUATION PROCESSES

E.8.1 Co-engineering team interview guide

The semi-directive interview scheme used to examine an Australian project team member's views on the LHEMP process is presented in Box E.1.

Box E.1 Oral semi-directive interview scheme used to examine a LHEMP project team member's views on the process

Leading questions

1. What do you see as the key strengths or advantages of the participatory workshop approach to action plan creation? And the disadvantages or weaknesses?
2. With your current knowledge of how the process has worked, what would you have changed if you were carrying it out again? How could the weaknesses be overcome?
3. Before the process started, you identified a couple of conflicts [. . .]. How do you think that the process has helped (or not) to manage these conflicts?
4. Did the process create or make any other conflicts visible, and how do you think these methods could have been better managed?
5. From your knowledge of past planning projects, how would you rate the efficiency of the process in terms of time, money and other resources?
6. Similarly, how would you rate the effectiveness of the process so far? For example, for developing a collectively agreed-upon action plan, for improving stakeholder relationships and for exchanging and understanding different points of view.
7. Do you think the results would have been different if other stakeholders had been involved?
8. To your knowledge, have there been any external effects of this process so far? For example, stakeholders taking a heightened interest in working with the council.
9. What feedback, positive or negative, have you received from the stakeholders about the process?
10. How are people to comment on the LHEMP once it goes on public exhibition? Will this be a standard or more participatory approach?
11. Do you have any thoughts on how the plan will be received by participating and non-participating stakeholders once it goes on public exhibition?
12. What do you like and dislike about the risk assessment and management approach? Do you think it will be effective in helping the estuary management to improve?
13. What do you see as the next major challenges for the estuary and its management?
14. What are the advantages of working with both consultants and researchers?
15. What is the most important thing that you have learnt from the process?

E.8.2 Participant evaluations

The following section outlines the responses to the final seven questions of the participant questionnaire. Further results and the full questionnaires are provided in Daniell (2007a; 2007b). The responses to questions presented here relate to the participants' overall thoughts and perceptions of the LHEMP process.

First of all, the question, '*What motivated you most to take part in this planning process?*' was posed. A number of participants replied that they had been motivated to take part in the process as it was their work, their responsibility to represent their group's interests, or that they were responsible for managing certain areas of the estuary and surrounding lands. However, it is noted that not all people in these positions with responsibility over the estuary's management, or interests to represent, attended the workshops, so what were some of the other underlying reasons for attending? Responses outlining some of these extra motivators included passion, desires and concern to help and improve the effective management of the estuary:

- *I believe it will make a difference to the environment and people who use the resource.*
- *Concern for estuaries and the chance to use my expertise.*
- *Because we want to work in the community and it is also my backyard.*
- *Previous studies or work with Hornsby Council on the river.*
- *A desire to participate effectively in the management of the natural resources of the area.*
- *Agreement that increased integration of estuary management will increase the likelihood of objectives being met.*
- *My passion for the estuary and contributing to the development of an ecologically sustainable plan: also contributing to the well-being of the community, which is impacted by the health of the river system.*
- *Mainly professional interests (planning, policy, ecology) and my concern for the lack of planning and management (or its implementation) on the river.*

One participant voiced this concern very strongly, stating simply that '*The river needs help.*' Adding a suggestion to this comment on how to help, the participant noted the opinion that, '*A river keeper is too mild – how about a River King with a band of knights as enforcers?*'

The next question, '*How do you think this process is helping to better manage the estuary?*', yielded a variety of responses from, '*Not sure it is helping yet – but give it time,*' to '*Community and agency involvement helps develop groundswell of support towards sustainable management concept.*' Some of the more hesitant responses included:

- *It may help a little but can't deal with the underlying growth factors that are the real problem (population + economic growth).*
- *The process provides a focus for the estuary, brings all these parties together to at least discuss and endeavour to try and plan to improve the estuary.*
- *Only time will tell.*
- *Will only help if it doesn't end in a report that isn't widely communicated and adopted.*
- *Hopefully, we will take some goodwill forward.*

On the more positive points, of how the process is helping to better manage the estuary, responses included:

- *Brings people who share similar concerns together.*
- *If implemented, especially into best practice and planning instruments, improved outcomes ought to result.*
- *Getting different groups (government + community) talking together and operating under agreed framework.*
- *This is an attempt to address estuary-wide issues, not site or community-specific issues.*
- *Incorporating all agencies and community or commercial representatives.*
- *A broad stakeholder involvement increases awareness of issues and includes many in creating solutions.*
- *Hornsby Council is a model other groups should follow.*

Finally, related to previous questionnaire comments and referring to the community representatives not being involved in the second workshop, one participant noted that, '*Disempowerment of the community in the process undermined a lot of commitment and work by many over the last years in preparation of the studies and plans.*' The context surrounding this comment has already been analysed in Section 7.3.2.

Following on from what the process may be able to achieve, the participants were asked about their own contributions to the process: '*Do you believe that your contribution to these workshops and planning process has been valued by the project team and other participants?*' All responses except one were a version of, '*Yes,*' or '*Hopefully,*' with the last one related to the aforementioned difficulties discussed in Section 7.3.2. Of the '*Yes*' responses, a few of the qualifying statements included:

- *Yes – always welcomed and comments encouraged.*
- *Yes, but hard to be sure.*
- *Yes – in proportion as one of many people.*
- *Yes. There seems to be material support and generally focused aims.*
- *Generally, yes, but greater knowledge of the waterway and its issues would have allowed greater input.*

As to whether the participants thought that outside stakeholder communities would accept the EMP resulting from the project process, responses to the question, '*Do you think the estuary management plan resulting from this process will be well accepted by the participants and outside stakeholder communities?*' were very varied. Comments resulting from this question appeared to indicate hope for successful outcomes, but concerns that the project was still not finished and a number of areas would still require further thought and attention before the final plan was produced and implemented. The responses received were:

- *Yes, as there have been extensive opportunities for participation.*
- *Maybe not by more extreme elements of the community as they felt excluded in the risk ranking.*
- *I hope so, but more tangible, grassroots action required before EMP nears completion.*
- *Probably – legislation and on-ground works – most people will ask whether their interests are accumulated (i.e. what is in it for stakeholders?).*
- *Not sure – is there a process for 'sign-off'? What happens next? Ask each participant to promote the multi-stakeholder process and contents to their organisation/networks. Provide a summary brochure etc. – promote it.*
- *Perhaps by participants. Or will it become yet another strategy or plan on the shelf?*
- *I do not know, it will be interesting to see what happens.*
- *Like any plan, it will satisfy some and not others but most will be indifferent unless it affects them directly, which is unlikely.*
- *Yes, although I am not convinced anything will change. It will at least be a benchmark.*
- *It should be if the results are communicated accurately and effectively.*
- *It should be. It has good representation from interested parties.*
- *Probably, because of the broad input.*
- *Possibly – if the objective, strategies and outcomes are clear, achievable and measurable within a timeframe and within reasonably expected resources.*
- *Depends how it is produced and distributed. Try and keep it simple but retain the power.*

As a follow-up to all the previous questions, the participants were asked, '*Overall, how do you think this estuary planning process could be improved?*' Most responses related to the contents of the workshops or improving the place of the project in the larger context. Related to the contents of the workshops, potential improvements included: adding a brief photographic presentation in the workshops or case study example '*To demonstrate the interrelatedness of river issues and the multiple benefits to all if appropriate action is taken;*' not using the categorisation of 'risks' in place of issues as '*They are a mixture of threats, pressures and management issues. Better classification and analysis would improve the process*;' identifying better whether the plan is '*Trying to operate at broad or specific level,*' as well as if '*A little more time in workshops and maybe pre-work where input is required*;' and installing a series of '*Interviews with participants to identify key strategies, plans, or major projects.*'

On process improvements of a contextual nature, short suggestions included: reducing '*the top-down approach which has come to dominate the process*'; and the probable need for '*further meetings of some people*' and '*more expertise in different areas*'. More elaborate suggestions on how the process could be improved included:

- *If this planning has the responsibility of higher level of government, or is supported by higher government levels;*
- *By placing it in a well-understood niche within the catchment management world. There are loads of catchment or estuary management forums and it's hard to pick the role or relationship of one to another; and*
- *Although it would have taken longer, maybe including all EMPs – require greater statutory weight, need greater consideration and enforcement.*

Finally, the participants were asked, '*Do you have any other comments or questions about this workshop or the overall project?*' To this question, half the questionnaire respondents left comments of varying note, from small remarks such as '*No. I look forward to the outcomes,*' '*It's frustrating that the real factors causing problems can't be dealt with at the local level,*' and '*Thank you Hornsby Council,*' to those of a more substantial nature related to the project's progression. These included:

- *Potentially a further workshop required for further discussion of actions*;
- *Work forward to end results – like to see a flexible document able to be adapted over time e.g. 5 years*; *and*
- *It would be valuable to be able to view the synthesis of the workshop notes in a table of some sort and comment on that. It will be easier to start pinning responsibilities when today's info is synthesised i.e. approx. 60–70 cards per sheet × 16 sheets = 1100–1200 cards for which to assign responsibility, i.e. impossible to be thorough or attentive and some ideas could easily be overlooked.*

A last observation on the project related to the previously mentioned issue of community representatives not taking part in the second workshop (see Section 7.3.2) was: '*I am sure that the process was well intentioned and I would like to see adoption of the plans by Government agencies and the community with a financial commitment by government and support by active participation by a management committee.*'

Appendix F Supplementary information on the Iskar Process, Bulgaria

This appendix supplements Chapter 8 by extending the brief co-engineering negotiation event summary table and example negotiation event in the form of an interpretative description of the detailed co-design and co-implementation processes. A table summarising the use of the decision-aiding process model through the intervention and the translation of the final Elin Pelin risk response plan are also provided.

F.1 DETAILED CO-DESIGN AND CO-IMPLEMENTATION

F.1.1 Preliminary interviews and Workshop 1: describing the context

The detailed co-design phase began in September 2006 in a meeting in Montpellier where the Bulgarian facilitator met and worked with the two French researchers (the Masters student's internship having been turned into a research contract) and the private research consultant from a German firm, who had been involved in the pre-assessment phase of the Iskar Basin and selection of the Local Public Stakeholder Forum (LPSF) for the AquaStress project. The preliminary questionnaires to be used for the policy makers' and citizens' interviews were designed, although not quite in the way that the researchers had intended. The researchers wanted to use cognitive mapping interviews but the Bulgarian facilitator did not feel comfortable enough with the technique, as she had not used it before, and preferred more traditional structured questionnaires, especially with the citizens (Popova, 2008, personal communication). She also thought that performing 100 such interviews with citizens would take too long, arguments that led to the French researchers finally agreeing to the text-based citizen interviews (Ribarova, 2008, personal communication). However, cognitive mapping interviews were still conducted with all of the policy makers, with the support and training provided in Bulgaria by the private research consultant, and the results were computerised in Germany (Hare, 2007). The French researchers simultaneously worked on how they might develop a semi-automatic data processing capability to create cognitive maps from the text response interviews (Ferrand, 2008, personal communication).

After this meeting, the interviews and participant selection were carried out by the facilitator, Bulgarian regional partner, a young student friend of the regional partner and a couple more university students who had participated in the Borowetz Summer School, with the interview structures being finalised in collaboration with the French researchers via email (Popova, 2008, personal communication). The results of these interviews were then processed by a group of three French researchers, an activity that proved to be more challenging and time-consuming than planned, and so was not entirely completed (Rougier, 2007). This challenge, along with the French researchers being slightly in disagreement with some of the participant selection methods for the workshop process, appeared to lead to a few small tensions emerging in the project team (Ferrand and Ribarova, 2008, personal communication). They considered that some of the citizen groups were not representative of the targeted populations in the region and were too close to the regional partners' own social networks, although they also understood the time and budgetary constraints placed on the Bulgarian work team and so respected their decisions (Rougier, 2007, personal communication; Ferrand, 2008, personal communication).

Despite a lack of complete interview processing being ready in time for the workshop process, the first workshop was considered a success from a number of points of view. The policy makers' workshop, which included national ministers, national government bureaucrats and national NGO representatives, was principally re-designed and planned by the private research consultant, the Bulgarian facilitator and the Bulgarian regional partner, with some support provided by the French contract researcher and the Bulgarian evaluator (Hare, 2007). Owing to constraints on some of these high-level policy makers participating in the workshop, some of them had to leave the workshop early. This meant that the organisers had to cope with reduced time for working. Despite this adjustment, the workshop was still implemented in a calm and well-organised manner in a pleasant hotel chosen by the Bulgarian regional partner (Hare, 2007). It appeared that most of the policy makers very much appreciated the opportunity for getting to know one another better and learning about each other's perspectives (Hare, 2007). For the members of the project team, this workshop helped to build their trust and personal relations with each other (Rougier, 2007, personal communication). However, although the relational objectives of this first workshop were well achieved, the method-based outcomes relative to the SAS model were found to be only partly satisfactory by the French researchers, as the elements of a water-related situation causal model of the region were not really obtained (Rougier and Ferrand, 2007, personal communication). The groups had instead created institutional or actor maps of regional water management, which the private research consultant considered some of the best that he had seen (Hare, 2007).

F.1.2 Workshops 2 and 3: visions and values

The second series of workshops' outcomes had similar issues of model content, since each of the facilitators' groups produced different types of cognitive map of the situation. As the French contract researcher mentioned when asked about this issue, it appeared that when the facilitators and groups appropriated the models as their own, they also adapted the syntax rules that had been developed. This meant that the groups felt strong ownership over the models that they had created but that the models could not be used as intended by the researchers (Rougier, 2007, personal communication).

This second set of workshops saw a change in the project team membership. The private research consultant's contract in the project had been completed, so he did not participate in any further workshops. The French research director took his place in the implementation finalisation at the last minute, as well as in-workshop re-design and organisation activities, along with the French contract researcher. During these workshops, there was also a need for more facilitators, so the regional partner's young student friend and the regional partner took on these roles. These workshops required a large amount of last-minute organisation; furthermore, there were a few small relational difficulties, not enough support for the untrained facilitators and too many in-workshop changes, which led to more tensions emerging in the project team (Rougier, 2007, personal communication; Popova, 2008, personal communication). Despite these challenges, the Bulgarian facilitation team dynamically adapted the pre-planned method relative to what they felt their facilitation capacity allowed and to the relational processes seen in the group (i.e. tiredness, confusion) (Ferrand, Ribarova and Popova, 2008, personal communication). This led to some changes through the workshop series and some aspects that were implemented as planned for all groups, the results of which were considered interesting by the French research director; especially the preference elicitation game developed at the last minute, which had worked well in all groups, and the actors–systems models that had been developed despite them having a range of model syntax that varied between the different facilitators' groups (Ferrand, personal communication). However, the French research director was frustrated at not being able to understand Bulgarian (as full translation was not ensured during the workshops) and thus understand how much the facilitators were driving or adding their own thoughts to the process (Ferrand, 2008, personal communication). In particular, he found the student facilitator too young for the target groups of stakeholders and did not yet have confidence in her (Ferrand and Ribarova, 2008, personal communication). The student facilitator realised this but also believed that she was capable of doing the job well and was proud of her own work, especially with one of the citizens' groups as it had been a very emotional experience for her and the citizens as they had shared and built a group model out of their own in-depth experiences of being recently flooded (Popova, 2008, personal communication).

At the same time, the Bulgarian regional partner was very stressed as the workshops had not been very well organised, by her standards, or planned sufficiently in advance and she was not overly pleased about having to spend so much time in roles that she had not envisaged taking on at the beginning of the project (Ribarova, 2008, personal communication), particularly since she knew that she could not be paid through the European project (Ferrand, 2008, personal communication). This led to her holding a meeting in which she stated that she wanted the other project team members to make more effort to organise the next workshops better in advance, and especially for the Bulgarian facilitator to take more responsibility over the process organisation, as had originally been planned in her contract (Ribarova, 2008, personal communication). The others seemed to understand her point of view and the situation improved in the next policy makers' workshop, where the process was successfully co-designed, planned and implemented predominantly between the Bulgarian facilitator and the French contract researcher, with some support from the Bulgarian regional partner and the French research director in the planning stage (Ribarova, 2008, personal communication). The French contract researcher, with one of his colleagues in France, had worked to process and format all the outputs of the previous workshops in order to be able to present some of their viewpoints to the policy makers; an effort that was respected by the policy makers and the regional partner. This workshop helped to rebuild some of the slightly damaged relations and led to the project team taking pride in their work (Rougier, 2007, personal communication), without pressure from the French research director to respect the research plan and to achieve the expected research-oriented outcomes required for the AquaStress project (Ferrand, 2008, personal communication) or intervention from the Bulgarian regional partner.

F.1.3 Mid-process project team changes

The period before the second series of interviews and fourth series of workshops turned out to be stressful, or simply full of changes for the project team members, and this would affect the future direction of the process and its outcomes. Just prior to the Workshop 4a series commencing, the project team found out that the Bulgarian facilitator was not likely to return from an overseas trip as planned, as she had an opportunity to take a job in the overseas country (Ribarova, 2008, personal communication). Although the project team members were very upset about this sudden personal and project loss, as well as the apparent disappearance of some of the data from the interviews, they mostly appeared to understand why she had left (Rougier, 2007, personal communication; Ferrand and Ribarova, 2008, personal communication). This left the Bulgarian regional partner very little time to decide how to reorganise the Bulgarian part of the project team. She finally managed to find two work colleagues, also working in the AquaStress project, to help her with the organisational aspects, such as contacting stakeholders, making bookings and other logistical aid (Ribarova, 2008, personal communication). She also decided to take on the rest of the facilitation herself with her student friend, as they already had gained sufficient experience and were at ease in the role. She thought it would be much easier and less risky for her, even if it would take her own time, rather than trying to find and train someone new who did not understand the project (Ribarova, 2008, personal communication).

On the French project team side, the French project director was also worried about what to do with the contract of the French contract researcher, which had already been extended once and for which there was no additional budget (Ferrand, 2008, personal communication). This was a difficult situation, especially since the project team claimed that it would be a great shame to lose him and his memory and understanding of the full process. However, considering the difficult-to-overcome financial issues and the French research director and contract researcher's trying relations, the French research director thought that I, his PhD student who had just returned to France, might be able to participate in the remaining phases of the process so that I could use the Bulgarian project as one of the case studies in my thesis and replace the contract researcher for the last workshop of the process.

F.1.4 Options interviews and Workshops 4a and 4b: strategy construction, merging and evaluation

The final lead-up work to the fourth series of workshops was remarkably calmly and effectively carried out despite all the project team and role changes. The Bulgarian student friend of the regional partner performed a very efficient and effective job of contacting and carrying out the second phase of cognitive mapping interviews with the policy makers, helped by the regional partner's supervisor (Popova, 2008, personal communication). The French research director, French contract researcher and I also worked well together in France to develop the list of options and the methods for use in the workshops for strategy building, despite this exercise originally being designed to be carried out with aid from the experts in another part of the AquaStress project. This issue, and the fact that some of the other group decisions made in France for the Workshop 4a implementation were not entirely in keeping with the original process to be tested for the AquaStress process, meant that the French research director upon his arrival in Bulgaria took the decision to try to adapt the implementation to make it closer to its original form, so as to limit the risk of not achieving the project's research objectives. However, process time constraints and the Bulgarian project team's general level of fatigue due to their increased level of participation in the process organisation, plus the additional pressures of the French research director who also had a difficult task to perform in fulfilling the European project's research requirements, led to a high level of stress in the project team (Ferrand and Ribarova, 2008, personal communication). Although the organisation of each workshop improved as the Bulgarian facilitators (the regional partner and her student friend) adapted the methods and worked to better fit the time allocations and to increase the interest of the process for the participants (Popova, 2008, personal communication), when the French contract researcher arrived after the first three workshops and I arrived for the last two workshops to meet the Bulgarians for the first time since the process began, we both found the Bulgarians very physically and mentally exhausted.

After observing how the Bulgarians had been implementing the workshops and helping to suggest further changes to improve them, the French contract researcher and I worked together with the Bulgarians to co-design the first workshop where all the groups' participants would work together for the first time in the process. As all the results from each of the Workshop 4a groups were formatted too differently to be of much comparative worth, and the results of the previous models had not been obtained and combined as planned by the French researchers, the French contract researcher and I took it upon ourselves to design what we thought would be of most interest to the stakeholders (but potentially less useful to the overall European project's research aims) and let them all meet each other and work effectively together. The chosen collaborative design was prepared ahead of schedule in the few days preceding the workshop, implemented as planned with the help of two more stakeholders from the process acting as facilitators and was commonly perceived to be a success, based on the objectives that we had set ourselves, and as evidenced by many comments in the debriefing session. However, this did leave us, and the French research director, with many more questions about how the results could be used later and specifically what could be done in the final workshop. This short-term success helped the new project team to bond together and to feel proud of their collective achievements in helping the stakeholders to appreciate each other's views and work well together. Nevertheless, it also made them all feel even more disappointed that the French contract researcher would probably not be able to participate any further in the process and they worried about how the last workshop would be designed to provide a positive end to the project that had so far built up so much hope in a number of the stakeholders for future improved management (Ribarova, Popova, Rougier, Vassileva and Daniell, 2008, collective Workshop 4b debriefing session).

F.1.5 Changing objectives and design of the final workshop

The description of this important stage of the process was outlined in the co-engineering negotiation event analysis in Section 7.3.2.

F.1.6 Workshop 5: creation of a flood risk response plan for Elin Pelin

The final workshop was well attended and ran smoothly, with the mixed groups of stakeholders working very effectively together on what appeared to be an equal level of understanding and competence (Popova, 2008, personal communication). In fact, it ran so smoothly that the activities were completed more quickly than expected and the risk response plan of 24 action sheets (projects) and their spatial mapping had been described, computerised, presented, debated, voted on and further discussed, and the workshop evaluation and a debriefing session carried out by lunchtime. By this stage, the facilitators and participants appeared to be at ease in using the types of group work and participatory methods that they had now been using for almost a year. Even two participants of the three who were asked to facilitate one of the groups seemed proficient in the role, which they had never previously practised. Almost everyone appeared very satisfied with the outcomes of this final workshop and the good relations that had been built with other stakeholders and the project team through the process. The evaluation results revealed that only one participant wondered how the results would materialise, and the French research director, although generally pleased, was still a little disappointed with some parts of the stakeholder analysis and the lack of stakeholders representing stronger interests in drought management at the final workshop. Otherwise, the process ended on a very positive note and the good personal relations of the adapted project team members had been significantly strengthened, including between the Bulgarians and me, who by then thought that we knew each other very well, both on a personal and work level, and between the French research director and the young facilitator.

F.2 DECISION-AIDING MODEL USE

In a more formal analysis of the intervention process, Table F.1 presents the elements of the Tsoukiàs (2007) decision-aiding process model and how members of these sets evolved through the process.

F.3 FURTHER SELECTED PROCESS RESULTS AND ANALYSES

F.3.1 Initial definition of actors and actions

In the second series of workshops (WS2), 23 joint actor–action cognitive maps were produced by the five groups of organised stakeholders and citizens, as outlined in Figure 8.5. The maps focused on flood and drought risk management by looking at actions required for prevention during times of crisis and for restoration and remediation, and the actors that could carry out these activities. The analysis of these maps was carried out conjointly between the Bulgarian facilitator in France and me, to ensure reasonable Bulgarian–English translations and count validations. In total, over 50 distinct actors, whole or named parts of institutional bodies, were counted on the maps. These were then classified as national government, including state-owned companies (48%), municipal government (7%), citizens and citizens' groups (7%), NGOs and international funders (13%) and private companies (24%). These results could be interpreted in a number of ways: firstly, that the citizens and municipalities believed that the state and other private companies (such as water supply companies) should take on more responsibility for ensuring their protection and aid before, during and after times of flood and drought, and that they themselves could do little; secondly, that the participants considered 'citizens or population' and 'municipality or municipal services' to form more coherent groups than some of the other types of actors, such as national government departments and different private businesses and NGOs, an idea which is supported by fact that both these categories received counts of over double many other actor types. However, elements of the first hypothesis also remain supported, as many of these mentioned were at the receiving end of aid, protection or information – only a few types of citizen action, such as reducing water consumption to relieve droughts and planting trees to reduce flood risks, were suggested. The top 10 out of 50 most highly counted actions or activities from the 23 maps are

Table F.1 *Decision-aiding process model element elicitation and evolution (the Australian researcher is the author)*

	Manifestation of model elements	Evolution through process
Representation of the problem situation: $\mathbf{P} = \langle \mathbf{A}, \mathbf{O}, \mathbf{R} \rangle$	**A** – set of actors: $\mathbf{A} = \mathbf{C} \cup \mathbf{K}, \mathbf{T} \subseteq \mathbf{A}$	
	C – *subset of core-participants*: approximately 60 members in the set over the whole process	Many more participants were involved in the first interview phase (100) than participated in the workshops – participants were chosen and invited primarily by the Bulgarian regional partners (who were themselves invited by the AquaStress project steering committee).
	K – *subset of associated stakeholders*: elicited through project team stakeholder analysis and through most of the interviews and workshops and in project team meeting discussions; present on cognitive maps of risk drivers and impacts, actors and actions maps, in option, strategy and project construction activities as 'responsibilities'	Approximately 50 discernible actors or actor groups were identified in the actors and actions mapping in WS2 (i.e. national government departments, municipal governments, private businesses, research groups, NGOs, community associations and user groups). Approximately 30 of these actor groups were invited or represented in the core participants – the actor networks linked to each of these representatives is thought to be large but was not specified entirely as part of the project. Main evolutions included increasing awareness that more parliamentary representatives (who would be present during the workshops) would help to increase the potential for real decisions to be made. Groups re-signalled in interviews and workshops not present in the workshops included the Ministry of Finance, the Roma communities and farmers.
	T – *subset of project team members*: approximately 30 from 20 institutions – university, government institution and private researchers, a diverse stakeholder steering group selected as part of the overall AquaStress Iskar test site (national ministry directors, NGOs, private businesses, research experts in water engineering, forestry and ecology, municipal government and citizen representatives) and a few additional citizen helpers	Membership different for development, design and implementation process. Twenty-one present in some part of the co-development, seven present in some part of the co-design and 15 present in some part of the co-implementation. five actors played the largest key co-engineering roles through the entire process: the regional partner (university researcher) and her chief, the French research director (government research institute), the Bulgarian facilitator (university student) and the Bulgarian external evaluator (university administrator). Others left and entered the process – Bulgarian facilitator (internationally qualified) left after WS3, French contract researcher left after WS4b, Australian researcher (privately funded PhD student) arrived prior to WS4a.
	O – *set of objects*: elicited as risk drivers and impacts, water system elements, expectations, visions in the first interviews, WS1, WS2 and WS3; others were present during discussions and time resources were present in WS4a and WS5	Hundreds of individually and group elicited objects were classified into a smaller number by the project team following the workshops, for example 18 risk driver categories and 15 impact categories. The six stakeholder groups' visions from WS2 and WS3 were synthesised by the project team into eight category 'values' and reintroduced into the last stage of the process, against which the potential impacts of the 24 chosen projects were evaluated (WS5).

Table F.1 (*cont.*)

	Manifestation of model elements	Evolution through process
	R – *set of resources*: elicited principally through the preliminary and secondary interviews. Some of these and some other resources (needs and requirements) were also elicited in WS2 and WS4a	The principal resources elicited in the first interviews were types of water resource, plus finances and social resources, such as information, competence and accountability. Many of these and other resources were treated implicitly through much of the process or just in oral discussion, such as the needs for 'expertise' and 'support'. Evolution of resources made explicit predominantly occurred during WS4a and WS5 when actions and projects were investigated as requiring certain resources for realisation i.e. finance and social and political coordination and will, although the categories were defined by the project team (time, infrastructure, citizens, institutions, other resources, cost).
Formulation of the problem and objectives: $\boldsymbol{\Gamma} = \langle \boldsymbol{\Pi}, \boldsymbol{A}, \mathbf{V} \rangle$	$\boldsymbol{\Pi}$ – *set of problem statements*: two risks formed the pseudo-problem statements – these were also divided into three problem categories each: prevention, during crisis and remediation	The problem statements were considered the risks requiring treatment: floods and droughts. In WS2, WS3 and WS4a they were analysed in terms of prevention, during crisis and remediation. This was an evolution from the original 'problem statement', which was to look only at crisis management (and not prevention and planning). In the final workshop, only the flood risk problem statement was treated in depth over the three categories.
	$\boldsymbol{A}$ – *set of potential actions*: these were elicited as 'actions' or options in the interviews, WS2 and WS3	Potential actions were brainstormed individually in the first set of interviews, and then laid out in the cognitive maps of WS2 and WS3. Especially at the beginning of the process, there was confusion or lack of attention to defining first which actions currently occur, and which ought to occur to cope with flood risks – some participants showed their confusion in the first interviews and WS2 maps.
	V – *set of potential points of view*: elicited from participants in interviews and cognitive maps. Some sourced through the visioning exercise in WS2 and WS3	Although explicated in the form of cognitive maps, the set of potential points of view that each actor was to use to analyse actions remained fairly tacit. This being the case, many of the points of view used to evaluate the options in WS4a and some in WS5 had been proposed by the researchers i.e. the impact categories: households, industry, agriculture, politics, nature, infrastructure, droughts, floods. The eight value categories extracted from the visions by the project team also became points of view for analysis in WS5.
Model exploration and options evaluation: $\mathbf{M} = \langle \boldsymbol{A}^*, \{\mathbf{D}, \mathbf{E}\}, \boldsymbol{H}, \boldsymbol{U}, \boldsymbol{F} \rangle$	$\boldsymbol{A}^*$ – *set of alternative sets of actions*: derived from full set of potential actions by the project team, with some redefinition in the second round of interviews, WS4a, WS4b and WS5	From the set of potential actions elicited from the participants and some others proposed by AquaStress project experts and the design team, over 40 'options' were provided and further defined in the second round of individual interviews and in WS4a as the 'set of alternative sets of actions'. Some of these were further refined in the strategies of WS4a, WS4b and WS5.

Table F.1 (*cont.*)

	Manifestation of model elements	Evolution through process
	D – *set of dimensions*: predominantly defined by the project team	Dimensions used in the large multi-criteria analysis matrices of strategies in WS4a looked at actor types, needs (i.e. resources) and impacts on the seven categories that were defined as the points of view above. Equally, the eight value categories, plus three more resource or constraint categories proposed by the project team were used to analyse the final projects in WS5.
	E – *set of corresponding scales*: predominantly defined by the project team and later adapted in collaboration with the stakeholder groups	For the matrices of WS4a, the scales varied somewhat between the facilitators, with a variety of coloured dots, ticks, crosses, numbers, pluses and minuses, as well as a few other comments. In fact, the scales often remained fairly cryptic to outsiders even after explanation from the facilitators. For WS5, the scales were much simpler with a '1' attributed to the value category on which the project would be perceived to have the most important positive impact, and a '2' for the second most important. Similarly, a '1' was attributed to the largest perceived constraint, and a '2' to the next most important perceived constraint.
	H – *set of preference criteria*: elicited as individuals and as groups in WS2 and WS3	Preferences were formally gathered as part of the money distribution game in WS2 and WS3, as outlined in Section 8.4.2. However, this was based on the categories defined by the project team and original project scoping: the sectors of agriculture, households, industry and nature in the regions of Samokov, Sofia and Elin Pelin. These were not really used in the process afterwards, except as part of the legitimisation for changing the scope of the final exercise to focus only on Elin Pelin. Preferences were also distributed in the form of votes on the final projects proposed in WS5.
	U – *an uncertainty structure*: uncertainty structure considered in WS4b as part of the extreme event analysis exercise	The structure of uncertainties was formally considered through the 'robustness analysis' that was carried out by subjecting the merged strategies of WS4b to three extreme events, each of which had been defined by the project team. Also, one category in the 'constraints' section of the project analysis in WS5 that was available to aid reflection was 'uncertainties in the execution'. Other uncertainties such as knowledge uncertainties were only discussed, often implicitly through the workshop, and were not systematically analysed.
	F – *set of operators*: most operators used in the workshops and for manipulating workshop outputs were defined by the project team. Other informal operators were developed or used intuitively by the participants throughout the workshops	Mostly very simple operators such as causal arrows and visualisation tools. Participants and facilitators often appropriated methods and changed the syntax (operators) en route through the workshops. Other implicit operators and arguments were used by the participants to

Table F.1 (*cont.*)

	Manifestation of model elements	Evolution through process
		synthesise, present or defend their views during the workshop series.
Final recommendations: Φ	Φ – *set of final recommendations*: given as 24 prioritised projects in the Elin Pelin risk response plan at the end of WS5	The construction of the final recommendations used many of the above sets, which resulted in 13 planning and prevention risk response projects, five projects to aid in times of crisis and six to aid current and future remediation efforts. The top seven projects received more than half the total votes, and contained three non-technical projects and four technical projects.

Table F.2 *Ten most required activities for flood and drought risk management as first perceived by the organised stakeholders and citizens*

Actions (provision of...)	Total count	Organised stakeholders' maps	Citizens' maps
Information	16	12	4
Money, finance, budget	15	11	4
Help, assistance	12	3	9
Management, control, organisation, accountability	11	7	4
Water supply system	6	3	3
Food, consumable supplies	5	4	1
Equipment	5	2	3
Dykes	4	2	2
Technical expertise, analysis	4	2	2
Medical help	4	1	3

outlined in Table F.2. More technical actions are indicated in normal font and the more non-technical actions in italic font.

These results demonstrate a certain bias towards the citizens' preference for direct help (medical or otherwise) and assistance, compared with that represented by the organised stakeholders in their cognitive maps. The organised stakeholders instead emphasised the management, information and financial issues required for risk management, in particular the coordination that is required between the different levels and departments of government, businesses and NGOs, and the information that is needed by municipalities and citizens to help them to plan, cope and restore their lives before, during and after flood and drought crises. These top 10 actions also show a tendency to preference the need for non-technical actions, and both citizens and organised stakeholders represented the technical options more consistently.

F.3.2 Final recommendations

The final projects defined and evaluated in the last workshop are shown in Figure F.1.

In Figure F.1, the less technical projects have been labelled in italic text, and the more technical projects in normal text. From these results, it can be seen that in the top five preferred projects were three technical and two non-technical projects. The first two projects were restoration activities, showing the difficulties the Bulgarians had in finding funding to maintain and restore their infrastructure following flood events. The next two projects were broad-scale education campaigns: one directed at the municipal government level about how to prepare and find funding for flood (and drought) risk management; and the other directed at the general population about how to prepare and cope with flood events more effectively. The final project of the top five is one to correct the current channel of the Iskar River to provide more control of flood drainage, another very 'hard' engineering solution. In total, 14 of the 24 final projects could be classified as largely technical and 10 as non-technical. The distribution of the different types of Iskar stakeholders' votes on the final project types is given in Figure F.2.

From Figure F.2, it can be observed that at the end of the process the policy makers preferred the more technical projects,

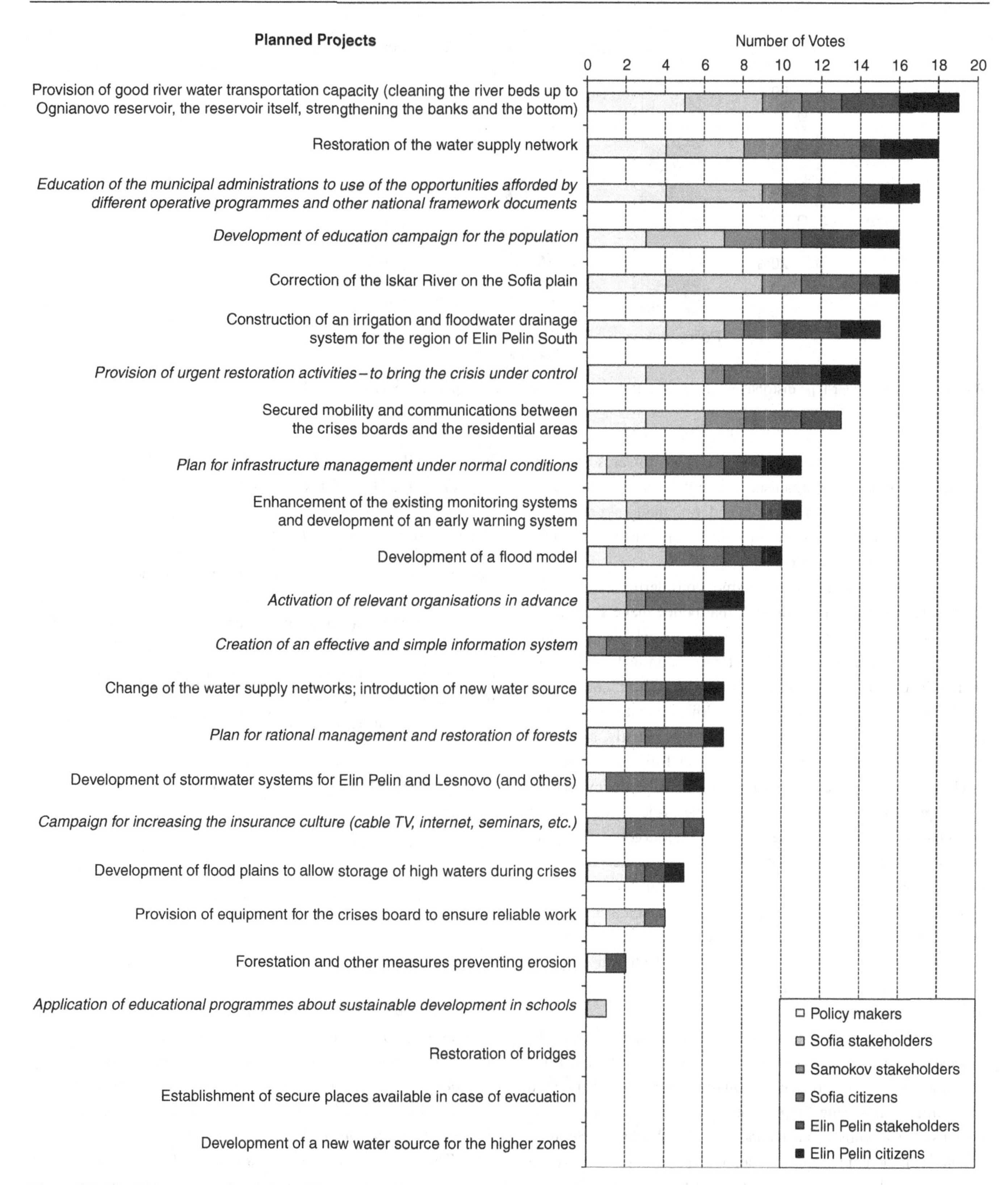

Figure F.1 Final votes on projects planned in the Elin Pelin risk management plan. Source: Daniell *et al.* (2011).

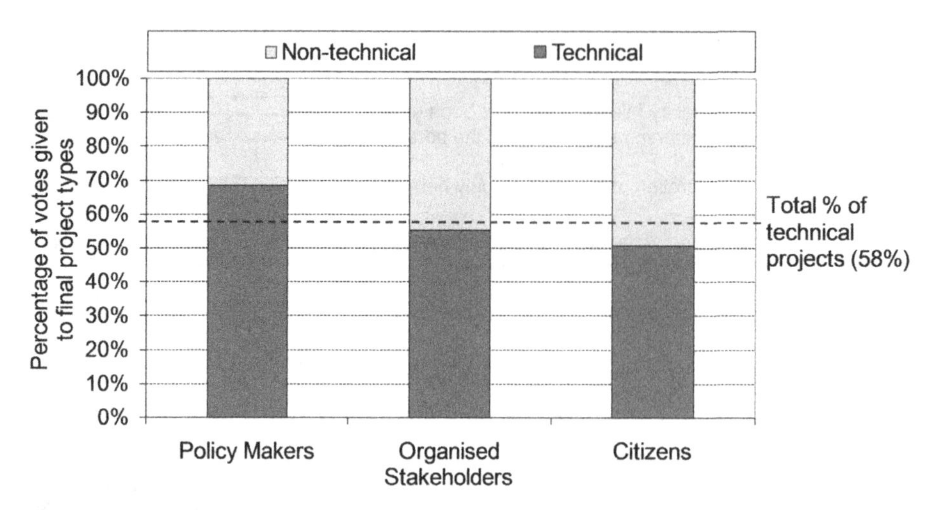

Figure F.2 Stakeholder-type distribution of votes on technical and non-technical projects. Source: Daniell *et al.* (2011).

while the citizens had an overall preference towards the non-technical solutions. The organised stakeholders were between the two other stakeholder types, but considering the total percentage of technical projects available for selection, chose slightly to the side of non-technical projects.

This is an interesting final outcome, considering the distribution of drivers of flood and drought that policy makers identified at the beginning of the workshop series, in which the policy makers outlined a large number of more non-technical socio-economic drivers for floods and droughts. There could be a number of reasons for this deviation, although just two potential hypotheses of explanation will be given here. Firstly, as the list of prioritised projects was to be targeting Bulgarian structural funds to finance projects, it is possible that the policy makers took a pragmatic stance of voting for the projects that had the best chances of being accepted – due to the largely 'infrastructural' nature of the funds. Secondly, the policy makers may have voted for the projects that they themselves would be able to run and fund, which were those that were more technically orientated. This may have equally been true for the citizens and municipalities voting for some of the non-technical projects, which could occur under their control or with which they could more easily be involved.

Whether this final voting underlies a strong appropriation of the process and personal willingness to continue to contribute to flood and drought management activities in the region is still difficult to determine. Considering the previous actions put forward by the citizens and organised stakeholders in the Workshop Series 2 (Table F.2), there do seem to be some changes in perspectives or preferences. The top voted project in the final workshop did not appear in the top ten mentioned actions of Workshop 2. In fact, it had been only mentioned twice by the same group – the Elin Pelin citizens. These issues of perspective change and learning will be further investigated in the evaluation results section.

F.4 FURTHER INFORMATION ON THE EVALUATION PROCESSES

F.4.1 Co-engineering team interview guide

The semi-directive interview scheme used to examine Bulgarian project team members' views on the Iskar process is presented in Box F.1.

This scheme was used in its entirety for one of the key Bulgarian facilitators and then reused in a simplified form to drive a cognitive mapping interview carried out jointly with the research and operational leaders of the Iskar process. Emphasis was placed in particular on points (1), (3), (5), (9) and (14).

F.4.2 Participant evaluation responses

This section supplements the information already provided in Sections 8.5.2 and 8.5.3.

Following the quantitative responses in Figure 8.13, qualitative evaluation responses of how the process helped stakeholders to work with and to communicate with others included:

- *'It helped me to realise that I am participant in a very useful and important civil activity.'*
- *'It provoked me to cope with different and unknown situations (during the workshops' games and activities).'*
- *'I had the opportunity to meet new people; this enlarged my possibilities, to ensure better coordination and relationships in my basic work, which is related to local crisis management and organisation.'*
- *'The long series of meetings and workshops gave me the opportunity to discuss the matters in detail and to achieve joint conclusions at the end of the process.'*
- *'The joint work had very positive influence upon all the participants. The discussions were open and straightforward, without confrontations or conflicts.'*

Box F.1 Oral semi-directive interview scheme used to examine Bulgarian project team members' views on the Iskar process

Guiding questions and information to elicit

For each process phase (interviews and workshop):

1. Objectives:
   a. What were the stated or perceived objectives of each project team member and any other visible process participants?
   b. To what extent were these achieved?
   c. Why were they achieved? If not, why not?
   d. When, why and how did any of these objectives change?
2. Methods and workshop or interviewing process:
   a. What were the advantages and disadvantages of what took place?
   b. Retrospective changes for improvement?
   c. What were the main lessons learnt?
3. Project team relationships:
   a. How would you describe the project team cohesion?
   b. Conflicts? Differences of opinion?
   c. To what extent were issues resolved or did they continue to cause problems?
   d. What negotiations took place over process or methods? What were the outcomes and did any major changes in planned activities result?
   e. Who controlled or led the process and decision making?
4. Participant relationships:
   a. How would you describe the relationships and cohesion between the participants in the workshops? Pre-existent relations?
   b. Conflicts of interests? Positive relations?
   c. How did these evolve?
5. Relationships between project team and participants:
   a. How would you describe the relationships and cohesion between the participants in the workshops? Pre-existent relations?
   b. Conflicts of interests? Positive relations?
   c. How did these evolve?
6. External effects:
   a. Were there any external effects of the workshop?
   b. Was any feedback (e.g. from workshop participants or other external stakeholders) obtained? What was it about?
7. Surprises:
   a. Did anything unexpected emerge from the workshop or interview process? Did this change the next stages of the process and in what way?

For the whole process:

8. Advantages or disadvantages of the process
9. Major lessons learnt
10. Any major changes in opinions or values?
11. Equity of participation for stakeholders and inside project team?
12. Major challenges seen in the next stages of the process (if there are follow-ups)?
13. Did the process highlight, lead to or manage any conflicts not already mentioned?
14. What are the major challenges of working in the intercultural participatory project team process? How do you think these could be better managed?
15. If project resources had been different, what would you have changed?
16. Any final comments or suggestions?

Table F.3 *Risk response plan with votes on costs and contributions to values*

No.	Activity	Geographic location	Why is it necessary?	Beginning	Duration	Administrative bodies taking decisions	Executors
1.1	Correction of the Iskar River – Sofia plain	The Iskar River – Sofia plain (Pancharevo – Chepinzi)	There is only partial correction of the river	2009	2 years	Ministry of Regional Development and Public Works (MRDPW) and Sofia City Municipality (SCM)	According to the Public Procurement Law
1.2	Construction of drainage system for the region of Elin Pelin – South	South of Lesnovska River	Drainage of slope waters and prevention of flooding	2008	3–4 years	Ministry of Agriculture and Forestry	According to the Public Procurement Law
1.3	Change the water supply networks; introduction of new water source	Elin Pelin municipality (the Iskar reservoir)	Old and amortised network, bad quality of the potable water; water supply problems	2009	3–4 years	MRDPW	According to the Public Procurement Law
1.4	Storm-water, sewerage for Elin Pelin and Lesnovo (and others)	The settlements along the Lesnovska River	Prevention of flooding	2008	2–3 years	MRDPW and the municipalities	According to the Public Procurement Law
2.1	Application of educational programmes about sustainable development in the schools	The schools in the Elin Pelin municipality	Bringing up responsible behaviour towards environment and development of skills for action in crises situation	2008, 2009	Permanently	Schools, municipalities	Teachers, municipal administration
2.2	Development of educational campaign for the population	All the settlements in the municipality	Enhanced culture for water consumption. Control of waste deposition on non-regulated places. Development of skills for action in crises. Formation of specialised groups of citizens for action in crises	2009	Permanently	The municipal administration	The civic protection and other structures at municipal level; the community centres and other structures of the civic society

Costs and infrastructure	Social and institutional	Uncertainties in the execution	To feel secure and healthy	To share our life	Sustainable economy	Sustainable agriculture	Preserved ecosystems	Effective management	Effective water supply	Treated potable water and treated wastewater
1			1							
1	2		1			2				
2	1								1	2
	1		1							
2		1	1	2						
1			1	1	2					

Table F.3 (*cont.*)

No.	Activity	Geographic location	Why is it necessary?	Beginning	Duration	Administrative bodies taking decisions	Executors
2.3	Education of the municipal administrations for utilisation of the opportunities afforded by different operative programmes and other national framework documents	The Elin Pelin municipality and the other settlements on the territory of the municipality	To develop competency in utilisation of the EU structural funds and to manage the preventive measures and the crisis intervention	2009	Permanently	MRDPW, the National Union of the Municipalities in the Republic of Bulgaria, the municipality	Different centres and organisations providing qualification at different administrative levels
2.4	Create effective and simple information system	The Elin Pelin municipality	The problem in this region is not education, but to inform the citizens			Municipality committee	Companies
3.1	Development of flooding model	The Elin Pelin municipality	Identification of the reasons causing floods and determination of the expected effects	2008	1 + 1 years	Ministry of State Policy for Disasters and Accidents	The Ministry of Environment and Water, the National Institute of Hydro-meteorology (NIHM), the municipality
3.2.	Enhancement of the existing monitoring systems and development of a system for early alert	The Elin Pelin municipality	The reliable database allows adequate decisions on time, in each situation	2009	2 years	МБА, civic protection	The NIHM, the civic protection, the municipality, the river basin directorate
3.3	Plan for management of the infrastructures at normal conditions	The Elin Pelin municipality	Who, what, when – for each infrastructural object; scheduled maintenance; operative service	2009	3 years	MRDPW, the municipality, the district, the Kremikovtzi	The municipality, Kremikovtzi
3.4	Plan for rational management and restoration of forests	The Elin Pelin municipality, Churek, Goljama Rakovitza	Limitation of the harmful effect of the slope waters	2008	3+ years	The forest agency, the municipality	The forest agency

Costs and infrastructure	Social and institutional	Uncertainties in the execution	To feel secure and healthy	To share our life	Sustainable economy	Sustainable agriculture	Preserved ecosystems	Effective management	Effective water supply	Treated potable water and treated wastewater
	1		2	2				1		
	1		1					2		
1			1					2		
1	2		1							
	2		1				2	2		

Table F.3 (*cont.*)

No.	Activity	Geographic location	Why is it necessary?	Beginning	Duration	Administrative bodies taking decisions	Executors
3.5	Development of flood plains for storage of high waters during crises	The Elin Pelin municipality	Clarifies the risk; affects the planning; adequate management, presumes (requires) insurance	2010	1 year	The municipality, the MRDPW, the municipal administration	The municipality
4.1	Equipment of the crises board for reliable work	The Elin Pelin municipality and all the settlements there	Reliable work of the crises board			The civic protection, the municipal council	The municipal administration, the civic protection
4.2	Secured mobility and communication among the crises boards and the residential areas	The Elin Pelin municipality and all the settlements there	The communication should not stop in the case of damage of the normal connections and roads			The civic protection, the municipal council	The civic protection, the municipal council
4.3	Create relevant organisations in advance	The Elin Pelin municipality	To be active and involved for time of crisis			The crisis board	Mayor, crisis board
4.4	Determination of secure places in case of evacuation.	The Elin Pelin municipality and all the settlements there	Saving people and techniques (equipment)		1 year	The crisis board	
4.5	Provision of urgent restoration activities – to bring the crisis under control	The Elin Pelin municipality and all the settlements there	Restriction of the consequences, spreading of the calamity			The crisis board	Pre-determined administrative bodies
5.1	Restoration of the water supply network	The village of Goljama Rakovitza	There is lack of potable water at the moment	2008	1 year	The mayor and the municipal councillor	Construction companies

Costs and infrastructure	Social and institutional	Uncertainties in the execution	To feel secure and healthy	To share our life	Sustainable economy	Sustainable agriculture	Preserved ecosystems	Effective management	Effective water supply	Treated potable water and treated wastewater
	1		1					2		
1	2		2				1			
1	2		2					1		
	1	2						1		
2	1		2					1		
1	2		2					1		
1			1						2	

Table F.3 (*cont.*)

No.	Activity	Geographic location	Why is it necessary?	Beginning	Duration	Administrative bodies taking decisions	Executors
5.2	Restoration of bridges	The village of Goljama Rakovitza	To make possible the communication in the village and among the neighbouring settlements	2008	1 year	The municipal council	Construction companies
5.3	Provision of good water transportation capacity of the rivers (cleaning of the river beds up to the Ognianovo reservoir and the reservoir itself, strengthening of the banks and the bottom)	The village of Goljama Rakovitza	To prevent and limit new flooding and damages	2008	2010 (2 years)	The district governor	Construction companies selected through competition
5.4	Campaign for increasing of the insurance culture (cable TV, internet, seminars, etc.)	The territory of the Elin Pelin municipality	For restoration of the damages and support of the population	2008	2011 (3 years)	The municipal council	Cable TV operators, insurance companies, the municipal administration
5.5	Forestation and other measures preventing erosion	The water catchment area of the village of Goljama Rakovitza	Decrease of the surface runoff and erosion – decrease of risk	2008	3 years	The owner of the forest	Companies selected through competition
5.6	Catching a new water source for the higher zones	The village of Goljama Rakovitza	To provide water source for the higher zones and to prevent the water potential of new flooding	2008	1 year	The municipal council	Companies selected through competition
							Total votes '1' **Total votes '2'** **Total**

Costs and infrastructure	Social and institutional	Uncertainties in the execution	To feel secure and healthy	To share our life	Sustainable economy	Sustainable agriculture	Preserved ecosystems	Effective management	Effective water supply	Treated potable water and treated wastewater
1				1				2		
1		2	2				1			
	1		1	2						
	1	2				1	1			
1				2					1	
12	**8**	**1**	**12**	**2**	**0**	**1**	**3**	**4**	**2**	**0**
3	**6**	**2**	**6**	**4**	**1**	**1**	**1**	**5**	**1**	**1**
15	**14**	**3**	**18**	**6**	**1**	**2**	**4**	**9**	**3**	**1**

This is an interesting final outcome, considering the distribution of drivers of flood and drought that policy makers identified at the beginning of the workshop series, in which the policy makers outlined a large number of more non-technical socio-economic drivers for floods and droughts. There could be a number of reasons for this deviation, although just two potential hypotheses of explanation will be given here. Firstly, as the list of prioritised projects was to be targeting Bulgarian structural funds to finance projects, it is possible that the policy makers took a pragmatic stance of voting for the projects that had the best chances of being accepted – due to the largely 'infrastructural' nature of the funds. Secondly, the policy makers may have voted for the projects that they themselves would be able to run and fund, which were those that were more technically orientated. This may have equally been true for the citizens and municipalities voting for some of the non-technical projects, which could occur under their control or with which they could more easily be involved.

Whether this final voting underlies a strong appropriation of the process and personal willingness to continue to contribute to flood and drought management activities in the region is still difficult to determine. Considering the previous actions put forward by the citizens and organised stakeholders in the Workshop Series 2 (Table F.2), there do seem to be some changes in perspectives or preferences. The top voted project in the final workshop did not appear in the top ten mentioned actions of Workshop 2. In fact, it had been only mentioned twice by the same group – the Elin Pelin citizens. These issues of perspective change and learning will be further investigated in the evaluation results section.

F.4 FURTHER INFORMATION ON THE EVALUATION PROCESSES

F.4.1 Co-engineering team interview guide

The semi-directive interview scheme used to examine Bulgarian project team members' views on the Iskar process is presented in Box F.1.

This scheme was used in its entirety for one of the key Bulgarian facilitators and then reused in a simplified form to drive a cognitive mapping interview carried out jointly with the research and operational leaders of the Iskar process. Emphasis was placed in particular on points (1), (3), (5), (9) and (14).

F.4.2 Participant evaluation responses

This section supplements the information already provided in Sections 8.5.2 and 8.5.3.

Following the quantitative responses in Figure 8.13, qualitative evaluation responses of how the process helped stakeholders to work with and to communicate with others included:

- *'It helped me to realise that I am participant in a very useful and important civil activity.'*
- *'It provoked me to cope with different and unknown situations (during the workshops' games and activities).'*
- *'I had the opportunity to meet new people; this enlarged my possibilities, to ensure better coordination and relationships in my basic work, which is related to local crisis management and organisation.'*
- *'The long series of meetings and workshops gave me the opportunity to discuss the matters in detail and to achieve joint conclusions at the end of the process.'*
- *'The joint work had very positive influence upon all the participants. The discussions were open and straightforward, without confrontations or conflicts.'*

From these last comments, which also have an operational or method-based content, it appears that the activities and new forms of interaction were appreciated by the stakeholders and helped them to work together.

More operational-based statements or issues of process were also given in these qualitative responses, including: *'I learnt more about the role of the different institutions in the field of water management. Actually I understood that the region of Iskar Basin is not ready yet to cope with these problems.'* and *'I understood that certain steps are being undertaken to manage with floods and droughts crises; this makes me feel satisfied!'* It is interesting to see that these two statements, although treating the operational management issues in the region, have both come from participants who gained different understandings of the progress, or lack of progress, that was being made – one understood that there was still much management work to be done, and the other one was satisfied to see that management steps were being undertaken.

Examining the changes induced by the process related to attitudes, practices or behaviours more directly: some stakeholders were not so sure that there had been many changes, e.g. *'I don't think that my participation made me change my attitude towards water very much. It was more likely that during this process I saw things that I've missed or escaped my notice before.'* On the other hand, other comments were on a very different level, suggesting that some stakeholders' attitudes or practices had undergone much more dramatic changes, as in: *'My attitude and behaviour towards water and water resources in general were totally changed, because I realised that drought as a disastrous event is not less serious and dangerous for the country and the population.'* Others were more in the middle of these two extremes; for example, *'To some extent I've already changed my attitude towards water issues and have become a more responsible citizen. I'm trying to pass my experience and knowledge to the other people as well – my family, colleagues.'* Therefore, it is evident that the process led to a range of changes for the stakeholders, from light attitude or knowledge construction to more dramatic mind frame shifts.

Whether all of the learning and changes can be considered as positive is another question. The final questionnaires also elicited some comments that could be interpreted as a little worrying. For example, one stakeholder stated, '*The most important thing I've learnt is that the problems can be solved.*' Although a very positive statement, exactly what is meant by this phrase is a little more of an issue. In fact, as has already been mentioned in Chapter 3, many complex problems cannot really be 'solved' but must be continuously managed. A similar comment of '*I learnt a lot about the best solutions of problems in case of flood and drought crises,*' is also worrying, as this stakeholder appeared to think that there are outright 'best solutions' to problems that can be learnt about, which was not one of the purposes of the process. Rather, the object was to help the participants appreciate that there might be many potentially good or useful options for managing problems. Of course, these interpretations of these comments might not be reasonable if the meaning in Bulgarian did not correspond to this English translation. The following could also be (mis-)interpreted in a similar way: '*The floods cannot be predicted but the risk and the bad impacts can certainly be prevented and the appropriate measures for their reduction can be undertaken in time.*'

Finally, some stakeholders seemed to become more afraid about things; again, this was not the purpose of the process: '*I changed my opinion about the floods; I was rather more scared of droughts.*' One of the issues in the co-engineering team that could have led to some of these potential misunderstandings, especially related to concept definitions, such as 'risk', was that the non-Bulgarian speaking researchers were unable to understand the subtleties of the content of the workshops being proposed and discussed. This was even the case with the translators present, as the meanings were likely to be filtered first in the understanding of the translators, and so many of the small differences in stakeholder understanding of the concepts was not transferred and could not be brought up for discussion.

It appears that at least the first steps towards developing coping capacity were aided by the participatory modelling process, as these participant comments on whether the process helped manage water in the Iskar Basin demonstrate:

- *'The knowledge and better understanding of the problems in the region are the key elements in the overall management.'*
- *'Without any doubt this process is helping the improvement of the whole area. It is a golden chance to discuss and identify the problems, and based on this analysis the most appropriate and suitable actions and activities can be undertaken.'*
- *'The project gives an excellent opportunity to bring around the table all stakeholders in the region – managers, common people and experts.'*
- *'If the state institutions take into consideration the outcomes and achievements of the project; this would lead to better management strategies and visions.'*
- *'The outcomes and achievements of the projects could help the Water Basin Directorates in their management.'*
- *'This process is a part of the overall management of the Iskar Basin. Undoubtedly the good solution of problems how to manage with floods and droughts will have a positive impact upon the general management.'*

In support of this view that the project had been innovative and new collective action described above, stakeholders mentioned in their optional evaluation comments that:

- *'The project is unique, ingenious and very interesting.'*
- *'New acquaintances and excellent opportunities for contacts with many new experts with rich experience and knowledge (from abroad); they showed us a new way of communicating with different people.'*
- *'I am really impressed by the support of the municipal and state institutions given to the project.'*
- *'It was very interesting to see representatives of different institutions, citizens and experts to come, step by step, to common solutions; actually I find this fact one of the most valuable achievements of the joint work. I learnt a lot.'*
- *'Nothing to add except my gratitude to the organisers. They managed to create an excellent team spirit and atmosphere during the meetings. I am glad that I was given an opportunity to enlarge my vision and enrich my knowledge.'*

This last comment brings up the idea of a 'team spirit' that also emerged within the workshops. This is of interest, especially considering the size and diversity of the final stakeholder groups in the workshops.

A stakeholder in the LPSF also outlined how constructive conflict management and resolution in the extended project team also led to the successful project outcomes:

> *'It is absolutely normal to expect some conflicts and different opinions in such a team with various people. Yes, there were some more discussed subjects and items; there were different points of view. All these were solved constructively by common decisions. But at the end, the teamwork brought about excellent results and decisions.'*

F.5 FINAL RISK RESPONSE PLAN

The full translation of the risk response plan created in the final workshop is provided in Table F.3.

References

Abbot, J., Chambers, R., Dunn, C. *et al.* (1998). Participatory GIS: opportunity or oxymoron? *PLA Notes*, **33**. London: IIED.

Abramovitz, J. N. (2001). Averting unnatural disasters. In *State of the World 2001*, ed. L. Starke. New York: Worldwatch Institute, pp 123–142. http://scidiv.bcc.ctc.edu/gj/AvertingUnnaturalDisasters.pdf.

ABS (1986). *The Snowy Mountains Hydro-Electric Scheme, 1301.0 – 1986 Special Article*. Canberra: Australian Bureau of Statistics. www.abs.gov.au/ausstats/ABS@.nsf/94713ad445ff1425ca25682000192af2/fde81ae268c76207ca2569de00274c14!OpenDocument.

ABS (2007). *1303.0 – Discover the ABS, 2006 to 2008: Release Date 26/09/2007*. Canberra: Australian Bureau of Statistics. www.ausstats.abs.gov.au/ausstats/subscriber.nsf/0/47FD15FA176D02B3CA257361001805D9/$File/13030_discover_the_abs_2006-2008.pdf.

Ackermann, F. and Eden, C. (2001). SODA and mapping in practice. In *Rational Analysis for a Problematic World Revisited: Problem Structuring Methods for Complexity, Uncertainty and Conflict*, revised edn, ed. J. Rosenhead and J. Mingers. Chichester, UK: John Wiley and Sons, pp 43–60.

Ackoff, R. L. (1960). Systems, organisations, and interdisciplinary research. *General Systems Yearbook, Society for General Systems Research*, **5**, 18.

Ackoff, R. L. (1974). The systems revolution. *Long Range Planning* **7**(6), 2–20.

Ackoff, R. L. (1979). The future of operational research is past. *Journal of the Operational Research Society*, **30**, 93–104.

Ackoff, R. L. (1999). *Ackoff's Best. His Classic Writings on Management*. New York: John Wiley and Sons.

Adair, J. (1983). *Effective Leadership*. London, UK: Pan Books.

Adair, J. (1986). *Effective Team Building*. Aldershot, UK: Gower.

Adams, D. and Carwardine, M. (1990). *Last Chance to See*. London, UK: William Heinemann.

Adams, W. M., Brockington, D., Dyson, J. and Vira, B. (2003). Managing tragedies: understanding conflict over common pool resources. *Science*, **302**, 1915–1916.

Adger, W. N., Hughes, T. P., Folke, C., Carpenter, S. R. and Rockström, J. (2005). Social-ecological resilience to coastal disasters. *Science*, **309**, 1036.

ADVISOR (2004). *Integrated Deliberative Decision Processes for Water Resources Planning and Evaluation, Guidance Document*. Europe: ADVISOR Project (Integrated Evaluation for Sustainable River Basin Governance).

Agarwal, C., Green, G. M., Grove, J. M., Evans, T. P. and Schweik, C. M. (2002). *A Review and Assessment of Land-Use Change Models: Dynamics of Space, Time, and Human Choice*, CIPEC Collaborative Report Series No 1. Bloomington, IN: Center for the Study of Institutions, Population and Environmental Change, Indiana University/USDA Forest Service. http://hero.geog.psu.edu/archives/AgarwalEtALInPress.pdf.

AGDHC (1975). *Operation and Maintenance Manual for South Tarawa Piped Water Supply System*. Canberra, Australia: AIDAB, Department of Housing and Construction, Australian Government.

Akkermans, H. (2001). Renga: a systems approach to facilitating inter-organisational network development. *System Dynamics Review*, **17**(3), 179–193.

Akkermans, H. A. and Vennix, J. A. M. (1997). Clients' opinions on group model-building: an exploratory study. *System Dynamics Review*, **13**(1), 3–31.

Albrow, M. (1970). *Bureaucracy*. London, UK: Pall Mall Press, Ltd.

Aldrich, H. and Whetten, D. (1981). Organization-sets, action-sets and networks: making the most of simplicity. In *Handbook of Organisational Design*, ed. P. C. Nystrom and W. H. Starbuck. Oxford, UK: Oxford University Press, pp 385–408.

Alexander, E. R. (1993). Interorganizational coordination: theory and practice. *Journal of Planning Literature*, **7**(4), 328–343.

Alexander, G. N., Morlat, G., O'Connell, P. E. *et al.* (1970). Heavy rainfall as a stochastic process. *Revue de l'Institut International de Statistique; Review of the International Statistical Institute*, **38**(1), 62–81 and 97–104.

Allan, C. (2007). Can adaptive management help us embrace the Murray-Darling Basin's wicked problems? In *Adaptive and Integrated Water Management: Coping With Complexity and Uncertainty*, ed. C. Pahl-Wostl, P. Kabat and J. Möltgen. Berlin: Springer-Verlag, pp 61–73.

Allan, J. A. (1997). *Virtual Water: A Long Term Solution for Water Short Middle Eastern Economies?* SOAS Water Issues Group, Occasional Paper No 10. London: School of Oriental and African Studies. http://www.soas.ac.uk/water/publications/papers/file38347.pdf.

Allen, W., Kilvington, M. and Horn, C. (2002). *Using Participatory and Learning-Based Approaches for Environmental Management to Help Achieve Constructive Behaviour Change*, Contract Report: LC0102/057. New Zealand: Landcare Research.

Allison, H. E., and Hobbs, R. J. (2006). *Science and Policy in Natural Resource Management: Understanding System Complexity*. Cambridge, UK: Cambridge University Press.

Althaus, C., Bridgeman, P. and Davis, G. (2007). *The Australian Policy Handbook*, 4th edn. Crows Nest, NSW, Australia: Allan & Unwin.

Andersen, D. F. and Richardson, G. P. (1997). Scripts for group model building. *System Dynamics Review*, **13**(2), 107–129.

Andersen, D. F., Richardson, G. P. and Vennix, J. A. M. (1997). Group model building: adding more science to the craft. *System Dynamics Review*, **13**(2), 187–201.

Angyal, A. (1941). A logic of systems. In *Foundations for a Science of Personality*, pp 243–261 (excerpt from Chapter 8). Cambridge, MA: Harvard University Press.

Ansoff, H. I. (1965). *Corporate Strategy*. New York: McGraw Hill.

Ansoff, H. I. (1984). *Implanting Strategic Management*. Edgewood Cliffs, NJ: Prentice-Hall.

ANZECC (2000). *Australian and New Zealand Guidelines for Fresh and Marine Water Quality*. Australian and New Zealand Environment and Conservation Council and Agriculture and Resource Management Council of Australia and New Zealand. www.mincos.gov.au/publications/australian_and_new_zealand_guidelines_for_fresh_and_marine_water_quality

Argyris, C. and Schön, D. A. (1996). *Organizational Learning II: Theory, Method and Practice*. Reading, MA: Addison-Wesley.

Argyris, C., Putnam, R. and MacLain-Smith, D. (1985). *Action Science: Concepts, Methods and Skills for Research and Intervention*. San Francisco: Jossey-Bass. http://actiondesign.com/resources/research/action-science.

Arnaud, G. (1996). Quelle strategie d'observation pour le chercheur en gestion? *Economie et Sociétés, série Sciences de Gestion*, **10**(22), 235–264.

Arnstein, S. R. (1969). A ladder of citizen participation. *Journal of American Institute of Planners*, **35**, 216–224.

Ashby, W. R. (1956). *An Introduction to Cybernetics*. London, UK: Chapman and Hall.

Asimov, I. (1978). *My Own View*. London, UK: Octopus Books.

Aslin, H. J. and Brown, V. A. (2004). *Towards Whole of Community Engagement: A Practical Toolkit.* Canberra: Murray Darling Basin Commission. http://publications.mdbc.gov.au/download/towards_whole_of_community_engagement_toolkit.pdf.

Astolfi, J. P., Darot, E., Ginsburger-Vogel, Y. and Toussaint, J. (1997). Mots-clés de la didactique des sciences: repères, définitions, bibliographies. De Boeck Université.

Australian Government (2008). *Water for the Future: The Australian Government's Plan for Water.* Canberra: Australian Government Department of the Environment, Water, Heritage and the Arts. www.environment.gov.au/water/publications/action/water-for-the-future.html.

Avenier, M.-J., Nourry, L. and Sweeney, M. (1999). Sciences of the artificial and knowledge production: the crucial role of intervention research in management sciences. *Design Issues*, **15**(2), 55–70.

Avery, J. (2005). *Science and Society*, 2nd edn. Copenhagen: H.C. Ørsted Institute. www.learndev.org/dl/ScienceAndSociety-Avery.pdf.

Axelrod, R. (1976). *The Structure of Decision: The Cognitive Maps of Political Elite.* Princeton, NJ: Princeton University Press.

Bachrach, P. and Baratz, M. S. (1962). Two faces of power. *The American Political Science Review*, **56**(4), 947–952.

Ball, P. (2001). *Life's Matrix: A Biography of Water.* Berkeley, CA: University of California Press.

Bammer, G. (2005). Integration and implementation sciences: building a new specialization. *Ecology and Society*, **10**(2), 6. www.ecologyandsociety.org/vol10/iss2/art6/.

Bammer, G. and Smithson, M. (eds) (2008). *Uncertainty and Risk: Multidisciplinary Perspectives.* London, UK: Earthscan.

Bandura, A. (1977). *Social Learning Theory.* Englewood Cliffs, NJ: Prentice-Hall.

Banxia Software (2008). *DecisionExplorer®.* www.banxia.com/dexplore.

Barbier, R. and Trepos, J.-Y. (2007). Humains et non-humains: un bilan d'étape de la sociologie des collectifs. *Revue d'anthropologie des connaissances*, **1**. www.ird.fr/socanco/article15.html.

Barnaud, C. (2008). *Équité, jeux de pouvoir et légitimité: les dilemmes d'une gestion concertée des ressources renouvelables – Mise à l'épreuve d'une posture d'accompagnement critique dans deux systèmes agraires des hautes terres du Nord de la Thaïlande.* PhD thesis, l'Université Paris X Nanterre, Paris.

Barreteau, O. (2003a). The joint use of role-playing games and models regarding negotiation processes: characterization of associations. *Journal of Artificial Societies and Social Simulation*, **6**(2). http://jasss.soc.surrey.ac.uk/6/2/3.html.

Barreteau, O. (2003b). Our companion modeling approach. *Journal of Artificial Societies and Social Simulation*, **6**(2). http://jasss.soc.surrey.ac.uk/6/2/1.html.

Barreteau, O., Bousquet, F. and Attonaty, J. M. (2001). Role-playing games for opening the black box of multi-agent systems: method and lessons of its application to Senegal River Valley irrigated systems. *Journal of Artificial Societies and Social Simulation*, **4**(2). http://jasss.soc.surrey.ac.uk/4/2/5.html.

Barreteau, O., Bots, P. W. G. and Daniell, K. A. (2010). A framework for clarifying 'participation' in participatory research to prevent its rejection for the wrong reasons. *Ecology and Society*, **15**(2). www.ecologyandsociety.org/vol15/iss2/art1/.

Basden, A. (2002). The critical theory of Herman Dooyeweerd? *Journal of Information Technology*, **17**(4), 257–269.

Bass, B. M. and Avolio, B. J. (eds.) (1994). *Improving Organizational Effectiveness through Transformational Leadership.* Thousand Oaks, CA: Sage Publications, Inc.

Bass, B., Huang, G. and Russo, J. (1997). Incorporating climate change into risk assessment using grey mathematical programming. *Journal of Environmental Management*, **49**(1), 107–123.

Batchelor, C. and Cain, J. (1999). Application of belief networks to water management studies. *Agricultural Water Management*, **40**(1), 51–57.

Bateson, G. (1972). *Steps to an Ecology of Mind.* New York: Ballantine Books.

Bawden, R. J. (1997). Learning to persist: a systemic view of development. In *Systems for Sustainability*, ed. F. A. Stowell, R. L. Ison, R. Armson, *et al.* New York: Plenum Press, pp 1–5.

Bayley, C. and French, S. (2008). Designing a participatory process for stakeholder involvement in a societal decision. *Group Decision and Negotiation*, **17**, 195–210.

Bearak, B. (14 July, 2002). Bangladeshis sipping arsenic as plan for safe water stalls. *New York Times (Late Edition (East Coast))*, p. 1.1.

Beaud, M. (2003). *L'art de la thèse.* Paris, France: La Découverte.

Beck, D. E. and Cowan, C. C. (1996). *Spiral Dynamics: Mastering Values, Leadership and Change – Exploring the New Science of Mimetics.* Oxford, UK: Blackwell Publishing.

Beck, U. (1992). *Risk Society: Towards a New Modernity.* London: Sage Publications. Translated from the German by M. Ritter.

Beder, S. (1993). From sewage farms to septic tanks: trials and tribulations in Sydney. *Journal of the Royal Australian Historical Society*, **79**(1 and 2), 72–95.

Beer, S. (1984). The viable system model: its provenance, development, methodology and pathology. *Journal of the Operational Research Society*, **35**(1), 7–25.

Beierle, T. C. and Cayford, J. (2002). *Democracy in Practice: Public Participation in Environmental Decisions.* Washington, DC: Resources for the Future.

Bellamy, J. A. (2007). Adaptive governance: the challenge for regional natural resource management. In *Federalism and Regionalism in Australia. New Approaches, New Institutions?* ed. A. J. Brown and J. A. Bellamy. Canberra: ANU E-Press, pp 95–118. http://epress.anu.edu.au/fra_citation.html.

Bellamy, J. A., Walker, D. H., McDonald, G. T. and Syme, G. J. (2001). A systems approach to the evaluation of natural resource management initiatives. *Journal of Environmental Management*, **63**, 407–423.

Bellenger, L. (1984). *La négociation.* Paris: Presses Universitaires de France.

Belton, V. and Stewart, T. J. (2002). *Multiple Criteria Decision Analysis: An Integrated Approach.* Norwell, MA: Kluwer Academic Publishers.

Bender, M. J. and Simonovic, S. P. (2000). A fuzzy compromise approach to water resource systems planning under uncertainty. *Fuzzy Sets and Systems*, **115**, 35–44.

Bennett, P. G. and Dando, M. R. (1979). Complex strategic analysis: a hypergame study of the fall of France. *Journal of the Operational Research Society*, **30**(1), 23–32.

Bennett, P. and Howard, N. (1996). Rationality, emotion and preference change: drama-theoretic models of choice. *European Journal of Operational Research*, **92**(3), 603–614.

Bennett, P., Cropper, S. and Huxham, C. (1989). Modelling interactive decisions: the hypergame focus. In *Rational Analysis for a Problematic World: Problem Structuring Methods for Complexity, Uncertainty and Conflict*, ed. J. Rosenhead, Chichester, UK: John Wiley and Sons.

Benson, J. K. (1975). The interorganizational network as a political economy. *Administrative Science Quarterly*, **20**(2), 229–249.

Berkes, F. and Folke, C. (eds) (1998). *Linking Social and Ecological Systems: Management Practices and Social Mechanisms for Building Resilience.* Cambridge, UK: Cambridge University Press.

Berry, M. (1995). Research and the practice of management: a French view. *Organization Science*, **6**(1), 104–116.

Berthelot, J.-M. (ed.) (2001). *Epistémologies des sciences sociales.* Paris, France: Presses Universitaires de France.

Bertrand, L. and Martel, J.-M. (2002). Une démarche participative multicritère en gestion integrée des forêts. *INFOR (Information systems and operational research)*, **40**(3), 223–239.

Billington, K. (2005). *River Murray Catchment Risk Assessment Project for Water Quality – Concepts and Method.* Murray Bridge, Australia: SA Environmental Protection Authority.

Biswas, A. K. (2004). From Mar del Plata to Kyoto: an analysis of global water policy dialogue. *Global Environmental Change Part A*, **14**(Supplement 1), 81–88.

Blackmore, S. (1999). *The Meme Machine.* Oxford, UK: Oxford University Press.

Blackstock, K. L., Kelly, G. J. and Horsey, B. L. (2007). Developing and applying a framework to evaluate participatory research for sustainability. *Ecological Economics*, **60**(4), 726–742.

BMT WBM (2007). *Lower Hawkesbury Estuary Synthesis Report.* NSW, Australia: Hornsby Shire Council.

BMT WBM (2008). *Lower Hawkesbury Draft Estuary Management Plan*, R.N1252.003.01.EMP: BMT WBM. NSW, Australia: Hornsby Shire Council.

Boin, A., t'Hart, P., Stern, E. and Sundelius, B. (2005). *The Politics of Crisis Management: Public Leadership Under Pressure.* Cambridge, UK: Cambridge University Press.

Bonner, G. M. (1993). Cognitive mapping as a technique for supporting international negotiation. *Theory and Decision*, **34**(3), 255–273.

Borgman, L. E., Amorocho, J., Morlat, G. *et al.* (1970). Some statistical problems in hydrology. *Revue de l'Institut International de Statistique/Review of the International Statistical Institute*, **38**(1), 84–104.

Borrini-Feyerabend, G., Farvar, M. T., Nguinguiri, J. C. and Ndangang, V. A. (2000). *Co-management of Natural Resources: Organising, Negotiating and Learning-by-Doing*. Heidelberg, Germany: GTZ and IUCN, Kasparek Verlag.

Borton, T. E. and Warner, K. P. (1971). Involving citizens in water resources planning: the communication-participation experiment in the Susquehanna River Basin. *Environment and Behavior*, **3**, 284–306.

Bots, P. W. G. (2007). Design in socio-technical system development: three angles in a common framework. *Journal of Design Research*, **5**(3), 382–396.

Bots, P. W. G., van Twist, M. J. W. and van Duin, R. (1999). Designing a power tool for policy analysts: dynamic actor network analysis. *Proceedings of the 32nd Hawaiian International Conference on System Sciences, 5–8 January 1999*. Maui, HI.

Bots, P., van Bueren, E., ten Heuvelhof, E. and Mayer, I. (2005). *Communicative tools in sustainable urban planning and building. Sustainable Urban Areas 5*. Delft, The Netherlands: Delft Centre for Sustainable Urban Areas, Delft University of Technology, DUP Science.

Bots, P., Bijlsma, R., von Korff, Y., van der Fluit, N. and Wolters, H. (2011). Supporting the constructive use of existing hydrological models in participatory settings: a set of 'rules of the game'. *Ecology and Society*, **16**(2), 16. www.ecologyandsociety.org/vol16/iss2/art16/.

Bouleau, G. (2003a). *Acteurs et circuits financiers de l'eau en France – Stakeholders and Their Financial Responsibilities in French Water Management*, 2nd edn. Paris, France: ENGREF – Ecole Nationale du Génie Rural des Eaux et des Forêts.

Bouleau, G. (2003b). Comment batir une prospective commune pour la gestion d'un fleuve transfrontalier? L'exemple de L'Escaut. *VertigO – La revue en sciences de l'environnement sur le WEB*, **4**(3). http://vertigo.revues.org/3747.

Bouleau, G., Snellen, B., Raymond, R. and Daniell, K. A. (2005). *Intermediate Report: Planning Theories and Water Stress.* D3.7.1. Europe: AquaStress IP, FP6.

Bousquet, F., Barreteau, O., Mullon, C. and Weber, J. (1996). *Modélisation d'accompagnement: systèmes multi-agents et gestion des resources renouvelables, 'Quel environnement au XXIème siècle? Environnement, maîtrise du long terme et démocratie'*. Paris, France: Abbaye de Fontevraud. *http://cormas.cirad.fr/pdf/accompagnement.pdf*.

Bousquet, F., Bakam, I., Proton, H. and Le Page, C. (1998). CORMAS: common-pool resources and multi-agent systems. *Proceedings of the 11th International Conference on Industrial and Engineering Applications of Artificial Intelligence and Expert Systems, Springer-Verlag*, **1416**, 826–838.

Bousquet, F., Barreteau, O., Le Page, C., Mullon, C. and Weber, J. (1999). An environmental modelling approach. The use of multi-agent simulation. In *Advances in Environmental and Ecological Modelling*, ed. F. Blasco and A. Weill. Oxford: Elsevier.

Bousquet, F., Barreteau, O., D'Aquino, P. *et al.* (2002). Multiagent systems and role games: collective learning processes for ecosystem management. In *Complexity and Ecosystem Management: The Theory and Practice of Multi-agent Systems*, ed. M. A. Janssen. Cheltenham, UK: Edward Elgar Publishers. pp 248–285.

Bouwen, R. and Taillieu, T. (2004). Multi-party collaboration as social learning for interdependence: developing relational knowing for sustainable natural resource management. *Journal of Community & Applied Social Psychology*, **14**, 137–153.

Bouwer, H. (2000). Integrated water management: emerging issues and challenges. *Agricultural Water Management*, **45**(3), 217–228.

Bouyssou, D., Marchant, T., Pirlot, M. *et al.* (2000). *Evaluation and Decision Models: A Critical Perspective*. Norwell, MA: Kluwer Academic Publishers.

Bouzon, A. (ed.) (2004). *La place de la communication dans la conception de systèmes à risques*. Paris, France: L'Harmattan.

Brans, J.-P. and Vincke, P. (1985). A preference ranking organisation method: the PROMETHEE method for multiple criteria decision-making. *Management Science*, **31**(6), 647–656.

Brans, J. P., Vincke, P. and Mareshal, B. (1986). How to select and to rank projects: the Promethee method. *European Journal of Operational Research*, **24**, 228–238.

Braybrooke, D. and Lindblom, C. E. (1963). *A Strategy of Decision: Policy Evaluation as a Social Process*. New York: Free Press.

Brews, P. J. and Hunt, M. R. (1999). Learning to plan and planning to learn: resolving the planning school/learning school debate. *Strategic Management Journal*, **20**, 889–913.

Brim, O. G., Glass, D. C., Lavin, D. E. and Goodman, N. (1962). *Personality and Decision Processes: Studies in the Social Psychology of Thinking*. Stanford, CA: Stanford University Press.

Brinkerhoff, D. W. and Goldsmith, A. A. (2006). Organising for mutual advantage: municipal associations in Bulgaria. *Public Administration and Development*, **26**(5), 373–382.

Brinkerhoff, J. M. (2002). Assessing and improving partnership relationships and outcomes: a proposed framework. *Evaluation and Program Planning*, **25**(3), 215–231.

Brocklesby, J. and Mingers, J. (1999). *The Cognitive Limits on Organisational Reframing: A Systems Perspective based on the Theory of Autopoiesis*. Paper presented at the International System Dynamics Conference, Wellington, New Zealand. www.systemdynamics.org/conferences/1999/PAPERS/PLEN4.PDF.

Brown, I. W. (2010). *Taking female Mud Crabs (Scylla Serrata): Assessment of Risks and Benefits*, Final report, FRDC 2009/031. Deception Bay, Queensland: Department of Employment, Economic Development and Innovation – Southern Fisheries Centre.

Brown, R. R., Ryan, R. and McManus, R. (2001). An Australian case study: why a transdisciplinary framework is essential for integrated urban stormwater planning, *UNESCO International Hydrological Program Symposium: Frontiers of Urban Water Management: Deadlock or Hope?* Marseilles, France, pp 251–259.

Brown, T. (2008). Design thinking. *Harvard Business Review*, June.

Bruen, M. (2007). Systems analysis – a new paradigm and decision support tools for the water framework directive. *Hydrology and Earth Systems Sciences Discussions*, **4**, 1491–1518. www.hydrol-earth-syst-sci-discuss.net/4/1491/2007.

Brugha, R. and Varvasovszky, Z. (2000). Stakeholder analysis: a review. *Health Policy and Planning*, **15**(3), 239–246.

Brugnach, M., Tagg, A., Keil, F. and de Lange, W. J. (2007). Uncertainty matters: computer models at the science-policy interface. *Water Resources Management*, **21**, 1075–1090.

Bucciarelli, L. L. (1994). *Designing Engineers*. Cambridge, MA: MIT Press.

Buchanan, M. (2002). *Small World: Uncovering Nature's Hidden Networks*. London: Weidenfeld & Nicholson.

Bunn, D. W. (1984). *Applied Decision Analysis*. New York: McGraw Hill.

Burby, R., Beatley, T., Berke, P. R. *et al.* (1999). Unleashing the power of planning to create disaster-resistant communities. *Journal of the American Planning Association*, **65**(3), 247–258.

Burchfield, J. (1998). Abandoned by the roadside: the long road ahead for collaborative stewardship. *Chronicle of Community*, **3**(1), 31–36.

Burke, E. (2005). *The Deep History of the Middle Eastern Environment, 1500 BCE–1500 CE*. UC World History Workshop. Essays and Positions from the World History Workshop. Paper 3. Santa Cruz.

Buzan, T. and Buzan, B. (1993). *The Mind Map Book: How to Use Radiant Thinking to Maximize Your Brain's Untapped Potential*. London, UK: BBC Books.

Cahill, C., Sultana, F. and Pain, R. (2007). Participatory ethics: politics, practices, institutions. *ACME: An International E-Journal for Critical Geographies*, **6**(3), 304–318.

Callon, M., Lascoumes, P. and Barthe, Y. (2001). *Agir dans un monde incertain: Essai sur la démocratie technique*. Paris, France: Seuil.

Capurso, E. and Tsoukiàs, A. (2003). Decision aiding and psychotherapy. *Bulletin of the EURO Working Group on MCDA*, **3**(8). www.inescc.pt/~ewgmcda/OpCapurso.html.

Carnes, S. A., Schweitzer, M., Peelle, E. B., Wolfe, A. K. and Munro, J. F. (1996). *Performance Measures for Evaluating Public Participation Activities in DOE's Office of Environmental Management*. ORNL-9605, Report prepared for the US Department of Energy. Oak Ridge, TN: Lockheed Martin Energy Research Group Corp., Oak Ridge National Laboratory.

Carpenter, C., Stubbs, J. and Overmars, M. (eds) (2002). *Proceedings of the Pacific Regional Consultation on Water in Small Island Countries in Sigatoka, Fiji Islands, 29 July–3 August 2002*. Suva, Fiji: Asian Development Bank and South Pacific Applied Geoscience Council.

Carpenter, D. O., Suk, W. A., Blaha, K. and Cikrt, M. (1996). Hazardous wastes in Eastern and Central Europe. *Environmental Health Perspectives*, **104**(3), 244–248.

CEAA (2008). *Public Participation Guide*. Canadian Environmental Assessment Agency, Canadian Government. www.acee-ceaa.gc.ca/default.asp?lang=En&n=46425CAF-1&offset=&toc=hide.

CEC (2007). *Towards Sustainable Water Management in the European Union: First stage in the Implementation of the Water Framework Directive 2000/60/EC*. SEC(2007) 362, COM(2007) 128 final. Brussels: Commission of the European Communities.

Cech, T. V. (2005). *Principles of Water Resources: History, Development, Management and Policy*, 2nd edn. Hoboken, NJ: John Wiley and Sons Inc.

Chadwick, E. (1843). *Report on the Sanitary Condition of the Labouring Population of Great Britain*. London: W. Clowes and Sons. www.archive.org/stream/reportonsanitary00chaduoft#page/n3/mode/2up.

Chambers, R. (2002). *Participatory Workshops: a Sourcebook of 21 Sets of Ideas and Activities*. London: Earthscan.

Chang, N.-B., Eh, S. C. Y. and Wu, G. C. (1999). Stability analysis of grey compromise programming and its application to watershed land-use planning. *International Journal of Systems Science*, **30**(6), 571–589. www.informaworld.com/10.1080/002077299292092.

Charnes, A., Cooper, W. W. and Ferguson, R. O. (1955). Optimal estimation of executive compensation by linear programming. *Management Science*, **1**(2), 138–151.

Chartres, C. J. (2006). *A Strategic Science Framework for the National Water Commission*. Canberra: National Water Commission.

Chau, K. (2004). River stage forecasting with particle swarm optimization. In *Innovations in Applied Artificial Intelligence. Proceedings of the 17th International Conference on Industrial and Engineering Applications of Artificial Intelligence and Expert Systems, IEA/AIE 2004, Ottawa, Canada, May 17–20, 2004*. Berlin: Springer-Verlag, pp 1166–1173.

Checkland, P. B. (1978). The origins and nature of hard systems thinking. *Journal of Applied Systems Analysis*, **5**, 99–110.

Checkland, P. B. (1981). *Systems Thinking, Systems Practice*. Chichester, UK: John Wiley.

Checkland, P. B. (1985). From optimizing to learning: a development of systems thinking for the 1990s. *Journal of the Operational Research Society*, **36**, 757–767.

Checkland, P. B. and Holwell, S. E. (1998). Action research: its nature and validity. *Systemic Practice and Action Research*, **11**(1), 9–21.

China Daily (10 June 2006). 'Great Wall' of water transport takes on new look. www.highbeam.com/doc/1P2-8851646.html.

Chown, E. (1999). Making predictions in an uncertain world: Environmental structure and cognitive maps. *Adaptive Behavior*, **7**(1), 17–33.

Churchman, C. W. (1968). *The Systems Approach*. New York: Dell Publishing Co.

Churchman, C. W. (1979). *The Systems Approach and Its Enemies*. New York: Basic Books.

CISEAU (2006). *Document introductif: salinisation induite par l'irrigation*. Paper presented at the Extension de la salinisation et stratégies de prévention et réhabilitation, Conférence électronique sur la salinisation. www.agrireseau.qc.ca/agroenvironnement/documents/Salinisation_irrigation.pdf.

Clarke, I. and Mackaness, W. (2001). Management 'intuition': an interpretative account of structure and content of decision schemas using cognitive maps. *Journal of Management Studies*, **38**(2), 147–172.

Clarke, R. (1993). *Water: The International Crisis*. Cambridge, MA: The MIT Press.

Coad, P., Wade, W., Letcher, R. A., Phillips, B. C. and Jakeman, A. J. (2006). *An Integrated Catchment Management System for the Shire of Hornsby, NSW*. Paper presented at the 30th Hydrology and Water Resources Symposium, 4–7 December 2006, Launceston, Australia.

Coad, P., Haines, P., Daniell, K. A., Guise, K. and Rollason, V. (2007). *Integration of Environmental Risk Assessment Within Estuary Management Planning for the Lower Hawkesbury-Nepean River, New South Wales*. Paper presented at the 16th NSW Coastal Conference, 7–9 November 2007, Yamba, NSW, Australia. www.coastalconference.com/2007/papers2007/Peter%20Coad.doc.

CoAG (2000). *Our Vital Resources: A National Action Plan for Salinity and Water Quality*: Council of Australian Governments. www.napswq.gov.au/publications/policies/pubs/vital-resources.pdf.

CoAG (2004). *Intergovernmental Agreement on a National Water Initiative: Between the Commonwealth of Australia and the Governments of New South Wales, Victoria, Queensland, South Australia, the Australian Capital Territory and the Northern Territory*. Coalition of Australian Governments. www.coag.gov.au/coag_meeting_outcomes/2004-06-25/docs/iga_national_water_initiative.pdf.

Coffman, L., Cheng, M.-S., Weinstein, N. and Clar, M. (1998). *Low-Impact Development Hydrologic Analysis and Design*. Paper presented at the Proceedings of the 25th Annual Conference on Water Resources Planning and Management, 7–10 June 1998, Water Resources and the Urban Environment, Chicago, IL.

Cohen, M. D., March, J. G. and Olsen, J. P. (1972). A garbage can model of organizational choice. *Administrative Science Quarterly*, **17**(1), 1–25.

Colebatch, H. K. (2006). *Beyond the Policy Cycle: the Policy Process in Australia*. Sydney, Australia: Allen & Unwin.

Colianese, C. (2002). *Is Satisfaction Success? Evaluating Public Participation in Regulatory Policymaking*. Regulatory Policy Program Working Paper RPP-2002–09. Cambridge, MA: Center for Business and Government, John F. Kennedy School of Government, Harvard University.

COMLAW (1997). *Natural Heritage Trust of Australia Act 1997: Act Compilation (current) – C2005C00211*. www.comlaw.gov.au/ComLaw/Legislation/ActCompilation1.nsf/0/D6613E7D95644C08CA256FC1001497AE?OpenDocument.

Commonwealth Of Australia (1900). *Constitution Act*. Section 100.

Connell, D. (2007). *Water Politics in the Murray Darling Basin*. Leichhart, NSW: The Federation Press.

Connick, S. and Innes, J. (2001). *Outcomes of Collaborative Water Policy Making: Applying Complexity Thinking to Evaluation*. Working Paper 2001–08, prepared for the Conference on Evaluating Environmental and Public Policy Dispute Resolution Programs and Policies, March 8–9, 2001, Washington, DC. Berkley, CA: Institute of Urban and Regional Development, University of California.

Corbett, J. and Rambaldi, G. (2009). 'Representing our reality': geographic information technologies, local knowledge and change. In *Qualitative GIS: Mixed Methods in Practice and Theory*, ed. M. Cope and S. Elwood. London: Sage Publications.

Corner, J., Buchanan, J. and Henig, M. (2001). Dynamic decision problem structuring. *Journal of Multi-Criteria Decision Analysis*, **10**, 129–141.

Cornwall, A. (1996). Towards participatory practice: participatory rural appraisal (PRA) and the participatory process. In *Participatory Research in Health: Issues and Experiences*, ed. K. de Koning and M. Marion. London: Zen Books Ltd., pp 94–107.

Cortner, H. J. and Moote, M. A. (1999). *The Politics of Ecosystem Management*. Washington, DC: Island Press.

Costanza, R. and Ruth, M. (1998). Using dynamic modeling to scope environmental problems and build consensus. *Environmental Management*, **22**(2), 183–195.

Couix, N. (1997). Evaluation «chemin faisant» et mise en acte d'une stratégie tâtonnante. In *La stratégie chemin faisant*, ed. M.-J. Avenier. Paris: Economica, pp 165–187.

Council of Europe (2008). *Council of Europe*. www.coe.int/lportal/web/coe-portal.

Covey, S. M. R. and Merrill, R. R. (2006). *The Speed of Trust*. London: Pocket Books.

Craig, W. J., Trevor, M. and Weiner, D. (2002). *Community Participation and Geographic Information Systems*. London, UK: Taylor and Francis.

Creighton, J. L. (2005). *The Public Participation Handbook: Making Better Decisions Through Citizen Involvement*. San Francisco, CA: Jossey-Bass.

Croke, B. F. W., Ticehurst, J. L., Letcher, R. A. *et al.* (2007). Integrated assessment of water resources: Australian experiences. *Water Resources Management*, **21**, 351–373.

CSIRO (2007). *Climate Change in the Hawkesbury Nepean Catchment*. Prepared for the NSW Government by the Commonwealth Scientific and Industrial Research Organisation (CSIRO). NSW, Australia: CSIRO.

CSIRO ASSERT (1992). *The Water Future of the ACT: A Community Discussion Document on the Major Issues*. Consultancy Report No. 92/45. Prepared for ACTEW by CSIRO ASSERT Social Science Unit, Division of Water Resources. Canberra, Australia: CSIRO.

CTGP (2008). *China Three Gorges Project*. www.ctg.com.cn/en.

Cu, E. (2000). Ecoterrorism as negotiating tactic. *The Christian Science Monitor*. www.csmonitor.com/2000/0721/p8s1.html.

Cumming, D. H. M. (2000). Drivers of resource management practices – fire in the belly? Comments on 'cross-cultural conflicts in fire management in northern Australia: not so black and white' by Alan Andersen. *Ecology and Society*, **4**(1), 4. www.consecol.org/vol4/iss1/art4/.

Cumming, G. S., Cumming, D. H. M. and Redman, C. L. (2006). Scale mismatches in social-ecological systems: causes, consequences, and solutions. *Ecology and Society*, **11**(1), 14. www.ecologyandsociety.org/vol11/iss1/art14/.

Cutts, B. B., White, D. D. and Kinzig, A. P. (2011). Participatory geographic information systems for the co-production of science and policy in an emerging boundary organization. *Environmental Science & Policy*, **14**(8), 977–985.

Cyert, R. M. and March, J. G. (1963). *A Behavioral Theory of the Firm*. Englewood Cliffs, NJ: Prentice-Hall.

D'Aquino, P. (2007). Empowerment and participation: how could the wide range of social effects of participatory approaches be better elicited and compared? *ICFAI Journal of Knowledge Management*, **5**(6), 76–87.

D'Aquino, P. (2008). *Fiches de travail pour construire une démarche participative*. www.lisode.com/images/MD/fiches_daquino.pdf.

D'Aquino, P., Barreteau, O., Etienne, M. *et al.* (2002a). *The Role Playing Games in an ABM Participatory Modeling Process: Outcomes from Five Different Experiments Carried out in the Last Five Years*. Paper presented at the Integrated Assessment and Decision Support, Proceedings of the First Biennial Meeting of the International Environmental Modelling & Software Society – iEMSs.

D'Aquino, P., Le Page, C. and Bousquet, F. (2002b). The SelfCormas Experiment: Aiding Policy and Land-Use Management by Linking Role-Playing Games, GIS, and ABM in the Senegal River Valley. In *Agent-Based Models of Land Use and Land Cover Change: Report and Review of an International Workshop, October 4–7, 2001*, ed. D. C. Parker, T. Berger and S. M. Manson. Irvine, CA, USA, pp 70–72.

D'Aquino, P., Le Page, C., Bousquet, F. and Bah, A. (2003). Using self-designed role-playing games and a multi-agent system to empower a local decision-making process for land use management: the SelfCormas experiment in Senegal. *Journal of Artificial Societies and Social Simulation*, **6**(3). http://jasss.soc.surrey.ac.uk/6/3/5.html.

Dahl, R. A. (1961). *Who Governs? Democracy and Power in an American City*. New Haven, CT: Yale University Press.

Daily, G. C. (2000). Management objectives for the protection of ecosystem services. *Environmental Science & Policy*, **3**(6), 333–339.

Damart, S. (2003). *Une étude de la contribution des outils d'aide à la décision aux démarches de concertation: le cas des décisions publiques de transport*. Université Paris IX Dauphine, Paris, France.

Dandy, G. C. and Warner, R. F. (1989). *Planning and Design of Engineering Systems*. Winchester, MA: Unwin Hyman.

Dandy, G. C., Warner, R. F., Daniell, T. M. and Walker, D. (2007). *Planning and Design of Engineering Systems: Revised Edition*. Abingdon, UK: Taylor & Francis Ltd.

Daniell, T. M. (1991). Neural networks. Applications in hydrology and water resources engineering. *National Conference Publication – Institute of Engineers Australia*, **3**, Part 22(91), 797–802.

Daniell, K. A. (2007a). Summary report: community workshop 1 for the Lower Hawkesbury Estuary Management Plan. In *BMT WBM, Lower Hawkesbury Estuary Synthesis Report*. NSW, Australia: Hornsby Shire Council, Appendix A.

Daniell, K. A. (2007b). *Summary Report: Stakeholder Workshops 2 & 3 for the Lower Hawkesbury Estuary Management Plan*, Report prepared for the Hornsby Shire Council and BMT WBM. Canberra, Australia: Australian National University.

Daniell, K. A. (2008). *Co-engineering Participatory Modelling Processes for Water Planning and Management, Co-ingénierie des processus de modélisation participative pour la planification et la gestion de l'eau* (2 volumes). PhD thesis, Australian National University, Canberra, Australia and AgroParisTech, Montpellier, France, (No 2008AGPT0071), 735 pp.

Daniell, K. A. (2011). Enhancing collaborative management in the Basin. In *Basin Futures: Water Reform in the Murray-Darling Basin*, ed. D. Connell and R. Q. Grafton. Canberra: ANU E-Press, pp 413–438. http://epress.anu.edu.au/basin_futures/pdf/ch26.pdf.

Daniell, T. M. and Daniell, K. A. (2006). Human impacts, complexity, variability and non-homogeneity: four dilemmas for the water resources modeller. In *Climate Variability and Change – Hydrological Impacts (Proceedings of the Fifth FRIEND. World Conference held at Havana, Cuba, November 2006)*. IAHS Publication No **308**: 10–15.

Daniell, K. A. and Ferrand, N. (2006). *Participatory Modelling for Water Resources Management and Planning*. Report D3.8.2. Europe: AquaStress IP, EU FP6.

Daniell, K. A., Kingsborough, A. B., Malovka, D. J. and Sommerville, H. C. (2004). *Assessment of the Sustainability of Housing Developments*. Honours Research Report, School of Civil and Environmental Engineering, University of Adelaide, Australia.

Daniell, K. A., White, I., Ferrand, N. *et al.* (2006a). *Towards an Art and Science of Decision Aiding for Water Management and Planning: A Participatory Modelling Process*. Paper presented at the Proceedings of the 30th Hydrology & Water Resources Symposium, 4–7 December 2006, Launceston, Australia.

Daniell, K A., Ferrand, N. and Tsoukiàs, A. (2006b). Investigating participatory modelling processes for group decision aiding in water planning and management. In *Proceedings of the Group Decision and Negotiation International Conference, 25-28 June 2006, Karlsruhe, Germany*.

Daniell, K. A., Coad, P., Ribarova, I. S. *et al.* (2008a). Participatory risk management approaches for water governance: insights from Australia and Bulgaria. In *Proceedings of the XIIIth World Water Congress: 'Global Changes and Water Resources', 1–4 September 2008, Montpellier, France*.

Daniell, K. A., Coad, P., Ferrand, N. *et al.* (2008b). Participatory values-based risk management for the water sector. In *Proceedings of the Water Down Under 2008 Conference, 14–17 April 2008, Adelaide, Australia*, pp 969–981.

Daniell, K. A., White, I. and Rollin, D. (2009). Ethics and participatory water planning. *Proceedings of the 32nd Hydrology and Water Resources Symposium: 'H2O 09', 30 November–3 December 2009*. Newcastle, Australia, pp 1476–1487.

Daniell, K. A., Mazri, C. and Tsoukiàs, A. (2010a). Real world decision-aiding: a case of participatory water management. In *e-Democracy: A Group Decision and Negotiation Perspective*, ed. D. Rios Insua and S. French. Dordrecht: Springer, pp 125–150.

Daniell, K. A., Ferrand, N., White, I. *et al.* (2010b). Co-engineering participatory water management processes: insights from Australia and Bulgaria. *Ecology and Society*, **15**(4), Art. 11. www.ecologyandsociety.org/vol15/iss4/art11/.

Daniell, K. A., Ribarova, I. S. and Ferrand, N. (2011). Collaborative flood and drought risk management in the Upper Iskar Basin, Bulgaria. In *Water Resources Planning and Management*, ed. R. Q. Grafton and K. Hussey. Cambridge University Press, pp 395–420.

Daniels, K., Markóczy, L. and de Chernatony, L. (1994). Techniques to compare cognitive maps. *Advances in Managerial Cognition and Organizational Information Processing*, **5**, 141–164.

Dart, J., Petheram, R. J. and Straw, W. (1998). *Review of Evaluation in Agricultural Extension*, RIRDC Publication no. 98/136. Barton, ACT, Australia: Rural Industries Research and Development Corporation. http://rufaelfassil.googlepages.com/goodmeformat.pdf.

David, A. (2000). La recherche-intervention, cadre général pour la recherche en sciences de gestion? In *Les nouvelles fondations des sciences en gestion*, ed. A. David, A. Hatchuel and R. Laufer. Paris: Vuibert, collection FNEGE, pp 193–213.

David, A. (2002). Décision, conception et recherche en sciences de gestion. *Revue française de gestion*, **3–4**(139), 173–185.

David, A. and Hatchuel, A. (2007). *Des connaissances actionnables aux théories universelles en sciences de gestion*. Paper presented at the AIMS XVIème Conférence Internationale de Management Stratégique, 6–9 June 2007, Montréal, Canada.

Davidoff, P. (1965). Advocacy and pluralism in planning. *Journal of American Institute of Planners*, **31**(4), 331–338.

Dawkins, R. (1976). *The Selfish Gene*. Oxford, UK: Oxford University.

Dayton, D. (2002). Evaluating environmental impact statements as communicative action. *Journal of Business and Technical Communication*, **16**(4), 355–405.

De Carvalho, R. C. and Magrini, A. (2006). Conflicts over water resource management in Brazil: a case study of inter-basin transfers. *Water Resources Management*, **20**(2), 193–213.

De Geus, A. P. (March–April, 1988). Planning as learning. *Harvard Business Review*, 70–74.

De Jong, A. (1996). Inter-organizational collaboration in the policy preparation process. In *Creating Collaborative Advantage*, ed. C. Huxham. London, UK: Sage Publications, pp 165–175.

De Marchi, B., Funtowicz, S., Lo Cascio, S. and Munda, G. (2000). Combining participative and institutional approaches with multicriteria evaluation: an empirical study for water issues in Troina, Sicily. *Ecological Economics*, **34**(2), 267–282.

De Raadt, J. D. R. (1997). A sketch for humane operational research in a technological society. *Systemic Practice and Action Research*, **10**(4), 421–441.

De Terssac, G. and Friedberg, E. (eds) (2002). *Coopération et Conception*, 2nd edn. Toulouse, France: Octares Editions.

Delli Priscoli, J. (2003). *Participation, Consensus Building, and Conflict Management Training Course (Tools for Achieving PCCP)*. IHP-IV Technical documents in Hydrology PC->CP Series, **No 22**. Paris, France: UNESCO-IHP prepared for the WWAP.

Delli Priscoli, J. and Wolf, A. T. (2009). *Managing and Transforming Water Conflicts*. Cambridge, UK: Cambridge University Press.

Deming, W. E. (1986). *Out of the Crisis*. Cambridge, MA: MIT Centre for Advanced Engineering Study.

Dennis, C. J. (1918). *Backblock Ballads and Later Verses*. Sydney, Australia: Angus and Robertson.

Derry, D. K. and Williams, T. I. (1960). *A Short History of Technology from Earliest Times to AD 1900*. Oxford: Oxford University Press.

DeSanctis, G. and Gallupe, R. B. (1987). A foundation for the study of group decision support systems. *Management Science*, **33**(5), 589–609.

Descartes, R. (1637). *Discours de la méthode*. www.hs-augsburg.de/~harsch/gallica/Chronologie/17siecle/Descartes/des_di00.html.

Dewey, J. (1910). *How We Think*. Boston: D.C. Heath & Co.

Dewey, J. (1927). *The Public and Its Problems*. New York: Holt.

DEWHA (2007). *National Water Quality Management Strategy*. www.environment.gov.au/water/quality/nwqms/index.html.

Dey, C., Berger, C., Foran, B. *et al.* (2007). Household environmental pressure from consumption: an Australian environmental atlas. In *Water Wind Art and Debate. How Environmental Concerns Impact on Disciplinary Research*, ed. G. Birch. Sydney University Press.

Diamond, J. (1999). *Guns, Germs, and Steel: The Fates of Human Societies*. New York: WW Norton & Company.

Dias, L. C. and Tsoukiàs, A. (2003). *On the Constructive and Other Approaches in Decision Aiding*. Paper presented at the Proceedings of the 56th meeting of the EURO MCDA working group, CCDRC, Coimbra, October 2004.

Dietz, T., Ostrom, E. and Stern, P. C. (2003). The struggle to govern the commons. *Science*, **302**, 1907–1912.

DIFD (2004). *Disaster Risk Reduction: A Development Concern*. London: Department for International Development, UK Government.

Digenti, D. (1999). Collaborative learning: a core capability for organizations in the new economy. *Reflections*, **1**(2), 45–57. http://mitpress.mit.edu/journals/SOLJ/Digenti.pdf.

Dikov, O., Cheshmedjiev, S., Tasseva, I. and Boneva, N. (2003). *Integrated Water Management in Bulgaria – Current State and National Priorities (Summary)*. Sofia, Bulgaria: Time Ecoprojects Foundation.

Dimitrova, A. (2002). Enlargement, institution-building and the EU's administrative capacity requirement. *West European Politics*, **25**(4), 171–190.

Dinesh Kumar, M. and Singh, O. P. (2005). Virtual water in food and water policy making: is there a need for rethinking? *Water Resources Management*, **19**, 759–789.

Dionnet, M., Daniell, K. A., Imache, A. *et al.* (2011). *Participatory Process Design and Testing for Natural Resources Management: Development and Evaluation of a Simulation Community of Practice*. Working Paper – Ref: 01/2010. Montpellier, France: Lisode.

Dixon, J. and Dogan, R. (2003). A philosophical analysis of management: improving praxis. *Journal of Management Development*, **22**(5/6), 458–482.

DLWC (2003). *Hawkesbury Lower Nepean Catchment Blueprint*. Australia: NSW Department of Land and Water Conservation.

Dodgson, J., Spackman, M., Pearman, A. and Phillips, L. (2000). *DTLR Multi-Criteria Analysis Manual*. National Economic Research Associates (NERA). http://iatools.jrc.ec.europa.eu/public/IQTool/MCA/DTLR_MCA_manual.pdf.

Donnelly, R. and Boyle, C. (2006). The catch-22 of engineering sustainable development. *Journal of Environmental Engineering, ASCE*, **132**(2), 149–155.

Dooyeweerd, H. (1958). *A New Critique of Theoretical Thought*. Vols I–IV. Philadelphia, PA: Presbyterian and Reformed Publisher Company.

Dorigo, M. and Stützle, T. (2004). *Ant Colony Optimization*. Cambridge, MA: MIT Press.

Dougherty, D. E. and Marryott, R. A. (1991). Optimal groundwater management: I. simulated annealing. *Water Resources Research*, **27**(10), 2493–2508.

Dovers, S. R. (1995). Information, sustainability and policy. *Australian Journal of Environmental Management*, **2**(3), 142–156.

Dovers, S. R. (2002). *Institutional and Policy Reform for Sustainability: Principles and Prospects for Australia*. Paper presented at 'Getting it right: what are the guiding principles for resource management in the 21st century?', 11–12 March 2002, Adelaide Convention Centre, Adelaide.

Dovers, S.R. (2005). *Environment and Sustainability Policy: Creation, Implementation, Evaluation*. Sydney: The Federation Press.

DPI (2006). *NSW Oyster Industry Sustainable Aquaculture Strategy – Final Report*. Australia: Fisheries Division, NSW Department of Primary Industries.

DPMC (2007). *A National Plan for Water Security*. Canberra: Department of the Prime Minister and the Cabinet, Australian Government. http://www.nalwt.gov.au/files/national_plan_for_water_security.pdf.

Dray, A., Perez, P., Le Page, C., D'Aquino, P. and White, I. (2005). *Companion Modelling approach: The ATOLLGAME Experience in Tarawa Atoll (Republic of Kiribati)*. Paper presented at the MODSIM 2005 International Congress on Modelling and Simulation, December 2005, Melbourne, Australia. www.mssanz.org.au/modsim05/papers/dray.pdf.

Dray, A., Perez, P., Jones, N. *et al.* (2006). The ATOLLGAME experience: from knowledge engineering to a computer-assisted role playing game. *Journal of Artificial Societies and Social Simulation*, **9**(1). http://jasss.soc.surrey.ac.uk/9/1/6.html.

Dray, A., Perez, P., Le Page, C., D'Aquino, P. and White, I. (2007). Who wants to terminate the game? The role of vested interests and metaplayers in the ATOLLGAME experience. *Simulation Gaming*, **38**, 494–511.

Dryzek, J. S. (1990). *Discursive Democracy: Politics, Policy and Political Science*. New York: Cambridge University Press.

Dryzek, J. S. (2010). *Foundations and Frontiers of Deliberative Governance*. Oxford, UK: Oxford University Press.

Durand, D. (1979). *La Systémique*, vol. 1795. Paris, France: Presses Universitaires de France.

Dwyer, A., Zoppou, C., Nielsen, O., Day, S. and Roberts, S. (2004). *Quantifying Social Vulnerability: A Methodology for Identifying Those at Risk to Natural Hazards*. Record 2004/14. Canberra: Geoscience Australia. www.ga.gov.au/image_cache/GA4267.pdf.

Earthbeat (2004). *Devil's Water* [Radio, recorded by A. de Blas, A. Page and J. May]. Australia: Radio National, Australian Broadcasting Corporation.

ecaabc (2004). *Agentlink – Agent Software*. www.agentlink.org/resources/agent-software.php.

Edelenbos, J. (1999). Design and management of participatory public policy making. *Public Management Review*, **1**(4), 569–576.

Eden, C. (1989). Using cognitive mapping for strategic options development and analysis. In *Rational Analysis for a Problematic World: Problem Structuring Methods for Complexity, Uncertainty and Conflict*, ed. J. Rosenhead. Chichester, UK: John Wiley and Sons, pp 21–42.

Eden, C. (2004). Analyzing cognitive maps to help structure issues or problems. *European Journal of Operational Research*, **159**(3), 673–686.

Eden, C. and Ackermann, F. (1998). *Making Strategy: The Journey of Strategic Management*. London, UK: Sage Publications.

Eden, C. and Ackermann, F. (2001). SODA – the principles. In *Rational Analysis for a Problematic World: Problem Structuring Methods for Complexity, Uncertainty and Conflict*, revised edn, ed. J. Rosenhead. Chichester, UK: John Wiley and Sons, pp 21–42.

Eden, C. and Ackermann, F. (2004). Cognitive mapping expert views for policy analysis in the public sector. *European Journal of Operational Research*, **152**(3), 615–630.

Eden, C. and Spender, J.-C. (eds) (1998). *Managerial and Organizational Cognition: Theory, Methods and Research*. London, UK: Sage Publications.

Eden, C., Ackermann, F. and Cropper, S. (1992). The analysis of cause maps. *Journal of Management Studies*, **29**(3), 309–324.

Eden, C., Ackermann, F., Bryson, J. M. *et al.* (2009). Integrating modes of policy analysis and strategic management practice: requisite elements and dilemmas. *Journal of the Operational Research Society*, **60**, 2–13.

Edwards, W. (1977). How to use multiattribute utility measurement for social decision making. *IEEE Transactions on Systems, Man and Cybernetics*, **7**(5), 326–340.

Edwards, W. and Newman, J. R. (1982). *Multiattribute Evaluation*. Beverly Hills, CA: Sage.

Eisenhardt, K. M., Kahwajy, J. L. and Bourgeois III, L. J. (1997). How management teams can have a good fight. *Harvard Business Review*, **75**(4), 77–85.

Elliott, J., Heesterbeek, S., Lukensmeyer, C. J. and Slocum, N. (2005). *Participatory Methods Toolkit. A Practitioner's Manual*. King Baudouin Foundation and the Flemish Institute for Science and Technology Assessment (viWTA). www.bchealthycommunities.ca/Groups/2009 Website Tools Resources/Public Participation Tool-kit (good).pdf.

Ellison, B. A. (2007). Public administration reform in Eastern Europe: a research note and a look at Bulgaria. *Administration and Society*, **39**(2), 221–232.

Elstub, S. (2008). *Towards a Deliberative and Associational Democracy*. Edinburgh, UK: Edinburgh University Press Ltd.

Emery, F. E. (ed.) (1969). *Systems Thinking*. Harmondsworth, UK: Penguin Books Ltd.

Enserink, B. and Monnikhof, R. A. H. (2003). Information management for public participation in co-design processes: evaluation of a Dutch example. *Journal of Environmental Planning and Management*, **46**(3), 315–344.

EPI (2006). *Eco-Economy Indicators*: Earth Policy Institute. www.earth-policy.org/publications/C39.

Estrella, M. and Gaventa, J. (1998). *Who Counts Reality? Participatory Monitoring and Evaluation: A Literature Review*, IDS Working Paper No 70. Brighton, UK: Institute of Development Studies.

Etienne, M. (ed.) (2010). *La modélisation d'accompagnement: une démarche participative en appui au développement durable*. Versailles: Editions Quae.

EU (2000). Directive 2000/60/EC of the European Parliament and of the Council, of 23 October 2000: establishing a framework for Community action in the field of water policy, L 327, 22.12.2000, EN. *Official Journal of the European Communities* 1–72.

EU (2002). *Guidance on Public Participation in Relation to the Water Framework Directive: Active involvement, Consultation, And Public Access to Information*. Final version after the Water Directors' meeting, December 2002. European Union.

EURO (2008). *What is Operational Research?* New York: The Association of European Operational Societies. www.euro-online.org/display.php?page=what_or.

EUROPA (2008). *The History of the European Union*. http://europa.eu/abc/history/index_en.htm.

European Community (1992). *Treaty on the European Union*. European Community.

Evan, W. M. and Manion, M. (2002). *Minding the Machines: Preventing Technological Disasters*. Upper Saddle River, NJ: Prentice Hall PTR.

Everingham, P. (2005). *Upper South East Water Quality Risk Management Strategy*. Adelaide, Australia: Department of Water, Land and Biodiversity Conservation, Government of South Australia.

Falkenmark, M. (1997). Meeting water requirements of an expanding world population. *Philosophical Transactions of the Royal Society of London*, **352**, 929–936.

Falkenmark, M. (2003). Freshwater as shared between society and ecosystems: from divided approaches to integrated challenges. *Philosophical Transactions: Biological Sciences*, **358**(1440), 2037–2049.

Falkenmark, M. and Lundqvist, J. (1998). Towards water security: Political determination and human adaptation crucial. *Natural Resources Forum*, **21**(1), 37–51.

Falkenmark, M. and Rockström, J. (2006). The new blue and green water paradigm: breaking new ground for water resources planning and management *Journal of Water Resources Planning and Management*, **132**(3), 129–132.

Falkenmark, M. and Widstrand, C. (1992). Population and water resources: a delicate balance. *Population Bulletin*, **47**(3), 1–36.

Falkenmark, M., Lundqvist, J. and Widstrand, C. (1989). Macro-scale water scarcity requires micro-scale approaches. Aspects of vulnerability in semi-arid development. *Natural Resources Forum*, **13**(4), 258–267.

Falkenmark, M., Gottschalk, L., Lundqvist, J. and Wouters, P. (2004). Towards integrated catchment management: increasing the dialogue between scientists, policy-makers and stakeholders. *International Journal of Water Resources Development*, **20**(3), 297–309.

Falkland, T. (2007). *Report on Water Investigations – Weno, Chuuk State, Federated States of Micronesia, November 2006 – February 2007: Chuuk Water Supply Follow-up Investigation*, RSC-C60725 (FSM): Loan Water Supply and Sanitation Project: Report prepared for the Asian Development Bank, March 2007.

Faltings, B. and Struss, P. (eds) (1992). *Recent Advances in Qualitative Physics*. Cambridge, MA: MIT Press.

FAO (2000). *New Dimensions in Water Scarcity: Water, Scarcity and Ecosystem Services in the 21th Century, AGL/MISC/25/2000*. Rome: Land and Water Development Division, Food and Agriculture Organization of the United Nations.

Fayesse, N. (2006). Troubles on the way: an analysis of the challenges faced by multi-stakeholder platforms. *Natural Resources Forum*, **30**, 219–229.

Fayol, H. (1918). *Administration industrielle et generale*. Paris: Durnod.

Feibleman, J. and Friend, J. W. (1945). The structure and function of organisation. *Philosophical Review*, **54**, 19–44.

Ferrand, N. (1997). *Modèles multi-agents pour l'aide à la décision et la négociation en aménagement du territoire*. Grenoble, France: UJF.

Ferrand, N. (2004a). *'ENCORE' Un modèle d'analyse, 'Intervention et Questions' from an unpublished report*. Montpellier, France: Cemagref.

Ferrand, N. (2004b). *Participative Tools in the WFD: What's for? – Needs and Requirements from HarmoniCOP and HarmoniQUA*. Paper presented at the HarmoniCA, November 16, 2004, Copenhagen, Denmark.

Ferrand, N. and Daniell, K. A. (2006). *Comment évaluer la contribution de la modélisation participative au développement durable?* Paper presented at the 'Séminaire interdisciplinaire sur le développement durable', 30 November 2006, Lille, France.

Ferrand, N., Nancarrow, B., Johnston, C. and Syme, G. (2005). *Simulation and Role-Playing Games For Social Justice Research*. Paper presented at the CABM-HEMA-SMAGET Joint Conference on Multi-Agent Modelling for Environmental Management, 21–25 March 2005, Bourg St Maurice, France. http://smaget.lyon.cemagref.fr/contenu/SMAGET%20proc/ABSTRACTS/ferrand.pdf.

Ferrand, N., Hare, M. and Rougier, J.-E. (2006). *Iskar Test Site Option Description Living with Flood and Drought. Methodological document to the Iskar Test Site*, Europe: AquaStress IP, FP6.

Ferrand, N., Ribarova, I. S., Daniell, K. A. *et al.* (2007). *Supporting a Multi-Levels Participatory Modelling Process For Floods and Droughts Co-Management*. Paper presented at the 'Journées de la Modélisation au Cemagref', 26–27 November 2007, Clermont-Ferrand, France.

Ferrand, N., Farolfi, S., Abrami, G. and du Toit, D. (2009a). *WAT-A-GAME: The Sand River Case*, Compte rendu de la Communauté Montpelliéraine de Participation, 07/09/09. Montpellier: Lisode. www.particip.fr/images/CR_2009/CR_cdp_test_JdR_Wat-A-Game_07.09.09.pdf.

Ferrand, N., Farolfi, S., Abrami, G. and du Toit, D. (2009b). *WAT-A-GAME: Sharing Water and Policies in Your Own Basin*. South Africa: AWARD Association for Water and Rural Development. www.award.org.za/file_uploads/File/ferrand-WAG-ISAGA-full.pdf.

Finaud-Guyot, P. (2005). *Development of an Evaluation Method for Public Involvement*. MSc thesis. Silsoe, UK: Institute of Water and Environment, Cranfield University.

Finlayson, B. L. and McMahon, T. A. (1988). Australia vs. the World: a comparative analysis of streamflow characteristics. In *Fluvial Geomorphology of Australia*, ed. R. F. Werner. Sydney, Australia: Academic Press, pp 17–40.

Fischer, F. (1990). *Technocracy and the Politics of Expertise: Critical Perspectives on the Managerial and Policy Sciences*. Newbury Park, CA: Sage Publications.

Fischer, F. (2000). *Citizens, Experts, and the Environment*. Durham: Duke University Press.

Fischer, F. (2003). *Reframing Public Policy: Discursive Politics and Deliberative Practices*. Oxford, UK: Oxford University Press.

Fisher, R. and Ury, W. (1981). *Getting to Yes: Negotiating Agreement Without Giving In*. Boston: Houghton Mifflin.

Flannery, T. (1994). *The Future Eaters: An Ecological History of the Australasian Lands and People*. Melbourne: Reed Books Australia.

Fleming, N. S. (1999). *Sustainability and Water Resources Management for the Northern Adelaide Plains, South Australia*. Adelaide: The University of Adelaide.

Fleming, N. S. (2005). *Systems Based Planning and Information Networks for Sustainability*. EIANZ Conference: working on the frontier – environmental sustainability in practice. Christchurch, New Zealand.

Fleming, N. S. and Daniell, T. M. (1996). *Matrix for Evaluation of Sustainability Achievement (MESA): An Aid to Water Resources Project Evaluation*. Paper presented at the Hydrology and Water Resources Symposium 1996: Water and the Environment; Preprints of Papers, No. 96/05, Australia: Barton, ACT.

Fletcher, W. J., Chesson, J. M. F., Sainsbury, K. J. and Hundloe, T. J. (2004). *National ESD Reporting Framework: The 'How To' Guide for Aquaculture. Version 1.1*. Canberra: FRDC.

Flood, R. L. (1995). Total systems intervention (TSI): a reconstitution. *Journal of the Operational Research Society*, **46**(2), 174–191.

Flood, R. L. (1998). Action research and the management and systems sciences. *Systemic Practice and Action Research*, **11**(1), 79–101.

Flood, R. L. and Jackson, M. C. (1991a). *Creative Problem Solving: Total Systems Intervention*. Chichester, UK: John Wiley and Sons.

Flood, R. L. and Jackson, M. C. (1991b). Total systems intervention: a practical face to critical systems thinking. *Systems Practice*, **4**(3), 197–213.

Florman, S. C. (1976). *The Existential Pleasure of Engineering*. New York: St. Martin's Press.

Foley, B. A., Daniell, T. M. and Warner, R. F. (2003). What is sustainability and can it be measured? *Australian Journal of Multidisciplinary Engineering*, **1**(1), 1–8.

Foran, B. (2008). Consuming the earth. In *Biodiversity: Integrating Conservation and Production*, ed. E. C. Lefroy, T. Lefroy, K. Bailey and T. Norton. Collingwood: CSIRO Publishing, pp 31–41.

Forester, J. (1989). *Planning in the Face of Power*. Berkeley, CA: University of California Press.

Forester, J. (1993). *Critical Theory, Public Policy, and Planning Practice: Towards a Critical Pragmatism*. Albany, NY: State University of New York.

Forester, J. (1999). *The Deliberative Practitioner: Encouraging Participatory Planning Processes*. Cambridge, MA: MIT Press.

Forrest, K. and Howard, A. (2004). *Central Coast: Regional Profile and Social Atlas*. NSW, Australia. http://catalogue.nla.gov.au/Record/3454948.

Forrester, J. W. (1961). *Industrial Dynamics*. Cambridge, MA: Productivity Press.

Forrester, J. W. (1968). *Principles of Systems*. Cambridge, MA: MIT Press.

Forrester, J. W. (1992a). Policies, decisions and information sources for modeling. *European Journal of Operational Research*, **59**(1), 42–63.

Forrester, J. W. (1992b). *Systems Dynamics, Systems Thinking, and Soft OR, D-4434–1, Road Maps: A Guide to Learning System Dynamics*. http://clexchange.org/curriculum/roadmaps/rm7.asp.

Forrester, J. W. (2007). System dynamic: a personal view of the first fifty years. *System Dynamics Review*, **23**(2–3), 345–358.

Forster, P., Ramaswamy, V., Artaxo, P. *et al.* (2007). Changes in atmospheric constituents and in radiative forcing. In *Climate Change 2007: The Physical Science Basis. Contribution of Working Group I to the Fourth Assessment Report of the Intergovernmental Panel on Climate Change*, ed. S. Solomon, D. Qin, M. Manning, *et al.* Cambridge, UK: Cambridge University Press.

Fotev, G. (2004). Bulgarian society and the water crisis: sociology and ethics. In *Drought in Bulgaria: A Contemporary Analog for Climate Change*, ed. C. G. Knight, I. Raev and M. P. Staneva. Aldershot, UK: Ashgate Publishing Ltd., pp 185–202.

Foucault, M. (1977). *Discipline and Punish: The Birth of the Prison*. London, UK: Penguin.

Foucault, M. (1980). *Power/Knowledge: Selected Interviews and Other Writings by Michel Foucault, 1972–1977*. Brighton, UK: Harvester.

Foucault, M. (1982). The subject and power. *Critical Inquiry*, **8**(4), 777–795.

Francis, M. J., Gulati, N. and Pashley, R. M. (2006). The dispersion of natural oils in de-gassed water. *Journal of Colloid and Interface Science*, **299**, 673–677.

Franco, L. A. (2008). Facilitating collaboration with problem structuring methods: a case study of an inter-organisational construction partnership. *Group Decision and Negotiation*, **17**(4), 267–286.

Frank, L. A. and Sigwarth, J. B. (2001). Detections of small comets with a ground-based telescope. *Journal of Geophysical Research – Space Physics*, **106**(3), 3665–3683.

Frank, L. A., Sigwarth, J. B. and Craven, J. D. (1986). On the influx of small comets into the earth's upper atmosphere II. Interpretation. *Geophysical Research Letters*, **13**(4), 307–310.

Freeman, H. E. (1977). The present status of evaluation research. In *Evaluation Studies Review Annual*, ed. M. Guttentag and S. Saar. London: Sage Publications, pp 17–51.

Freeman, R. E. (1984). *Strategic Management: A Stakeholder Approach*. Boston, MA: Pitman.

Frey, F. W. (1993). The political context of conflict and cooperation over international river basins. *Water International*, **18**(1), 54–68.

Friedmann, J. (1973). *Retracking America: A Theory of Transactive Planning*. Garden City, NY: Anchor Books.

Friedmann, J. (1993). Toward a non-Euclidian mode of planning. *Journal of the American Planning Association*, **59**(4), 482–485.

Friend, J. (1993). Searching for appropriate theory and practice in multi-organizational fields. *Journal of the Operational Research Society*, **44**(6), 585–598.

Friend, J. (2001). The strategic choice approach. In *Rational Analysis for a Problematic World Revisited: Problem Structuring Methods for Complexity, Uncertainty and Conflict*, revised edn, ed. J. Rosenhead and J. Mingers. Chichester, UK: John Wiley and Sons, pp 115–149.

Friend, J. and Hickling, A. (1987). *Planning Under Pressure: The Strategic Choice Approach*. Oxford, UK: Pergamon Press.

Fuchs, C. (2004). Knowledge management in self-organizing social systems. *Journal of Knowledge Management Practice*, **5**. www.tlainc.com/articl61.htm.

Fung, A. (2006). Varieties of participation in complex governance. *Public Administration Review*, **66**(s1), 66–75.

Funtowicz, S. O. and Ravetz, J. R. (1993). Science for the post-normal age. *Futures*, **25**(7), 739–755.

Gaillard, I. and Lamonde, F. (2001). Ingénierie concourante et conception collective: point de vue de l'ergonomie. In *Compétances collectives au travail*, ed. D. R. Kouabenan. Paris, France: L'Harmattan, pp 149–164.

Gardner, A. and Bowmer, K. H. (2007). Environmental water allocations and their governance. In *Managing Water For Australia: The Social and Institutional Challenges*, ed. K. Hussey and S. Dovers. Vol. 43–57. Collingwood, Australia: CSIRO Publishing.

Gass, S. I. (1983). Decision-aiding models: validation, assessment, and related issues for policy analysis. *Operations Research*, **31**(4), 603–631.

Gellman, B. (27 June 2002). Cyber-attacks by Al Qaeda feared. *The Washington Post*, p. A01. www.cartome.org/aq-cyberattacks.htm.

Gennari, J., Musen, M. A., Fergerson, R. W. *et al.* (2002). The evolution of protégé: an environment for knowledge-based systems development. *International Journal of Human–Computer Studies*, **58**(1), 89–123.

Ghassemi, F., Jakeman, A. J. and Nix, H. A. (1995). *Salinisation of Land and Water Resources: Human Causes, Extent, Management and Case Studies*. Sydney: CAB International and UNSW Press.

GHD (2007). *Using Recycled Water for Drinking: An Introduction*, Waterlines Occasional Paper No 2. Canberra: GHD.

Gidley, J. (2007). The evolution of consciousness as a planetary imperative: an integration of integral views *Integral Review*, **5**, 7–226.

Ginestet, A. (16 January 2008). La politique de l'eau européenne s'est développée en trois phases. *Le Journal de l'Environnement*, **10**, 25.

Giordano, R., Passarella, G., Uricchio, V. F. and Vurro, M. (2005). Fuzzy cognitive maps for issue identification in a water resources conflict resolution system. *Physics and Chemistry of the Earth*, **30**(6–7), 463–469.

Giordano, R., Passarella, G., Uricchio, V. F. and Vurro, M. (2007). Integrating conflict analysis and consensus reaching in a decision support system for water resource management. *Journal of Environmental Management*, **84**(2), 213–228.

Glaser, B. and Strauss, A. L. (1967). *The Discovery of Grounded Theory: Strategies for Qualitative Research*. Chicago: Adline Publishers.

Gleick, J. (1987). *Chaos: Making a New Science*. New York: Viking Penguin.

Gleick, P. H. (1990). Vulnerability of water systems. In *Climate Change and US Water Resources*, ed. P. E. Waggoner. New York: John Wiley and Sons Inc.

Gleick, P. H. (1998). *The World's Water 1998–1999: The Biennial Report on Freshwater Resources*. Washington DC: Island Press.

Gleick, P. H. (2000a). The changing water paradigm, a look at twenty-first century water resources development. *Water International*, **25**(1), 127–138.

Gleick, P. H. (2000b). *The World's Water 2000–2001: The Biennial Report on Freshwater Resources*. Washington, DC: Island Press.

Gleick, P. H. (2006). Environment and security: water conflict chronology version 2006–2007. In *The World's Water 2006–2007: The Biennial Report on Freshwater Resources*, ed. P. H. Gleick, H. Cooley *et al.* Washington DC: Island Press, pp 189–218.

Gleick, P. H. and Lane, J. (2005). Large international water meetings: time for a reappraisal. *Water International*, **30**(3), 410–414.

Gleick, P. H., Burns, W. C. G., Chalecki, E. L. *et al.* (2002). *The World's Water 2002–2003: The Biennial Report on Freshwater Resources*. Washington DC: Island Press.

Gleick, P. H., Cain, N. L., Haasz, D. *et al.* (eds) (2004). *The World's Water 2004–2005: The Biennial Report on Freshwater Resources*. Washington, DC: Island Press.

Gleick, P. H., Cooley, H., Katz, D. *et al.* (2006). *The World's Water 2006–2007: The Biennial Report on Freshwater Resources*. Washington DC: Island Press.

Global Water Partnership (2000). *Integrated Water Resources Management*, TAC Background Papers No 4, Technical Advisory Committee. Stockholm, Sweden: Global Water Partnership.

Glover, F. and Laguna, M. (1997). *Tabu Search*. Boston: Kluwer Academic Publishers.

Goicoechea, A., Hansen, D. R. and Duckstein, L. (1982). *Multiobjective Decision Analysis with Engineering and Business Applications*. New York: John Wiley & Sons.

Goldberg, D. E. (1989). *Genetic Algorithms in Search, Optimization and Machine Learning*. Boston, MA: Addison-Wesley.

Gourlay, S. (2006). Conceptualizing knowledge creation: a critique of Nonaka's theory. *Journal of Management Studies*, **43**(7), 1415–1436. http://onlinelibrary.wiley.com/doi/10.1111/j.1467-6486.2006.00637.x/full.

Government of the Republic of Kiribati (2008a). *National Water Resources Implementation Plan: Sustainable Water Resource Management, Use, Protection and Conservation – A 10 Year Plan, November 2008*. Government of the Republic of Kiribati. www.sprep.org/att/IRC/eCOPIES/Countries/Kiribati/95.pdf.

Government of the Republic of Kiribati (2008b). *National Water Resources Policy: Water for Healthy Communities, Environments and Sustainable Development*. Government of the Republic of Kiribati. www.sprep.org/att/IRC/eCOPIES/Countries/Kiribati/94.pdf.

Grabow, S. and Heskin, A. (1973). Foundations for a radical concept of planning. *Journal of the American Planning Association*, **39**(2), 106–114.

Gratton, L. and Erickson, T. J. (2007). 8 ways to build collaborative teams. *Harvard Business Review*, **85**(11), 100–109.

Gratton, L., Voigt, A. and Erickson, T. J. (2007). Bridging faultlines in diverse teams. *MIT Sloan Management Review*, **48**(4), 22.

Gray, B. (1989). *Collaborating: Finding Common Ground For Multiparty Problems*. San Francisco, CA: Jossey-Bass.

Greenberg, I. (5 April 2006). A vanished sea reclaims its form in Central Asia. *International Herald Tribune*. www.nytimes.com/2006/04/05/world/asia/05iht-sea.html?scp=1&sq=A vanished sea reclaims its form in Central Asia&st=cse.

Gregory, W. J. (1996). Discordant pluralism: a new strategy for critical systems thinking. *Systems Practice*, **9**(6), 605–625.

Grimble, R. and Wellard, K. (1997). Stakeholder methodologies in natural resource management: a review of principles, contexts, experiences and opportunities. *Agricultural Systems*, **55**(2), 173–193.

GRMA (2006). *The Great Man-Made River Project Libya*. Paper presented at the Mege Projects MENA, 20–21 November 2006, Dubai, UAE. http://www.gmmra.org/en/.

Guba, E. G. and Lincoln, Y. S. (1989). *Fourth Generation Evaluation*. Newbury Park, CA: Sage Publications Inc.

Guitouni, A. and Martel, J.-M. (1998). Tentative guidelines to help choosing an appropriate MCDA method. *European Journal of Operational Research*, **109**(2), 501–521.

Gunderson, L. H. and Holling, C. S. (eds) (2002). *Panarchy: Understanding Transformations in Human and Natural Systems*. Washington DC: Island Press.

Habermas, J. (1972). *Knowledge and Human Interests*. New York: Beacon Press.

Habermas, J. (1984). *The Theory of Communicative Action*. Vols I and II, 3rd edn. Oxford, UK: Polity Press.

Habermas, J. (1996). *Between Facts and Norms*. Cambridge, UK: Polity Press.

Hajkowicz, S., Young, M., Wheeler, S., MacDonald, D. H. and Young, D. (2000). *Supporting Decisions: Understanding Natural Resource Management Assessment Techniques, A Report to the Land and Water Resources Research and Development Corporation*. Glen Osmond, Australia: CSIRO Land and Water. www.clw.csiro.au/publications/consultancy/2000/support_decisions.pdf.

Haken, H. (1981). *The Science of Structure: Synergetics*. New York: Van Nostrand Reinhold Company. (Translated by F. Bradley. This edition first published in 1984.)

Hämäläinen, R. P., Kettunen, E., Marttunen, M. and Ehtamo, H. (2001). Evaluating a framework for multi-stakeholder decision support in water resources management. *Group Decision and Negotiation*, **10**(4), 331–353.

Handmer, J. W., Smith, D. I. and Dorcey, A. H. J. (1991). The Australian context for mediation in water management. In *Negotiating Water: Conflict Resolution in Australian Water Management*, ed. J. W. Handmer, A. H. J. Dorcey and D. I. Smith. Canberra, Australia: Centre for Resource and Environmental Studies, Australian National University, pp 3–19.

Hardy, C. and Phillips, N. (1998). Strategies of engagement: lessons from the critical examination of collaboration and conflict in an interorganizational domain. *Organization Science*, **9**(2), 217–230.

Hare, M. (2006). *Evaluation of Process and Next Steps for the Iskar River Basin Test Site Within the AquaStress Project*, Seecon Report, Seecon09/2006. Osnabrück, Germany: Seecon Deutschland GmbH.

Hare, M. (2007). *Policy Makers' Interviews and Report on the 1st Policy Makers' Workshop of Case Study 3 of the Iskar River Basin Test Site within the AquaStress Project*, Seecon Report # Seecon01/2007. Osnabrück, Germany: Seecon Deutschland GmbH.

Hare, M. and Pahl-Wostl, C. (2002). Stakeholder categorisation in participatory integrated assessment processes. *Integrated Assessment*, **3**(1), 50–62.

Hare, M., Letcher, R. A. and Jakeman, A. J. (2003). Participatory natural resource management: a comparison of four case studies. *Integrated Assessment*, **4**(2), 62–72.

Hargroves, K. J. and Smith, M. H. (eds) (2005). *The Natural Advantage of Nations: Business Opportunities, Innovation and Governance in the 21st Century*. London: Earthscan.

HarmoniCOP (2005a). *Learning Together to Manage Together: Improving Participation in Water Management*. Europe: 'Harmonising Collaboration Planning' Project. http://waterwiki.net/index.php/HarmoniCOP_2005:_Learning_together_to_manage_together_-_Improving_participation_in_water_management.

HarmoniCOP (2005b). *Sustainability Learning for River Basin Management and Planning in Europe*, Integration report. Europe: HarmoniCOP WP6. http://www.harmonicop.uni-osnabrueck.de/_files/_down/WP6 Integration reportFINAL.pdf.

Hashimoto, T., Loucks, D. P. and Stedinger, J. R. (1982a). Robustness of water resources systems. *Water Resources Research*, **18**(1), 21–26.

Hashimoto, T., Stedinger, J. R. and Loucks, D. P. (1982b). Reliability, resiliency, and vulnerability criteria for water resource system performance criteria. *Water Resources Research*, **18**(1), 14–20.

Hatchuel, A. (1994). Les savoirs de l'intérvention en entreprise. *Entreprises et histoire*, **7**, 59–75.

Hatchuel, A. (2000). Quel horizon pour les sciences de gestion? Vers une théorie de l'action collective. In *Les nouvelles fondations des sciences de gestion*, ed. A. David, A. Hatchuel and R. Laufer. Paris, France: Vuibert/FNEGE, pp 7–43.

Hatchuel, A. (2004). Du débat public à la conception collective: qu'est-ce qu'une expertise démocratique? In *Expertises et projet urbain*, ed. T. Evette. Paris, France: Editions de La Villette.

Hatchuel, A. (2005). Towards an epistemology of collective action: management research as a responsive and actionable discipline. *European Management Review*, **2**, 36–47.

Hatchuel, A. and Molet, H. (1986). Rational modelling in understanding and aiding human decision-making: about two case studies. *European Journal of Operational Research*, **24**, 178–186.

Held, I. M. and Soden, B. J. (2006). Robust responses of the hydrological cycle to global warming. *Journal of Climate*, **19**, 5686–5699.

Helfgott, A. (2008). *Situating Strength-Based Approaches to Community Development in a Systems Thinking Context: The Cambodia Tales*. Paper presented at the UK Systems Society Conference 2008 – *Building Resilience: Responding to a Turbulent World*, 1–3 September, Oxford, UK.

Heron, J. (1981). Experiential research methodology. In *Human Inquiry: a Sourcebook of New Paradigm Research*, ed. P. Reason and J. Rowan. Chichester, UK: John Wiley and Sons.

Heron, J. (1996). *Co-operative Inquiry: Research into the Human Condition*. London: Sage.

Heyer, R. (2004). *Understanding Soft Operations Research: The Methods, Their Application and its Future in the Defence Setting*, Report No DSTO-GD-0411. Edinburgh, Australia: Command and Control Division Information Sciences Laboratory, Defence Science and Technology Organisation, Australian Government. www.dsto.defence.gov.au/publications/3451/DSTO-GD-0411.pdf.

Hickling, A. (2001). Gambling with frozen fire? In *Rational Analysis for a Problematic World Revisited: Problem Structuring Methods for Complexity, Uncertainty and Conflict*, revised edn, ed. J. Rosenhead and J. Mingers. Chichester, UK: John Wiley and Sons Ltd, pp 151–180.

High Performance Systems (1992). *Stella II: An Introduction to Systems Thinking*. Nahover, NH: High Performance Systems, Inc.

Hipel, K. W. and Ben-Haim, Y. (1999). Decision making in an uncertain world: information-gap modeling in water resources management. *IEEE Transactions on Systems, Man and Cybernetics – Part C: Applications and Reviews*, **29**(4), 506–517.

Hipel, K. W., Ragade, R. K. and Unny, T. E. (1974). Metagame analysis of water resources conflicts. *ASCE Journal of the Hydraulics Division*, **100**, 1437–1455.

Hjern, B. and Porter, D. O. (1981). Implementation structures: a new unit of administrative analysis. *Organization Studies*, **2**(3), 211–227.

Hjortsø, C. N. (2004). Enhancing public participation in natural resource management using Soft OR: an application of strategic option development and analysis in tactical forest planning. *European Journal of Operational Research*, **152**(3), 667–683.

HNCMA (2005). *Hawkesbury-Nepean Draft Catchment Action Plan 2006–2015*. NSW, Australia: Hawkesbury Nepean Catchment Management Authority.

Hobbs, B. F., Ludsin, S. A., Knight, R. L. *et al.* (2002). Fuzzy cognitive mapping as a tool to define management objectives for complex ecosystems. *Ecological Applications*, **12**(5), 1548–1565.

Holling, C. S. (1973). Resilience and stability of ecological systems. *Annual Review of Ecological Systems*, **4**, 1–23.

Holling, C. S. (ed.) (1978). *Adaptive Environmental Assessment and Management*. New York: John Wiley.

Holmes, T. and Scoones, J. (2000). *Participatory Environmental Policy Processes: Experiences From North and South*, IDS working paper 133. Brighton: Institute of Development Studies.

Holz, L., Kuczera, G. and Kalma, J. (2004). *Sustainable Urban Water Resource Planning in Australia: A Decision Sciences Perspective*. Paper presented at the International Conference on Water Sensitive Urban Design, Adelaide, Australia.

Hong, L. and Page, S. E. (2004). Groups of diverse problem solvers can outperform groups of high-ability problem solvers. *Proceedings of the National Academy of Sciences of the United States of America*, **101**, 16385–16389.

Howard, N. (1989). The manager as a politician and general: the metagame approach to analysing cooperation and conflict. In *Rational Analysis for a Problematic World: Problem Structuring Methods for Complexity, Uncertainty and Conflict*, ed. J. Rosenhead. Chichester, UK: John Wiley and Sons.

Howell, D. T. (1989). *Beyond Optimality: A Survey of Ideas About Choosing the Best*. Paper presented at the Proceedings of the National Workshop on Planning and Management of Water Resource Systems: Risk and Reliability, AWRC Conference Series No 17.

Hristov, T., Nikolova, R., Yancheva, S. and Nikolova, N. (2004). Water resource management during the drought. In *Drought in Bulgaria – A Contemporary Analog for Climate Change*, ed. C. G. Knight, I. Raev and M. P. Staneva. Aldershot, UK: Ashgate Publishing Limited. pp 241–276.

HSC (2002). *Berowra Creek Estuary Management Study and Management Plan*. NSW, Australia: Hornsby Shire Council. www.hornsby.nsw.gov.au/media/documents/environment-and-waste/water-catchments/estuary-management/reports/berowra-creek/Berowra-Creek-Estuary-Management-Study-and-Plan.pdf.

HSC (2004a). *Community Sustainability Indicators Project*. NSW, Australia: Hornsby Earthwise, Hornsby Shire Council.

HSC (2004b). *Water Management Plan*. NSW, Australia: Hornsby Shire Council.

HSC (2005a). *2004/05 State of the Environment Report*. NSW, Australia: Hornsby Earthwise, Hornsby Shire Council. www.hornsby.nsw.gov.au/media/documents/about-council/corporate-documents-and-reports/soe-reports/SOE-Report-2004-2005.pdf.

HSC (2005b). *Management Plan*. NSW, Australia: Hornsby Shire Council.

HSC (2005c). *Water Quality Monitoring Program: 2004–2005 Annual Report: Water Catchments Team*. NSW, Australia: Hornsby Shire Council. www.hornsby.nsw.gov.au/media/documents/environment-and-waste/water-catchments/water-quality/Water-Quality-Annual-Report-2004-2005.pdf.

HSC (2006a). *Q26/2006 Lower Hawkesbury Estuary Management Plan – Tender Document*. NSW, Australia: Water Catchments Team, Hornsby Shire Council.

HSC (2006b). *Summary of the 2006 State of the Environment Report*. Hornsby, NSW, Australia: Environmental Health and Protection Team, Hornsby Shire Council.

HSC (2009). *Estuary Management Program: 2008–2009 Annual Report, Hawkesbury Estuary Program*. Hornsby, NSW: Hornsby Shire Council.

HSC (2010). *Estuary Management Program: 2009–2010 Annual Report, Hawkesbury Estuary Program*. Hornsby, NSW: Hornsby Shire Council.

HSC/WBM (2006). *Brooklyn Estuary Management Plan: WBM Oceanics*. NSW, Australia: Hornsby Shire Council.

Huang, G. H., Baetz, B. W. and Patry, G. G. (1992). A grey linear programming approach for municipal solid waste management planning under uncertainty. *Civil Engineering and Environmental Systems*, **9**(4), 319–335.

Hubert, B. (2002). Sustainable development: think forward and act now – agricultures and sustainable development: the stakes of knowledge and research attitudes, *Dossier No 22, INRA faced with Sustainable Development: Landmarks for the Johannesburg Conference*. Johannesburg, South Africa. www.inra.fr/dpenv/huberd22e.htm.

Hugo, G. (2007). *Recent Population Change in Australia, Fenner Conference on the Environment: Water, Population and Australia's Future*. Canberra: Australian Academy of Sciences.

Huitema, D. and Meijerink, S. (eds) (2009). *Water Policy Entrepreneurs: A Research Companion to Water Transitions Around the Globe*. Cheltenham: Edward Elgar Publishing.

Hussey, K. and Dovers, S. (eds) (2007). *Managing Water for Australia: The Social and Institutional Challenges*. Collingwood, Australia: CSIRO Publishing.

Hutton, D. and Connors, L. (1999). *A History of the Australian Environmental Movement*. Cambridge, UK: Cambridge University Press.

Huxham, C. (ed.) (1996). *Creating Collaborative Advantage*. London, UK: Sage Publications.

ICWE (1992). *The Dublin Statement on Water and Sustainable Development*. International Conference on Water and the Environment. Dublin, Ireland.

IEA (1993). *At What Price Data? An Initiative of the National Committee on Coastal and Ocean Engineering*. Canberra, Australia: The Institution of Engineers, Australia.

IEA (2000). *Code of Ethics*: Institution of Engineers, Australia. www.engineersaustralia.org.au/sites/default/files/shado/About%20Us/Overview/Governance/CodeOfEthics2000.pdf.

Imache, A. (2008). *Construction de la demande en eau agricole au niveau régional en intégrant le comportement des agriculteurs: application aux exploitations agricoles collectives de la Mitidja-Ouest (Algerie)*. Doctoral thesis, l'Institut des Sciences et Industries du Vivant et de l'Environnement (Agro Paris Tech), Montpellier.

Imache, A., Le Goulven, P., Bouarfa, S., Chabaca, M. and Daniell, K. A. (2008). Farmers' behaviours and trends in agricultural water demand: the case of the irrigated Mitidja Plain (Algeria), *Proceedings of Water Down Under 2008: Incorporating the 31st Hydrology Symposium and the 4th Conference on Water Resources and Environment Research*. Adelaide: Engineers Australia, pp 2703–2710.

Imache, A., Bouarfa, S., Dionnet, M. *et al.* (2009a). Les arrangements de proximité sur les terres publiques: un choix délibéré ou une «question de survie» pour l'agriculture irriguée en Algérie? In *Actes du quatrième atelier régional du projet Sirma, 'Economies d'eau en systèmes irrigués au Maghreb', Mostaganem, Algérie, 26–28 mai 2008*, ed. T. Hartani, A. Douaoui and M. Kuper. Montpellier: Cirad.

Imache, A., Bouarfa, S., Kuper, M., Hartani, T. and Dionnet, M. (2009b). Integrating 'invisible' farmers in a regional debate on water productivity: the case of informal water and land markets in the Algerian Mitidja Plain. *Irrigation and Drainage*, **58**, S264–S272.

Imache, A., Dionnet, M., Bouarfa, S. *et al.* (2009c). Scénariologie participative: une démarche d'apprentissage social pour appréhender l'avenir de l'agriculture irriguée dans la Mitidja (Algérie). *Cahiers agricultures*, **18**(5), 417–423.

Imache, A., Hartani, T., Bouarfa, S. and Kuper, M. (eds) (2010). *La Mitidja 20 ans après. Réalités agricoles aux portes d'Alger*. Algiers: Editions Alpha.

Ingram, H. and Schneider, A. (1999). Science, democracy, and water policy. *Water Resources Update*, **133**, 21–28.

Islam, S., Oki, T., Kanae, S. *et al.* (2007). A grid-based assessment of global water scarcity including virtual water trading. *Water Resources Management*, **21**, 19–33.

Ison, R. L., Maiteny, P. T. and Carr, S. (1997). Systems methodologies for sustainable natural resources research and development. *Agricultural Systems*, **55**(2), 257–272.

Ison, R. L., Steyaert, P., Roggero, P. P., Hubert, B. and Jiggins, J. (eds) (2004). *The SLIM (Social Learning for the Integrated Management and Sustainable Use of Water at Catchment Scale) Final Report*. Europe: Prepared for the European Commission, FP5.

Ivanova, M. (2007). Inequality and government policies in Central and Eastern Europe. *East European Quarterly*, **41**(2), 167–204.

IWR (2007). *Shared Vision Planning: Shared Vision Models*. Institute for Water Resources, US Army Corps of Engineers. www.sharedvisionplanning.us/models.cfm.

Jackson, M. C. (1988). Some methodologies for community operational research. *Journal of the Operational Research Society*, **39**(8), 715–724.

Jackson, M. C. (1990). Beyond a system of systems methodologies. *Journal of the Operational Research Society*, **41**(8), 657–668.

Jackson, M. C. (2000). *Systems Approaches to Management*. Boston, MA: Kluwer Academic Publishing.

Jackson, M. C. (2003). *Systems Thinking: Creative Holism for Managers*. Chichester: Wiley.

Jackson, M. C. and Keys, P. (1984). Towards a system of systems methodologies. *Journal of the Operational Research Society*, **35**(6), 473–486.

Jackson, P. M. and Stainsby, L. (2000). Managing public sector networked organisations. *Public Money and Management*, **20**(1), 11–16.

Jackson, S., Storrs, M. and Morrison, J. (2005). Recognition of Aboriginal rights, interests and values in river research and management: Perspectives from northern Australia. *Ecological Management and Restoration*, **6**(2), 105–110.

Jaeger, C. C., Renn, O., Rosa, E. A. and Webler, T. (2001). *Risk, Uncertainty, and Rational Action*. London: Earthscan Publications Ltd.

Jakeman, A. J., Letcher, R. A. and Norton, J. P. (2006). Ten iterative steps in development and evaluation of environmental models. *Environmental Modelling & Software*, **21**(5), 602–614.

Janssen, M. A. (ed.) (2002). *Complexity and Ecosystem Management: The Theory and Practice of Multi-Agent Approaches*. Northampton, MA: Edward Elgar Publishers.

Jeffery, P. and Muro, M. (2005). *Review of Test and Evaluation Protocols*, Report D5.2–1. Europe: AquaStress IP, EU FP6.

Jensen, F. V. (2001). *Bayesian Networks and Decision Graphs*. New York: Springer-Verlag.

Johnson-Laird, P. N. (1983). *Mental Models: Towards a Cognitive Science of Language, Inference and Consciousness*. Cambridge, UK: Cambridge University Press.

Jones, N. A. (2007a). *Lower Hawkesbury Estuary Management Plan (LHEMP) On-going Evaluation Report*, Report prepared for the ADD-ComMod project (La modélisation d'accompagnement: une pratique de recherche en appui au développement durable, un project sous le programme 2005 du Agriculture et Developpement Durable en France). Canberra, Australia: ANU/CSIRO.

Jones, N. A. (2007b). *AtollGame: Ex-post Evaluation Report*, Report prepared for the ADD-ComMod project (La modélisation d'accompagnement: une pratique de recherche en appui au développement durable, un project sous le programme 2005 du Agriculture et Developpement Durable en France). Canberra, Australia: ANU/CSIRO.

Jones, N. A., Perez, P., Measham, T. G. *et al*. (2009). Evaluating participatory modeling: developing a framework for cross-case analysis. *Environmental Management*, **44**(6), 1180–1195.

Kahneman, D. and Tversky, A. (1979). Prospect theory: an analysis of decision under risk. *Econometrica*, **47**(2), 263–291.

Kain, J., Kuruppu, I. and Billing, R. (2003). *Australia's National Competition Policy: Its Evolution and Operation*, E-brief: online only: Economics, Commerce and Industrial Relations Group, Parliament of Australia, Parliamentary Library. www.aph.gov.au/library/intguide/econ/ncp_ebrief.htm.

Kant, I. (1781). *Critique of Pure Reason*. Basingstoke: Palgrave MacMillan. Translated by N. K. Smith.

Katz, D. (2006). Going with the flow: preserving and restoring instream water allocations. In *The World's Water: The Biennial Report on Freshwater Resources*, ed. P. H. Gleick, H. Cooley, D. Katz *et al*. Washington DC: Island Press, pp 29–49.

Katz, D. and Kahn, R. L. (1966). *The Social Psychology of Organisations*. New York: Wiley.

Katzenbach, J. R. and Smith, D. K. (2002). *The Discipline of Teams: A Mindbook-Workbook for Delivering Small Group Performance*. New York: John Wiley and Sons.

Kazakçi, A. O. and Tsoukiàs, A. (2005). Extending the C-K design theory: a theoretical background for personal design assistants. *Journal of Engineering Design*, **16**(4), 399–411.

Keeney, R. (1992). *Value Focused Thinking: A Path to Creative Decision Making*. Cambridge, MA: Harvard University Press.

Keeney, R. and Raiffa, H. (1976). *Decision with Multiple Objectives: Preferences and Value Trade-offs*. New York: John Wiley and Sons.

Kelly, D. (2001). *Community Participation in Rangeland Management*, RIRDC Publication No 01/118. Gatton, QLD, Australia: Rural Industries Research and Development Corporation.

Kelly, G. A. (1955). *The Psychology of Personal Constructs*, Vols 1 and 2. New York: Norton.

Kennedy, J. and Eberhart, R. (1995). *Particle Swarm Optimization*. Paper presented at the Proceedings of the IEEE International Conference on Neural Networks, 27 Nov–1 Dec 1995, Perth, Australia.

Ker Rault, P. A. (2008). *Public Participation in Integrated Water Management: A Wicked Concept for a Complex Societal Problem – Which Type of Public Participation for Which Type of Water Management Challenges in the Levant*. PhD thesis, Cranfield University, UK.

Ker Rault, P. A. and Jeffrey, P. J. (2008). Deconstructing public participation in the Water Framework Directive: implementation and compliance with the letter or with the spirit of the law? *Water and Environment Journal*, **22**(4), 241–249.

Kerr, T. (2004). As if bunyips mattered... cross-cultural mytho-poetic beasts in Australian subaltern planning. *Journal of Australian Studies*, **80**, 14–27.

Khisty, C. J. (2000). Citizen involvement in the transportation planning process: what is and what ought to be. *Journal of Advanced Transportation*, **34**(1), 125–142.

Khisty, C. J. (2006). A fresh look at the systems approach and an agenda for action: peeking through the lens of Churchman's aphorisms. *Systemic Practice and Action Research*, **19**(1), 3–25.

Killingsworth, M. J. and Palmer, J. S. (1992). The environmental impact statement and the rhetoric of democracy. In *Environmental Discourse and Practice: A Reader*, ed. L. M. Benton and J. R. Short. Oxford, UK: Blackwell Publishing, pp 156–160.

Kimmerikong (2005). *Hawkesbury–Nepean River Estuary Management – Scoping Study – Final Report*. NSW, Australia: Kimmerikong Pty Ltd Natural Resource Management.

Kirby, M. W. (2007). Paradigm change in operations research: thirty years of debate. *Operations Research*, **55**(1), 1–13.

Kirkpatrick, S., Gerlatt Jr., C. D. and Vecchi, M. P. (1983). Optimization by simulated annealing. *Science*, **220**(4598), 671–680.

Kirst, M. W. (1984). Policy issue networks: their influence on state policy-making. *Policy Studies Journal*, **13**(2), 247–264.

Kittel, N. (Friday, 22 April 2005). The seven engineering wonders – the Ord River irrigation scheme. *ABC Northern Territory*. www.abc.net.au/nt/stories/s1349572.htm.

Klinke, A. and Renn, O. (2002). A new approach to risk evaluation and management: risk-based, precaution-based, and discourse-based strategies. *Risk Analysis*, **22**(6), 1071–1094.

KNAW (2003). *Hydrology: A Vital Component of Earth System Science: Preliminary Foresight Study on Hydrological Science: Preliminary Foresight Committee on Hydrological Science in The Netherlands*. Royal Netherlands Academy of Arts and Sciences (KNAW).

Knight, C. G., Raev, I. and Staneva, M. P. (eds) (2004). *Drought in Bulgaria: A Contemporary Analog for Climate Change*. Aldershot, UK: Ashgate Publishing Limited.

Knight, D. W., Sullivan, J. F., Poole, S. J. and Carlson, L. E. (2002). *Skills Assessment in Hands-On Learning and Implications for Gender Differences in Engineering Education*. Paper presented at the Proceedings of the 2002 American Society for Engineering Education Annual Conference & Exposition. http://itll-www.colorado.edu/images/uploads/about_us/publications/Papers/ASEE02%20SkillsAssmnt%20FINAL%202430.pdf.

Kohonen, T. (1988). *Self-organization and Associative Memory*, 2nd edn. Berlin: Springer-Verlag.

Kolfschoten, G. L., Briggs, R. O., Appelman, J. H. and de Vreede, G.-J. (2004). Thinklets as building blocks for collaboration processes: a further conceptualization. In *CRIWG 2004, LNCS*, Vol. 3198, ed. G.-J. de Vreede, L. A. Guerrero and G. M. Raventos, Berlin, Germany: Springer-Verlag, pp 137–152.

Kolkman, M. J., Kok, M. and van der Veen, A. (2005). Mental model mapping as a new tool to analyse the use of information in decision-making in integrated water management. *Physics and Chemistry of the Earth*, **30**, 317–332.

Korfmacher, K. S. (2001). The politics of participation in watershed modeling. *Environmental Management*, **27**(2), 161–176.

Kosko, B. (1986). Fuzzy cognitive maps. *International Journal of Man–Machine Studies*, **24**(1), 65–75.

Krane, D. (2001). Disorderly progress on the frontiers of policy evaluation. *International Journal of Public Administration*, **24**(1), 95–123.

Krastev, I., Dorosiev, R. and Ganev, G. (2005). *Nations in Transit – Bulgaria (2005)*. Washington DC: Freedom House Inc. www.freedomhouse.org/template.cfm?page=47&nit=359&year=2005.

Kuczera, G. (2008). At-site flood frequency analysis (draft). In *Australian Rainfall and Runoff*, ed. NCWE. National Committee on Water Engineering, Engineers Australia.

Kuczera, G. and Parent, E. (1998). Monte Carlo assessment of parameter uncertainty in conceptual catchment models: the metropolis algorithm. *Journal of Hydrology*, **211**(1–4), 69–85.

Kuhn, T. S. (1962). *The Structure of Scientific Revolutions*. Chicago, IL: University of Chicago Press.

Kuhnert, K. W. (1994). Transforming leadership: developing people through delegation. In *Improving Organisational Effectiveness through Transformational Leadership*, ed. B. M. Bass and B. J. Avolio. Thousand Oaks, CA: Sage Publications, Inc., pp 10–25.

Kundzewicz, Z. W. and Schellnhuber, H.-J. (2004). Floods in the IPCC TAR perspective. *Natural Hazards*, **31**(1), 111–128.

Kundzewicz, Z. W. and Takeuchi, K. (1999). Flood protection and management: quo vadimus? *Hydrological Sciences Journal*, **44**(3), 417–432.

Kuper, M., Errahj, M., Faysse, N. *et al.* (2009). Autonomie et dépendance des irrigants en grande hydraulique: observations de l'action organisée au Maroc et en Algérie. *Natures sciences sociétés* **17**(3), 248–256.

Lachapelle, P. R., McCool, S. F. and Patterson, M. E. (2003). Barriers to effective natural resource planning in a 'messy' world. *Society & Natural Resources*, **16**(6), 473–490.

Landry, M. (2000). *Repères pour la formulation des problèmes organisationnels complexes*, Document de travail 2000–006. Canada: Centre de recherche en modélisation, information et décision (CERAMID), Université Laval.

Landry, M., Malouin, J. L. and Oral, M. (1983). Model validation in operations research. *European Journal of Operational Research*, **14**(3), 207–220.

Landry, M., Banville, C. and Oral, M. (1996). Model legitimisation in operational research. *European Journal of Operational Research*, **92**(3), 443–453.

Langfield-Smith, K. and Wirth, A. (1992). Measuring differences between cognitive maps. *Journal of the Operational Research Society*, **43**(12), 1135–1150.

Lankford, B. A. (2008). Integrated, adaptive and domanial water resources management. In *Adaptive and integrated water management: coping with complexity and uncertainty*, ed. C. Pahl-Wostl, P. Kabat and J. Möltgen. Berlin: Springer Verlag, pp 39–59.

Lao Tzu (600BC–531BC). *Lao Tzu Quotes*. http://thinkexist.com/quotation/be_careful_what_you_water_your_dreams_with-water/340213.html.

Lasut, A. (2005). *Creative Thinking and Modelling for the Decision Support in Water Management*, NRM 81.2005. Milano, Italy: Fondazione Eni Enrico Mattei. http://papers.ssrn.com/sol3/papers.cfm?abstract_id=740293.

László, E. (2004). *Science and the Akashic Field: An Integral Theory of Everything*. Rochester, VT: Inner Traditions.

László, E. (2006). *The Chaos Point: The World at the Crossroads*. London, UK: Piatkus Books Ltd.

Lauriol, J. (1998). Les représentations sociales dans la décision. In *Penser la stratégie: fondements et perspectives*, ed. H. Laroche and J.-P. Nioche. Paris, France: Vuibert, pp 320–348.

Lazega, E. (1998). *Réseaux sociaux et structures relationnelles*, Vol. 3399. Paris, France: Presses Universitaires de France.

Le Bars, M. and Ferrand, N. (2004). *The ENCORE Paradigm, Unpublished Report: CEMAGREF*, Montpellier, France.

Le Moigne, J.-L. (1977). *La théorie du système général: théorie de la modélisation*. Paris, France: Presses Universitaires de France.

Le Moigne, J.-L. (1999). *Les épistémologies constructivistes*, 2nd edn. Paris: Presses Universitaires de France.

Leach, G. and Wallwork, J. (2003). *Enabling Effective Participation, Negotiation, Conflict Resolution and Advocacy in Participatory Research: Tools and Approaches for Extension Professionals*. Paper presented at the APEN 2003 Forum, 26–28 November 2003, Hobart, Australia. www.regional.org.au/au/apen/2003/refereed/083leachgwallworkj.htm.

Lebel, L. and Garden, P. (2008). Deliberation, negotiation and scale in the governance of water resources in the Mekong region. In *Adaptive and integrated Water Management: Coping With Complexity and Uncertainty*, ed. C. Pahl-Wostl, P. Kabat and J. Möltgen, Vol. 205–225. Berlin: Springer Verlag.

Lee, K. N. (1999). Appraising adaptive management. *Conservation Ecology*, **3**(2), 3. www.consecol.org/vol3/iss2/art3/.

Leeuwis, C. (2000). Reconceptualizing participation for sustainable rural development: towards a negotiation approach. *Development and Change*, **31**(5), 931–959.

Letcher, R. A. (2002). *Issues in Integrated Assessment and Modelling for Catchment Management*. Canberra: Australian National University.

Letki, N. (2004). Socialization for participation? Trust, membership, and democratization in East-Central Europe. *Political Research Quarterly*, **57**(4), 665–679.

Levin, S. A. (1998). Ecosystems and the biosphere as complex adaptive systems. *Ecosystems*, **1**, 431–436.

Levin-Rozalis, M. (2004). Searching for the unknowable: a process of detection – abductive research generated by projective techniques. *International Journal of Qualitative Methods*, **3**(2), Art.1. www.ualberta.ca/~iiqm/backissues/3_2/pdf/rozalis.pdf.

Levrel, H. and Bouamrane, M. (2008). Instrumental learning and indicators efficiency: outputs from co-construction experiments in West African biosphere reserves. *Ecology and Society*, **13**(1), 28. www.ecologyandsociety.org/vol13/iss1/art28/.

Lewicki, R., Saunders, D., Minton, J. and Barry, B. (2001). *Essentials of Negotiation*: McGraw-Hill.

Lewin, K. (1951). *Field Theory in Social Science*. New York: Harper and Row.

Li, X., Harbottle, G., Zhang, J. and Wang, C. (2003). The earliest writing? Sign use in the seventh millennium BC at Jiahu, Henan Province, China. *Antiquity*, **77**(295), 31–44.

Libicki, M. C. and Pfleeger, S. L. (2004). *Collecting the Dots: Problem Formulation and Solution Elements, Occasional Paper. USA*. RAND Science and Technology, RAND Corporation.

Lienhard, J. H. (2000). *The Engines of Our Ingenuity: An Engineer Looks at Technology and Culture*. New York: Oxford University Press.

Lilien, G. L. (1975). Model relativism: a situational approach to model building. *Interfaces*, **5**(3), 11–18.

Lingiari Foundation (2002). *Onshore Water Rights Discussion Booklet*. Broome, Australia: Lingiari Foundation Inc.

Linnerooth-Bayer, J., Vári, A. and Mechler, R. (2005). Designing a disaster insurance pool: participatory and expert approaches in Hungary and Turkey. In *Catastrophic Risks and Insurance*. Paris, France: OECD Publishing, pp 267–290.

Linstone, H. A. and Turoff, M. (2002). *The Delphi Method: Techniques and Applications*. Newark, NJ: Information Systems Department, New Jersey Institute of Technology. www.is.njit.edu/pubs/delphibook/.

Lipshitz, R. and Bar-Ilan, O. (1996). How problems are solved: reconsidering the phase theorem. *Organizational Behavior and Human Decision Processes*, **65**(1), 48–60.

Lombardi, P. and Brandon, P. (2007). The multimodal system approach to sustainability planning evaluation. In *Sustainable Urban Development Volume 2: The Environmental Assessment Methods*, ed. M. Deakin, G. Mitchell, P. Nijkamp and R. Vreeker. London: Routledge, pp 47–64.

Lord, W. B., Adelman, L., Wehr, P. *et al.* (1979). *Conflict Management in Federal Water Resource Planning*. Washington, DC: Office of Water Research and Technology.

Lorenz, E. N. (1963). Deterministic nonperiodic flow. *Journal of the Atmospheric Sciences*, **20**, 130–141.

Loubier, S. and Rinaudo, J.-D. (2003). *L'eau et les habitants du bassin versant de l'Hérault: résultats d'une enquête sur les pratiques et perceptions*, Rapport intermédiaire, BRGM/RP-52584-FR. France: BRGM.

Loucks, D. P. (1992). Water resource systems models: their role in planning. *Journal of Water Resources Planning and Management*, **118**(3), 214–223.

Loucks, D. P. (1998). *Watershed Planning: Changing Issues, Processes and Expectations, Water Resources Update, Universities' Council on Water Resources*, Issue No 111, Spring Carbondale, IL, pp 38–45.

Loucks, D. P., Kindler, J. and Fedra, K. (1985). Interactive water resources modeling and model use: an overview. *Water Resources Research*, **21**(2), 95–102.

Low, A. (2002). *Creating Consciousness: A Study of Consciousness, Creativity, Evolution, and Violence*. Ashgate, OR: White Cloud Press.

Luckman, J. (1967). An approach to the management of design. *Operational Research Quarterly*, **18**(4), 345–358.

Luna-Reyes, L. F., Martinez-Moyano, I. J., Pardo, T. A. *et al.* (2006). Anatomy of a group model-building intervention: building dynamic theory from case study research. *System Dynamics Review*, **22**(4), 291–320.

LWA (2006). *Water Perspectives: Knowledge for Managing Australian Landscapes: Outcomes of an Expert Workshop Scoping Social and Institutional Research Questions in Support of the Implementation of the National Water Initiative*. Canberra: Land and Water Australia, Australian Government. http://lwa.gov.au/land-and-water-australia/knowledge-managing-australian-landscapes.

Lynham, T., de Jong, W., Sheil, W., Kusumanto, T. and Evans, K. (2007). A review of tools for incorporating community knowledge, preferences and values into decision making in natural resource management. *Ecology and Society*, **12**(1), Art. 5. www.ecologyandsociety.org/vol12/iss1/art5/.

MacKenzie, K. D. (1986). Virtual positions and power. *Management Science*, **32**(5), 622–642.

Magnusson, R. J. (2001). *Water Technology in the Middle Ages: Cities, Monasteries and Waterworks after the Roman Empire*. Baltimore, MD: The John Hopkins University Press.

Maier, H. R. (2007). Meeting the challenges of engineering education via online roleplay simulations. *Australasian Journal of Engineering Education*, **13**(1), 31–39.

Maier, H. R. and Ascough II, J. C. (July 2006). Uncertainty in environmental decision-making: issues, challenges and future directions. In *Proceedings of the iEMSs Third Biennial Meeting: 'Summit on Environmental Modelling & Software'*, ed. A. Voinov, A. J. Jakeman and A. E. Rizzoli. Burlington, USA: International Environmental Modelling and Software Society, 2006. CD ROM. www.iemss.org/iemss2006/sessions/all.html.

Maier, H. R., Simpson, A. R., Zecchin, A. C. *et al.* (2003). Ant colony optimization for design of water distribution systems. *Journal of Water Resources Planning and Management*, **129**(3), 200–209.

Maier, H. R., Baron, J. and McLaughlan, R. G. (2007). Using online roleplay simulations for teaching sustainability principles to engineering students. *International Journal of Engineering Education*, **23**(6), 1162–1171.

Malmqvist, D. P.-A., Heinicke, G., Karrman, E., Stenstrom, T. A. and Svensson, G. (eds) (2006). *Strategic Planning of Sustainable Urban Water Management*. London, UK: IWA Publishing.

Mara, D. D. (2003). Water, sanitation and hygiene for the health of developing nations. *Public Health*, **117**(6), 452–456.

March, J. G. (1978). Bounded rationality, ambiguity, and the engineering of choice. *The Bell Journal of Economics*, **9**(2), 587–608.

March, J. G. and Olsen, J. P. (1976). *Ambiguity and Choice in Organizations*. Bergen: Universitetsforlaget.

Marks, J. (2004). *Negotiating Change in Urban Water Management: Attending to Community Trust in the Process*. Paper presented at the Water Sensitive Urban Design: Cities as Catchments International Conference, Adelaide, Australia, pp 203–215.

Marsh, R., Rowe, G. and Frewer, L. J. (2001). *A Toolkit for Evaluating Public Participation Exercises, Report to the Department of Health and Safety*. Norwich, UK: Institute of Food Research.

Martin, E. (2003). *La participation publique dans le domaine de la santé au Canada: la régionalisation comme agent de démocratisation?* Québec: Université Laval.

Maslow, A. H. (1943). A theory of human motivation. *Psychological Review*, **50**, 370–396.

Mason, R. (2002). Assessing values in conservation planning: methodological issues and choices. In *Assessing the Values of Cultural Heritage: Research Report*, ed. M. de la Torre. Los Angeles, CA: The Getty Conservation Institute.

Mason, R. O. and Mitroff, I. I. (1981). *Challenging Strategic Planning Assumptions: Theory, Cases, and Techniques*. Chichester, UK: John Wiley and Sons.

Matondo, J. I. (2002). A comparison between conventional and integrated water resources management and planning. *Physics and Chemistry of the Earth*, **27**, 831–838.

Matthews, D. (2004). *Rethinking Systems Thinking: Towards a Postmodern Understanding of the Nature of Systemic Inquiry*. PhD thesis, School of Mathematical Sciences, The University of Adelaide, Australia.

Maurel, P. (2001). Les représentations spatiales: concepts de base et éléments de typologie. In *Représentations spatiales et développement territorial*, ed. S. Lardon, P. Maurel and V. Piveteau. Paris: Hermes, pp 75–108.

Maurel, P. (2003). *Public Participation and the European Water Framework Directive. Role of Information and Communication Tools*, WP3 report of the HarmoniCOP project, deliverable No. 3a. Europe: EC contract No EVK1-CT-2002-00120.

Maurel, P., Cernesson, F., Ferrand, N., Craps, M. and Valkering, P. (2004). *Some Methodological Concepts to Analyse the Role of IC-tools in Social Learning Processes*. Paper presented at the International Environmental Modelling and Software Society, iEMSs 2004 International Conference: Complexity and Integrated Resources Management, 14–17 June 2004, University of Osnabrück, Germany. www.iemss.org/iemss2004/pdf/dss2/maursome.pdf.

Mayer, I. S., Van Daalen, C. E. and Bots, P. W. G. (2004). Perspectives on policy analyses: a framework for understanding and design. *International Journal of Technology, Policy and Management*, **4**(2), 169–191.

Mayer, P. (1986). *Vers un renouvellement méthodologique et épistemologique en recherche en gestion*. Paper presented at the Conference on 'La qualité des informations scientifiques en gestion', Paris, France.

Mays, L. W. (2005). *Water Resources Engineering: 2005 Edition*. New York: John Wiley and Sons Inc.

Mazri, C. (2007). *Apport méthodologique pour la structuration de processus de décision publique en contexte participatif. Le cas des risques industriels majeurs en France*. Paris: Université Paris Dauphine.

McDaniels, T., Gregory, R. and Fields, D. (1999). Democratizing risk management: successful public involvement in an electric utility water management decision. *Risk Analysis*, **19**(3), 491–504.

McIntyre-Mills, J. J. (2006). *Rescuing the Enlightenment from Itself: Critical and Systemic Implications for Democracy*. Vol. 1. New York: Springer.

McKinnon, K. (2007). Postdevelopment, professionalism, and the politics of participation. *Annals of the Association of American Geographers*, **97**(4), 772–785.

McMahon, T. A. (1988). Droughts and arid zone hydrology. *Civil Engineering Transactions (Institution of Engineers Australia)*, **CE30**(4), 175–186.

MDBC (2004). *The Cap: Providing Security for Water Users and Sustainable Rivers*. Canberra: Murray Darling Basin Commission.

MEA (2005). *Ecosystems and Human Well-Being: A Framework for Assessment*. Washington DC: Island Press.

Meijerink, S. and Huitema, D. (2010). Policy entrepreneurs and change strategies: lessons from sixteen case studies of water transitions around the globe. *Ecology and Society*, **15**(2), 21. www.ecologyandsociety.org/vol15/iss2/art21/.

Mermet, L. (1992). *Stratégies pour la gestion de l'environnement*. Paris: L'Harmattan.

Meyer-Sahling, J.-H. (2004). Civil service reform in post-communist Europe: the bumpy road to depoliticisation. *West European Politics*, **27**(1), 71–103.

Meyn, S. P. and Tweedie, R. L. (1993). *Markov Chains and Stochastic Stability*. London: Springer-Verlag.

Micklin, P. (2007). The Aral Sea disaster. *Annual Review of Earth and Planetary Sciences*, **35**(1), 47–72. http://arjournals.annualreviews.org/doi/abs/10.1146/annurev.earth.35.031306.140120.

Midgley, G. (1997a). Developing the methodology of TSI: from oblique use of methods to creative design. *Systems Practice*, **10**, 305–319.

Midgley, G. (1997b). Mixing methods: developing systemic intervention. In *Multimethodology: The Theory and Practice of Combining Management Science Methodologies*, ed. J. Mingers and A. Gill. Chichester, UK: Wiley, pp 250–290.

Midgley, G. (2000). *Systemic Intervention: Philosophy, Methodology, and Practice*. New York: Kluwer Academic/Plenum Publishers.

Miller, B. M. and van Senden D. C. (eds) (2003). *Brooklyn Estuary Processes Study*. Manly Vale, NSW, Australia: Water Research Laboratory, The University of New South Wales.

Milly, P. C. D., Betancourt, J., Falkenmark, M. *et al.* (2008). Stationarity is dead: whither water management? *Science*, **319**(5863), 573–574.

Mingers, J. (2001). Multimethodology – mixing and matching methods. In *Rational Analysis for a Problematic World Revisited: Problem Structuring Methods for Complexity, Uncertainty and Conflict*, revised edn, ed. J. Rosenhead and J. Mingers. Chichester, UK: John Wiley and Sons, pp 289–309.

Mingers, J. (2003). A classification of the philosophical assumptions of management science methods. *Journal of the Operational Research Society*, **54**(6), 559–570.

Mingers, J. and Gill, A. (eds) (1997). *Multimethodology: Theory and Practice of Combining Management Science Methodologies*. Chichester, UK: Wiley.

Mintzberg, H. (1979a). Beyond implementation: an analysis of the resistance to policy analysis. In *Operational Research '78*, ed. K. B. Haley. Amsterdam: North Holland, pp 106–162.

Mintzberg, H. (1979b). An emerging strategy of 'direct' research. *Administrative Science Quarterly*, **24**(4), 582–589.

Mintzberg, H. (1989). *Mintzberg on Management. Inside Our Strange World of Organisations*. New York: Free Press.

Mintzberg, H. (1994). *The Rise and Fall of Strategic Planning*. London: Prentice Hall.

Mintzberg, H., Raisinghani, D. and Théorêt, A. (1976). The structure of 'unstructured' decision processes. *Administrative Science Quarterly*, **21**, 246–275.

Miser, H. J. and Quade, E. S. (1988). *Handbook of Systems Analysis: Craft Issues and Procedural Choices*. New York: Wiley.

Mitchell, R. K., Agle, B. R. and Wood, D. J. (1997). Toward a theory of stakeholder identification and salience: defining the principle of who and what really counts. *The Academy of Management Review*, **22**(4), 853–886.

Moellenkamp, S., Lamers, M. and Ebenhoeh, E. (2008). Institutional elements for adaptive water management regimes. Comparing two regional water management regimes in the Rhine basin. In *Adaptive and Integrated Water Management. Coping with Complexity and Uncertainty*, ed. C. Pahl-Wostl, P. Kabat and J. Moeltgen. Berlin: Springer Verlag, pp 147–166.

Moellenkamp, S., Lamers, M., Huesmann, C. *et al.* (2010). Informal participatory platforms for adaptive management. insights into niche-finding: collaborative design and outcomes from a participatory process in the Rhine basin. *Ecology and Society*, **15**(4), 41. www.ecologyandsociety.org/vol15/iss4/art41/.

Moglia, M., Perez, P. and Burn, S. (2008). Urbanization and water development in the Pacific Islands. *Development*, **51**, 49–55.

Mohammed, S., Klimoski, R. and Rentsch, J. R. (2000). The measurement of team mental models: we have no shared schema. *Organizational Research Methods*, **3**, 123–165.

Moisdon, J. C. (1984). Recherche en gestion et intervention. *Revue française de gestion*, (Sept.–Oct.), 61–73.

Montibeller, G. N., Ackermann, F., Belton, V. and Ensslin, L. (2001). *Reasoning Maps for Decision Aid: A Method to Help Integrated Problem Structuring and Exploring of Decision Alternatives.* Paper presented at the ORP3 2001, Paris, 26–29 September.

Moore, R. E. (1979). *Method and Application of Interval Analysis.* Philadelphia: SIAM.

Morin, E. (1990). *Introduction à la pensée complexe.* Paris, France: ESF éditeur.

Mostert, E. (2003a). The European Water Framework Directive and water management research. *Physics and Chemistry of the Earth*, **28**(12–13), 523–527.

Mostert, E. (2003b). *Public Participation and the European Water Framework Directive. A Framework for Analysis: Inception Report of the HarmoniCOP Project*, EC contract No EVK1-CT-2002–00120, Deliverable No 1. www.harmonicop.uni-osnabrueck.de/_files/_down/HarmoniCOPinception.pdf.

Muller-Merbach, H. (2002). The cultural roots linking Europe. *European Journal of Operational Research*, **140**(2), 212–224.

Munda, G. (2004). Social multi-criteria evaluation (SMCE): methodological foundations and operational consequences. *European Journal of Operational Research*, **158**(3), 662–677.

Murray, P. (2000). Evaluating participatory extension programs: challenges and problems. *Australian Journal of Experimental Agriculture*, **40**(4), 519–526.

Mustajoki, J., Hämäläinen, R. P. and Marttunen, M. (2004). Participatory multicriteria decision support with Web-HIPRE: a case of lake regulation policy. *Environmental Modelling & Software*, **19**(6), 537–547.

Myšiak, J. and Brown, J. D. (2006). Environmental policy aid under uncertainty. In *Proceedings of the iEMSs Third Biennial Meeting: 'Summit on Environmental Modelling and Software'*, ed. A. Voinov, A. J. Jakeman and A. E. Rizzoli. International Environmental Modelling and Software Society, Burlington, USA, July 2006. CD ROM. www.iemss.org/iemss2006/sessions/all.html.

Myšiak, J., Giupponi, C. and Rosato, P. (2005). Towards the development of a decision support system for water resource management. *Environmental Modelling & Software*, **20**(2), 203–214.

Nakova, K. (2007). Energy efficiency networks in Eastern Europe and capacity building for urban sustainability: experience of two municipal networks. *Indoor and Built Environment*, **16**(3), 248–254.

Nancarrow, B. (2005). *When the Modeller Meets the Social Scientist or Vice-Versa.* Paper presented at the MODSIM 2005 International Congress on Modelling and Simulation. Modelling and Simulation Society of Australia and New Zealand, December 2005, Melbourne, Australia. www.mssanz.org.au/modsim05/papers/nancarrow.pdf.

Nandalal, K. D. W. and Simonovic, S. P. (2003). *State-of-the-Art Report on Systems Analysis Methods for Resolution of Conflicts in Water Resources Management.* A report prepared for division of water sciences UNESCO: UNESCO-IHP.

National Driller (1 December 2002). Building from the past: the history of the treatment of drinking water. *National Driller Magazine* www.nationaldriller.com/CDA/Archives/c9275bc6c6197010VgnVCM100000f932a8c0____.

Needham, J. (1959). Review of KA Wittfogel's oriental despotism: a comparative study of total power. *Science and Society*, **23**, 58–65.

NeWater (2009). *Rhine, Wupper: Results of Workshops: NeWater: New Approaches to Adaptive Water Management under Uncertainty.* www.newater.uos.de/index.php?pid=1102.

Nichols, D., Stewart, T. and von Hippel, D. (2000). *Planning Approaches, Thematic Review V.1, Prepared as an Input to the World Commission on Dams.* Cape Town, South Africa.

Nicholson-Cole, S. A. (2005). Representing climate change futures: a critique on the use of images for visual communication. *Computers, Environment and Urban Systems*, **29**, 255–273.

Nonaka, I. and Takeuchi, H. (1995). *The Knowledge-Creating Company: How Japanese Companies Create the Dynamics of Innovation.* New York: Oxford University Press.

Nonaka, I. and Toyama, R. (2003). The knowledge-creating theory revisited: knowledge creation as a synthesizing process. *Knowledge Management Research & Practice*, **1**(1), 2–10.

Novak, J. D. and Cañas, A. J. (2006). *The Theory Underlying Concept Maps and How to Construct Them.* Technical Report IHMC CmapTools 2006–01: Florida Institute for Human and Machine Cognition. http://cmap.ihmc.us/Publications/ResearchPapers/TheoryCmaps/TheoryUnderlyingConceptMaps.htm.

NSW Government (1992). *Estuary Management Manual.* NSW, Australia: NSW Government.

Nutt, P. C. (1984). Types of organizational decision processes. *Administrative Science Quarterly*, **29**(3), 414–450.

Nutt, P. C. (2005). Decision aiding: search during decision making. *European Journal of Operational Research*, **160**, 851–876.

NWC (2007). *Australian Water Resources 2005: A Baseline Assessment of Water Resources for the National Water Initiative.* National Water Commission, Australian Government. www.water.gov.au/Publications/AWR2005_Level_2_Report_May07.pdf.

OCDE (2001). *Des citoyens partenaires: manuel de l'OCDE sur l'information, la consultation et la participation à la formulation des politiques publiques* (English version: *OECD Handbook on Citizens as Partners: Information, Consultation and Participation in Policy-Making*). Paris, France: Organisation de Coopération et de Développement Économiques. www.keepeek.com/Digital-Asset-Management/oecd/governance/des-citoyens-partenaires_9789264295575-fr.

Ohlsson, L. (2000). Water conflicts and social resource scarcity. *Physics and Chemistry of the Earth, Part B*, **25**(3), 213–220.

Olsson, P., Folke, C. and Berkes, F. (2004). Adaptive co-management for building resilience in socio-ecological systems. *Environmental Management*, **34**, 75–90.

Olsson, P., Gunderson, L. H., Carpenter, S. R. *et al.* (2006). Shooting the rapids: navigating transitions to adaptive governance of social-ecological systems. *Ecology and Society*, **11**(1), 18. www.ecologyandsociety.org/vol11/iss1/art18/.

Ostanello, A. and Tsoukiàs, A. (1993). An explicative model of 'public' interorganizational interactions. *European Journal of Operational Research*, **70**, 67–82.

Ostrom, E. (1990). *Governing the Commons: The Evolution of Institutions for Collective Action.* New York: Cambridge University Press.

Ostrom, E. (1996). Crossing the great divide: coproduction, synergy, and development. *World Development*, **24**(6), 1073–1087.

Ostrom, E. (2005). *Understanding Institutional Diversity.* Princeton, NJ: Princeton University Press.

O'Toole Jr., L. J. (1986). Policy recommendations for multi-actor implementation: an assessment of the field. *Journal of Public Policy*, **6**(2), 181–210.

Page, S. E. (2007). *The Difference: How the Power of Diversity Creates Better Groups, Firms, Schools and Societies.* Princeton, NJ: Princeton University Press.

Pahl-Wostl, C. (2002). Towards sustainability in the water sector – the importance of human actors and processes of social learning. *Aquatic Sciences*, **64**, 394–411.

Pahl-Wostl, C. (2007). Transitions towards adaptive management of water facing climate and global change. *Water Resources Management*, **21**, 49–62.

Pahl-Wostl, C. and Hare, M. (2004). Processes of social learning in integrated resource management. *Journal of Community & Applied Social Psychology*, **14**(3), 193–206.

Pahl-Wostl, C., Kabat, P. and Möltgen, J. (eds) (2008). *Adaptive and Integrated Water Management: Coping with Complexity and Uncertainty.* Berlin: Springer Verlag.

Palmer, R. N., Keyes, A. M. and Fisher, S. (1993). *Empowering Stakeholders Through Simulation in Water Resources Planning.* Paper presented at the Water management in the '90s: a time for innovation, ASCE Annual Conference, 1–5 May 1993, Seattle, WA.

Parker, D. C., Manson, S. M., Janssen, M. A., Hoffmann, M. J. and Deadman, P. (2003). Multi-agent systems for the simulation of land-use and land-cover change: a review. *Annals of the Association of American Geographers*, **93**(2), 314–337.

Pateman, C. (1970). *Participation and Democratic Theory*. Cambridge, UK: Cambridge University Press.

Pawlak, Z. (1991). *Rough Sets: Theoretical Aspects of Reasoning about Data*. Dordrecht: Kluwer Academic Publishers.

Pearce, F. (2004). *Keepers of the Spring: Reclaiming Our Water in an Age of Globalisation*. Washington DC: Island Press.

Pease, A. and Pease, B. (2005). *The Definitive Book of Body Language: How to Read Others' Thoughts by their Gestures*. London: Orion Publishing Group.

Pease, B. and Pease, A. (2000). *Why Men Don't Listen and Women Can't Read Maps: How We're Different and What to Do about It*. New York: Welcome Rain.

Peev, E. (1995). Separation of ownership and control in transition: the case of Bulgaria. *Europe–Asia Studies*, **47**(5), 859–875.

Penman, R. (1988). *Evaluation of ACT Water Conservation Campaign, Summer 1988*. Communication Research Institute of Australia Incorporated.

Perez, P. and Batten, D. (eds) (2006). *Complex Science for a Complex World: Exploring Human Ecosystems with Agents*. Canberra, Australia: ANU E-Press.

Perrow, C. (1984). *Normal Accidents: Living with High Risk Technologies*. New York: Basic Books.

Pfeffer, J. and Salancik, G. R. (1974). Organizational decision making as a political process: the case of a university budget. *Administrative Science Quarterly*, **19**(2), 135–151.

Piaget, J. (1967a). *Biologie et connaissance*. Paris: Gallimard.

Piaget, J. (1967b). *Logique et connaissance scientifique*. Paris: Gallimard.

Pimentel, D., Houser, J., Preiss, E. *et al.* (1997). Water resources: agriculture, the environment, and society. *Bioscience*, **47**(2), 97–106.

Pinkney, D. H. (1955). Napoleon III's transformation of Paris: the origins and development of the idea. *Journal of Modern History*, **27**(2), 125–134.

PMSU (2004). *Strategy Survival Guide*. London, UK: Prime Minister's Strategy Unit, UK Government. http://interactive.cabinetoffice.gov.uk/strategy/survivalguide/index.htm.

Polya, G. (1957). *How to Solve It: A New Aspect of Mathematical Method*. Princeton, NJ: Princeton University Press.

Pomerol, J.-C. (1997). Artificial intelligence and human decision making. *European Journal of Operational Research*, **99**(1), 3–25.

Poncelet, E. C. (2001). Personal transformation in multistakeholder environmental partnerships. *Policy Sciences*, **34**, 273–301.

Poteete, A. R., Janssen, M. A. and Ostrom, E. (2010). *Working Together: Collective Action, the Commons, and Multiple Methods in Practice*. Princeton, NJ: Princeton University Press.

Poujol, S. (2004). *Conception d'une infrastructure de modélisation participative par des systèmes multi agents – ou du tricotage de modèles par nos grand mères*. Rapport du stage (work experience research report). Montpellier, France: Cemagref.

Poussin, J.-C. (1987). Notions de système de modèle. *Cahier des sciences humaines*, **23**(3–4), 439–441.

Powell, J. M. (2002). Environment and institutions: three episodes in Australian water management, 1880–2000. *Journal of Historical Geography*, **28**(1), 100–114.

Pretty, J. (2003). Social capital and the collective management of resources. *Science*, **302**(5652), 1912–1914.

Priscoli, J. D. (1990). *Public Involvement; Conflict Management; and Dispute Resolution in Water Resources and Environmental Decision Making*. Alternative Dispute Resolution Series, working paper No 2. Fort Belvoir, VA: Army Engineer Institute for Water Resources.

Priscoli, J. D. (1998). *Water and Civilisation: Conflict, Cooperation and the Roots of a New Eco-Realism*, A keynote address for the 8th Stockholm Water Symposium, 10–13 August 1998. Stockholm, Sweden.

Pruitt, D. G. and Rubin, J. Z. (1986). *Social Conflict: Escalation, Stalemate, and Settlement*. New York: Random House.

Quaddus, M. A. and Siddique, M. A. B. (2001). Modelling sustainable development planning: a multicriteria decision conferencing approach. *Environment International*, **27**(2–3), 89–95.

Rahaman, M. M., Varis, O. and Kajander, T. (2004). EU water framework directive vs. integrated water resources management: the seven mismatches. *Water Resources Development*, **20**(4), 565–575.

Ramirez, R. (1999). Stakeholder analysis and conflict management. In *Cultivating Peace: Conflict and Collaboration in Natural Resource Management*, ed. D. Buckles. Ottawa, Canada: International Development Research Centre/World Bank Institute, pp 101–126.

Raskin, P., Gleick, P. H., Kirshen, P., Pontius, G. and Strzepek, K. (1997). *Comprehensive Assessment of the Freshwater Resources of the World. Water Futures: Assessment of Long-Range Patterns and Problems*. Stockholm: Stockholm Environment Institute.

Ravetz, J. R. (1971). *Scientific Knowledge and its Social Problems*. Oxford, UK: Oxford University Press.

Rayner, S. (2007). The rise of risk and the decline of politics. *Environmental Hazards*, **7**(2), 165–172.

Reisner, M. (1986). *Cadillac Desert: The American West and its Disappearing Water*. New York: Viking Penguin Inc.

Renn, O., Webler, T. and Wiedemann, P. (eds) (1995). *Fairness and Competence in Citizen Participation. Evaluating Models for Environmental Discourse*. Dordrecht: Kluwer.

Ribarova, I., Assimacopoulos, D., Balzarini, A. *et al.* (2006). *AquaStress Case Study Iskar*. Report of the JWT. Presented to the PSG on the 29th of June, 2006. Brussels, Belgium.

Ribarova, I., Assimacopoulos, D., Jeffery, P. *et al.* (2008a). Research-supported participatory planning and its application for water stress mitigation. *Journal of Environmental Planning and Management*, **54**(2), 283–300.

Ribarova, I., Ninov, P. I., Daniell, K. A., Ferrand, N. and Hare, M. (2008b). *Integration of Technical and Non-Technical Approaches for Flood Identification*. Paper presented at the Proceedings of the Water Down Under 2008 International Conference, 14–17 April, 2008, Adelaide, Australia.

Richard-Ferroudji, A. (2008). *L'appropriation des dispositifs de gestion locale et participative de l'eau: composer avec une pluralité de valeurs, d'objectifs et d'attachements*. Doctoral thesis, Ecole des Hautes Etudes en Sciences Sociales, Paris, France.

Richardson, G. P., Andersen, D. F., Rohrbaugh, J. W. and Steinhurst, W. (1992). Group model building. *Proceedings of the 1992 International System Dynamics Conference*, System Dynamics Society: Utrecht, Albany, NY.

Rijsberman, F. R. (2006). Water scarcity: fact or fiction? *Agricultural Water Management*, **80**(1–3), 5–22.

Rinaudo, J.-D. and Garin, P. (2003). *An Operational Methodology to Analyse Conflicts Over Water Used at the River Basin*, 54th International Executive Council of ICID, 20th ICID European Conference. Montpellier, France.

Rittel, H. and Webber, M. (1973). Dilemmas in a general theory of planning. *Policy Sciences*, **4**, 155–169.

Rivas, J. M. (19 June 2007). Poll suggests that Hurricane Katrina may overshadow 9/11 events for 2008 election. *Associated Content*. www.associatedcontent.com/article/286388/poll_suggests_that_hurricane_katrina.html.

Rixon, A., Robinson, P. and Boschetti, F. (2006). *The Meme as a Design Pattern for Social Learning in Agent-Based Models*. www.per.marine.csiro.au/staff/Fabio.Boschetti/papers/memes.pdf.

Rocha, E. M. (1997). A ladder of empowerment. *Journal of Planning Education and Research*, **17**(1), 31–44.

Rose, D. B. (1996). *Nourishing Landscapes: Australian Aboriginal Views of Landscape and Wilderness*. Canberra: National Heritage Commission.

Rosenhead, J. (1980). Planning under uncertainty: II. A methodology for robustness analysis. *Journal of the Operational Research Society*, **31**(4), 331–341.

Rosenhead, J. (2001). Robustness analysis: keeping your options open. In *Rational Analysis for a Problematic World Revisited: Problem Structuring Methods for Complexity, Uncertainty and Conflict*, revised edn, ed. J. Rosenhead and J. Mingers. Chichester, UK: John Wiley and Sons, Ltd, pp 181–207.

Rosenhead, J. and Mingers, J. (2001a). An overview of related methods: VSM, system dynamics, and decision analysis. In *Rational Analysis for a Problematic World Revisited: Problem Structuring Methods for Complexity, Uncertainty and Conflict*, revised edn, ed. J. Rosenhead and J. Mingers. Chichester, UK: John Wiley and Sons, pp 267–288.

Rosenhead, J. and Mingers, J. (eds) (2001b). *Rational Analysis for a Problematic World Revisited: Problem Structuring Methods for Complexity, Uncertainty and Conflict*, revised edn. Chichester, UK: John Wiley and Sons, Ltd.

Rothschild, L. J. and Mancinelli, R. L. (2001). Life in extreme environments. *Nature*, **409**(6823), 1092–1101.

Rougier, J.-E. (2006). *Quelles modalités de participation des acteurs à la gestion locale de l'eau? Réflexion sur trois cas européens.* Montpellier, France: Thèse Professionnelle ISIGE6.

Rougier, J.-E. (2007). *Living with Floods and Drought: AquaStress Project Bulgarian Test Site – Case Study 3.* Internal AquaStress Project meeting presentation and report, 18 April 2007, Montpellier, France.

Rousseau, D., Verdonck, F., Moerman, O. *et al.* (2001). Development of a risk assessment based technique for design/retrofitting of WWTPs. *Water Science and Technology*, **43**(7), 287–294.

Rousseau, L. and Deffuant, G. (2005). Gestion des territories: aider à la formulation collective de problèmes. *Natures Sciences Sociétés*, **13**, 21–33.

Rowe, A. J. and Davis, S. A. (1996). *Intelligent Information Systems: Meeting the Challenge of the Knowledge Era.* Westport, CT: Greenwood Publishing Group.

Rowe, G. and Frewer, L. J. (2000). Public participation methods: a framework for evaluation. *Science, Technology and Human Values*, **25**(1), 3–29.

Rowe, G. and Frewer, L. J. (2004). Evaluating public-participation exercises: a research agenda. *Science, Technology and Human Values*, **29**(4), 512–556.

Roy, B. (1985). *Méthodologie multicritere d'aide à la decision.* Paris: Economica.

Roy, B. (1993). Decision science or decision-aid science? *European Journal of Operational Research*, **66**(2), 184–203.

Roy, B. (2006). A process model of integral theory. *Integral Review*, **3**, 118–152.

Russo, J. E. and Schoemaker, P. J. H. (1989). *Decision Traps: The Ten Barriers to Decision-Making & How to Overcome Them.* New York: Fireside.

Ryan, S. (31 December 2007). Farmers' aid flows on after drought. *The Australian.*

SA Water (2008). *Water Systems – Pipelines.* Adelaide: SA Water. www.sawater.com.au/SAWater/Education/OurWaterSystems/Pipelines.htm.

Saaty, T. L. (1980). *The Analytic Hierarchy Process.* New York: McGraw-Hill.

Sabatier, P. A. and Jenkins-Smith, H. C. (eds) (1993). *Policy Change and Learning: An Advocacy Coalition Approach.* Boulder, CO: Westview Press.

SAGE LMEP (1999). *Présentation du périmetre du SAGE Lez-Mosson-Etangs-Palavasiens.* Région Languedoc-Roussillon, France: Conception – Commission Locale de l'Eau/Anne Roux, Réalisation – SIEE/Dom. www.languedoc-roussillon.ecologie.gouv.fr/eau/sage/lez/intro.htm.

Saha, M. and Alex, B. (2004). Oceanographic survey for Kalpasar. *Project Monitor: India's First Newspaper on Projects.* www.projectsmonitor.com/detailnews.asp?newsid=7473.

SCEH (2000). *Co-ordinating Catchment Management: Report of the Inquiry into Catchment Management.* Canberra: The Parliament of the Commonwealth of Australia, House of Representatives, Standing Committee on Environment and Heritage.

Schein, E. H. (1987). *The Clinical Perspective in Fieldwork.* London, UK: Sage Publications.

Schein, E. H. (1999). *Process Consultation Revisited: Building the Helping Relationship.* Reading, MA: Addison Wesley Longman, Inc.

Schneider, A. and Ingram, H. (2007). *Ways of Knowing: Implications for Public Policy*, Paper presented at the annual meeting of the American Political Science Association, 29 August – 2 September 2007, Chicago.

Schoemaker, P. J. H. (1982). The expected utility model: its variants, purposes, evidence and limitations. *Journal of Economic Literature*, **20**, 529–563.

Scholl, H. J. (2001). *Agent-Based Versus System Dynamics Modeling.* Proceedings of the 34th Hawaiian International Conference on System Sciences (HICSS-34), January 2001, Maui, HI, USA.

Scholl, H. J. (July 2004). *Can SD Models Have Greater Relevance to Practice When Used Within Participatory Action Research Designs?* Paper presented at the ISDC 2004 22nd International System Dynamics Conference, Oxford, UK.

Schon, D. A. (1987). *Educating the Reflective Practitioner: Toward a New Design for Teaching and Learning in the Professions.* San Francisco, California: Jossey-Bass.

Schumacher, E. F. (1977). *A Guide for the Perplexed.* New York: Harper & Row.

Seckler, D., Amarasinghe, U., Molden, D., de Silva, R. and Barker, R. (1998). *World Water Demand and Supply, 1990 to 2025: Scenarios and Issues*, Research report No 19. Colombo, Sri Lanka: International Water Management Institute.

Seecon Deutschland GmbH (2008). *Partizipative Planung von Maßnahmen zur Gewässerentwicklung. Ergebnisdokument der Workshops, Untere Dhünn', 7 April 2008.* Osnabrück/Wuppertal: Seecon Deutschland GmbH, Institut für Umweltsystemforschung der Universität Osnabrück (USF), and Wupperverband (WV). http://www.wupperverband.de/internet/wupperverbandwys.nsf/2d475fe384f2de43c1256f3c002e68be/4867509108a5e5efc1257426002e64e2/$FILE/ErgebnisdokumentWorkshopsUntereDhuenn_final.pdf.

Selznick, P. (1948). Foundations of the theory of organisations. *American Sociological Review*, **13**, 25–35.

Senge, P. M. (1990). *The Fifth Discipline: The Art and Practice of The Learning Organisation.* New York: Currency Doubleday.

Sexton, M. (2003). *Silent Flood: Australia's Salinity Crisis.* Sydney, Australia: ABC Books.

Shalev, Z. (January 1992). *Draft 10 Year National Water Master Plan.* Betio, Tarawa:, Ministry Works and Energy and UNDTCP.

Sharifi, M. A. (2003). Integrated planning and decision support systems for sustainable water resources management: concepts, potentials and limitations, *Seminar on Water Resources Management for Sustainable Agricultural Productivity.* Lahore, Pakistan.

Shaw, A., Sheppard, S., Burch, S. *et al.* (2009). Making local futures tangible: synthesizing, downscaling, and visualizing climate change scenarios for participatory capacity building. *Global Environmental Change Part A*, **19** (4), 447–463.

Sheitanov, N. (1994). Сексуална философия на българина, В: Защо сме такива? В търсне на българската културна идентичност. Съст. 4., В. Еленков, Р. Димитров. Изд. Просвета. София стр. 282.

Shiklomanov, I. A. (1999). *World Water Resources and Their Use.* http://webworld.unesco.org/water/ihp/db/shiklomanov/.

Siebenhüner, B. and Barth, V. (2005). The role of computer modelling in participatory integrated assessments. *Environmental Impact Assessment Review*, **25**(4), 367–389.

Simeon, R. (1976). Studying public policy. *Canadian Journal of Political Science/Revue Canadienne de science politique*, **9**(4), 548–580.

Simon, H. A. (1954). A behavioural model of rational choice. *Quarterly Journal of Economics*, **69**, 99–118.

Simon, H. A. (1960). *The New Science of Management Decision.* New York: Harper and Row.

Simon, H. A. (1973). The structure of ill-structured problems. *Artificial Intelligence*, **4**, 181–201.

Simon, H. A. (1977). *The New Science of Management Decision, Revised Edition.* Englewood Cliffs, NJ: Prentice-Hall.

Simon, H. A. (1996). *The Sciences of the Artificial*, 3rd edn. Cambridge, MA: MIT Press.

Simpson, A. R., Murphy, L. J. and Dandy, G. C. (1993). *Pipe Network Optimisation Using Genetic Algorithms.* Paper presented at the Proceedings of the ASCE Water Resources Planning and Management Division's 20th Anniversary Conference, 'Water Management in the'90s: A Time for Innovation', 1–5 May 1993, Seattle, WA.

SIRMA (2011). *La Mitidja.* www.eau-sirma.net/la_recherche/les-terrains/algerie/la-mitidja.

SJB Planning (2005). *Hornsby Shire Waterways Review.* Sydney: SJB Planning Pty, Ltd.

SKM (2004). *Monitoring, Evaluation and Reporting Framework for Gippsland NRM*, Report for the West Gippsland Catchment Management Authority. SKM Ref: WC02773. Melbourne: Sinclair Knight Merz.

SLA (29 March 2007). *DECISIONARIUM: Global Space for Decision Support.* www.decisionarium.tkk.fi/index-new.html.

Slovic, P. (2000). *The Perception of Risk.* London: Earthscan Publications, Ltd.

Smith, D. I. and Handmer, J. W. (1991). Water conflict and resolution: a case study of dams in southwest Tasmania. In *Negotiating Water: Conflict Resolution in Australian Water Management*, ed. J. W. Handmer, A. H. J. Dorcey and D. I. Smith. Canberra, ACT: Centre for Resource and Environmental Studies, The Australian National University, pp 47–62.

Soncini-Sessa, R., Castelletti, A. and Weber, E. (24–27 June 2002). Participatory decision making in reservoir planning. *Proceedings of the First Biennial Meeting of the International Environmental Modelling and Software Society on 'Integrated Assessment and Decision Support'*, Vol. 1. Lugano, Switzerland. www.iemss.org/iemss2002/proceedings/pdf/volume%20tre/soncini.pdf.

SP AusNet (2006). *Risk Management Framework.*

Speil, K., Rotter, S., Interwies, E. and Moellenkamp, S. (2008). Systematische Gestaltung eines Partizipationsprozesses: Ziele, Methoden und

Herausforderungen am Bespiel der 'Workshops Untere Dhünn'. UVP-Report. *Special Issue on Water Management, Umweltprüfung und wasserwirtschaftliche Planung*, **22**(1/2), 36–41.

Spicer, D. P. (1998). Linking mental models and cognitive maps as an aid to organisational learning. *Career Development International*, **3**(3), 125–132.

Srdjevic, B. and Medeiros, Y. D. P. (2007). Fuzzy AHP assessment of water management plans. *Water Resources Research*. www.springerlink.com/content/c6077063v78462j4/.

Staddon, C. (1999). Localities, natural resources and transition in Eastern Europe. *Geographical Journal*, **165**(2), 200–208.

Staddon, C. and Cellarius, B. (2002). Paradoxes of conservation and development in postsocialist Bulgaria: recent controversies. *European Environment*, **12**, 105–116.

Standards Australia (2004). *AS/NZS 4360:2004, Risk Management; and Companion Handbook HB 436:2004*. Sydney, Australia: Standards Australia.

Standards Australia (2006). *HB 203:2006, Environmental risk management – Principles and processes*. Sydney, Australia: Standards Australia.

Stanners, D. and Bourdeau, P. (eds) (1995). *Europe's Environment: The Dobris Assessment*: European Environmental Agency.

Steiner, G. A. (1969). *Top Management Planning*. New York: Macmillan.

Stephenson, K. (2003). *The What and Why of Shared Vision Planning for Water Supply*. Paper prepared for the panel session 'Collaborative Water Supply Planning: A Shared Vision Approach for the Rappahannock River Basin', Universities Council on Water Resources, Water Security in the 21st Century Conference, 30 July 2003, Washington, DC.

Sterman, J. D. (1989). Modeling managerial behavior: misperceptions of feedback in a dynamic decision making experiment. *Management Science*, **35**(3), 321–339.

Stern, P. J. and Fineberg, H. V. (eds) (1996). *Understanding Risk: Informing Decisions in a Democratic Society*. Washington, DC: National Academy Press.

Stevens, L. A. (1974). *Clean Water – Nature's Way to Stop Pollution*. New York: EP Dutton.

Stewart, T. A. (ed.) (Spring 2008). Leading high-performance teams. *Harvard Business Review OnPoint*.

Stewart, W. J. (1994). *Introduction to the Numerical Solution of Markov Chains*. Princeton, NJ: Princeton University Press.

Steyaert, P. and Ollivier, G. (2007). The European Water Framework Directive: how ecological assumptions frame technical and social change. *Ecology and Society*, **12**(1), 25. www.ecologyandsociety.org/vol12/iss1/art25/.

Stirling, A. (1999). Risk at a turning point? *Journal of Environmental Medicine*, **1**, 119–126.

Stradspan Ltd (2007). *Strategic Decision Support*. www.btinternet.com/~stradspan/.

Stringer, J. (1967). Operational research for 'multi-organizations'. *Operational Research Quarterly*, **18**(2), 105–120.

Strömgren, O. (2003). *The Importance of Evolutionary Epistemologies for Exploring the Limits of the Asian-Pacific Nation State*. Ritsumeikan Asia Pacific Conference. 27–28 November 2003.

Sullivan, C. A. (2001). The potential for calculating a meaningful water poverty index. *Water International*, **26**, 471–480.

Sullivan, C. A. and Meigh, J. (2007). Integration of the biophysical and social sciences using an indicator approach: Addressing water problems at different scales. *Water Resources Management*, **21**, 111–128.

Sullivan, C. A., Manez, M., Schmidt, S. *et al.* (2007). *Report on Indicators for Water Stress*, D 2.1–3. Europe: AquaStress IP, FP6.

Sultana, F. (2007). Reflexivity, positionality and participatory ethics: negotiating fieldwork dilemmas in international research. *ACME: An International E-Journal for Critical Geographies*, **6**(3), 374–385.

Sundar, N. (2000). Unpacking the 'joint' in joint forest management. *Development and Change*, **31**(1), 255–279.

Susman, G. I. and Evered, R. D. (1978). An assessment of the scientific merits of action research. *Administrative Science Quarterly*, **23**(4), 582–603.

Svejnar, J. (2002). Transition economies: performance and challenges. *Journal of Economic Perspectives*, **16**(1), 3–28.

Syme, G. J. and Sadler, B. S. (1994). Evaluation of public involvement in water resources planning: a researcher-practitioner dialogue. *Evaluation Review*, **18**(5), 523–542.

Taket, A. R. (1992). Review of 'Creative problem solving: total systems intervention'. *Journal of the Operational Research Society*, **43**(10), 1013–1016.

Taket, A. R. and White, L. A. (1993). After OR: an agenda for postmodernism and poststructuralism in OR. *Journal of the Operational Research Society*, **44**(9), 867–881.

Taket, A. R. and White, L. A. (1996). Pragmatic pluralism: an explication. *Systems Practice*, **9**(6), 571–586.

Talbot, M. (1993). *Mysticism and the new physics: revised edition*. London, UK: Penguin Arkana.

Tan, K. S. and Rhodes, B. G. (2008). *Implications of the 1997–2006 Drought on Water Resources Planning for Melbourne*, Water Down Under 2008. Adelaide Convention Centre, Adelaide.

Tan, P. L., Jackson, S., Oliver, P. *et al.* (2008). *Collaborative Water Planning: Context and Practice Literature Review*, TRaCK: Land and Water Australia. http://lwa.gov.au/files/products/track/pn21213/pn21213.pdf.

Tanev, T. A. (2001). Emerging from post-communist chaos: the case of Bulgaria. *International Journal of Public Administration*, **24**(2), 235–248.

Taylor, F. W. (2003). *Scientific Management*. New York: Routledge.

Tharenou, P., Donohue, R. and Cooper, B. (2007). *Research Management Methods*. Port Melbourne, Australia: Cambridge University Press.

Theesfeld, I. and Boevsky, I. (2005). Reviving pre-socialist cooperative traditions: the case of water syndicates in Bulgaria. *Sociologia Ruralis*, **45**(3), 171–186.

Thomas, E. J. and Rothman, J. (eds) (1994). *Intervention Research: Design and Development for Human Service*. New York: The Harworth Press Incorporated.

Thomas, I. and Elliott, M. (2005). *Environmental Impact Assessment in Australia: Theory and Practice*, 4th edn. Leichhardt, NSW: The Federation Press.

Thomas, K. W. (1976). Conflict and conflict management. In *Handbook of Industrial and Organizational Psychology*, ed. M. D. Dunnette. Chicago: Rand McNally, pp 889–935.

Thomas, K. W. (1992). Conflict and conflict management: reflections and update. *Journal of Organizational Behavior*, **13**(3), 265–274.

Thomas, R. L. (2004). Management of freshwater systems: the interactive roles of science, politics and management, and the public. *Lakes and Reservoirs: Research and Management*, **9**(1), 65–73.

Thomasson, F. (2004). *Water and Local Conflict: A Brief Review of the Academic Literature and Other Sources*. Stockholm, Sweden: Conflict and Water Group, Swedish Water House. www.swedishwaterhouse.se/swh/resources/20050425162906Water_and_Local_Conflict.pdf.

Thomson, A. M. and Perry, J. L. (2006). Collaboration processes: inside the black box. *Public Administration Review*, **66**, 20–32.

Ticehurst, J. L., Rissik, D., Letcher, R. A., Newham, L. H. T. and Jakeman, A. J. (2005). Development of decision support tools to assess the sustainability of coastal lakes. *Proceedings of MODSIM conference, Melbourne, 12–15 December 2005*.

Ticehurst, J. L., Newham, L. T. H., Rissik, D., Letcher, R. A. and Jakeman, A. J. (2007). A Bayesian network approach for assessing the sustainability of coastal lakes in New South Wales, Australia. *Environmental Modelling & Software*, **22**(8), 1129–1139.

Toeman, Z. and Thompson, L. (1950). Action research. *Scientific Monthly*, **70**(5), 345–346. www.jstor.org/pss/19961.

Torbert, W. R. (1976). *Creating a Community of Inquiry: Conflict, Collaboration, Transformation*. London: Wiley. http://eric.ed.gov/ERICWebPortal/contentdelivery/servlet/ERICServlet?accno=ED103551.

Tordjman, G. (1996). Une morale de l'inadvertance. In *Sun Tzu – L'art de la guerre: les treize articles*. Vol. 122. Paris: Editions mille et une nuits. Translated from Chinese by Le Père Amiot.

Torrieri, F., Concilio, G. and Nijkamp, P. (2002). a scenario approach to risk management of the Vesuvio Area in Naples, Italy. *Decision Support Tools for Urban Contingency Policy*, **10**(2), 95–112.

Travis, J. (2005). Hurricane Katrina: scientists' fears come true as hurricane floods New Orleans (news article). *Science*, **309**, 1656–1659.

Tsoukas, H. (1993). The road to emancipation is through organizational development: a critical evaluation of total systems intervention. *Systems Practice*, **6**(1), 53–70.

Tsoukiàs, A. (2007). On the concept of decision aiding process: an operational perspective. *Annals of Operations Research*, **154**(3), 3–27.

Tung, C.-P. and Chou, C.-A. (2002). Application of tabu search to ground water parameter zonation. *Journal of the American Water Resources Association*, **38**(4), 1115–1125.

UEC (2006). *Mt Arthur Coal Environmental Assessment for Proposed Exploration Adit: Umwelt Environmental Consultants*. www.umwelt.com.au/mtarthurcoal-adit/.

Ulrich, W. (1983). *Critical Heuristics of Social Planning: A New Approach to Practical Philosophy*. Berne, Switzerland: Haupt Academic Publishers.

Ulrich, W. (1991). Critical heuristics of social systems design. In *Critical Systems Thinking: Directed Readings*, ed. R. L. Flood and M. C. Jackson. Chichester, UK: John Wiley and Sons, pp 103–115.

United Nations (2005). *World Urbanization Prospects: The 2005 Revision*. www.un.org/esa/population/publications/WUP2005/2005wup.htm.

UNDP (2003). *Human Development Report 2003*. New York: United Nations Development Programme.

UNDP (2006). *Beyond Scarcity: Power, Poverty and the Global Water Crisis, Human Development Report 2006*. New York: United Nations Development Programme.

UNESCO (2006). *Non-Renewable Groundwater Resources: A Guidebook on Socially-Sustainable Management for Water Policy-Makers*. IHP-IV, Series on Groundwater No 10. Paris, France: UNESCO.

UNESCO-WWAP (2003). *Water for People, Water for Life – UN World Water Development Report*. Paris, France: UNESCO/World Water Assessment Program.

UNITAR (2005). *Guidance on Action Plan Development for Sound Chemicals Management: Draft Guidance Document for UNITAR/UNDP/GEF Action Plan Project Review*. Geneva: United Nations Institute for Training and Research.

United Nations (1977). *Report of the United Nations Water Conference*, Mar del Plata, March 14–25 1977, E.77.II.A.12. New York: United Nations.

United Nations (1982). *Les organismes des nations unies et l'eau: note d'information à l'intention des coordinnateurs residents/représentants résidents, des représentants dans les pays et des directeurs de projets associés à diverses organisations, SC.83/WS/61*, Conférence des Nations Unies sur l'Eau 1977. New York: Groupe intersecrétariats du Comité d'administratif de coordination pour les ressources en eau, Organisation des Nations Unies.

United Nations (1992). *Agenda 21, The Earth Summit: The United Nations Conference on Environment and Development*. Rio de Janerio, Brazil: United Nations.

United Nations (1997). *Comprehensive Assessment of the Freshwater Resources of the World*. New York: Commission on Sustainable Development, United Nations.

United Nations (2000). *Ministerial Declaration of the Hague on Water Security in the 21st Century*. Agreed to on Wednesday 22 March 2000 in The Hague, Netherlands.

United Nations (2002). *Substantive Issues Arising in the Implementation of the International Covenant on Economic, Social and Cultural Rights, E/C.12/2002/11*. New York.

United Nations (2003). *Ministerial Declaration of the 3rd World Water Forum: Message from Lake Biwa & Yoda River Basin*. Agreed to on 23 March 2003 in Kyoto, Japan.

United Nations (2006). *4th World Water Forum Declaration*. Agreed to on 22 March 2006, in Mexico City, Mexico.

United Nations (2010). *A/RES/64/292 The Human Right to Water and Sanitation: United Nations General Assembly*. www.scribd.com/doc/38690215/A-RES-64-292-The-Human-Right-to-Water-and-Sanitation.

US Government (23 February 2006). *The Federal Response to Hurricane Katrina: Lessons Learned*. White House, Washington DC: US Government.

UTS/SKM (2005). *Sustainable Total Water Cycle Management Strategy: Institute for Sustainable Futures, University of Technology Sydney and Sinclair Knight Merz*. NSW, Australia: Hornsby Shire Council. www.hornsby.nsw.gov.au/media/documents/environment-and-waste/water-catchments/sustainable-total-water-cycle/Sustainable-Total-Water-Cycle-Management-Strategy-Vol-1-Summary-Document.pdf.

van Asselt, M. B. A., Mellors, J., Rijkens-Klomp, N. *et al.* (2001). *Building Blocks for Participation in Integrated Assessment: A Review of Participatory Methods*. ICIS Working Paper I01-E003. Maastricht, The Netherlands: ICIS.

van Ast, J. A. and Boot, S. P. (2003). Participation in European water policy. *Physics and Chemistry of the Earth*, **28**, 555–562.

van den Belt, M. (ed.) (2004). *Mediated Modeling: A System Dynamics Approach to Environmental Consensus Building*. Washington, DC: Island Press.

van der Brugge, R., Rotmans, J. and Loorbach, D. (2005). The transition in Dutch water management. *Regional Environmental Change*, **5**, 164–176.

van der Sluijs, J. P., Janssen, P. H. M., Petersen, A. C. *et al.* (2004). *RIVM/MNP Guidance for Uncertainty Assessment and Communication: Tool Catalogue for Uncertainty Assessment*, Report No NWS-E-2004-37. Utrecht/Bilthoven: Copernicus Institute & RIVM. www.nusap.net/downloads/toolcatalogue.pdf.

van Rooy, P. J. T. C., de Jong, J., Jagtman, E., Hosper, S. H. and Boers, P. C. M. (1998). Comprehensive approaches to water management. *Water Science and Technology*, **37**(3), 201–208.

Vance, B. W. (2007). Zen and 5 steps to ERM. *Risk Management Magazine*, **54**(4), 36–40.

Varady, R. G. (2004). *Global Water Initiatives: Preliminary Observations on Their Evolution and Significance*. International Specialty Conference: 'Good Water Governance for People & Nature: What Roles for Law, Institutions, Science & Finance?', Dundee, Scotland.

Varela, F. and Maturana, H. (1973). Mechanism and biological explanation. *Philosophy of Science*, **39**, 378–382.

Vasileva, S. (2007). *Technical Evaluation Report (for the Iskar Test Site, Bulgaria)*. AquaStress IP, FP6, Europe.

Vasileva, S. (2008). *Final Report on the Evaluation Activities of the Participatory Processes in Iskar Case Study – Bulgaria*. Europe, AquaStress IP, FP6.

Vassilev, R. (2006). Bulgaria's population implosion. *East European Quarterly*, **40**(1), 71–87.

Vennix, J. A. M. (1996). *Group Model Building: Facilitating Team Learning Using System Dynamics*. Chichester, UK: John Wiley.

Ventana Systems (1995). *Vensim Users' Guide*. Belmont, MA: Ventana Systems Inc.

Vickers, A. (1991). The emerging demand-side era in water management. *American Water Works Association*, **83**(10), 38–43.

Viessman Jr., W. (1998). Water policies for the future: bringing it all together. *Water Resources Update*, Issue No 111(Spring), 104–110. Carbondale, IL: Universities' Council on Water Resources.

Vinck, D. (1999). Les objets intermédiaires dans les réseaux de coopération scientifique: contribution à la prise en compte des objets dans les dynamiques sociales. *Revue française de sociologie*, **XL-2**, 385–414.

Vinck, D. and Jeantet, A. (1995). Mediating and commissioning objects in the sociotechnical process of product design: a conceptual approach. In *Designs, Networks and Strategies*. Vol. 2, ed. D. Maclean, P. Saviotti and D. Vinck. Brussels: EC Directorate General Science R&D.

Vlachos, E. C. (1998). Practicing hydrodiplomacy in the 21st century. *Water Resources Update*, Issue No 111(Spring). Carbondale, IL: Universities' Council on Water Resources.

Vogel, R. M., Tsai, Y. and Limbrunner, J. F. (1998). The regional persistence and variability of annual streamflow in the United States. *Water Resources Research*, **34**(12), 3445–3459.

Voinov, A. and Bousquet, F. (2010). Modelling with stakeholders. *Environmental Modelling & Software*, **25**, 1268–1281.

von Bertalanffy, L. (1950). An outline of general system theory. *British Journal for the Philosophy of Science*, **1**(2), 134–165.

von Glasersfeld, E. (1989). Cognition, construction of knowledge, and teaching. *Synthese*, **80**(1), 121–140. www.vonglasersfeld.com/118.

von Korff, Y., d'Aquino, P., Daniell, K. A. and Bijlsma, R. (2010). Designing participation processes for water management and beyond. *Ecology and Society*, **15**(3), 1. www.ecologyandsociety.org/vol15/iss3/art1/.

WA Water Corporation (2006). *Yearly Streamflow for Major Surface Water Sources: Reduced Inflow to Dams*, Western Australian Water Corporation. www.watercorporation.com.au/D/dams_streamflow.cfm.

Wackernagel, M., Monfreda, C. and Deumling, D. (2002). *Ecological Footprint of Nations: How Much Nature Do They Use? How Much Do They Have?* Sustainability issue brief. Oakland, CA: Redefining Progress.

Walker, B., Carpenter, S., Anderies, J. *et al.* (2002). Resilience management in socio-ecological systems: a working hypothesis for a participatory approach. *Conservation Ecology*, **6**(1), 14. www.consecol.org/vol6/iss1/art14/print.pdf.

Walker, B., Holling, C. S., Carpenter, S. R. and Kinzig, A. (2004). Resilience, adaptability and transformability in social-ecological systems. *Ecology and Society*, **9**(2), 5. www.ecologyandsociety.org/vol9/iss2/art5/.

Walker, J. (2008). Where Goyder made his mark. *The Australian*.

Wang, R. and Li, F. (2008). Eco-complexity and sustainability in China's water management. In *Adaptive and Integrated Water Management: Coping with Complexity and Uncertainty*, ed. C. Pahl-Wostl, P. Kabat and J. Möltgen. Berlin, Germany: Springer Verlag, pp 23–38.

Warner, J. F. and Johnson, C. L. (2007). 'Virtual water' – real people: useful concept or prescriptive tool? *Water International*, **32**(1), 63–77.

Warren, R. L. (1967). The interorganizational field as a focus for investigation. *Administrative Science Quarterly*, **12**(3), 396–419.

Warshall, P. (1995). The morality of molecular water. *Whole Earth Review*, (Spring). http://findarticles.com/p/articles/mi_m1510/is_n85/ai_16816218.

Watson, D. J. (2000). The international resource cities program: building capacity in Bulgarian local governments. *Public Administration Review*, **60**(5), 457–463.

Watson, S. R. and Buede, D. M. (1987). *Decision Synthesis: The Principles and Practice of Decision Analysis*. Cambridge, UK: Cambridge University Press.

Watzlawick, P., Weakland, J. and Fisch, R. (1974). *Change: Principles of Problem Formation and Problem Resolution*. New York: WW Norton & Company Inc.

WCD (2000). *Dams and Development: A New Framework. The Report of the World Commission on Dams*. London: Earthscan Publications, Ltd.

Webber, L. M. and Ison, R. L. (1995). Participatory rural appraisal design: conceptual and process issues. *Agricultural Systems*, **47**(1), 107–131.

Webler, T. (1995). 'Right' discourse in citizen participation: an evaluative yardstick. In *Fairness and Competence in Citizen Participation: Evaluating Models for Environmental Discourse*, ed. O. Renn, T. Webler and P. Wiedemann. Dordrecht: Kluwer Academic Publishers, pp 35–86.

Webler, T., Kastenholz, H. G. and Renn, O. (1995). Public participation in impact assessment: a social learning perspective. *Environmental Impact Assessment Review*, **15**(5), 443–463.

Webler, T., Tuler, S. and Krueger, R. (2001). What is a good public participation process? Five perspectives from the public. *Environmental Management*, **27**(3), 435–450.

Weiss, G. (1999). *Multi-Agent Systems: A Modern Approach to Distributed Artificial Intelligence*. Cambridge, MA: MIT Press.

Wenger, E. (1998). *Communities of Practice: Learning, Meaning, and Identity*. Cambridge, UK: Cambridge University Press.

Wenk Jr., E. and Kuehn, T. J. (1977). Interinstitutional networks in technology delivery systems. In *Science and Technology Policy*. Vol. 153–166, ed. J. Haberer, Lexington, MA: Lexington Books.

Wentworth Group (2006). *Australia's Climate is Changing Australia: the State of Australia's Water*. Wentworth Group of Concerned Scientists. www.wentworthgroup.org/docs/Australias_Climate_is_Changing_Australia.pdf.

Werick, W. (2000). *The Future of Shared Vision Planning*. Paper presented at the ASCE 2000 Joint Conference in Water Resources Engineering and Water Resources Planning and Management, July 30 – August 2, 2000, Section 42, Chapter 4, Minneapolis, MN, USA.

Werick, W. J. and Whipple Jr., W. (1994). *Managing Water for Drought: National Study of Water Management During Drought*. IWR Report 94-NDS-8: US Army Corps of Engineers, Water Resources Support Centre, Institute for Water Resources. www.iwr.usace.army.mil/docs/iwrreports/94nds8.pdf.

Wesselink, A. J. (2007). Flood safety in the Netherlands: the Dutch response to Hurricane Katrina. *Technology in Society*, **29**(2), 239–247.

Whish-Wilson, P. (2002). The Aral Sea environmental health crisis. *Journal of Rural and Remote Environmental Health*, **1**(2), 29–34. www.jcu.edu.au/jrtph/vol/v01whish.pdf.

White, I. (2007). *A Whole-of-Government Approach to Water Policy and Planning*, Final Report, September 2007, Republic Of Kiribati Pilot Project, EU-SOPAC Pacific Programme for Water Governance. Canberra, Australia: Australian National University.

White, I. and Falkland, A. (2010). Management of freshwater lenses on small Pacific islands. *Hydrogeology Journal*, **18**(1), 227–246.

White, I., Falkland, A., Perez, P. *et al.* (2002). An integrated approach to groundwater management and conflict reduction in low coral islands, *Proceedings of the International Symposium on Low-Lying Coastal Areas: Hydrology and Integrated Coastal Zone Management*. Paris: UNESCO, pp 249–256.

White, I., Melville, M., Macdonald, B. *et al.* (2007). From conflicts to wise practice agreement and national strategy: cooperative learning and coastal stewardship in estuarine floodplain management. *Journal of Cleaner Production*, **15**(16), 1545–1558.

White, I., Falkland, A., Metutera, T. *et al.* (2008). Safe water for people in low, small island pacific nations: the rural-urban dilemma. *Development*, **51**(2), 282–287.

White, L. A. and Taket, A. R. (1998). Experience in the practice of one tradition of multimethodology. *Systemic Practice and Action Research*, **11**(2), 153–168.

WHO (2000). *Towards an Assessment of the Socioeconomic Impact of Arsenic Poisoning in Bangladesh*, No WHO/SDE/WSH/00.4. World Health Organisation. www.who.int/water_sanitation_health/dwq/arsenic2/en/index5.html.

WHO (2003). *Guidelines for Safe Recreational Environments, Volume 1: Coastal and Fresh Water*. World Health Organisation. www.who.int/water_sanitation_health/bathing/srwe1/en/.

Whyte, W. F. (ed.) (1991). *Participatory Action Research*. Newbury Park, CA: Sage Publications.

Wiedemann, P. M. and Femers, S. (1993). Public participation in waste management decision making: analysis and management of conflicts. *Journal of Hazardous Materials*, **33**, 355–368.

Wiedemann, P. M., Clauberg, M., Gray, P. C. R. *et al.* (2000). *Risk Communication for Companies: Thriving and Surviving in an Age of Risk*. Germany: MUT.

Wien, J. J. F., Otjens, A. J. and van der Wal, T. (2003). *ICT Tools for Participatory Planning*. Paper presented at the EFITA 2003 Conference, Debrecen, Hungary.

Wilber, K. (2000). *A Theory of Everything: An Integral Vision From Business, Politics, Science and Spirituality*. Boston, MA: Shambhala Publications Incorporated.

Wild River, S. and Healy, S. (2006). *Guide to Environmental Risk Management*. Sydney, Australia: CHH Aust. Ltd.

Witte, E. (1972). Field research on complex decision making processes – the phase theory. *International Studies of Management and Organization*, (Fall), 156–182.

Wittfogel, K. A. (1957). *Oriental Despotism: A Comparative Study of Total Power*. London: Oxford University Press.

Wittgenstein, L. (1953). *Philosophical Investigations*. Oxford, UK: Blackwell.

Wolf, A. T. (2002a). *Atlas of International Freshwater Agreements: United Nations Environment Programme*. www.transboundarywaters.orst.edu/publications/atlas/atlas_zipped.zip.

Wolf, A. T. (ed.) (2002b). *Conflict Prevention and Resolution in Water Systems*. Cheltenham, UK: Elgar.

Wolf, A. T., Kramer, A., Carius, A. and Dabelko, G. D. (2005). Managing water conflict and cooperation. In *State of the World 2005: Redefining World Security*. New York: Worldwatch Institute, pp 80–95.

Wolfe, S. and Brooks, D. B. (2003). Water scarcity: an alternative view and its implications for policy and capacity building. *Natural Resources Forum*, **27**, 99–107.

Wolff, G. H. and Gleick, P. H. (2002). The soft path for water. In *The World's Water: The Biennial Report on Freshwater Resources 2002–2003*, ed. P. H. Gleick, W. C. G. Burns, E. L. Chalecki, *et al.* Washington DC: Island Press.

Wong, P. (2008). *Water for the Future*. Speech to the 4th Annual Australian Water Summit, Sydney Convention and Exhibition Centre, 29–30 April 2008. www.climatechange.gov.au/~/media/Files/minister/previous%20minister/wong/2008/major-speeches/April/sp20080429.pdf.

Wong, T. H. F. and Eadie, M. L. (2000). *Water Sensitive Urban Design: A Paradigm Shift in Urban Design*. The International Water Resources Association for the Xth World Water Congress, 12–16 March 2000, Melbourne.

Woolley, R. N. and Pidd, M. (1981). Problem structuring: a literature review. *Journal of the Operational Research Society*, **32**(3), 197–206.

World Bank (2007). *World Development Indicators*, 3. Environment: World Bank. http://siteresources.worldbank.org/DATASTATISTICS/Resources/WDI07section3-intro.pdf.

Wupperverband (2011). *Planungseinheit Dhünn*. www.wupperverband.de/aufgaben/gewaesser/planungseinheit.dhuenn.html.

WWAP (2007). *Milestones, 1972–1996: From Stockholm to Mexico*, http://www.unesco.org/water/wwap/milestones/.

WWC (2009). *Global Water Framework: Outcomes of the 5th World Water Forum*. Istanbul: World Water Council. www.worldwaterforum5.org/fileadmin/WWF5/Final_Report/GWF.pdf.

Yammarino, F. J. (1994). Indirect leadership: transformational leadership at a distance. In *Improving Organizational Effectiveness through Transformational Leadership*, ed. B. M. Bass and B. J. Avolio. Thousand Oaks, CA: Sage Publications Inc., pp 26–47.

Yeh, W. W. G. (1985). Reservoir management and operations models: a state of the art review. *Water Resources Research*, **21**(12), 1797–1818.

Yin, R. K. (2003). *Case Study Research: Design and Methods*, 3rd edn. Thousand Oaks, CA: Sage Publications.

Yoffe, S. B., Wolf, A. T. and Giordano, M. (2001). Conflict and cooperation over international freshwater resources: Indicators and findings of the basins at risk. *Journal of American Water Resources Association*, **39**(5), 1109–1126.

Yoveva, A., Gocheva, B., Voykova, G., Borissov, B. and Spassov, A. (2000). *Sofia: Urban Agriculture in an Economy in Transition*. RUAF Foundation. www.ruaf.org/system/files?file=Sofia.PDF.

Yu, J. E. (2004). Reconsidering participatory action research for organizational transformation and social change. *Journal of Organisational Transformation and Social Change*, **1**(2–3), 111–141.

Yu, S. (2000). *Ngapa Kunangkul: Living Water. Report on the Aboriginal Cultural Values of Groundwater in the La Grange Sub-Basin*, 2nd edn. Nedlands, Western Australia: Centre for Anthropological Research, University of Western Australia.

Yunkaporta, T. (11 June 2006). *Linear vs. Circular Logic: Conflict Between Indigenous And Non-Indigenous Logic Systems*. http://aboriginalrights.suite101.com/article.cfm/linear_vs_circular_logic.

Yunkaporta, T. (18 September 2007a). *Indigenous Cosmology and Science: Extra-terrestrial life – Indigenous and Western Scientific Viewpoint*. http://australian-indigenous-peoples.suite101.com/article.cfm/indigenous_cosmology_and_science.

Yunkaporta, T. (31 July 2007b). *Indigenous Knowledge Systems: Comparing Aboriginal and Western Ways of Knowing*. http://aboriginalrights.suite101.com/article.cfm/indigenous_knowledge_systems.

Zadeh, L. A. (1978). Fuzzy sets as a basis for a theory of possibility. *Fuzzy Sets and Systems*, **1**, 3–28.

Zagonel, A. A. (2002). Model conceptualization in group model building: a review of the literature exploring the tension between representing reality and negotiating a social order, July 28–August 1 2002. *Proceedings of the International Conference of the System Dynamics Society, Palermo, Italy*.

Zeitoun, M. and Warner, J. (2006). Hydro-hegemony: a framework for analysis of transboundary water conflicts. *Water Policy*, **8**, 435–460.

Zeleny, M. (1973). Compromise programming. In *Multiple Criteria Decision Making*, ed. J. L. Cochrane and M. Zeleny. Columbia, SC: University of South Carolina Press, pp 262–301.

Zhang, L. (1999). *Social Impacts of Large Dams: The China Case, Prepared for Thematic Review I*. 1. Social impacts of large dams equity and distributional issues – contributing paper for the World Commission on Dams, China.

Zhu, Z. (1998). Conscious mind, forgetting mind: two approaches in multimethodology. *Systemic Practice and Action Research*, **11**(6), 669–690.

Zhu, Z. (2010). Theorizing systems methodologies across cultures. *Systems Research and Behavioral Science*, **27**, 208–223.

Zwaan, R. A. and Radvansky, G. A. (1998). Situation models in language comprehension and memory. *Psychological Bulletin*, **123**(2), 162–185.

Index

Page numbers in *italics* are located in Appendices. **Bold** page numbers indicate Tables or Figures.

Printed in the United States
by Baker & Taylor Publisher Services